GENETIC ABERRANCIES AND NEURODEGENERATIVE DISORDERS

Volume 3 • 1999

ADVANCES IN CELL AGING AND GERONTOLOGY

GENETIC ABERRANCIES AND NEURODEGENERATIVE DISORDERS

VOLUME 3 • 1999

ADVANCES IN CELL AGING AND GERONTOLOGY

Series Editors: PAOLA S. TIMIRAS
Department of Molecular and Cell Biology
University of California—Berkeley

E. EDWARD BITTAR
Department of Physiology
University of Wisconsin—Madison

Volume Editor: MARK P. MATTSON
Laboratory of Neurosciences
National Institute on Aging

JAI PRESS INC.
Stamford, Connecticut

100 Prospect Street
Stamford, Connecticut 06904

ISBN: 0-7623-0405-7

Transferred to digital printing 2006

CONTENTS

LIST OF CONTRIBUTORS

Robert B. Bell	Department of Clinical Neurosciences Health Sciences Centre University of Calgary Calgary, Alberta, Canada
Robert J. Ferrante	ENR Memorial Veteran's Hospital Bedford, Massachusetts
Pierluigi Gambetti	Department of Pathology Indiana University School of Medicine Indianapolis, Indiana
Bernardino Ghetti	Department of Pathology Indiana University School of Medicine Indianapolis, Indiana
Gordon Glazner	Sanders-Brown Center on Aging University of Kentucky Lexington, Kentucky
Nina Horn	The John F. Kennedy Institute Glostrup, Denmark
R.D. Jolly	Institute of Veterinary Animal and Biomedical Science Massey University Palmerston North, New Zealand
Edward J. Kasarskis	Neurology Service Veterans Affairs Medical Center Lexington, Kentucky
A. Kohlschütter	Universitätkinderklinik University of Hamburg Hamburg, Germany

Neil W. Kowall	ENR Memorial Veteran's Hospital Bedford, Massachusetts
Stephan Kummerle	ENR Memorial Veteran's Hospital Bedford, Massachusetts
Douglas Lanska	Department of Neurology University of Kentucky Lexington, Kentucky
Mark P. Mattson	Laboratory of Neurosciences National Institute on Aging Baltimore, Maryland
D.N. Palmer	Department of Animal and Veterinary Science Lincoln University Canterbury, New Zealand
LaRoy Penix	Neuroscience Institute Morehouse School of Medicine Atlanta, Georgia
Asuri N. Prasad	Pediatrics and Neurology University of Manitoba Winnipeg, Manitoba, Canada
Chitra Prasad	Pediatrics and Human Genetics University of Manitoba Winnipeg, Manitoba, Canada
P. Hemachandra Reddy	Molecular Neurogenetics Section Genetics and Molecular Biology Branch National Human Genome Research Institute National Institutes of Health Bethesda, Maryland
John T. Slevin	Department of Neurology University of Kentucky Lexington, Kentucky
Daret K. St. Clair	Neurology Service Veterans Affairs Medical Center Lexington, Kentucky

Carl E. Stafstrom
Neurology and Pediatrics
University of Wisconsin-Madison
Madison, Wisconsin

Danilo A. Tagle
Molecular Neurogenetics Section
Genetics and Molecular Biology Branch
National Human Genome Research Institute
National Institutes of Health
Bethesda, Maryland

Zeynep Tümer
Department of Medical Genetics
University of Copenhagen
Copenhagen, Denmark

S.U. Walkley
Department of Neuroscience
Albert Einstein College of Medicine
Bronx, New York

PREFACE

Recent advances in epidemiological investigations, molecular genetics, and methods in molecular biology have led to compelling evidence that gene mutations and polymorphisms make major contributions to most, if not all, neurodegenerative disorders. The purpose of organizing this volume was to bring together, under one cover, fundamental information concerning the roles of inherited traits in the pathogenesis of different neurodegenerative disorders. In addition to providing a catalogue of the known genetic alterations that are linked to specific neurodegenerative disorders, we have attempted to convey the current state of understanding of the cellular and biochemical mechanisms whereby the genetic aberrancies lead to neuronal dysfunction and degeneration. This latter realm of investigation is only in its infancy and, indeed, in most cases we have very little clue as to the precise sequence of events that lead from a genetic defect to nerve cell degeneration. Nevertheless, the emerging pictures of each disorder, painted by pathological, biochemical, and molecular brushes, share several important features, including increased levels of oxidative stress, perturbed ion homeostasis, mitochondrial dysfunction, apoptotic proteolytic cascades, and protein aggregation. The existence of these common themes of the neurodegenerative process provides the opportunity to design experiments (in cultured nerve cells and transgenic mice, for example) that can establish the precise pathogenic mechanism of a specific mutation or genetic risk factor. The value of this approach is exemplified by recent studies of how mutations in Cu/Zn-superoxide dismutase cause amyotrophic lateral sclerosis

(ALS) and how presenilin mutations result in early-onset Alzheimer's disease (AD).

Some neurodegenerative disorders arise solely from a genetic defect, with Huntington's disease (HD) being an example in which the disorder is caused by trinucleotide expansions in the huntingtin gene. The genetic contributions to many other prominent neurodegenerative disorders are quite complex, with some cases being caused by gene mutations and other cases being influenced by polymorphisms that can be considered susceptibility or risk factors. The major risk factor for many of the disorders covered in this volume is increasing age. The aging process is still not well understood, but clearly involves both environmental factors such as progressive accumulation of free radical-mediated damage to cellular constituents and genetic factors such as apolipoprotein E genotype. It is of considerable interest and importance that events that occur during aging predispose neurons to genetic aberrancies that promote degenerative cascades, and that specific genetic defects exert their influence on certain populations of neurons in a disorder-specific manner. Why do presenilin mutations results in age-related degeneration of neurons in the entorhinal cortex and hippocampus, whereas polyglutamine expansions in Huntington render striatal neurons vulnerable and Cu/Zn SOD mutations afflict mainly lower motor neurons? The answer to such questions has proven elusive because of the fact that, in most cases, the defective gene is expressed at similar levels in both vulnerable and nonvulnerable populations of neurons. It is hoped that the chapters in this volume will stimulate readers to generate new hypotheses concerning the pathogenic mechanisms of genetic aberrancies that can be experimentally tested.

The scope of the subject of this book is certainly too vast to fully cover in detail in one volume. However, I expect the reader will find each chapter information-rich and will appreciate the relative lack of redundancy that often occurs in such edited volumes. The organization of the book is rather simple—each chapter covers a given disorder or class of disorders. AD is an intriguing disorder because its genetics are quite complex, and the cellular and molecular neurodegenerative mechanisms that appear to be operative involve a set of regulatory systems that normally function in neuronal development and adaptive synaptic plasticity. In the chapter on AD, I have attempted to provide a view of how genetic causal and risk factors interact with age-related changes in the brain to promote degenerative biochemical cascades. The trinucleotide repeat disorders are covered by P. Hemachandra Reddy and Danilo Tagle. The latter have several intriguing features including "anticipation," in which the number of trinucleotide repeats in the affected gene increase in successive generations resulting in enhance voracity of the clinical phenotype. The fact that such trinucleotide repeats are responsible for degeneration of different neuronal populations depending upon the disorder provides a novel opportunity to address basic mechanisms underlying selective neuronal death. Neil Kowall and colleagues cover the seemingly straightforward genetics of HD, and the less straightforward mechanisms whereby polyglutamine repeats in the huntingtin protein lead to degeneration of striatal neurons. Recent findings suggesting that aggregation of abnormal huntingtin may be involved in its neurodegenerative effects, and may provide mechanistic links between this disease and AD and PD, two other disorders that exhibit abnormal aggregations of amyloid β-peptide and synuclein, respectively. The identification of mutations in Cu/Zn SOD as causal for some cases of familial amyotrophic lateral sclerosis (ALS) brought the free radical theory of neurodegenerative disorders to center stage; Ed Kasarskis and Daret St. Clair cover the genetics of ALS and possible mechanisms underlying the pathogenic actions of Cu/Zn SOD mutations, which appear not to be as simple as originally thought. Mutations in prion proteins result in formation and amplification of amyloid protein aggregations, which appear to be the key events in the infective feature of these disorders. Bernardino Ghetti and Pierluigi Gambetti describe the genetic alterations responsible for several different types or prion disorders and emphasize the role of these mutations in protein conformation and protein-protein interactions.

Epilepsy has a remarkably complex genetics, probably because there are so many factors that can alter neuronal excitability. Carl Stafstrom, John Slevin, and coauthors provide a view of this complexity and its implications for mechanisms of disease pathogenesis. As described by LaRoy Penix and Doug Lanska, stroke is a leading cause of disability and death that has important genetic predisposition and causal factors. The inherited disorder CADASIL was recently shown to be caused by mutations in Notch-3, which raises very interesting questions concerning its pathogenic mechanism, and suggests possible mechanistic links between AD and stroke, since presenilin-1 appears to be involved in the Notch signaling pathway. Damage to oligodendrocytes resulting in demyelination in the central nervous system is a central feature of multiple sclerosis (MS). While accumulating data suggest an autoinflammatory response plays a role in MS, the causes are unkown. Robert Bell presents emerging evidence that is revealing genes which predispose to MS. Because many other neurodegenerative disorders involve inflammation-like processes, a better understanding of genetic influences on cytokine cascades and leukocyte physiology is clearly of interest.

Mitochondrial DNA mutations are increasingly recognized as playing important roles in an array of age-related neurodegenerative disorders. Gordon Glazner describes the central role of mitochondria in free radical metabolism and calcium homeostasis, and provides a view of how disruption of these processes may be a central consequence of mitochondrial DNA mutations. The pivotal role of mitochondria in apoptosis, and the likely involvement of this form of cell death in many different neurodegenerative disorders, emphasizes the importance of understanding how abberancies of the mitochondrial genome arise and how they lead to nerve cell degeneration. There exists a quite remarkable array of hereditary neurodegenerative disorders that arise from alterations in metabolic pathways. Zeynep Tümer and Nina Horn describe inherited disorders of copper metabolism, including their unusual clinical manifestations and their biochemical bases. Robert Jolly and colleagues present the lysosomal storage diseases classified as ceroid-lipofuscinoses, a complex set of related disorders often characterized by electroencephalogram abnormalities and retinal degeneration. The genetic alterations responsible for the latter disorders are beginning to be identified, and it is expected that definition of the protein alterations that lead to perturbed lysosomal metabolism will soon follow.

We have attempted to capture the excitement and optimism of the current era of genetic and molecular biology as applied to neurodegenerative disorders, and hope that this volume spurs interactive research efforts aimed at identifying mechanisms that are shared among neurodegenerative disorders.

MARK P. MATTSON

ACKNOWLEDGEMENTS

We, as authors, are indebted to Sally Malley who has been the key person involved in organizing and editing the manuscripts for this volume. We also thank the many outstanding coworkers who contributed to the basic research described in the chapters.

Chapter 1

Genetic Contributions to the Pathogenesis of Alzheimer's Disease

MARK P. MATTSON

INTRODUCTION

Alzheimer's disease (AD) is a progressive degenerative disorder characterized by nerve cell dysfunction and death in brain regions involved in learning and memory processes, including the hippocampus, entorhinal cortex, and basal forebrain. Examination of postmortem brain tissue from AD patients reveals striking abnormalities including degenerated neurons and dystrophic neurites containing abnormal accumulations of insoluble straight and twisted filaments comprised of a cytoskeletal protein called tau (see Selkoe, 1991, for review). Tau normally functions in the modulation of microtubule polymerization, thereby regulating adaptive

Advances in Cell Aging and Gerontology
Volume 3, pages 1–31

ISBN: 0-7623-0405-7

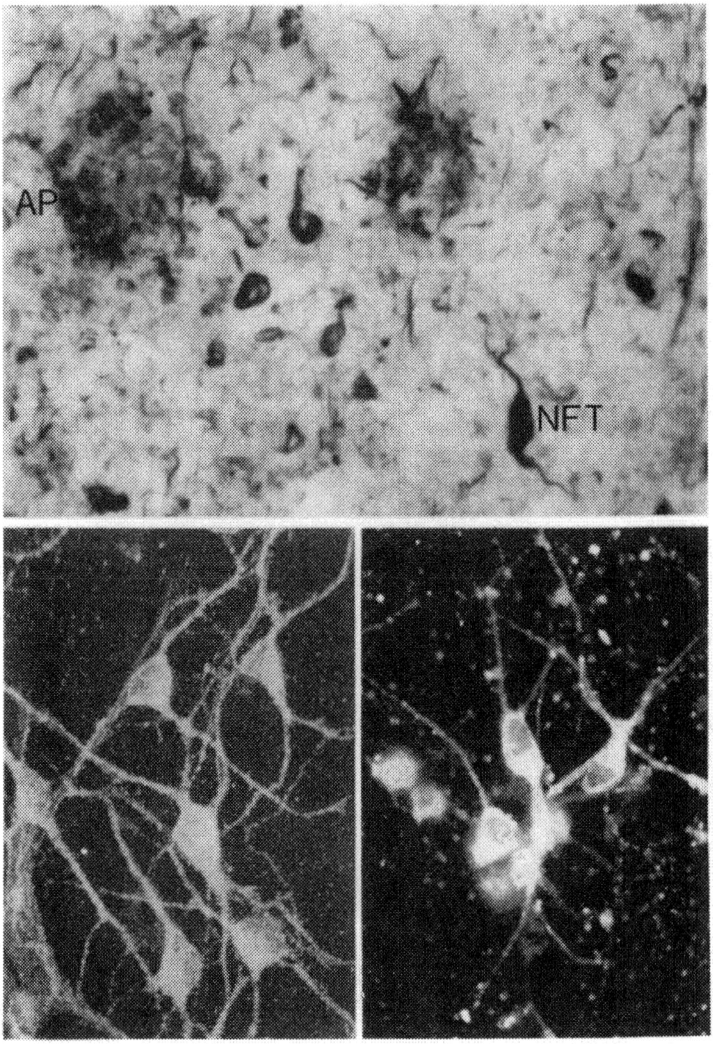

Figure 1. Association of amyloid deposition with neuronal degeneration in Alzheimer's disease (AD). The upper panel shows the two overt abnormalities observed upon microscopic examination of brain tissue from AD patients—the neurofibrillary tangle (NFT) consisting of intracellular accumulations of insoluble filaments of the microtubule-associated protein tau in the nerve cell body, and amyloid plaques (AP) comprised of extracellular aggregations of amyloid β-peptide (Aβ) often associated with degenerated neurites. The lower panels show an example of the damaging effect of Aβ on cultured hippocampal neurons—one culture was exposed to a control peptide (Control) and the other to the fibril-forming Aβ. The cells were then reacted with an antibody that recognizes phosphorylated tau (white color). Aβ caused accumulation of phosphorylated tau filaments and neuron degeneration.

changes in neuronal structure and physiology. In AD tau becomes hyperphosphorylated and self-aggregates, and microtubules depolymerize which might contribute to neurite dysfunction and degeneration. Immunohistochemical and ultrastructural analyses have shown that extensive synapse loss occurs in the relatively early stages of AD (see Lassman, 1996, for review). Another prominent abnormality in AD brain is the accumulation of extracellular amyloid plaques which are spherical structures comprised mainly of a protein called amyloid β-peptide (Aβ) (Figure 1). Aβ is a 40- to 42-amino acid peptide generated as a proteolytic product of a much larger amyloid precursor protein (APP) (see Mattson, 1997a, for review). Plaques manifest as either a diffuse form in which Aβ is in an unaggregated state not associated with neuronal degeneration, and a compact form in which Aβ forms antiparallel fibrils with a β-pleated sheet structure that exhibit birefringence under polarized light. The fibrillar Aβ deposits are often associated with degenerated neurites. Cognitive deficits are strongly correlated with density of neurofibrillary tangles, amyloid burden, and synapse loss suggesting that neurodegeneration is responsible for cognitive dysfunction and that amyloid accumulation is linked to the neurodegenerative process.

In addition to the neuronal degeneration and Aβ deposition present in brains of AD victims, there are numerous cellular and biochemical alterations that suggest the presence of an inflammation-like process (see McGeer and McGeer, 1995, for review). Reactive astrocytes and microglia are associated with neuritic plaques, with astrocytes surrounding the plaques and microglia being concentrated within the plaques. Local increases in several cytokines have been described in association with neuritic plaques including interleukin-1β, interleukin-6, and tumor necrosis factor-α. Moreover, immunohistochemical studies indicate the presence of complement proteins such as C1q in association with neuritic plaques. By analogy with inflammatory responses in other tissues, the glial and immune alterations in AD brain most likely represent a secondary response to a primary neurodegenerative process. Nevertheless, such secondary responses may accelerate the neurodegenerative process. Indeed, recent epidemiological data suggest that nonsteroidal anti-inflammatory drugs may be beneficial in delaying the onset of the symptoms of AD (Breitner et al., 1994). Although genetic contributions to AD may act at the level of the primary neurodegenerative process, it is important to consider the roles of inflammatory processes in the progression of the disease.

MUTATIONS LINKED TO EARLY-ONSET INHERITED ALZHEIMER'S DISEASE

There are families in which AD is inherited in an autosomal dominant manner such that all affected family members develop AD symptoms at an early age, usually when they are in their 30s, 40s, and 50s. Such familial AD (FAD) cases account for approximately 15% of all AD cases (see Finch and Tanzi, 1997, for review). The remaining 85% of cases are not caused by a specific genetic defect, and are

Table 1. Genetic Causal and Risk Factors for Alzheimer's Disease

	Gene	*Chromosome*	*Age of Onset (yrs)*
Causal Factor	Amyloid precursor protein	21	45–65
	Presenilin-1	14	28–50
	Presenilin-2	1	40–55
Risk Factor	Apolipoprotein E4	19	65–85
	α2-macroglobulin	12	65–85
	Bleomycin hydrolase	17	65–85

characterized by a relatively late age of onset, typically in the range of 65 to 85 years of age; the sporadic forms of AD are, however, influenced by genetic polymorphisms that can be considered susceptibility or risk factors. During the past 10 years, tremendous progress has been made in identifying the genetic defects responsible for FAD (Table 1). At least five different chromosomes harbor defective genes including chromosomes 1, 12, 14, 17, and 21. The first gene linked to FAD was the β-amyloid precursor protein (APP) located on chromosome 21 (see Mullan and Crawford, 1993, for review). Three years ago, two homologous genes were identified as harboring mutations linked to the most vigorous (earliest age of onset) forms of AD. The genes are now called presenilin-1 (chromosome 14) and presenilin-2 (chromosome 1) (see Hardy, 1997, for review). The defective genes located on chromosomes 12 and 17 have yet to be identified, although recent findings suggest that α2-macroglobulin or the low-density lipoprotein–related receptor (LRP), or both, may be the culprit(s) on chromosome 12 (Blacker et al., 1998) and that tau is the affected gene on chromosome 17 (Poorkaj et al., 1998).

APP MUTATIONS RESULT IN ABERRANT PROTEOLYTIC PROCESSING OF APP, LEADING TO OXIDATIVE STRESS AND PERTURBED CALCIUM REGULATION IN NERVE CELLS

APP is a large transmembrane protein that is expressed in neurons and glial cells throughout the nervous system, as well as in many non-neural tissues including vascular smooth muscle and endothelial cells (see Mattson, 1997a, for review). In neurons APP is axonally transported and accumulates in presynaptic terminals and growth cones. APP mutations in FAD cases are missense in nature resulting in either a single amino acid change at codon 717 ("London" mutation) or a two amino acid substitution at codons 670 and 671 ("Swedish" mutation). The mutations are located immediately N- (Swedish mutation) or C-terminal (V717F) to the Aβ sequence (Figure 2). Additional families harbor mutations at codons 692 ("Flemish" mutation) or 693 ("Dutch" mutation), which lie within the Aβ sequence. A cleavage of APP in the middle of the Aβ sequence, effected by an enzyme called

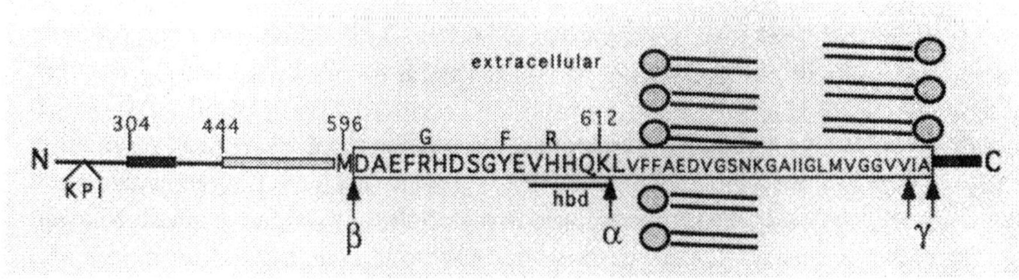

Figure 2. Structure, proteolytic cleavage sites, and functional domains of human APP. APP consists of a large extracellular domain of approximately 616 amino acids, a hydrophobic transmembrane domain, and a short cytoplasmic C-terminus. The numbering of amino acids in the diagram is based on the 695 amino acid isoform; additional isoforms of APP (APP751 and APP770) contain a kunitz protease inhibitor (KPI) domain near the N-terminus. The amino acid sequence of Aβ is indicated in the boxed portion of the diagram. APP is proteolytically cleaved at at least four different sites. α-Secretase cleaves between amino acids 612 and 613, which lies within the Aβ sequence (this cleavage releases sAPPα from the cell surface). β-Secretase cleaves at the N-terminus of Aβ (between amino acids 596 and 597), releasing sAPPβ from the cell surface and leaving a C-terminal membrane-associated fragment containing intact Aβ. γ-Secretase cleaves at the C-terminus of Aβ in at least two different sites, resulting in release of intact Aβ1-40 or Aβ1-42. There are several functional domains in sAPPα including a region just N-terminal to the β-secretase cleavage site that is involved in modulating neuronal excitability and survival, and a heparin-binding domain (hbd) at the C-terminus. Within the Aβ domain, the amino acid sequences of human and rodent Aβ differ by three amino acids (G, F, and R are present in rat Aβ as indicated in the diagram).

α-secretase, results in release of a secreted form of APP called sAPPα from axon terminals. This secretory cleavage prevents production of intact, and therefore potentially amyloidogenic, Aβ. Interestingly, the secretory cleavage of APP is induced when neurons are electrically active and studies of the effects of sAPPα on neuronal activity and synaptic function suggest that sAPPα may play an important role in learning and memory processes (Doyle et al., 1990; Huber et al., 1993; Roch et al., 1994; Furukawa et al., 1996a, 1996b; Ishida et al., 1997; Furukawa and Mattson, 1998). A striking biological activity of sAPPα is its ability to protect neurons from being damaged and killed by conditions relevant to the pathogenesis of AD including exposure oxidative and metabolic insults (Mattson et al., 1993; Smith-Swintosky et al., 1994; Furukawa et al., 1996b). An alternative cleavage of APP at the N-terminus of the Aβ sequence (by β-secretase activity) leaves a membrane-associated fragment that contains intact Aβ and can be further cleaved at the C-terminus (by γ-secretase activity) resulting in the release of Aβ. APP mutations result in altered APP processing in a manner that increases production of Aβ and decreases production of sAPPα.

There are at least two mechanisms whereby APP mutations may promote neuronal degeneration in AD (see Mattson, 1997a, for review). The first involves increased Aβ production leading to excessive accumulation of fibrillar Aβ, which causes cell damage and death. The second mechanism involves reduced production of sAPPα; it has been shown that sAPPα can prevent death of cultured neurons and of brain neurons in adult rodents exposed to metabolic and oxidative insults relevant to the pathogenesis of AD, suggesting that APP mutations may result in loss of a neuroprotective function of APP.

When mutated forms of APP are expressed in cultured cells and transgenic mice, there is increased production of Aβ, particulary the longer 42 amino acid form of the peptide (Aβ1-42). Aβ1-42 exhibits an increased propensity to self-aggregate and form amyloid fibrils; this property of Aβ1-42 is correlated with increased toxic activity toward cultured neurons. Exposure of cultured hippocampal neurons to Aβ results in an increase in levels of various oxyradicals, and consequent free radical–mediated damage to membrane lipids, proteins, and DNA. Aβ itself may generate free radicals upon interaction with certain metals such as iron (Fe^{2+}); such generation of peptide-associated radicals may play an important role in covalent crosslinking of Aβ to form amyloid fibrils (Dyrks et al., 1992; Hensley et al., 1994; Mattson, 1995). Alternatively (or coincidentally) Aβ may induce oxidative stress by engaging receptor-mediated pathways. For example, data suggest that Aβ binds to the receptor for advanced glycation end products (RAGE), and that this interaction results in increased nitric oxide production in cells such as microglia that express RAGE (Mattson and Rydel, 1996; Yan et al., 1996).

Membrane lipid peroxidation appears to be a critical early event that results in a cascade of events induced by Aβ that increases the vulnerability of neurons to degeneration (Mattson, 1998). Lipid peroxidation causes release of an aldehyde called 4-hydroxynonenal (HNE), which covalently binds to proteins at cysteine, lysine, and histidine residues. Studies of cultured primary neurons, astrocytes, and synaptosomes have shown that Aβ can impair the function of membrane proteins involved in the regulation of ion homeostasis and energy metabolism, and that 4-hydroxynonenal plays a key role in these actions of Aβ (Figure 3). Exposure of cultured rat hippocampal neurons or human cortical synaptosomes to Aβ impairs the function of the Na^+/K^+-adenosine triphosphatase (ATPase) and the Ca^{2+}-ATPase, two membrane enzymes critical for maintenance of resting membrane potential and intracellular calcium levels (Mark et al., 1995). The latter effects of Aβ are blocked by antioxidants that suppress membrane lipid peroxidation, and are mimicked by 4-hydroxynonenal (Mark et al., 1995, 1997a; Keller et al., 1997a, 1997b). Additional studies have shown that both Aβ and HNE can impair the function of neuronal and synaptosomal glucose transporters (Mark et al., 1997b; Keller et al., 1997a), and astrocytic glutamate transporters (Keller et al., 1997b; Blanc et al., 1998). The discovery that Aβ impairs the function of these membrane transporters via an oxyradical-mediated mechanism may explain why only neurons, and not glial cells, degenerate and die in AD. Neurons that degenerate in AD express

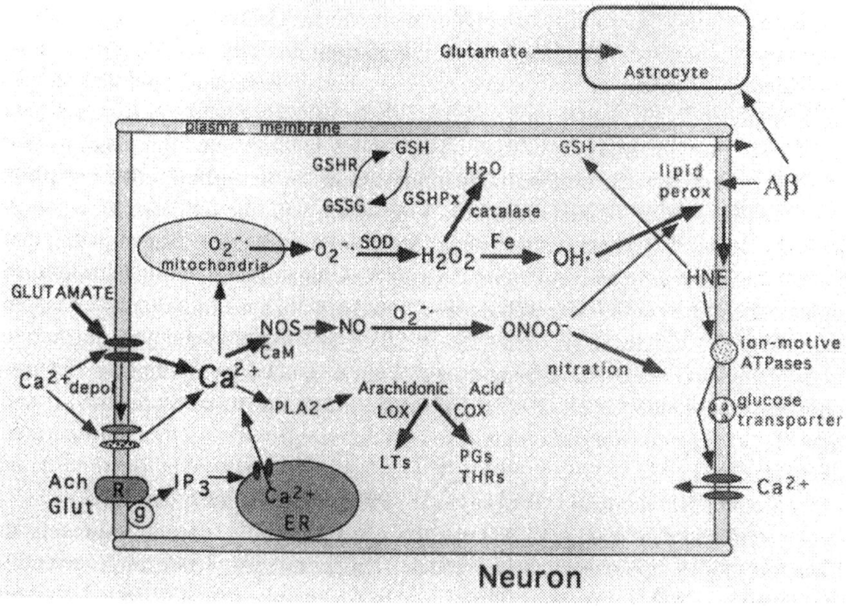

Figure 3. Mechanisms whereby Aβ and activation of glutamate receptors induce oxidative stress and disrupt ion homeostasis in neurons. Aβ can induce membrane lipid peroxidation (MLP), resulting in the production of 4-hydroxynonenal (HNE) an aldehyde that covalently modifies membrane transporters (Na^+/K^+-ATPase, Ca^{2+}-ATPase, glucose transporter, and glutamate transporter) and thereby impairs their functions. These adverse effects of MLP promote membrane depolarization and excessive activation of glutamate receptors resulting in excitotoxicity. Oxidative stress also perturbs ion homeostasis in endoplasmic reticulum (ER) and mitochondria. Activation of glutamate receptors results in calcium influx which, in turn, promotes oxyradical production in several different ways including compromising mitochondrial calcium homeostasis and membrane potential resulting in increased production of superoxide anion radical (O_2^-·). Superoxide dismutases (SOD) convert O_2^-· to H_2O_2 which, in the presence of Fe^{2+}, generates OH·. O_2^-· also interacts with nitric oxide (NO) to form peroxynitrite. Both OH· and peroxynitrite induce MLP. Calcium also promotes arachidonic acid production the activities of cyclooxygenases (COX) and lipoxygenases (LOX) with resulant generation of O_2^-·.

high levels of receptors for the excitatory neurotransmitter glutamate. Neurons that express glutamate receptors are vulnerable to being killed by a mechanism termed "excitotoxicity" in which activation of glutamate receptors under adverse conditions (e.g., when the cells are subjected to oxidative and metabolic stress) results in massive calcium influx and disruption of various structural components and metabolic pathways in the neurons. Exposure of cultured hippocampal neurons to Aβ (Mattson et al., 1992) and 4-hydroxynonenal (Mark et al., 1997a) greatly increases their vulnerability to glutamate toxicity. Central to the mechanism whereby Aβ increases neuronal vulnerability to excitotoxicity is lipid peroxidation-mediated impairment of ion-motive ATPases, and glucose and glutamate transporters (Keller et al., 1997b; Mark et al., 1997a, 1997b). Such experimental data are consistent with analyses of the human brain which revealed that neurons that degenerate in AD, such as those in the hippocampus and entorhinal cortex, express very high levels of glutamate receptors. Consistent with the oxidative stress–perturbed calcium hypothesis of neuronal degeneration in AD are data showing that insults that induce oxidative stress and disrupt calcium homeostasis in neurons also induce alterations in the microtubule-associated protein tau similar to those seen in neurofibrillary tangles in AD (Mattson, 1990). 4-Hydroxynonenal may play a role in the latter process by promoting crosslinking of tau and preventing its dephosphorylation (Mattson et al., 1997a). Although it remains to be established if and how tau hyperphosphorylation and crosslinking contributes to the neuronal cell death process in AD, the recent finding that mutations in tau account for some cases of frontotemporal dementia (Poorkaj et al., 1998) suggests this possibility.

A prominent abnormality in AD patients is that the ability of their brain cells to transport glucose is severely compromised. This abnormality has been repeatedly documented in brain imaging studies in which the uptake of radiolabeled glucose into brain cells is quantified. Moreover, a deficit in glucose transport can be detected prior to clinical symptoms in patients at risk for AD. Oxidative stress resulting from Aβ deposition and age-related changes likely plays an important role in the impairment of glucose uptake in neurons (Mark et al., 1997b). Aβ also impairs glucose uptake in vascular endothelial cells (Blanc et al., 1997b); these cells provide the main route of transport of glucose from blood to brain. The adverse effect of Aβ on glucose transport can be prevented by treating the neurons and vascular endothelial cells with antioxidants such as vitamin E, glutathione ethyl ester, and 17β-estradiol (Blanc et al., 1997b; Keller and Mattson, 1997; Mark et al., 1997b).

Exposure of cultured cortical neurons to sublethal levels of Aβ resulted in impaired muscarinic cholinergic signaling, analogous to the cholinergic alterations documented in brain tissue from AD patients (Kelly et al., 1996). Detailed analyses indicated that the defect in the signaling pathway involved impaired coupling of the muscarinic receptors to the guanosine triphosphate (GTP)-binding protein G_{q11}. The adverse effect of Aβ on cholinergic signal transduction was mimicked by exposure of cells to Fe^{2+}, and prevented in cells treated with vitamin E, suggesting a role for lipid peroxidation (Kelly et al., 1996). 4-Hydroxynonenal may mediate lipid peroxidation-induced impairment of muscarinic signal transduction, possibly by covalently crosslinking G_{q11} (Blanc et al., 1997a). Other signaling pathways involving GTP-binding proteins such as those activated by metabotropic glutamate receptors (Blanc et al., 1997a) and thrombin receptors (Mattson and Begley, 1996) may also be adversely affected by Aβ via a lipid peroxidation-mediated mechanism.

Deficits in mitochondrial function and increased oxidative damage to mitochondria have been documented in studies of AD patients (see Benzi and Moretti, 1997;

Mattson, 1997b, for review). Exposure of cultured rat hippocampal neurons or cortical synaptosomes to Aβ results in a decrease in mitochondrial reducing potential, mitochondrial membrane depolarization, and accumulation of reactive oxygen species in the mitochondria (Keller and Mattson, 1997; Keller et al., 1997b). Overexpression of Mn-SOD in cultured neural cells results in a preservation of mitochondrial function and makes the cells resistant to apoptosis induced by Aβ, indicating a major role for mitochondrial failure in the neurotoxic action of Aβ (Keller et al., 1998b). Further evidence that mitochondrial failure plays a role in neurotoxic cascades induced by Aβ comes from studies showing that cyclosporin A, an inhibitor of mitochondrial membrane permeability transition, protects cultured hippocampal neurons against apoptosis induced by Aβ (Figure 4). Finally, Aβ induces alterations in the cytoskeleton of cultured neurons similar to those seen in the neurofibrillary tangles of AD, including hyperphosphorylation of the microtubule-associated protein tau; these cytoskeletal alterations appear to result from increased oxidative stress and calcium overload (Mattson, 1990; Mattson et al., 1997a). The collective data showing that APP mutations increase Aβ production on the one hand, and that Aβ damages and kills neurons in a manner consistent with

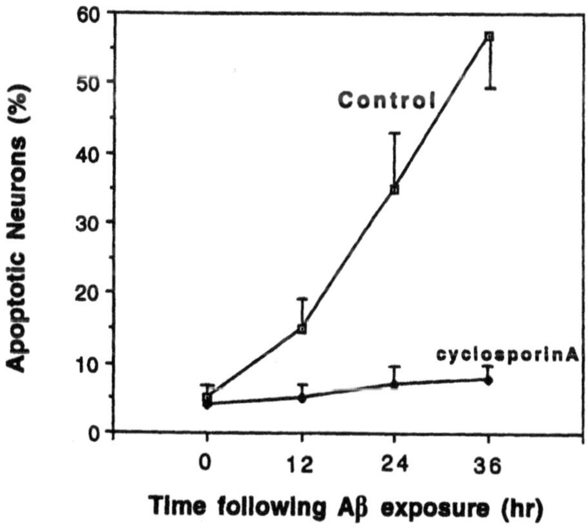

Figure 4. Cyclosporin A, an inhibitor of mitochondrial membrane permeability transition, protects cultured hippocampal neurons against Aβ toxicity. Rat hippocampal cultures were pretreated with vehicle (Control) or 1 μM cyclosporin A for 1 hour, and were then exposed to 10 μM Aβ25–35 for the indicated time periods. Cultures were fixed and stained with the fluorescent DNA-binding dye Hoecsht 33342, and the percentage of neurons exhibiting apoptotic nuclei in each culture was quantified. Values are the mean and SEM of determinations made in four separate cultures.

the neuronal degeneration in AD patients on the other hand, strongly suggests an important role for Aβ deposition in the pathogenesis of AD.

The decreased levels of sAPPα that result from APP mutations are likely to contribute greatly to the neurodegenerative process in AD. The α-secretase cleavage of APP is induced by activity in neurons and, indeed, sAPPα is released from hippocampal slices during stimulation at frequencies that induce long-term potentiation, a cellular correlate of learning and memory (Nitsch et al., 1993). Studies in embryonic rat hippocampal and human cortical cell cultures showed that picomolar concentrations of sAPPα can suppress glutamate-induced Ca^{2+} influx (Mattson et al., 1993). This effect of sAPPα on neuronal calcium homeostasis appears to play a role in modulation of dendrite outgrowth and cell survival in developing hippocampal neurons (Mattson, 1994). Both sAPPα695 and sAPPα751 exhibited similar effects on calcium responses to glutamate, indicating that the kunitz protease inhibitor domain was not involved in these actions of sAPPα. Recordings of whole-cell, and single channel ion currents in cultured rat hippocampal neurons showed that sAPPα activates high-conductance, charybdotoxin-sensitive potassium channels (Furukawa et al., 1996a). Activation of the potassium channels was correlated with membrane hyperpolarization and suppression of ongoing, glutamate receptor-mediated, synaptic activity. The potency of sAPPβ in activating potassium channels was 100-fold lower than that of sAPPα (Furukawa et al., 1996b). In addition to activating potassium channels, sAPPα may affect neuronal excitability by modulating the activity of excitatory amino acid receptors. Treatment of cultured hippocampal neurons with sAPPα resulted in a marked and selective decrease in currents induced by N-methyl-D-aspartate (Furukawa and Mattson, 1998). The suppressive effect of sAPPα on NMDA current was rapid (sec) and reversible upon washout of the sAPPα. These suppressive effects of sAPPα on neuronal excitability likely contributes to its excitoprotective actions (Figure 5).

The effects of sAPPα on NMDA currents and potassium channels may provide an explanation for the evidence that APP plays a role in learning and memory. A recent study showed that exposure of hippocampal slices to sAPPα alters the frequency-dependence of long-term depression (LTD), and enhances LTP in area CA1 (Ishida et al., 1997). When slices were pretreated with 1 to 10 nM sAPPα for 1 to 2 hours, a higher stimulation frequency (10 Hz) was required to induce LTD, while the magnitude of increase in the postsynaptic response following high-frequency stimulation was enhanced in slices treated with sAPPα. The role of sAPPα in synaptic plasticity suggests that reduced levels of sAPPα resulting from APP mutations may contribute to the cognitive impairment in AD patients.

An action of sAPPα of particular importance for understanding how APP mutations lead to neuronal degeneration in AD is its ability to protect neurons against excitotoxic, metabolic and oxidative insults (Mattson et al., 1993; Smith-Swintosky et al., 1994; Furukawa et al., 1996b). Expression of βAPP in the nervous system increases following various insults (excitotoxic, ischemic, oxidative), suggesting a role for βAPP in neuronal responses to the injury. Pretreatment of cultured

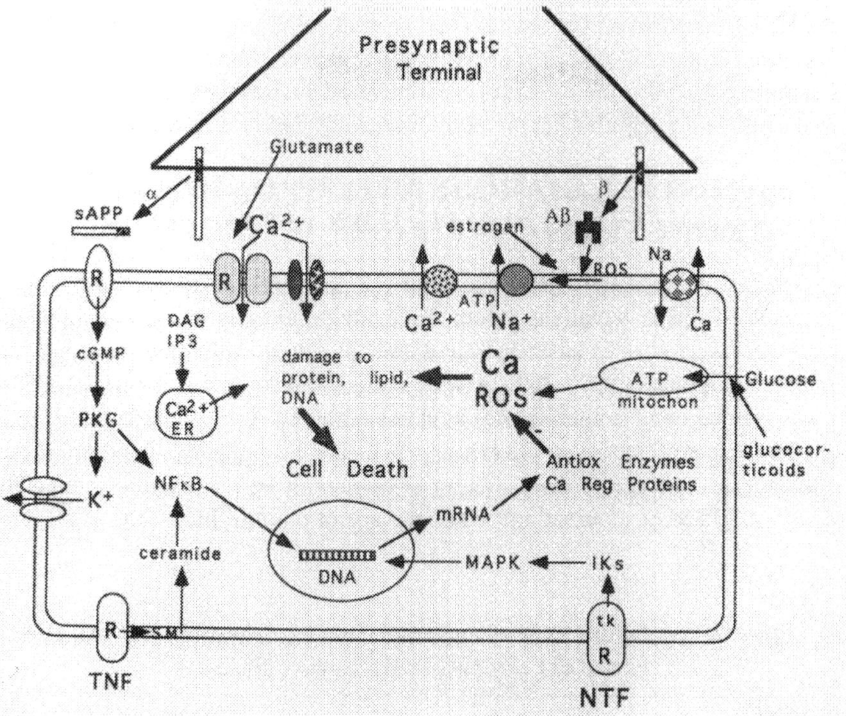

Figure 5. Modulation of neuronal excitability and vulnerability to excitotoxicity by APP metabolites, neurotrophic factors, and cytokines. See text for description.

hippocampal neurons with sAPPα increases their resistance to excitotoxicity and glucose deprivation–induced injury by a mechanism involving stabilization of calcium homeostasis (Mattson et al., 1993). Further studies of cultured rat hippocampal cells showed that sAPPαs can protect neurons against Aβ toxicity and other oxidative insults (Goodman and Mattson, 1994). The mechanism whereby sAPPα protects neurons against metabolic, excitotoxic, and oxidative insults appears to involve both rapid effects on ion channel function, and delayed transcription-dependent processes involving the transcription factor NF-κB (Barger and Mattson, 1996). Indeed, recent studies have shown that activation of NF-kβ in hippocampal neurons confers resistance to excitotoxic, metabolic, and oxidative insults that induce apoptosis, including exposure to Aβ (Barger et al., 1995; Mattson et al., 1997b). Infusion of sAPPα into the lateral ventricles of adult rats immediately prior to administration of transient global forebrain ischemia significantly reduced degeneration of hippocampal CA1 neurons (Smith-Swintosky et al., 1994), demonstrating a neuroprotective action of sAPPα in vivo. Moreover, transgenic mice overexpressing human βAPP exhibited resistance to excitotoxic injury induced by the HIV-1 envelope protein gp120 (Mucke et al., 1995). In light of evidence for decreased levels of sAPPα in brain tissue of AD patients (Van Nostrand et al., 1992; Lannfelt et al., 1995), these kinds of experimental data strongly suggest that reduced levels of sAPPα contribute to the neurodegenerative process in AD.

PRESENILIN MUTATIONS ALTER CELLULAR CALCIUM HOMEOSTASIS AND PERTURB APP PROCESSING

Mutations in presenilin-1(PS-1) account for up to 8% of all AD cases, while presenilin-2 (PS-2) mutations account for many fewer cases. The presenilin mutations are transmitted in an autosomal dominant pattern with 100% penetrance in most cases (Hardy, 1997). To date, 45 mutations in PS-1 and 2 mutations in PS-2 have been reported in familial AD kindreds (Table 2). With one exception, all of the PS-1 and PS-2 mutations are missense in nature, resulting in a single amino acid substitution. The age of disease onset of carriers of PS-1 mutations is typically between 30 and 50 years of age, while the age of onset in PS-2 cases is 50 to 65

Table 2. Mutations in PS-1 and PS-2 Linked to Autosomal Dominant Forms of Early-Onset Alzheimer's Disease[a]

Mutation	*Domain*	*Onset Age (yrs)*
PS-1		
A79V, V82L, V96F	TM1	45–55
Y115C, Y115H, E120K, E120D	TM1/2	35–40
N135D, M139T, M139V, M139I, I143F, I143T, M146V, M146L	TM2	35–49
1H163Y, H163R	TM2/3	47
G209V, I213T	TM4	45
A231V, A231T, L235P	TM5	50–55
A246E, L250S, A260V	TM6	50–55
C263R, P264L, P267S, R269H, R269G	TM6/HP	35–50
E280A, E280G, A285V, L286V, 290-319del, E318G	HP/loop	42–55
G384A, G392V	TM7	35–46
C410Y, A426P, P436S	TM8	45–50
PS-2		
N141I	TM2	50–65
M239V	TM5	50

Notes: TM, transmembrane. HP, hydrophobic domains.

[a]Domains are based on an eight transmembrane domain structure of PSs (see Figure 6). The nomenclature for defining mutants is such that the number refers to the amino acid location in the protein, the letter preceding the number refers to the amino acid normally present at that position and the succeeding number refers to the amino acid present at that position in the mutant protein. These data were compiled from the following references: Campion et al., 1995; Cruts et al., 1995; Levy-Lahad et al., 1995; Perez-Tur et al., 1995; Rogaev et al., 1995; Sherrington et al., 1995; Boteva et al., 1996; Campion et al., 1996; Doan et al., 1996; Kamino et al., 1996; Tanahashi et al., 1996.

years. The mutations, which occur in or adjacent to transmembrane domains, affect amino acids that are conserved in PS-1 and PS-2. Two regions of the presenilin protein in which mutations are clustered are putative transmembrane domain 2 (15 mutations) and a region immediately adjacent to the hydrophilic loop domain (11 mutations).

PS-1 and PS-2 are widely expressed in the nervous system (Levy-Lahad et al., 1995; Sherrington et al., 1995), wherein neurons appear to express them at higher levels than glial cells. PS-1 is concentrated in cell bodies and dendrites of hippocampal and cortical neurons, with lower levels present in axons (Cook et al., 1996; Elder et al., 1996). Immunocytochemical analyses of AD and control brains have shown that PS-1 is present in both nonvulnerable and vulnerable neurons (Cribbs et al., 1996; Murphy et al., 1996; Uchihara et al., 1996; Busciglio et al., 1997; Giannakopoulos et al., 1997). Changes in PS-1 expression and proteolytic processing may occur during brain ontogeny, suggesting a developmental role for PS-1 in the brain (Hartmann et al., 1997). Confocal and electron microscope analyses indicate that presenilins are localized primarily in the endoplasmic reticulum (ER) of cultured cell lines and primary neurons (Cook et al., 1996; Guo et al., 1996; Kovacs et al., 1996; Lah et al., 1997). Hydrophobicity plots and topological analyses suggest that presenilins are transmembrane proteins with six or eight transmembrane domains, and both the C- and N-termini on the same (cytosolic) side of the ER membrane (Figure 6) (Lehmann et al., 1997).

The normal function(s) of presenilins are not known, but clues are accumulating. Presenilins have considerable homology to two *C. elegans* genes called *spe*-4 and *sel*-12, which function in spermatogenesis and egg laying (Levitan and Greenwald, 1995; Levitan et al., 1996). Interestingly, *sel*-12 mutants can be rescued by human

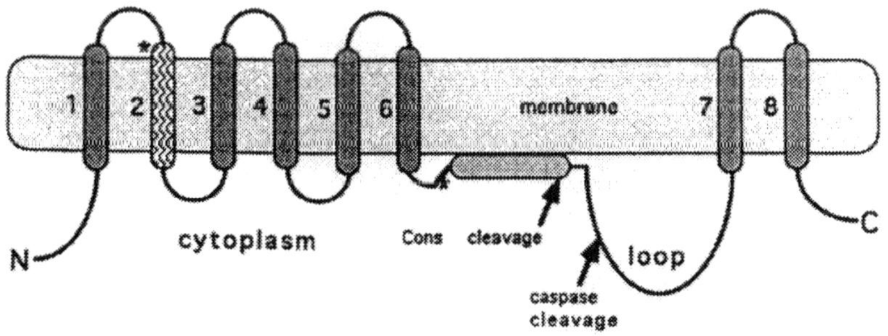

Figure 6. Structure, proteolytic cleavages sites, and sites of mutations in presenilin-1. Presenilin-1 is believed to have eight transmembrane domains with both the N- and C-termini, and a hydrophobic domain and a loop region residing on the cytoplasmic side of the endoplasmic reticulum membrane. An enzymatic cleavage site is located near the loop region, and a caspase cleavage site is located in the loop region. Missense mutations are clustered in and immediately adjacent to transmembrane 2 and adjacent to the loop region (*).

presenilins (Levitan et al., 1996), demonstrating a conserved function for these two genes. *Sel*-12 functions in the Notch signaling pathway, and the phenotypes of Notch and PS-1 knockout mice are identical (defective somite formation and embryonic lethality), indicating a critical role for presenilins in development (Conlon et al., 1995; Shen et al., 1997; Wong et al., 1997). Additional studies suggest a role for presenilins in the regulation of cellular calcium homeostasis (Guo et al., 1996, 1998a).

Two major alterations in cells expressing presenilin mutations have been identified. One alteration involves aberrant processing of APP, resulting in increased levels of Aβ1-42 (Borchelt et al., 1996; Duff et al., 1996; Scheuner et al., 1996). Aβ1-42 more readily forms fibrils than does Aβ1-40, and also exhibits increased neurotoxicity. It is unclear how presenilin mutations increase Aβ1-42 production. Presenilins might directly interact with APP (Weidemann et al., 1997; Xia et al., 1997), or might indirectly affect APP processing by increasing levels of ER "stress" (Guo et al., 1997).

A second adverse effect of presenilin mutations is to perturb calcium regulation in the ER, and thereby increase neuronal vulnerability to various metabolic and oxidative insults, thereby promoting cell degeneration and death (Mattson et al., 1998). Calcium imaging studies showed that agonist-induced calcium release from ER is enhanced in neural cells expressing mutant PS-1 (Guo et al., 1996). Calcium release in response to thapsigargin (an inhibitor of the ER Ca^{2+}-ATPase) was also enhanced in cells expressing mutant PS-1, suggesting an increased ER calcium pool. Whether wild-type presenilins serve a normal function in regulating intracellular calcium levels is unclear, but in light of the central roles of calcium in regulating developmental processes, and in neurodegenerative disorders (see Mattson, 1992, for review), it seems reasonable to consider roles for presenilins in regulation of cellular calcium homeostasis. Recent data suggest that PS-1 mutations can alter NGF-induced differentiation of PC12 cells, which is associated with alterations in cellular calcium homeostasis and transcription factor AP-1 activation (Furukawa et al., 1998). Because of calcium's involvement in the regulation of neuronal development and synaptic plasticity, perturbations of calcium homeostasis may be an important consequence of PS-1 mutations that results in age-related synaptic dysfunction and neuronal degeneration.

Apoptosis is a form of cell death in which the cells shrink and exhibit nuclear chromatin condensation and DNA fragmentation, but maintain membrane integrity. Two general mechanisms responsible for apoptosis are induction of the expression of "death genes" and reduced activation of antiapoptotic signaling pathways (Mattson and Furukawa, 1996). Studies of postmortem brain tissue from AD patients (Su et al., 1994; Smale et al., 1995) and of the neurotoxic actions of Aβ in cultured neurons (Loo et al., 1993; Kruman et al., 1997) suggest a role for apoptosis in AD. Overexpression of PS-1 mutations (L286V and M146V) in cultured PC12 cells increases their vulnerability to apoptosis induced by exposure to Aβ or trophic factor withdrawal (Guo et al., 1996, 1997, 1998a, 1998b). PS-2 mutations may also result in increased vulnerability to cells to apoptosis (Wolozin et al., 1996). PC12 cells expressing the antiapoptotic gene product *Bcl*-2 were resistant to the apoptosis-enhancing action of mutant PS- 1 (Figure 7). The apoptosis-enhancing action of presenilin mutations may involve perturbation of ER calcium signaling and calcium overload because the adverse effect of PS-1 mutations is suppressed by dantrolene and nifedipine (compounds that block calcium release from ER and calcium influx through plasma membrane voltage-dependent channels, respectively), and by overexpression of the calcium-binding protein calbindin D28k (Guo et al., 1996, 1997, 1998a). Moreover, calcium imaging studies showed that elevations of $[Ca^{2+}]_i$ induced by agonists that induce calcium release from ER, and by Aβ, were enhanced in PC12 cells expressing mutant PS-1 (Guo et al., 1996, 1997). In addition, it was recently shown that hippocampal neurons in PS1 mutant knockin mice exhibit perturbed calcium homeostasis and increased vulnerability to apoptosis and excitotoxic necrosis (Guo et al., 1999a; 1999b). Levels of cellular oxidative stress (accumulation of superoxide anion radical, hydrogen peroxide and peroxynitrite) following exposure to Aβ or trophic factor withdrawal were greatly increased in PC12 cells expressing mutant PS-1 (Guo et al., 1997, 1998a, 1998b). PC12 cells expressing mutant PS-1 were exquisitely sensitive to mitochondrial membrane depolarization and metabolic failure following exposure to Aβ or the mitochondrial toxin 3-nitropropionic acid, suggesting a widespread defect in calcium handling and oxyradical metabolism (Guo et al., 1998a; Keller et al., 1998a). Thus, presenilin mutations may promote neuronal death by enhancing levels of oxidative stress and disrupting mitochondrial function (Figure 8). It remains to be established whether the same pathogenic mechanism of action applies to PS-2.

A novel gene product that may participate in the apoptosis-enhancing effect of PS-1 mutations was recently identified. *Par*-4 (prostate apoptosis response 4) was identified by differential screening of genes induced in prostate tumor cells undergoing apoptosis, and was shown to function as a death-promoting signal in those cells (Sells et al., 1997). Levels of *Par*-4 mRNA and protein are greatly increased in AD brain tissue, and immunohistochemical analysis indicates that *Par*-4 is localized primarily in neurofibrillary tangle-bearing neurons (Guo et al., 1998c). Levels of *Par*-4 are increased in cultured primary rat hippocampal neurons following exposure to Aβ1-42, and treatment of these cells with *Par*-4 antisense oligodeoxynucleotides protects them against apoptosis induced by Aβ1-42 (Guo et al., 1998c). When PC12 cells overexpressing PS-1 mutations are exposed to apoptotic insults (trophic factor withdrawal and amyloid β-peptide), increases in *Par*-4 expression are exacerbated (Guo et al., 1998c). The apoptosis-enhancing action of PS-1 mutations can be blocked by overexpression of a dominant-negative *Par*-4 leucine zipper domain, indicating that *Par*-4 participates in the pathogenic mechanism of the PS-1 mutations (Guo et al., 1998c).

The altered proteolytic processing of APP resulting from presenilin mutations might be explained by a direct interaction of presenilins and APP (Weidemann et al., 1997). On the other hand, perturbed ER calcium homeostasis and an ER stress

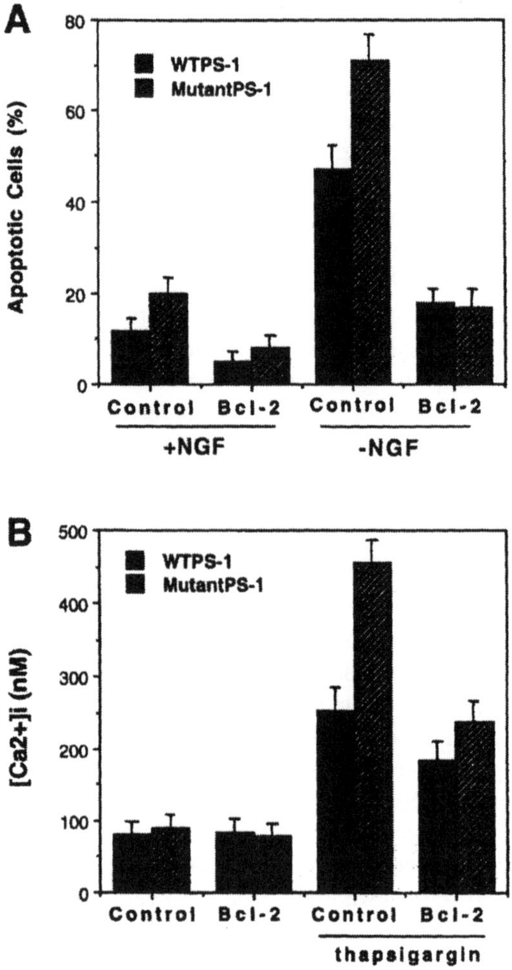

Figure 7. A presenilin-1 mutation increases the vulnerability of differentiated PC12 cells to NGF withdrawal-induced apoptosis, and enhances calcium release from endoplasmic reticulum. PC12 cells overexpressing either wild-type human PS-1 (WTPS-1) or the PS-1L286V mutation (MutantPS-1), were co-transfected with empty vector (Control) or Bcl-2. **A**. The different clones were differentiated into a neuron-like phenotype by chronic exposure to NGF and were then incubated for 48 hours in serum-free medium containing or lacking NGF and the percentage of cells exhibiting nuclear condensation and fragmentation was quantified. Values are the mean and SEM of determinations made in at least four separate cultures. **B**. The different clones were incubated in serum-free medium and basal $[Ca^{2+}]_i$, and the peak $[Ca^{2+}]_i$ following exposure to 1 μM thapsigargin was quantified. Values are the mean and SEM of determinations made in at least four separate cultures. Modified from Guo et al. (1997).

response might lead, secondarily, to altered APP processing (Guo et al., 1997; Mattson et al., 1998). Indeed, it has been shown that calcium overload and metabolic impairment can increase Aβ production in cultured cells (Gabuzda et al., 1994; Querfurth and Selkoe, 1994).

Interestingly, sAPPα can protect cultured neural cells against the pro-apoptotic action of PS-1 mutations (Guo et al., 1998b). Treatment of cultured cells with sAPPα largely abolished the enhancement of apoptosis, intracellular calcium levels, oxidative stress, and mitochondrial dysfunction normally observed in cells expressing mutant PS-1 following exposure of cells to Aβ (Figure 9). The mechanism whereby sAPPα protects cells against the adverse effects of PS-1 mutations appears to involve activation of the transcription factor NF-κB because treatment of cells with κB decoy DNA (which selectively blocks activation of NF-κB) abolishes the protective effect

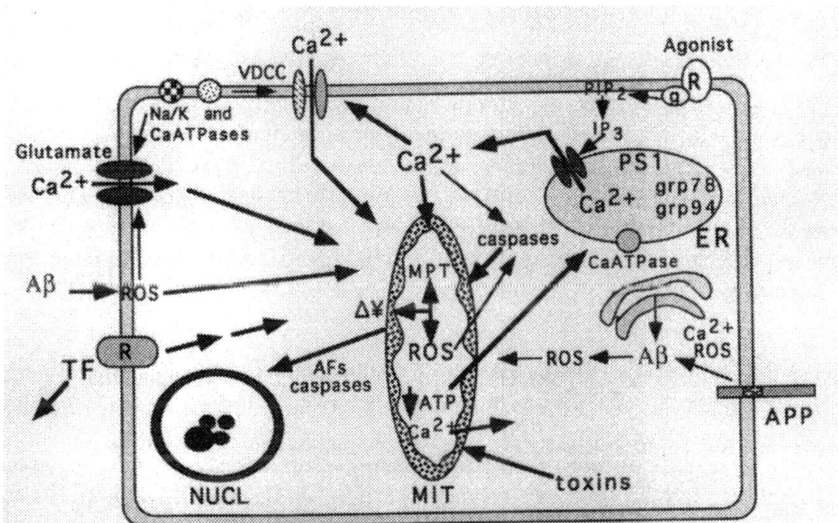

Figure 8. Putative mechanisms whereby PS-1 mutations increase neuron vulnerability to mitochondrial dysfunction and apoptotis. PS-1 mutations perturb calcium homeostasis in a manner that leads to enhanced calcium release from the endoplasmic reticulum when cells are exposed to various insults. Impairment of mitochondrial function, as may occur during aging, leads to ATP depletion, further destabilization of calcium homeostasis, and production of reactive oxygen species (ROS). The ROS, in turn, damage mitochondrial membranes and proteins resulting in membrane depolarization (Δ¥) and permeability transition (MPT), and ultimately release of apoptotic factors (AFs) and nuclear apoptosis. Perturbed calcium homeostasis and oxyradical metabolism promotes aberrant processing of APP resulting in increased production of amyloidogenic forms of Aβ and decreased production of neuroprotective sAPPα. In addition, increased levels of intracellular calcium and oxyradicals lead to activation of caspases which contribute to various aspects of the apoptotic cell death process. Modified from Keller et al. (1998a).

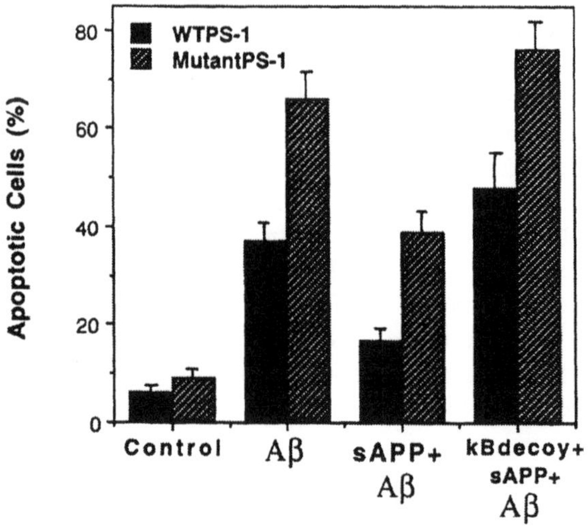

Figure 9. sAPPα protects neural cells against the apoptosis-enhancing activity of presenilin-1 mutations by a mechanism involving activation of NF-κB. The indicated lines of PC12 cells were pretreated for 24 hours with sAPPα (10 nM) or κB decoy DNA (25 μM), or for 1 hour with dantrolene (10 μM) or nifedipine (1 μM), and were then exposed to Aβ (50 μM) for 48 hours; apoptosis was then quantified. Values are the mean and SEM of determinations made in four to six separate cultures. Modified from Guo et al. (1998b).

of sAPPα (Guo et al., 1998b). Because presenilin mutations enhance Aβ production, and may decrease sAPPα production (Ancolio et al., 1997), the latter findings suggest a role for reduced sAPPα levels in the pathogenic action of the mutations.

Links between activation of caspases, a family of proteases that play major roles in apoptosis, and the apoptosis-enhancing action of presenilin mutations is suggested by several recent findings. Both PS-1 and PS-2 are substrates for caspases, and caspase-induced cleavage of presenilins is increased in cells expressing mutant forms of the presenilins (Kim et al., 1997; Loetscher et al., 1997). The consequences of such presenilin cleavage for the cell death process are unclear, but a recent study suggests that a carboxy-terminal caspase cleavage product of PS-2 inhibits apoptosis (Vito et al., 1997).

In addition to sensitizing neurons to apoptosis, presenilin mutations may perturb physiological signaling pathways in neurons. Levels of choline acetyltransferase (ChAT), the enzyme responsible for the synthesis of the neurotransmitter acetylcholine, were greatly decreased in PC12 cells expressing mutant PS-1 compared to control PC12 cell lines and to lines overexpressing wild-type PS-1 (Pedersen et al., 1997). The adverse effect of mutant PS-1 on ChAT levels may be relevant to the well-documented deficits in ChAT and acetylcholine in basal forebrain cholinergic neurons and their cortical and hippocampal targets in AD (Bartus et al., 1982).

Another signaling pathway altered in cells expressing mutant PS-1 involve NF-κB (Guo et al., 1998b), a transcription factor previously linked to prevention of neuronal apoptosis (Mattson et al., 1997b). Presenilins have been shown to interact with β-catenin, a cytoplasmic protein involved in the *Wg/Wnt* signaling pathway (Zhou et al., 1997). Because members of the *Wnt* family are localized in the ER, the latter findings suggest a link between *Wnt* signaling and the pathogenesis of AD. Finally, a recent report demonstrated an interaction between presenilins and members of the filamin family of actin-binding proteins (Zhang et al., 1998), suggesting a role for presenilins in modulating cytoskeletal behaviors.

LINKS BETWEEN DOWN SYNDROME AND ALZHEIMER'S DISEASE

Essentially all individuals with Down syndrome develop classic AD pathology, including Aβ deposition, neuritic plaques, neurofibrillary tangles, and synapse loss (Cutler et al., 1985). Epidemiological data suggest an increased incidence of AD in mothers of adults with Down syndrome (Schupf et al., 1994), suggesting an underlying genetic predisposition to AD. Down syndrome is caused by trisomy 21, and efforts have therefore focused on determining which gene(s) on that chromosome is responsible for the abnormal phenotype (see Schellenberg et al., 1992, for review). Two likely candidates include the gene encoding APP and the gene encoding Cu/Zn superoxide dismutase (SOD). Overexpression of APP in transgenic mice does not result in formation of Aβ deposits or neuronal degeneration, suggesting that an increase in APP gene dosage is unlikely to explain the pathogenesis of Down syndrome. In a recent study, van Leeuwen et al. (1998) reported evidence for the existence of frameshift mutations in APP and ubiquitin-B in the brains of Down syndrome and AD patients. Frameshift mutations arise post-transcriptionally and could be an important factor in non-FAD forms of the disease, although the mechanisms whereby such acquired mutations might promote neuronal degeneration are unknown.

Mice overexpressing human Cu/Zn-SOD exhibit some alterations similar to those seen in Down syndrome patients, but do not exhibit amyloid deposition or neurofibrillary degeneration in brain (Ceballos et al., 1991). Primary cortical neurons in cell cultures established from Down syndrome fetuses exhibit spontaneous apoptosis associated with increased levels of oxidative stress (Busciglio and Yankner, 1995). As appears to be the case in AD, perturbed cellular calcium homeostasis and increased levels of oxidative stress appear to be fundamental alterations in Down syndrome. As evidence, glial cells from trisomy 16 mice (mice with phenotypic similarities to Down patients) exhibit increased rest levels of cytoplasmic calcium, and markedly increased calcium release from ER in response to agonists that activate the IP_3 pathway (Bambrick et al., 1997). The latter findings are of considerable interest in light of the evidence that mutations in APP and presenilins also perturb cellular calcium homeostasis (see earlier discussion), suggesting a common degenerative cascade of events involving dysregulation of calcium in Down syndrome and AD.

Perhaps the most reasonable interpretation of the Down phenotype is that multiple alterations arising from an extra chromosome 21 result in increased cellular oxidative stress, perturbed calcium homeostasis, and altered APP processing. The latter explanation is consistent with the fact that the organ systems most severely affected in Down syndrome, namely the heart and brain, are comprised of cells with high metabolic demands and high levels of oxidative stress and calcium mobilization. Approaches aimed are slowing or halting the pathogenic process in both AD and Down syndrome should therefore include treatments that suppress oxyradical accumulation and stabilize calcium homeostasis.

GENETIC RISK FACTORS IN ALZHEIMER'S DISEASE

An increasing number of genetic risk factors are being identified that predispose individuals to developing late-onset AD. One risk factor involves polymorphisms in the genes encoding apolipoprotein E (Saunders et al., 1993). There are three isoforms of apolipoprotein E (E2, E3, and E4), and persons with one or two copies of the E4 isoform have an increased risk for AD. It is unclear why E4 increases the risk of AD, but several relevant activities of the apolipoproteins have been identified. Perhaps the most straightforward explanation relates to the well-established role of apolipoproteins in transporting cholesterol in the blood and in mediating its uptake into cells in the liver. It has been known for many years that individuals with E4 are at increased risk for developing atherosclerosis (Hixon, 1991), suggesting that vascular abnormalities may contribute to the increased incidence of AD in individuals with E4. The latter explanation is parsimonious with the considerable evidence that there are alterations in cerebral blood vessels in the AD brain that appear to be associated with the neurodegenerative process (see de La Torre, 1997, for review). Indeed, data suggest an increased prevalence of AD in patients with hypertension or atherosclerosis, or both, and feeding rabbits a high-cholesterol diet may promote amyloid deposition in the brain (Sparks et al., 1994). Damage to endothelial cells in cerebral vessels may result in reduced nutrient availability to neurons, consistent with positron emission tomography imaging studies showing that levels of glucose uptake into brain cells is decreased in AD patients (Hoyer et al., 1991; Jagust et al., 1991).

Apolipoproteins may also have direct effects on neurons and glial cells that may influence the pathogenic process in AD. Apolipoprotein E affects neurite outgrowth and cell survival in cultured neurons (Nathan et al., 1994), and astrocytes produce apolipoprotein E and its expression is increased during nerve cell degeneration and regeneration (Poirier, 1994), suggesting a role in the brain's response to injury. The presence of apolipoprotein E in cerebral amyloid plaques in AD, and interactions of apolipoprotein E isoforms with Aβ suggest that apolipoprotein E may play a direct role in Aβ deposition or clearance, or both, from the brain (Castano et al.,

1995; Evans et al., 1995). It has also been reported that apolipoprotein E can protect cultured neurons against Aβ toxicity (Whitson et al., 1994). The latter effect of apolipoprotein E might possibly be explained by antioxidant effects of the Apo E (Miyata and Smith, 1996). Apolipoprotein E has also been shown to enhance the neuroprotective action of sAPPα, with E2 and E3 exhibiting a greater activity than E4 (Barger and Mattson, 1996). Apolipoprotein E2 and E3 isoforms might, therefore, protect against neurodegeneration in AD by promoting trophic activities of sAPPα. It has was reported that levels of lipoprotein receptor–related protein (LRP) are increased in reactive astrocytes and senile plaques in AD (Rebeck et al., 1993), suggesting a role for this receptor in the neurodegenerative process. Clearly, the actions of apolipoprotein E in the brain are complex and considerable further work will be required to establish which actions are central to the increased risk for AD in individuals with the E4 allele.

Two additional genes in which genetic variability may be linked to increased risk for AD are those encoding bleomycin hydrolase on chromosome 17 (Montoya et al., 1998) and α2-macroglobulin on chromosome 12 (Blacker et al., 1998). Bleomycin hydrolase is a cysteine protease in the papain superfamily that exhibits aminopeptidase and endopeptidase activities. Analyses of frequencies of a polymorphism at codon 1450 (A to G substitution) revealed a strong correlation with incidence of late onset AD. Interestingly, the increased risk for AD was confined to individuals lacking the Apo E4 allele. In the case of α2-macroglobulin, a previously recognized five-base deletion in exon 2 was shown to confer increased risk for AD (Blacker et al., 1998). The latter findings are intriguing because α2-macroglobulin is a protease inhibitor, is a major ligand for LRP, and may play roles in modulating Aβ aggregation and clearance (Narita et al., 1997; Du et al., 1998). Moreover, it is conceivable that alterations in α2-macroglobulin may enhance vascular dysfunction and damage associated with lipoprotein metabolism or Aβ production (Blanc et al., 1997b).

Aside from polymorphisms in nuclear DNA, alterations in mitocondrial DNA may also play a role in the pathogenesis of late-onset AD. Mitochondria are the energy factory of cells; they "burn" glucose and produce ATP, which cells use for a variety of vital processes. During the electron transport process oxyradicals are produced. Some mitochondrial proteins are encoded by genes in the cell nucleus, whereas others are encoded by DNA located within the mitochondria. Because mitochondrial DNA is bombarded with oxyradicals, and because mitochondria have relatively inefficient DNA repair mechanisms, mutations to mitochondrial DNA accumulate progressively during a lifetime. Recent analyses of mitochondrial DNA from AD patients and age-matched control subjects, have revealed an increased frequency of specific missense mutations in mitochondrial DNA in AD patients (Davis et al., 1997). However, subsequent studies have not confirmed such an association (Hutchin et al., 1997; Wallace et al., 1997). Mitochondrial DNA mutations are transferred from generation to generation in a manner that is quite different than nuclear DNA mutations. The fertilized egg contains only mitochondria from the mother (no mitochondria are contributed by the sperm), and so mitochondrial defects are passed on only from the mother. Some data suggest that there is a "maternal" genetic factor in AD, although it is apparently not a major factor.

HORMONAL MODIFIERS OF ALZHEIMER'S DISEASE RISK

Postmenopausal women who receive estrogen replacement therapy have a greatly reduced risk for developing AD (Henderson et al., 1994 ; Tang et al., 1996). Estrogen has a profound beneficial impact on many organ systems susceptible to age-related oxidative damage, including the cardiovascular system and nervous system. The mechanism whereby estrogen protects against AD may involve direct actions in neurons. Recent studies have shown that estrogens can protect cultured hippocampal neurons against insults relevant to the pathogenesis of AD including exposure to Aβ, and excitotoxic and metabolic insults (Goodman et al., 1996). The mechanism whereby estrogen protects neurons appears to involve inherent antioxidant activity involving the phenol group of the first steroid ring (Goodman et al., 1996; Green et al., 1997). Estrogen suppresses membrane lipid peroxidation and thereby preserves the function of key membrane transport systems that would otherwise be adversely affected by oxidative stress (Goodman et al., 1996; Keller and Mattson, 1997). Treatment of cultured neural cells expressing mutant PS-1 with estrogen largely abolishes the increased sensitivity of the cells to apoptosis induced by Aβ and trophic factor withdrawal (Mattson et al., 1997c). Data in the latter study indicated that the mechanism whereby estrogen protects against the adverse effects of presenilin mutations is by suppressing oxidative stress and preserving mitochondrial function.

Glucocorticoids, steroid hormones released from cells in the adrenal cortex in response to physical and psychological stressors, may promote neuronal degeneration in some brain regions in aging and AD (see Wise et al., 1997, for review). The mechanism whereby glucocorticoids endanger neurons appears to involve suppression of glucose transport resulting in metabolic compromise (Sapolsky, 1994). Exposure of cultured hippocampal neurons to glucocorticoids increases their vulnerability to oxidative and excitotoxic insults, and to death induced by Aβ (Goodman et al., 1996). The latter study provided evidence that glucocorticoids enhance disruption of calcium homeostasis and oxyradical production in neurons. There is evidence that AD patients have perturbed regulation of glucocorticoid production resulting in increased levels of circulating glucocorticoids, which could contribute to the neurodegenerative process.

DIETARY MODIFIERS OF ALZHEIMER'S DISEASE RISK

Increasing evidence suggests that AD may be forestalled through appropriate dietary measures. The same dietary risk factors that increase risk of other age-related degenerative conditions such as cardiovascular disease and diabetes also appear to increase risk for AD. These risk factors include high caloric intake (combined with low level of exercise) and low antioxidant intake. The ability of a low-calorie diet to extend life span and forestall the development of age-related diseases is well-known (Sohal and Weindruch, 1996) and appears to apply also to brain aging (Finch and Morgan, 1997). The beneficial effect of vitamin E supplementation in slowing the progression of AD in a recent clinical trial (Sano et al., 1997) suggests that antioxidants may provide protection against the neurodegenerative process. In addition, a recent epidemiological study showed that ethnic groups known to have diets high in calories and low in antioxidants have an increased incidence of AD (Tang et al., 1998). Moreover, recent studies have shown that maintenance of adult rats and mice on a dietary restriction regimen results in increased resistance of neurons in the hippocampus, substantia nigra, striatum, and cerebral cortex to insults relevant to the pathogenesis of AD and other age-related neurodegenerative disorders (Bruce-Keller et al., 1999; Duan and Mattson, 1999; Yu and Mattson, 1999).

REFERENCES

Ancolio, K., Marambaud, P., Dauch, P. & Checler, F. (1997). α-Secretase–derived product of β-amyloid precursor protein is decreased by presenilin-1 mutations linked to familial Alzheimer's disease. J. Neurochem. 69, 2494–2499.

Bambrick, L.L., Golovina, V.A., Blaustein, M.P., Yarowsky, P.J. & Krueger, B.K. (1997). Abnormal calcium homeostatis in astrocytes from the trisomy 16 mouse. Glia 19, 352–358.

Barger, S.W. & Mattson, M.P. (1996). Induction of neuroprotective κB-dependent transcription by secreted forms of the Alzheimer's β-amyloid precursor. Mol. Brain Res. 40, 116–126.

Barger, S.W., Horster, D., Furukawa, K., Goodman, Y., Krieglstein, J. & Mattson, M.P. (1995). Tumor necrosis factors α and β protect neurons against amyloid β-peptide toxicity: evidence for involvement of a κB-binding factor and attenuation of peroxide and Ca^{2+} accumulation. Proc. Natl. Acad. Sci. U.S.A. 92, 9328–9332.

Bartus, R.T., Dean, R.L., Beer, B. & Lippa, A.S. (1982). The cholinergic hypothesis of geriatric memory dysfunction. Science 214, 408–417.

Benzi, G. & Moretti, A. (1997). Contributions of mitochondrial alterations to brain aging. Adv. Cell Aging Gerontol. (JAI Press, Greenwich, CT) 2, 129–160.

Blacker, D., Wilcox, M.A., Laird, N.M., Rodes, L., Horvath, S.M., Go, R.C.P., Perry, R., Watson, B., Bassett, S.S., McInnis, M.G., Albert, M.S., Hyman, B.T. & Tanzi, R.E. (1998). Alpha-2 macroglobulin is a novel Alzheimer's disease gene on chromosome 12. Nat. Genet. In press.

Blanc, E.M., Kelly J. F., Mark, R.J. & Mattson, M.P. (1997a). 4-Hydroxynonenal, an aldehydic product of lipid peroxidation, impairs signal transduction associated with muscarinic acetylcholine and metabotropic glutamate receptors: possible action on $G\alpha_{q/11}$. J. Neurochem. 69, 570–580.

Blanc, E.M., Toborek, M., Mark, R.J., Hennig, B. & Mattson, M.P. (1997b). Amyloid β-peptide disrupts barrier and transport functions and induces apoptosis in vascular endothelial cells. J. Neurochem. 68, 1870–1881.

Blanc, E.M., Keller, J.N., Fernandez, S. & Mattson, M.P. (1998). 4-Hydroxynonenal, a lipid peroxidation product, inhibits glutamate transport in astrocytes. Glia 22, 149–160.

Borchelt, D.R., Thinakaran, G., Eckman, C.B., Lee, M.K., Davenport, F., Ratovitsky, T., Prada, C-M., Kim, G., Seekins, S., Yager, D., Slunt, H.H., Wang, R., Seeger, M., Levey, A.I., Gandy, S.E., Copeland, N.G., Jenkins, N.A,. Price, D.L., Younkin, S.G. & Sisodia, S.S. (1996). Familial

Alzheimer's disease-linked presenilin 1 variants elevate Aβ1-42/1-40 ratio in vitro and in vivo. Neuron 17, 1005–1013.

Boteva, K., Vitek, M., Mitsuda, H., de Silva, H., Xu, P.T., Small, G. & Gilbert, J.R. (1996). Mutation analysis of presenillin 1 gene in Alzheimer's disease. Lancet 347, 130–131.

Breitner, J.C.S., Gau, B.A., Welsh, K.A., Plassman, B.L., McDonald, W.M., Helms, M.J. & Anthony, J.C. (1994). Inverse association of anti-inflammatory treatments and Alzheimer's disease: initial results of a co-twin control study. Neurology 44, 227–232.

Bruce-Keller, A.J., Umberger, G., McFall, R., & Mattson, M.P. (1999). Food restriction reduces brain damage and improves behavioral outcome following excitotoxic and metabolic insults. Ann. Neurol. 45, 8–15.

Busciglio, J. & Yankner, B.A. (1995). Apoptosis and increased generation of reactive oxygen species in Down's syndrome neurons in vitro. Nature 378, 776–779.

Busciglio, J., Hartmann, H., Lorenzo, A., Wong, C., Baumann, K., Sommer, B., Staufenbiel, M. & Yankner, B.A. (1997). Neuronal localization of presenilin-1 and association with amyloid plaques and neurofibrillary tangles in Alzheimer's disease. J. Neurosci. 17, 5101–5107.

Campion, D., Flaman, J.M., Brice, A., Hannequin, D., Dubois, B., Martin, C., Moreau, V., Charbonnier, F., Didierjean, O. & Tardieu, S. (1995). Mutations of the presenilin I gene in families with early-onset Alzheimer's disease. Hum. Mol. Genet. 4, 2373–2377.

Campion, D., Brice, A., Dumanchin, C., Puel, M., Baulac, M., De La Sayette, V., Hannequin, D., Duyckaerts, C., Michon, A., Martin, C., Moreau, V., Penet, C., Martinez, M., Clerget-Darpoux, F., Agid, Y. & Frebourg, T. (1996). A novel presenilin 1 mutation resulting in familial Alzheimer's disease with an onset age of 29 years. Neuroreport 7, 1582–1584.

Castano, E., Prelli, F., Wisniewski, T., Golabek, A., Kumar, R., Soto, C. & Frangione, B. (1995). Fibrillogenesis in Alzheimer's disease of amyloid β peptides and apolipoprotein E. Biochem. J. 306, 599–604.

Ceballos, I., Nicole, A., Briand, P., Grimber, G., Delacourte, A., Flament, S., Blouin, J.L., Thevenin, M., Kamoun, P. & Sinet, P.M. (1991). Expression of human Cu-Zn superoxide dismutase gene in transgenic mice: model for gene dosage effect in Down syndrome. Free Rad. Res. Commun. 12–13, 581–589.

Conlon, R.A., Reaume, A.G.& Rossant, J. (1995). Notch1 is required for the coordinate segmentation of somites. Development 121, 1533–1545.

Cook, D.B., Sung, J.C., Golde, T.E., Felsenstein, K.M., Wojczyk, B.S., Tanzi, R.E., Trojanowski, J.Q., Lee, V.M.Y. & Doms, R.W. (1996). Expression and analysis of presenilin 1 in a human neuronal system: localization in cell bodies and dendrites. Proc. Natl. Acad. Sci. U.S.A. 93, 9223–9228.

Cribbs, D.H., Chen, L., Bendle, S.M. & La Ferla, F.M. (1996). Widespread neuronal expression of the presenilin-1 early-onset Alzheimer's disease in the murine brain. Am. J. Pathol. 148, 1797–1806.

Cruts, M., Backhovens, H., Wang, S.Y., Gassen, G.V., Theuns, J., De Jonghe, C.D., Wehnert, A., De Voecht, J., De Winter, G. & Cras, P. (1995). Molecular genetic analysis of familial early-onset Alzheimer's disease linked to chromosome 14q24.3. Hum. Mol. Genet. 4, 2363–2371.

Cutler, N.R., Heston, L.L., Davies, P., Haxby, J.V. & Schapiro, M.B. (1985). NIH Conference. Alzheimer's disease and Down's syndrome: new insights. Ann. Intern. Med. 103, 566–578.

Davis, R.E., Miller, S., Herrnstadt, C., Ghosh, S.S., Fahy, E., Shinobu, L.A., Galasko, D., Thal, L.J., Beal, M.F., Howell, N. & Parker, W.D. Jr. (1997). Mutations in mitochondrial cytochrome c oxidase genes segregate with late-onset Alzheimer disease. Proc. Natl. Acad. Sci. U.S.A. 94, 4526–4531.

de la Torre, J.C. (1997). Cerebrovascular changes in the aging brain. In: The Aging Brain (Mattson, M.P. & Geddes, J.W., eds.). Adv. Cell Aging Gerontol. (JAI Press, Greenwich, CT) 2, 77–107.

Doan, A., Thinakaran, G., Borchelt, D.R., Slunt, H.H., Ratovistsky, T., Podlisny, M., Selkoe, D.J., Seeger, M., Gandy, S.E., Price, D.L. & Sisodia, S.S. (1996). Protein topology of presenilin 1. Neuron 17, 1023–1030.

Doyle, E., Bruce, M.T., Breen, K.C., Smith, D.C., Anderton, B. & Regan, C.M. (1990). Intraventricular infusions of antibodies to amyloid β-protein precursor impair the acquisition of a passive avoidance response in the rat. Neurosci. Lett. 115, 97–102.

Du, Y., Bales, K.R., Dodel, R.C., Liu, X., Glinn, M.A., Horn, J.W., Little, S.P. & Paul, S.M. (1998). α2-Macroglobulin attenuates β-amyloid peptide 1-40 fibril formation and associated neurotoxicity of cultured fetal rat cortical neurons. J. Neurochem. 70, 1182–1188.

Duan, W. & Mattson, M.P. (1999). Dietary restriction and 2-deoxyglucose administration improve behavioral outcome and reduce degeneration of dopaminergic neurons in models of Parkinson's disease. J. Neurosci. Res. 57, 195–206.

Duff, K., Eckman, C., Zehr, C., Yu, X., Prada, C.-M., Perez-tur, J., Hutton, M., Buee, L., Harigaya, Y., Yager, D., Morgan, D., Gordon, M.N., Holcomb, L., Refolo, L., Zenk, B., Hardy, J. & Younkin, S. (1996). Increased amyloid-β42(43) in brains of mice expressing mutant presenilin 1. Nature 383, 710–713.

Dyrks, T., Dyrks, E., Hartmann, T., Masters, C. & Beyreuther, K. (1992). Amyloidogenicity of βA4 and βA4-bearing amyloid protein precursor fragments by metal-catalyzed oxidation. J. Biol. Chem. 267, 18210–18217.

Elder, G.A., Tezapsidis, N., Carter, J., Shioi, J,. Bouras, C., Li, D., Johnston, J.M., Efthimiopoulos, S., Friedrich, V.L. & Robakis, N.K. (1996). Identification and neuron specific expression of the S182/presenilin 1 protein in human and rodent brains. J. Neurosci. Res. 45, 308–320.

Evans, K.C., Berger, E.P., Cho, C.G., Wiesbraber, K.H. & Lansbury, P.T. (1995). Apolipoprotein E is a kinetic but not a thermodynamic inhibitor of amyloid formation: implications for the pathogenesis and treatment of Alzheimer's disease. Proc. Natl. Acad. Sci. U.S.A. 92, 763–767.

Finch, C.E. & Morgan, T.E. (1997). Food restriction and brain aging. In: The Aging Brain (Mattson, M.P. and Geddes, J.W., eds.). Adv. Cell Aging Gerontol. (JAI Press, Greenwich, CT) 2, 279–297.

Finch, C.E. & Tanzi, R.E. (1997). Genetics of aging. Science 278, 407–411.

Furukawa, K. & Mattson, M.P. (1998). Secreted amyloid precursor protein α selectively suppresses NMDA receptor currents in hippocampal neurons: involvement of cyclic GMP. Neuroscience 83, 429–438.

Furukawa, K., Barger, S.W., Blalock, E. & Mattson, M.P. (1996a). Activation of K^+ channels and suppression of neuronal activity by secreted β-amyloid precursor protein. Nature 379, 74–78.

Furukawa, K., Sopher, B., Rydel, R.E., Begley, J.G., Martin, G.M. & Mattson, M.P. (1996b). Increased activity-regulating and neuroprotective efficacy of α-secretase–derived secreted APP is conferred by a C-terminal heparin-binding domain. J Neurochem. 67, 1882–1896.

Furukawa, K., Guo, Q., Schellenberg, G.D. & Mattson, M.P. (1998). Presenilin-1 mutation alters NGF-induced neurite outgrowth, calcium homeostasis, and transcription factor (AP-1) activation in PC12 cells. J. Neurosci. Res. 52, 618–624.

Gabuzda, D., Busciglio, J., Chen, L., Matsudaira, P. & Yankner, B.A. (1994). Inhibition of energy metabolism alters the processing of amyloid precursor protein and induces a potentially amyloidogenic derivative. J. Biol. Chem. 269, 13623–13628.

Giannakopoulos, P., Bouras, C., Kovari, E., Shioi, J., Tezapsidis, N., Hof, P.R. & Robakis, N.K. (1997). Presenilin-1 immunoreactive neurons are preserved in late-onset Alzheimer's disease. Am. J. Pathol. 150, 429–436.

Goodman, Y. & Mattson, M.P. (1994). Secreted forms of β-amyloid precursor protein protect hippocampal neurons against amyloid β-peptide–induced oxidative injury. Exp. Neurol. 128, 1–12.

Goodman, Y., Bruce, A.J., Cheng, B. & Mattson, M.P. (1996). Estrogens attenuate and corticosterone exacerbates excitotoxicity, oxidative injury and amyloid β-peptide toxicity in hippocampal neurons. J. Neurochem. 66, 1836–1844.

Green, P.S., Gordon, K. & Simpkins, J.W. (1997). Phenolic A ring requirement for the neuroprotective effects of steroids. J. Steroid Biochem. Mol. Biol. 6, 229–235.

Guo, Q., Furukawa, K., Sopher, B.L., Pham, D.G., Xie, J., Robinson, N., Martin, G.M. & Mattson, M.P. (1996). Alzheimer's PS-1 mutation perturbs calcium homeostasis and sensitizes PC12 cells to death induced by amyloid β-peptide. Neuroreport 8, 379–383.

Guo, G., Sopher, B.L., Pham, D.G., Furukawa, K., Robinson, N., Martin, G.M. & Mattson, M.P. (1997). Alzheimer's presenilin mutation sensitizes neural cells to apoptosis induced by trophic factor withdrawal and amyloid β-peptide: involvement of calcium and oxyradicals. J. Neurosci. 17, 4212–4222.

Guo, Q., Christakos, S., Robinson, N. & Mattson, M.P. (1998a). Calbindin blocks the pro-apoptotic actions of mutant presenilin-1: reduced oxidative stress and preserved mitochondrial function. Proc. Natl. Acad. Sci. U.S.A. 95, 3227–3232.

Guo, Q., Robinson, N. & Mattson, M.P. (1998b). Secreted APPα counteracts the pro-apoptotic action of mutant presenilin-1 by activation of NF-κB and stabilization of calcium homeostasis J. Biol. Chem. 273, 12341–12351.

Guo, Q., Fu, W., Xie, J., Luo, H., Sells, S.F., Geddes, J.W., Bondada, V., Rangnekar, V.M. & Mattson, M.P. (1998c). Par-4 is a mediator of neuronal degeneration associated with the pathogenesis of Alzheimer's disease. Nat. Med. 4, 957–962.

Guo, Q., Fu, W., Sopher, B.L., Miller, M.W., Ware, C.B., Martin, G.M. & Mattson, M.P. (1999a). Increased vulnerability of hippocampal neurons to excitotoxic necrosis in presenilin-1 mutant knock-in mice. Nature Med. 5, 101–106.

Guo, Q., Sebastian, L., Sopher, B.L., Miller, M.W., Ware, C.B., Martin, G.M., & Mattson, M.P. (1999b). Increased vulnerability of hippocampal neurons from presenilin-1 mutant knock-in mice to amyloid beta-peptide toxicity: Central roles of superoxide production and caspase activation. J. Neurochem. 72, 1019–1029.

Hardy, J. (1997). Amyloid, the presenilins and Alzheimer's disease. Trends Neurosci. 20, 154–159.

Hartmann, H., Busciglio, J., Baumann, K.H., Staufenbiel, M. & Yankner, B.A. (1997). Developmental regulation of presenilin-1 processing in the brain suggests a role in neuronal differentiation. J. Biol. Chem. 272, 14505–14508.

Henderson, V.W., Paganini-Hill, A., Emanuel, C.K., Dunn, M.E., &Buckwalter, J.G. (1994). Estrogen replacement therapy in older women. Comparisons between Alzheimer's disease cases and nondemented control subjects. Arch. Neurol. 51, 896–900.

Hensley, K., Carney, J.M., Mattson, M.P., Aksenova, M., Harris, M., Wu, J.F., Floyd, R. & Butterfield, D.A. (1994). A model for β-amyloid aggregation and neurotoxicity based on free radical generation by the peptide: relevance to Alzheimer's disease. Proc. Natl. Acad. Sci. U.S.A. 91, 3270–3274.

Hixon, J.E. (1991). Apolipoprotein E polymorphisms affect atherosclerosis in young males. Arterioscler. Thromb. 11, 1237–1244.

Hoyer, S., Nitsch, R. & Oesterreich, K. (1991). Predominant abnormality in cerebral glucose utilization in late-onset dementia of the Alzheimer type: a cross-sectional comparison against advanced late-onset and incipient early-onset cases. J. Neural Transm. 3, 1–14.

Huber, G., Martin, J.R., Loffler, J. & Moreau, J.L. (1993). Involvement of amyloid precursor protein in memory formation in the rat: an indirect antibody approach. Brain Res. 603, 348–352.

Hutchin, T.P., Heath, P.R., Pearson, R.C. & Sinclair, A.J. (1997). Mitochondrial DNA mutations in Alzheimer's disease. Biochem. Biophys. Res. Commun. 241, 221–225.

Ishida, A., Furukawa, K., Keller, J.N. & Mattson, M.P. (1997). Secreted form of β-amyloid precursor protein shifts the frequency dependence for induction of LTD, and enhances LTP in hippocampal slices. Neuroreport 8, 2133–2137.

Jagust, W.J., Seab, J.P., Huesman, R.H., Valk, P.E., Mathis, C.A., Reed, B.R., Coxson, P.G. & Budinger, T.F. (1991). Diminished glucose transport in Alzheimer's disease: dynamic PET studies. J. Cereb. Blood Flow Metab. 11, 323–330.

Kamino, K., Sato, S., Sakaki, Y., Yoshiiwa, A., Nishiwaki, Y., Takeda, M., Tanabe, H., Nishimura, T., Ii, K., St George-Hyslop, P.H., Miki, T. & Ogihara, T. (1996). Three different mutations of presenilin 1 gene in early-onset Alzheimer's disease families. Neurosci. Lett. 208, 195–198.

Keller, J.N. & Mattson, M.P. (1997). 17β-Estradiol attenuates oxidative impairment of synaptic Na^+/K^+-ATPase activity, glucose transport and glutamate transport induced by amyloid β-peptide and iron. J. Neurosci. Res. 50, 522–530.

Keller, J.N., Mark, R.J., Bruce-Keller, A.J., Blanc, E.M., Rothstein, J.D., Uchida, K. & Mattson, M.P. (1997a). 4-Hydroxynonenal, an aldehydic product of membrane lipid peroxidation, impairs glutamate transport and mitochondrial function in synaptosomes. Neuroscience 80, 685–696.

Keller, J.N., Pang, Z., Geddes, J.W., Begley, J.G., Germeyer, A., Waeg, G. & Mattson, M.P. (1997b). Impairment of glucose and glutamate transport and induction of mitochondrial oxidative stress and dysfunction in synaptosomes by amyloid β-peptide: role of the lipid peroxidation product 4-hydroxynonenal. J. Neurochem. 69, 273–284.

Keller, J.N., Guo, Q., Holtsberg, F.W., Bruce-Keller, A.J. & Mattson, M.P. (1998a). Increased sensitivity to mitochondrial toxin-induced apoptosis in neural cells expressing mutant presenilin-1 is linked to perturbed calcium homeostasis and enhanced oxyradical production. J. Neurosci. 18, 4439–4450.

Keller, J.N., Kindy, M.S., Holtsberg, F.W., St Clair, D.K., Yen, H.C., Germeyer, A., Steiner, S.M., Bruce-Keller, A.J., Hutchins, J.B. & Mattson, M.P. (1998b). Mitochondrial MnSOD prevents neural apoptosis and reduces ischemic brain injury: suppression of peroxynitrite production, lipid peroxidation and mitochondrial dysfunction. J. Neurosci. 18, 687–697.

Kelly, J., Furukawa, K., Barger, S.W., Mark, R.J., Rengen, M.R., Blanc, E.M., Roth, G.S. & Mattson, M.P. (1996). Amyloid β-peptide disrupts carbachol-induced muscarinic cholinergic signal transduction in cortical neurons. Proc. Natl. Acad. Sci. U.S.A. 93, 6753–6758.

Kim, T.-W., Pettingell, W.H., Jung, Y.-K., Kovacs, D.M. & Tanzi, R.E. (1997). Alternative cleavage of Alzheimer-associated presenilins during apoptosis by a caspase-3 family protease. Science 277, 373–376.

Kovacs, D.M., Fausett, H.J., Page, K.J., Kim, T-W., Moir, R.D., Merriam, D.E., Hollister, R.D., Hallmark, O.G., Mancini, R., Felsenstein, K.M., Hyman, B.T., Tanzi, R.E. & Wasco, W. (1996). Alzheimer-associated presenilins 1 and 2: neuronal expression in brain and localization to intracellular membranes in mammalian cells. Nat. Med. 2, 224–229.

Kruman, I., Bruce-Keller, A.J., Bredesen, D.E., Waeg, G. & Mattson, M.P. (1997). Evidence that 4-hydroxynonenal mediates oxidative stress-induced neuronal apoptosis. J. Neurosci. 17, 5097–5108.

Lah, J.J., Heilman, C.J., Nash, N.R., Rees, H.D., Yi, H., Counts, S.E. & Levey, A.I. (1997). Light and electron microscopic localization of presenilin-1 in primate brain. J Neurosci. 17, 1971–1980.

Lannfelt, L., Basun, H., Wahlund, L.O., Rowe, B.A. & Wagner, S.L. (1995). Decreased α-secretase-cleaved amyloid precursor protein as a diagnostic marker for Alzheimer's disease. Nat. Med. 1, 829–832.

Lassmann, H. (1996). Patterns of synaptic and nerve cell pathology in Alzheimer's disease. Behav. Brain Res. 78, 9–14.

Lehmann, S., Chiesa, R. & Harris, D.A. (1997). Evidence for a six-transmembrane domain structure of presenilin 1. J. Biol. Chem. 272, 12047–12051.

Levitan, D. & Greenwald, I. (1995). Facilitation of lin-12-mediated signalling by sel-12, a *Caenorhabditis elegans* S182 Alzheimer's disease gene. Nature 377, 351–354.

Levitan, D., Doyle, T.G., Brousseau, D., Lee, M.K., Thinakarin, G., Slunt, H.H., Sisodia, S.S. & Greenwald, I. (1996). Assessment of normal and mutant human presenilin function in *Caenorhabditis elegans*. Proc. Natl. Acad. Sci. U.S.A. 93, 14940–14944.

Levy-Lahad, E., Wasco, W., Poorkaj, P., Romano, D.M., Oshima, J., Pettingell, W.H., Yu, C-E., Jondro, P.D., Schmidt, S.D., Wang, K., Crowley, A.C., Fu, Y-H., Guenette, S.Y., Galas, D., Nemens, E., Wijsman, E.M., Bird, T.D., Schellenberg, G.D. and Tanzi, R.E. (1995). Candidate gene for the chromosome 1 familial Alzheimer's disease locus. Science 269, 973–977.

Loetscher, H., Deuschle, U., Brockhaus, M., Reinhardt, D., Nelboeck, P., Mous, J., Grunberg, J., Haass, C. & Jacobsen, H. (1997). Presenilins are processed by caspase-type proteases. J. Biol. Chem. 272, 20655–20659.

Loo, D., Copani, A., Pike, C., Whittemore, E., Walencewicz, A. & Cotman, C.W. (1993). Apoptosis is induced by β-amyloid in cultured central nervous system neurons. Proc. Natl. Acad. Sci. U.S.A. 90, 7951–7955.

Mark, R.J., Hensley, K., Butterfield, D.A. & Mattson, M.P. (1995). Amyloid β-peptide impairs ion-motive ATPase activities: evidence for a role in loss of neuronal Ca^{2+} homeostasis and cell death. J. Neurosci. 15, 6239–6249.

Mark, R.J., Lovell, M.A., Markesbery, W.R., Uchida, K. & Mattson, M.P. (1997a). A role for 4-hydroxynonenal in disruption of ion homeostasis and neuronal death induced by amyloid β-peptide. J. Neurochem. 68, 255–264.

Mark, R.J., Pang, Z., Geddes, J.W. & Mattson, M.P. (1997b). Amyloid β-peptide impairs glucose uptake in hippocampal and cortical neurons: involvement of membrane lipid peroxidation. J. Neurosci. 17, 1046–1054.

Mattson, M.P. (1990). Antigenic changes similar to those seen in neurofibrillary tangles are elicited by glutamate and calcium influx in cultured hippocampal neurons. Neuron 4, 105–117.

Mattson, M.P. (1992). Calcium as sculptor and destroyer of neural circuitry. Exp.Gerontol. 27, 29–49.

Mattson, M.P. (1994). Secreted forms of β-amyloid precursor protein modulate dendrite outgrowth and calcium responses to glutamate in cultured embryonic hippocampal neurons. J. Neurobiol. 25, 439–450.

Mattson, M.P. (1995). Untangling the pathophysiochemistry of β-amyloid. Nat. Struct. Biol. 2, 926–928.

Mattson, M.P. (1997a). Cellular actions of β-amyloid precursor protein, and its soluble and fibrillogenic peptide derivatives. Physiol. Rev. 77, 1081–1132.

Mattson, M.P. (1997b). Mother's legacy: Mitochondrial DNA mutations and Alzheimer's disease. Trends Neurosci. 20, 373–375.

Mattson, M.P. (1998). Modification of ion homeostasis by lipid peroxidation: roles in neuronal degeneration and adaptive plasticity. Trends Neurosci. 21, 53–57.

Mattson, M.P. & Begley, J.G. (1996). Amyloid β-peptide alters thrombin-induced calcium responses in cultured human neural cells. Amyloid 3, 28–40.

Mattson, M.P. & Furukawa, K. (1996). Programmed cell life: anti-apoptotic signaling and therapeutic strategies for neurodegenerative disorders. Retorative Neurol. Neurosci. 9, 191–205.

Mattson, M.P. & Rydel, R.E. (1996). Amyloid ox-tox transducer. Nature 382, 674–675.

Mattson, M.P., Cheng, B., Davis, D., Bryant, K., Lieberburg, I. & Rydel, R.E. (1992). β-Amyloid peptides destabilize calcium homeostasis and render human cortical neurons vulnerable to excitotoxicity. J. Neurosci. 12, 376–389.

Mattson, M.P., Cheng, B., Culwell, A., Esch, F., Lieberburg, I. & Rydel, R.E. (1993). Evidence for excitoprotective and intraneuronal calcium-regulating roles for secreted forms of β-amyloid precursor protein. Neuron 10, 243–254.

Mattson, M.P., Fu, W., Waeg, G. & Uchida, K. (1997a). 4-Hydroxynonenal, a product of lipid peroxidation, inhibits dephosphorylation of the microtubule-associated protein tau. Neuroreport 8, 2275–2281.

Mattson, M.P., Goodman, Y., Luo, H., Fu, W. & Furukawa, K. (1997b). Activation of NF-κB protects hippocampal neurons against oxidative stress-induced apoptosis: evidence for induction of Mn-SOD and suppression of peroxynitrite production and protein tyrosine nitration. J. Neurosci. Res. 49, 681–697.

Mattson, M.P., Robinson, N. & Guo, Q. (1997c). Estrogens stabilize mitochondrial function and protect neural cells against the pro-apoptotic action of mutant presenilin-1. Neuroreport 8, 3817–3821.

Mattson, M.P., Guo, Q., Furukawa, K. & Pedersen, W.A. (1998). Presenilins, the endoplasmic reticulum, and neuronal apoptosis in Alzheimer's disease. J. Neurochem. 70, 1–14.

McGeer, P.L. & McGeer, E.G. (1995). The inflammatory response system of brain: implications for therapy of Alzheimer and other neurodegenerative diseases. Brain Res. Rev. 21, 195–218.

Miyata, M. & Smith, J.D. (1996). Apolipoprotein E allele-specific antioxidant activity and effects on cytotoxicity by oxidative insults and beta-amyloid peptides. Nat. Genet. 14, 55–61.

Montoya, S.E., Aston, C.E., DeKosky, S.T., Kamboh, M.I., Lazo, J.S. & Ferrell, R.E. (1998). Bleomycin hydrolase is associated with risk of sporadic Alzheimer's disease. Nat. Gen. 18, 211–212.

Mucke, L., Abraham, C., Ruppe, M., Rockenstein, E., Togas, S., Mallory, M., Alford, M. & Masliah, E. (1995). Protection against HIV-1 gp120-induced brain damage by neuronal expression of human amyloid precursor protein. J. Exp. Med. 181, 1551–1556.

Mullan, M. & Crawford, F. (1993). Genetic and molecular advances in Alzheimer's disease. Trends Neurosci. 16, 398–403.

Murphy, G.M., Forno, L.S., Ellis, W.G., Nochlin, D., Levy-Lahad, E., Poorkaj, P., Bird, T.D., Jiang, Z. & Cordell, B. (1996). Antibodies to presenilin proteins detect neurofibrillary tangles in Alzheimer's disease. Am. J. Pathol. 149, 1839–1846.

Narita, M., Holtzman, D.M., Schwartz, A.L. & Bu, G. (1997). α2-Macroglobulin complexes with and mediates the endocytosis of β-amyloid peptide via cell surface low-density lipoprotein receptor-related protein. J. Neurochem. 69, 1904–1911.

Nathan, B.P., Bellosta, S., Sanan, D.A., Weisgraber, K. H., Mahle, R.W. & Pitas, R.E. (1994). Differential effects of apolipoproteins E3 and E4 on neuronal growth in vitro. Science 264, 850–852.

Nitsch, M.R., Farber, A.S., Growdon, H.J. & Wurtman, J.R. (1993). Release of amyloid β-protein precursor derivatives by electrical depolarization of rat hippocampal slices. Proc. Natl. Acad. Sci. U.S.A. 90, 5191–5193.

Pedersen, W., Guo, Q., Hartman, B.K. & Mattson, M.P. (1997). NGF-independent reduction in choline acetyltransferase activity in PC12 cells expressing mutant presenilin-1. J. Biol. Chem. 272, 22397–22400.

Perez-Tur, J., Froelich, S., Prihar, G., Crook, R., Baker, M., Duff, K., Wragg, M., Busfield, F., Lendon, C., Clark, R.F., Roques, P., Fuldner, R.A., Johnston, J., Cowburn, R., Forsell, C., Axelman, K., Lilius, L., Houlden, H., Karran, E., Roberts, G.W., Rossor, M., Adams, M.D., Hardy, J., Goate, A., Lannfelt, L. & Hutton, M. (1995). A mutation in Alzheimer's disease destroying a splice acceptor site in the presenilin-1 gene. Neuroreport 7, 297–301.

Poirier, J. (1994). Apolipoprotein E in animal models of CNS injury and in Alzheimer's disease. Trends Neurosci. 17, 525–530.

Poorkaj, P., Bird, T.D., Wijsman, E., Nemens, E., Garruto, R.M., Anderson, L., Andreadis A., Wiederholt, W.C., Raskind, M. & Schellenberg, G.D. (1998). Tau is a candidate gene for chromosome 17 frontotemporal dementia. Ann. Neurol. 43, 815–825.

Querfurth, H.W. & Selkoe, D.J. (1994). Calcium ionophore increases amyloid β peptide production by cultured cells. Biochemistry 33, 4550–4561.

Rebeck, G.W., Reiter, J.S., Strickland, D.K. & Hyman, B.T. (1993). Apolipoprotein in Alzheimer's disease: allelic variation and receptor interactions. Neuron 11, 575–580.

Roch, J.-M., Masliah, E., Roch-Levecq, A., Sundsmo, M., Otero, D.A., Veinbergs, I. & Saitoh, T. (1994). Increase of synaptic density and memory retention by a peptide representing the trophic domain of the amyloid β/A4 protein precursor. Proc. Natl. Acad. Sci. U.S.A. 91, 7450–7454.

Rogaev, E.I., Sherrington, R., Rogaeva, E.A., Levesque, G., Ikeda, M., Liang, Y., Chi, H., Lin, C., Holman, K., Tsuda, T., Mar, L., Sorbi, S., Nacmias, B., Piacentini, S., Amaducci, L., Chumakov, I., Cohen, D., Lannfelt, L., Fraser, P.E., Rommens, J.M. & St. George-Hyslop, P.H. (1995). Familial Alzheimer's disease in kindreds with missense mutations in a gene on chromosome 1 related to the Alzheimer's disease type 3 gene. Nature 376, 775–778.

Sano, M., Ernesto, C., Thomas, R.G., Klauber, M.R., Schafer, K., Grundman, M., Woodbury, P., Growdon, J., Cotman, C.W., Pfeiffer, E., Schneider, L.S. & Thal, L.J. (1997). A controlled trial of selegiline, alpha-tocopherol, or both as treatment for Alzheimer's disease. The Alzheimer's Disease Cooperative Study. N. Engl. J. Med. 336, 1216–1222.

Sapolsky, R.M. (1994). The physiological relevance of glucocorticoid endangerment of the hippocampus. Ann. N.Y. Acad. Sci. 746, 294–304.

Saunders, A.M., Strittmatter, W.J., Schmechel, D., St. George-Hyslop, P.H., Pericak, V.M.A., Joo, S.H., Rosi, B.L., Gusella, J.F., Crapper-MacLachlan, D.R., Alberts, M.J., Hulette, C., Crain, B., Goldgaber, D. & Roses, A.D. (1993). Association of apolipoprotein E allele e4 with late-onset familial and sporadic Alzheimer's disease. Neurology 43,1467–1472.

Schellenberg, G.D., Kamino, K., Bryant, E.M., Moore, D. & Bird, T.D. (1992). Genetic heterogeneity, Down syndrome, and Alzheimer disease. Prog. Clin. Biol. Res. 379, 215–226.

Scheuner, D., Eckman, C., Jensen, M., Song, X., Citron, M., Suzuki, N., Bird, T.D., Hardy, J., Hutton, M., Kukull, W., Larson, E., Levy-Lahad, E., Viitanen, M., Peskind, E., Poorkaj, P., Schellenberg, G., Tanzi, R., Wasco, W., Lannfelt, L., Selkoe, D. & Younkin, S. (1996). The amyloid β protein deposited in the senile plaques of Alzheimer's disease is increased *in vivo* by the presenilin 1 and 2 and APP mutations linked to familial Alzheimer's disease. Nat. Med. 2, 864–870.

Schupf, N., Kapell, D., Lee, J.H., Ottman, R. & Mayeux, R. (1994). Increased risk of Alzheimer's disease in mothers of adults with Down's syndrome. Lancet 344, 353–356.

Selkoe, D.J. (1991). The molecular pathology of Alzheimer's disease. Neuron 6, 487–498.

Sells, S.F., Han, S.-S., Muthukkumar, S., Maddiwar, N., Johnstone, R., Boghaert, E., Gillis, D., Liu, G., Nair, P., Monnig, S., Collini, P., Mattson, M.P., Sukhatme, V.P., Zimmer, S.G., Wood, D.P., McRoberts, J.W., Shi, Y. & Rangnekar, V.M. (1997). Expression and function of the leucine zipper protein PAR-4 in apoptosis. Mol. Cell. Biol. 17, 3823–3832.

Shen, J., Bronson, R.T., Chen, D. F., Xia, W., Selkoe, D.J. & Tonegawa, S. (1997). Skeletal and CNS defects in Presenilin-1-deficient mice. Cell 89, 629–639.

Sherrington, R., Rogaev, E.I., Liang, Y., Rogaeva, E.A., Levesque, G., Ikeda, M., Chi, H., Lin, C., Li, G., Holman, K., Tsuda, T., Mar, L., Foncin, J-F., Bruni, A.C., Montesi, M.P., Sorbi, S., Rainero, I., Pinessi, L., Nee, L., Chumakov, I., Pollen, D., Brookes, A., Sansequ, P., Polinsky, R.J., Wasco, W., Da Silva, H.A.R., Haines, J.L., Pericak-Vance, M.A., Tanzi, R.E., Roses, A.D., Fraser, P.E., Rommens, J.M. & St. George-Hyslop, P.H. (1995). Cloning of a gene bearing missense mutations in early-onset familial Alzheimer's disease. Nature 375, 754–760.

Smale, G., Nichols, N.R. & Brady, D.R. (1995). Evidence for apoptotic cell death in Alzheimer's disease. Exp. Neurol. 133, 225–230.

Smith-Swintosky, V.L., Pettigrew, L.C., Craddock, S.D., Culwell, A.R., Rydel, R E. & Mattson, M. P. (1994). Secreted forms of β-amyloid precursor protein protect against ischemic brain injury. J. Neurochem. 63, 781–784.

Sohal, R.S. & Weindruch R. (1996). Oxidative stress, caloric restriction, and aging. Science 273, 59–63.

Sparks, D.L., Scheff, S.W., Hunziker, J.C., Liu, H., Landers, T. & Gross, D.R. (1994). Induction of Alzheimer-like beta-amyloid immunoreactivity in the brains of rabbits with dietary cholesterol. Exp. Neurol. 126, 88–94.

Su, J.H., Anderson, A.J., Cummings, B. & Cotman, C.W. (1994). Immunocytochemical evidence for apoptosis in Alzheimer's disease. Neuroreport 5, 2529–2533.

Tanahashi, H., Kawakatsu, S., Kaneko, M., Yamanaka, H., Takahashi, K. & Tabira, T. (1996). Sequence analysis of presenilin-1 gene mutation in Japanese Alzheimer's disease patients. Neurosci. Lett. 218, 139–141.

Tang, M.X., Jacobs, D., Stern, Y., Marder, K., Schofield, P., Gurland, B., Andrews, H. & Mayeux, R. (1996). Effect of oestrogen during menopause on risk and age at onset of Alzheimer's disease. Lancet 348, 429–432.

Tang, M.X., Stern, Y., Marder, K., Bell, K., Gurland, B., Lantigua, R., Andrews, H., Feng, L., Tycko, B. & Mayeux, R. (1998). The Apo-E4 allele and risk of Alzheimer's disease among African Americans, whites and hispanics. JAMA. 279, 751–755.

Uchihara, T., el Hachimi, H.K., Duyckaerts, C., Foncin, J.F., Fraser, P.E., Levesque, L., St. George-Hyslop, P.H. & Hauw, J.J. (1996). Widespread immunoreactivity of presenilin in neurons of normal and Alzheimer's disease brains: double-labeling immunohistochemical study. Acta Neuropathol. 92, 325–330.

van Leeuwen, F.W., de Kleijn, D.P., van den Hurk, H.H., Neubauer, A., Sonnemans, M.A., Sluijs, J.A., Koycu, S., Ramdjielal, R.D.J., Salehi, A., Martens, G.J.M., Grosveld, F.G., Peter, J., Burbach, H. & Hol, E.M. (1998). Frameshift mutants of beta amyloid precursor protein and ubiquitin-B in Alzheimer's and Down patients. Science 279, 242–247.

Van Nostrand, W.E., Wagner, S.L., Shankle, W.R., Farrow, J.S., Dick, M., Rozemuller, J.M., Kuiper, M.A., Wolters, E.C., Zimmerman, J., Cotman, C.W. & Cunningham, D. (1992). Decreased levels of soluble amyloid β-protein precursor in cerebrospinal fluid of live Alzheimer disease patients. Proc. Natl. Acad. Sci. U.S.A. 89, 2551–2555.

Vito, P., Ghayur, T. & D'Adamio, L. (1997). Generation of anti-apoptotic presenilin-2 polypeptides by alternative transcription, proteolysis, and caspase-3 cleavage. J. Biol. Chem. 272, 28315–28320.

Wallace, D.C., Stugard, C., Murdock, D., Schurr, T. & Brown, M.D. (1997). Ancient mtDNA sequences in the human nuclear genome: a potential source of errors in identifying pathogenic mutations. Proc. Natl. Acad. Sci. U.S.A. 94, 14900–14905.

Weidemann, A., Paliga, K., Durrwang, U., Czech, C., Evin, G., Masters, C.L. & Beyreuther, K. (1997). Formation of stable complexes between two Alzheimer's disease gene products, presenilin-2 and β-amyloid precursor protein. Nat. Med. 3, 328–332.

Whitson, J., Mims, M., Strittmatter, W., Yamaki, T., Morrisett, J. & Appel, S. (1994). Attenuation of the neurotoxic effect of Aβ amyloid peptide by apolipoprotein E. Biochem. Biophys. Res. Commun. 199, 163–170.

Wise, P.M., Herman, J.P. & Landfield, P. (1997). Neuroendocrine aspects of the aging brain. In: The Aging Brain (Mattson, M.P. and Geddes, J.W., eds.). Adv. Cell Aging Gerontol. (JAI Press, Greenwich, CT) 2, 193–241.

Wolozin, B., Iwasaki, K., Vito, P., Ganjei, J.K., Lacana, E., Sunderland, T., Zhao, B., Kusiak, J.W., Wasco, W. & D'Adamio, L. (1996). Participation of presenilin 2 in apoptosis: enhanced basal activity conferred by an Alzheimer mutation. Science 274, 1710–1713.

Wong, P.C., Zheng, H., Chen, H., Becher, M.W., Sirinathsinghji, D.J.S., Trumbauer, M.E., Chen, H.Y., Price, D.L., Van der Ploeg, L.H.T. & Sisodia, S.S . (1997). Presenilin 1 is required for *Notch*1 and *Dll*1 expression in the paraxial mesoderm. Nature 387, 288–292.

Xia, W., Zhang, J., Perez, R., Koo, E.H. & Selkoe, D.J. (1997). Interaction between amyloid precursor protein and presenilins in mammalian cells: implications for the pathogenesis of Alzheimer's disease. Proc. Natl. Acad. Sci. U.S.A. 94, 8208–8213.

Yan, S.D., Chen, X., Fu, J., Chen, M., Zhu, H., Roher, A., Slattery, T., Zhao, L., Nagashima, M., Morser, J., Migheli, A., Nawroth, P., Stern, D. & Schmidt, A.M. (1996). RAGE and amyloid-beta peptide neurotoxicity in Alzheimer's disease. Nature 382, 685–691.

Yu, Z.F. & Mattsoin, M.P. (1999). Dietary restriction and 2-deoxyglucose administration reduce focal ischemic brain damage and improve behavioral outcome: Evidence for a preconditioning mechanism. J. Neurosci. Res. 57, 830–839.

Zhang, W., Han, S.W., McKeel, D.W., Goate, A. & Wu, J.Y. (1998). Interaction of presenilins with the filamin family of actin-binding proteins. J. Neurosci. 18, 914–922.

Zhou, J., Liyangage, U., Medina, M., Ho, C., Simmons, A.D., Lovett, M. & Kosik, K.S. (1997). Presenilin 1 interaction in the brain with a novel member of the Armadillo family. Neuroreport 8, 1489–1494.

Chapter 2

The Biology of Trinucleotide Repeat Disorders

P. HEMACHANDRA REDDY and DANILO A. TAGLE

Advances in Cell Aging and Gerontology
Volume 3, pages 33–79

ISBN: 0-7623-0405-7

INTRODUCTION

The identification of a novel class of mutation, referred to as dynamic mutations (Richards and Sutherland, 1992) or tandem repeat expansions (Tables 1 and 2), has revolutionized our understanding of how genetic diseases can arise in humans. Dynamic mutations arise from genetic instability of short tandem DNA repeat sequences that are generally polymorphic. Instability generally favors the tendency toward expansion, though on some occasions, contractions in repeat lengths have been observed. The repeat tract expands in patients beyond the normal range found in the general population and becomes inherited as unstable DNA sequences. Dynamic mutations now provide an explanation for what in the past several decades has been an enigma in the field of genetics; that is, for some dominantly inherited diseases, the age of onset manifests at an earlier age and with increasing severity in successive generations, a phenomenon most commonly referred to as genetic anticipation. Classic Mendelian genetics failed to explain this phenomenon, which has often been ascribed to biases in clinical ascertainment of patients (Penrose, 1948).

Dynamic mutations involving dinucleotide (primarily $[CA]_n$ and $[AT]_n$) repeat sequences have been shown to be associated with certain cancers (Ionov et al., 1993; Thibodeau et al., 1996, 1998), as well as trinucleotide repeat instability in diseases of the central nervous system. Dinucleotide repeat instabilities tend to be small contractions or expansions, are quite widespread in the genome, and arise through defects in the mismatch repair system (Modrich, 1991). By contrast, trinucleotide repeats involved in inherited neurological disorders correlate with expansion of the repeat sequence, occur less frequently and usually within the context of the genome organization of genes, and arise primarily from replication errors during meiosis.

Currently, dynamic mutations involving trinucleotide repeat tracts have been implicated in at least 13 inherited neurological diseases in humans. These triplet repeat mutations can be effectively categorized into expansions occurring within noncoding (Figure 1) and coding sequences (Figure 2). The former includes $(CGG)_n$ expansions resulting in chromosome fragile sites (FRAXA, FRAXE and FRA11B) associated with mental retardation (Verkerk et al., 1991; Knight et al., 1993; Jones et al., 1994), $(CTG)_n$ expansions found in myotonic dystrophy (Brook et al., 1992; Fu et al., 1992; Mahadevan et al., 1992), and $(GAA)_n$ expansions seen in Friedreich's ataxia. Two additional fragile sites, FRAXF and FRA16A, are associated with $(CGG)_n$ triplet repeat expansions but have not been implicated in any human disease (Nancarrow et al., 1994; Parrish et al., 1994). Dynamic mutations in coding regions involving $(CAG)_n$ expansions observed in patients with spinal and bulbar muscular atrophy (SBMA) (La Spada et al., 1991); Huntington's disease (HD) (Huntington's Disease Collaborative Research Group,1993); spinocerebellar ataxia

Table 1. Summary of Noncoding Part of Trinucleotide Repeat Associated Neurological Disorders

Disease	*Gene Location*	*Type of Repeat*	*Location of Repeats*	*Repeat Normal*	*Length Premutation*	*Mutation*	*Clinical Hallmarks*
Myotonic distrophy	19q13.3	CTG	3′ UTR	5–37	50–180	200–2000	Muscle weakness, wasting myotonia, cataracts, and mental retardation
Fragile X syndrome (FRAXA)	Xq27.3	CGG	5′ UTR	6–52	60–200	200–2000	Mental retardation
Fragile x syndrome (FRAXE)	Xq28	GCC	??	6–25	43–200	> 200	Mental retardation
Fredreich's ataxia	9q13	GAA	Intron	7–22	—	200–900	Progressive gait, limb ataxia, and dysarthria

Table 2. Summary of CAG-Repeat Associated Neurological Disorders

Disease	*Gene Location*	*Transcript Size (kb)*	*Repeat Normal*	*Length Affected*	*Localization of Normal Protein*	*Clinical Hallmarks*	*Pathology*
Spinobulbar mus-cular atrophy	Xq11-12	3.5	11–34	40–62	??	Muscle weakness and atrophy	Loss of bulbar neurons
Dentatorubral-pal-lidoluysian atrophy	12q23-24	4.5	7–25	49–88	??	Choreoathetosis, ataxia, and dementia	Globus pallidus, dentatorubral, and subthalamus
Huntington's dis-ease	4p16.3	10.5	6–35	40–120	Cytoplasm	Chorea, dementia, and intellectual impairment	Striatum and cerebral cortex
Spinocerebellar ataxia type 1	6p22-23	10.0	6–39	41–81	Nucleus	Ataxia, dysarthria, and dysmetria	Cerebellar cortex, dentate nucleus, and brain stem
Spinocerebellar ataxia type 2	12q23-24.1	4.2	15–29	35–59	Cytoplasm	Ataxia and dysarthria	Cerebellum, pontine nuclei, and substantia nigra

Spinocerebellar ataxia type 3	14q32.1	1.2	13–36	68–79	Cytoplasm	Ataxia, dystonia, and opthalmoplegia	Substantia nigra, globus pallidus, pontine nucleus, and caudate nucleus
Spinocerebellar ataxia type 6	19p13	8.5	4–16	21–27	Not determined	Ataxia, dysarthria, nystagmus, and vibratory sensory loss	Cerebellar and mild brain stem atrophy
Spinocerebellar ataxia type 7	3p12-13	7.5	7–17	38–130	Cytoplasm	Ataxia dysarthria, pyramidal and extrapyramidal signs, deep sensory loss, or dementia	Cerebellar cortex, basis pontis, inferior olive, and retinal ganglial cells

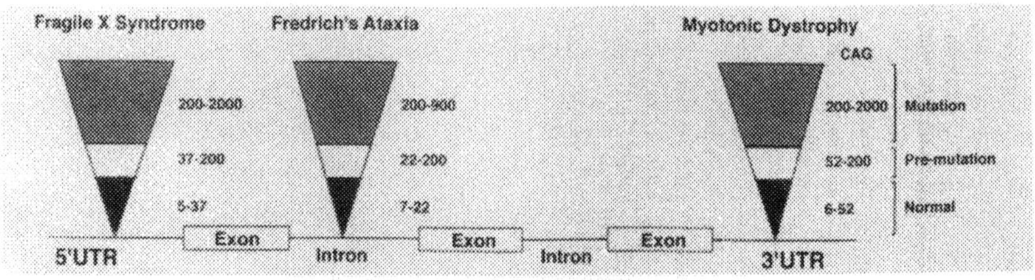

Figure 1. Schematic representation of noncoding part of trinucleotide repeat associated diseases, showing gene location and number of repeat expansions within a model gene. The number and type of repeats are shown on right side of a triangle for each disease gene.

type 1 (SCA1) (Orr et al., 1993); dentatorubral-pallidoluysian atrophy (DRPLA), which is allelic to Haw River syndrome (Burke et al., 1994; Koide et al., 1994; Nagafuchi et al., 1994a, 1994b; Potter et al., 1996); Machado-Joseph disease, which is allelic to spinocerebellar ataxia type 3 (SCA3) (Kawaguchi et al., 1994); and more recently spinocerebellar ataxia type 2 (SCA2) (Imbert et al., 1996) and spinocerebellar ataxia type 6 (SCA6) (Zhuchenko et al., 1997). Instability of triplet repeats in coding sequences are generally characterized by relatively small expansions between 40 and 138 copies. On the other hand, repeat instability involving noncoding sequences are generally quite large (in the order of thousands of copies)

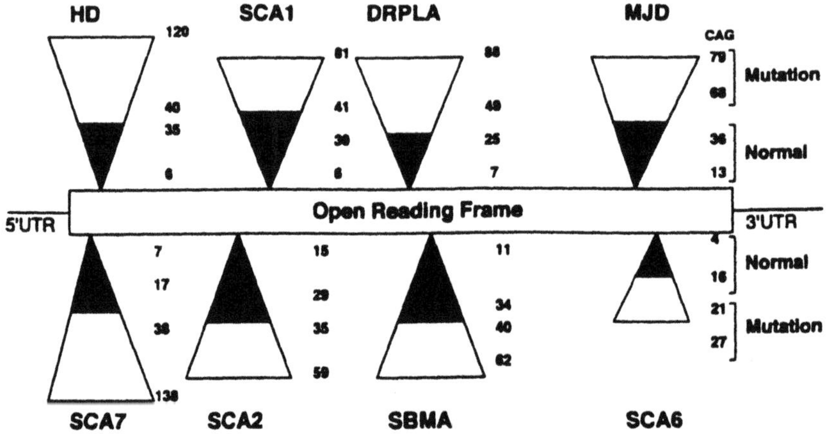

Figure 2. Schematic representation of coding part of CAG trinucleotide repeat associated diseases, showing gene location and number of repeat expansions within the model gene. Normal and mutation range is shown on right side of a triangle for each disease gene.

and in some cases are associated with fragile sites in chromosomes. In this chapter a survey is presented of diseases caused by triplet repeat expansions, as well as an overview of mechanisms involved in their pathogenesis.

NEUROLOGICAL DISORDERS CAUSED BY TRIPLET REPEAT INSTABILITY

Of the 3^4 possible permutations for triplet repeats, the number of unique trinucleotide repeat combinations can be effectively reduced to ten (Table 3) when considering the tandem nature of the repeat and the complementary strand of the DNA molecule. In the past, a nomenclature system has been proposed (Sutherland and Richards, 1993) based on the system proposed for microsatellite repeats at the Human Genome Mapping Conference 10.5 (Williamson, 1990). Using this system, the four possible nucleotide bases of DNA that make up the tandem repeat are written in alphabetical order from 5′ to 3′ direction; thus, for example, the $(CAG)_n$ and $(CTG)_n$ repeats found in coding sequences and myotonic dystrophy, respectively, would be represented by $(AGC)_n$. However, the repeats have been and continue to be described according to the codon of the open reading frame of the specific gene that they are found in, or in the order of the first obvious tandem repeat. It is also becoming apparent that categorizing triplet repeat diseases based on this nomenclature would be too simplistic as the mechanism of pathogenesis may be different based on the type of repeat sequence, its location within the target gene, protein function, and other molecules it interacts with. For example, clear distinctions should be made between diseases caused by $(CTG)_n$ and $(CAG)_n$ instability, as discussed in more detail later in this chapter.

Table 3. Ten Possible Triplet Repeat Sequence Combinations

Sequence	*Complement*	*Disease Caused*
AAT	ATT	
ATC	GAT	
CGG	CCG	FRAXA, FRAXE
CTG	CAG	DM, SBMA, HD, DRPLA, SCA 1,2,3,6,7
GTA	TAC	
GTT	AAC	
CAC	GTG	
AGG	CCT	
TCG	CGA	
AAG	CTT	FRDA

REPEAT INSTABILITY IN NONCODING SEQUENCES

Fragile X Syndrome

Clinical Features

X-linked mental retardation is the most common form of inherited mental retardation in humans. FRAXA (for FRAgile site, X chromosome, A site) is a form of mental retardation associated with a folate-sensitive fragile site at Xq27.3 seen in metaphase spreads from cultures of peripheral blood lymphocytes grown in folate-deprived media (Lubs, 1969; Gerald, 1980). FRAXA, also known as fragile X syndrome, is characterized by moderate to severe mental retardation, macro-orchidism, facial abnormalities (narrow face with prominent forehead, enlarged ears, prominent jaw), and high-pitched jocular speech (Giraud et al., 1976; Harvey et al., 1977; Sutherland, 1977). Expression is variable, with mental retardation being the most common feature. FRAXA mental retardation affects approximately 1 in 1,250 males and approximately 1 in 2,000 females. Longitudinal observations indicate a deterioration of IQ level with age. An age-related increase in the volume of the hippocampus and an age-related decrease in the volume of the superior temporal gyrus was observed in fragile X patients compared to controls (Reiss et al., 1994).

Genetics

An unusual genetic aspect of this disorder can be observed in which 20% of obligatory carrier males are phenotypically and cytogenetically normal, yet can transmit the disease. The heterozygous daughters of such males are never mentally retarded and have few or no fragile sites, and by contrast in the next generation, a third of heterozygous females are affected (Sherman et al., 1985). This phenomenom, which is a form of genetic anticipation, indicates that the risk of mental retardation is much higher in the grandchildren of a normal transmitting male than in his siblings and has come to be known as the Sherman paradox (Fu et al., 1991).

The cloning and characterization of the gene involved in FRAXA helped unravel the mystery of this paradox. The gene, designated *FMR1* for *f*ragile site *m*ental *r*etardation-*1*, expresses a 4.8-kb message in human brain (Verkerk et al., 1991), which encodes an RNA-binding protein (Ashley et al., 1993). The *FMR1* gene consists of 17 exons spanning 38 kb of genomic sequence (Eichler et al., 1993). In a large number of patients with fragile X syndrome, the phenotype correlates with the expansion of $(CGG)_n$ repeat in the 5′ untranslated region (5′ UTR) of the *FMR1* gene. In unaffected individuals, the repeat size range is from 6 to 52 repeats with a majority of alleles having 30 CGGs. In contrast, patients with fragile X syndrome have expansions of 230 to more than 2,000 repeats. Carriers have an intermediate range of 60 to 230 repeats (sometimes referred to as a premutation) (see Table 1). The premutation exhibits meiotic instability toward large increases in the number of repeats, particularly when transmitted maternally. Quite often, affected females inherit the expanded allele from female carriers but not from fathers with the premutation. All males who carry the full mutation are clinically affected, whereas only 50% of females who are heterozygous for the full mutation are affected, reflecting the random pattern of X inactivation. Since the size of the $(CGG)_n$ repeat increases in subsequent generations, the penetrance or severity of the disease also tends to follow this increase in repeat size.

The expansion of the $(CGG)_n$ repeat to the full mutation range results in hypermethylation of the 5′ end of the *FMR1* gene, thus rendering the gene transcriptionally inactive. However, the 5′ region of the *FMR1* gene remains unmethylated when it carries the premutation repeat range of 50 to 230 CGGs. There are very few mutations in the *FMR1* gene that can lead to fragile X syndrome outside of the $(CGG)_n$ repeat expansion. An exception would be 1.6-kb deletion proximal to the CGG repeat of the *FMR1* gene in one family with fragile X syndrome (Meijer et al., 1994). In the latter case, the *FMR1* transcript was not detectable in the affected males, suggesting deletion of the *FMR1* promoter sequences due to rearrangement and loss of part of the $(CGG)_n$ repeats. The deletion patients had approximately 45 CGG repeats in the *FMR1* gene; these were not interrupted by AGG triplets that are usually present in both normal and expanded repeats (Meijer et al., 1994). Most RNA-binding proteins contain a highly conserved 50 amino acid sequence, the K homology (KH) domain, originally described in the pre-mRNA–binding heterogeneous nuclear ribonucleoprotein (hnRNP) K protein. A single point mutation causing an Ile367Asn (isoleucine to asparagine change at position 367) mutation in a conserved residue of the KH domain in the FMR1 protein product can lead to severe fragile X syndrome (De Boulle et al., 1993; Siomi et al., 1994), further indicating the correlation between loss of RNA-binding activity in the *FMR1* gene and fragile X syndrome.

Transcript and Protein

In situ hybridization studies indicated that *FMR1* is expressed from an early stage in proliferating and migrating cells of the nervous system and retina as well as several non-nervous tissues (Abitbol et al., 1993). Expression is highest in cholinergic neurons of the nucleus basalis magnocellularis and in pyramidal neurons of the hippocampus in the brain of 25-week-old fetuses. The FMR1 protein is found to be cytoplasmic in cells from normal individuals and from those with the premutation but is not detectable by immunoblotting of cell lysates from patients (Devys et al., 1993; Siomi et al.,1993; Verheij et al., 1993). The highest levels of expression can be found in neurons, while glial cells contained very low levels. In epithelial tissues, levels of FMR1 were higher in dividing layers. In adult testes, FMR1 was detected only in spermatogonia.

Animal Model

In situ hybridization studies in adult mice also demonstrated that expression of the mouse homologous gene, *Fmr-1*, was localized to several areas of the brain and the tubules of the testes (Hinds et al., 1993). A nullizygous mouse model for fragile X syndrome (Dutch-Belgian Fragile X Consortium, 1994) demonstrated features similar to the human disease: macroorchidism, learning deficits, and hyperactivity.

Fragile X Site-E Mental Retardation

Clinical Features

A few patients with mental retardation and fragile X cytogenetic abnormality did not show the $(CGG)_n$ repeat expansion in the FMR1 gene. In patients with this rare form of mental retardation, a second folate-sensitive fragile site was identified approximately 600 kb telomeric to the FRAXA site at Xq28 (Sutherland and Baker, 1992). In addition to the mild mental retardation, FRAXE patients exhibit nonspecific developmental delay and variable microcephaly, and were either of normal stature or unusually tall but otherwise were unremarkable physically (Knight et al., 1996; Abrams et al., 1997).

Genetics

A gene, *FMR-2*, associated with FRAXE mental retardation was subsequently identified where a repeat sequence $(GCC)_n$ in the 5′ UTR of the gene was found to be highly unstable (Knight et al., 1993, 1994; Gu et al., 1996; Gecz et al., 1997). The repeat size in normal alleles ranged from 6 to 35 copies of the GCC repeat, whereas patients with the FRAXE-fragile site have greater than 200 copies of the repeat sequence (Knight et al., 1993, 1994; Chakrabarti et al., 1996). As in FRAXA, expansion of the repeat results in hypermethylation and transcriptional downregulation of the gene (Gu et al., 1996; Barnicoat et al., 1997).

Transcript and Protein

The *FMR2* gene is expressed as an 8.7-kb transcript in placenta and adult brain. It is a large gene spanning 22 exons within a 500-kb genomic interval on Xq28. The FMR2 protein shows homology to transcription factors involved in cancer, and lymphocyte differentiation (Gu et al., 1996, Gecz et al., 1997).

Myotonic Dystrophy

Clinical Features

Myotonic dystrophy (DM) is an autosomal dominant disorder characterized by muscle wasting in the face and extremities, myotonia affecting cardiac and smooth muscles, early cataract formation, immunoglobulin abnormalities, hypogonadism, frontal balding, and often mild mental retardation (Caughey and Myrianthopoulos, 1963; Harper, 1989). DM is the most common form of inherited neuromuscular disease in adults and, unlike other muscular dystrophies, it is the only one involving myotonia and early involvement of the distal rather than the proximal muscles of the extremities. Though the disease has been mapped to a single locus on chromosome 19q13.3, the disease manifestation is quite pleiotropic both within and between families. The mildest form of the disease has a late-onset presentation and is characterized by cataracts with little or no muscle involvement whereas the more common form of the disease presents in early adult life and is characterized by myotonia and muscle weakness. Congenital DM is observed in a small proportion of cases with features of neonatal hypotonia accompanied by respiratory distress, mental retardation, and facial diplegia (Harper, 1975) and is commonly inherited maternally (Harper and Dyken, 1972; Tanaka et al., 1981).

Genetics

The most striking example of genetic anticipation is found in studies of DM pedigrees. Before the discovery of the genetic defect responsible for DM, there was no biological explanation for anticipation and it was generally attributed to ascertainment bias of patients. The identification of highly variable genomic length clones (Aslanidis et al., 1992; Buxton et al., 1992; Harley et al., 1992) led to the finding of an amplified trinucleotide $(CTG)_n$ repeat in the 3′ UTR of a protein kinase gene on chromosome 19, which explained many of the unusual features of this disorder (Brook et al., 1992).

A correlation between severity of DM and the number of repeats was found whereby normal individuals have repeat lengths from 5 to 37 copies, mildly affected persons possessing 50 to 80 repeats, and severely affected individuals have 1,000 or more copies. Repeat instability is observed in successive generations with the largest expansions being inherited through the maternal line; thus explaining the occurrence of the severe congenital form almost exclusively in the offspring of affected women (Mahadevan et al., 1992; Tsilfidis et al., 1992). The largest repeat sizes were seen in patients with congenital DM, while the less affected patients had minimal repeat sizes.

Transcript and Protein

The DM gene contains 15 exons distributed over about 13 kb of genomic DNA (Fu et al., 1993; Shaw et al., 1993). It encodes a protein, DM protein kinase (*DMPK*), of 624 amino acids with an N-terminal domain highly homologous to cyclic adenosine monophosphate (cAMP)-dependent serine-threonine protein kinases, an intermediate domain with a high alpha-helical content and weak similarity to various filamentous proteins, and a hydrophobic C-terminal segment (Fu et al., 1992). Human and mouse show homology in the *DMPK* 3′ UTR sequence but the unstable (CTG)5-30 motif is found uniquely in humans (Mahadevan et al., 1993). In both species another active gene, called *DMR-N9*, was found in close proximity to the DM-kinase gene. *DMR-N9* transcripts, mainly expressed in brain and testis, possess a single large open reading frame (Jansen et al., 1992). The function of the protein product is unknown. It is possible that clinical manifestations of DM caused by the expanded CTG-repeat may compromise the expression of DM-kinase or DMR-N9 proteins. Fu et al. (1993) demonstrated that decreased levels of the mRNA and protein expression are associated with the adult form of DM, suggesting that the autosomal dominant nature of the disease is due to a DMPK dosage deficiency. The CTG repeat acts to both reduce the levels of primary DM transcripts and impair the processing of these transcripts (Carango et al., 1993).

Animal Models

Mouse models wherein *Dmpk* expression was knocked out showed subtle abnormalities during all phases of mouse development and aging. In these *Dmpk*$^{-/-}$mice, a late-onset progressive skeletal myopathy characterized by decreased force generation, increased fiber degeneration and regeneration, and loss of sarcomeric organization was noted (Jansen et al., 1996; Reddy et al., 1996). Myotubes of *Dmpk*$^{-/-}$mice (Jansen et al., 1996) exhibited a higher resting intracellular calcium concentration and smaller and slower calcium responses than did myotubes of wild-type mice after triggering with either acetylcholine or high external potassium (Benders et al., 1997).

On the other hand, transgenic mice overexpressing the *DMPK* gene showed cardiomyopathy, but this condition was dependent on the transgene copy number (Jansen et al., 1996). Additional transgenic mouse models expressing either a truncated version of the *DMPK* gene with 162 CTG repeats in the 3′ UTR (Monckton et al., 1997) or a genomic clone containing the entire *DMPK* gene with 55 CTG repeats (Gourdon et al., 1997) showed intergenerational and somatic cell instability of the CTG trinucleotide repeats.

Friedreich's Ataxia

Clinical Features

Friedreich's ataxia (FRDA) is an autososmal recessive disorder generally characterized by incoordination of limb movements, dysarthria, nystagmus, diminished or absent tendon reflexes, Babinski sign, impairment of position and vibratory senses, scoliosis, pes cavus, and hammer toe. The triad of hypoactive knee and ankle jerks, signs of progressive cerebellar dysfunction, and preadolescent onset is commonly regarded as sufficient for diagnosis. Pathologically, FRDA involves the spinocerebellar tracts, dorsal columns, pyramidal tracts, and to a lesser extent, the cerebellum and medulla. A variant form of the disease was identified with later onset (older than 20 years), a more slowly progressive peripheral involvement (muscle weakness and loss of vibratory perception), a lower incidence or absence of cardiomyopathy, and a slightly longer life span than the most common form of FRDA (Barbeau et al., 1984; De Michele et al., 1994).

Genetics

The FRDA1 locus was assigned to the human 9q12-q13 region by genetic linkage analysis (Fujita et al., 1989; Hanauer et al., 1990) and by in situ hybridization studies (Raimondi et al., 1990; Shaw et al., 1990). A second FRDA locus has been

Table 4. Point Mutations Associated with Friedreich's Ataxia

Nucleotide Change	*Protein Change*	*Reference*
3 G → T	Met1Ile	Cossee et al., 1997
316 T → G	Leu106Ter	Campuzano et al., 1996
385 -2 A → G	Loss of splice acceptor	Campuzano et al., 1996
389 G → T	Gly130Val	Bidichandani et al., 1998
460 A → T	Ile154Phe	Campuzano et al., 1996

identified (FRDA2) in two patients who do not show linkage to chromosome 9 by haplotype analysis despite being clinically indistinguishable from FRDA1 patients. A 150-kb interval on the long arm of chromosome 9 was subsequently defined (Montermini et al., 1995) as the FRDA1 candidate region from which the *FRDA* gene was identified using cDNA selection (Campuzano et al., 1996). The majority (94%) of FRDA patients were homozygous for an expansion of a GAA repeat in the first intron of the gene (Campuzano et al., 1996). In a small number of patients who were heterozygous for the GAA repeat expansion, point mutations (Table 4) within the gene were also identified in conjunction with the repeat expansion. The repeat expansions in FRDA patients were typically between 120 and 1,186 copies compared to 7 to 22 copies in the normal population (Campuzano et al., 1996; Durr et al., 1996; Filla et al., 1996). As with other triplet repeat diseases, the larger expansions correlated with earlier age of onset and shorter times to loss of ambulation. In FRDA, the size of the GAA expansions, especially in the repeat length of the shorter allele, was associated with the frequency of cardiomyopathy and loss of reflexes in the upper limbs. Though the FRDA GAA repeat is highly unstable during meiosis, unlike other triplet repeat diseases contractions outnumber expansions and there appears to be no parental bias in towards the tendency to contract or expand (Pianese et al., 1997). In addition, clinical variability in FRDA is related to the size of the expanded repeat where the milder forms of the disease (i.e., late-onset FRDA and FRDA with retained reflexes) are associated with shorter expansions, especially with the shorter of the two expanded alleles (Monroe et al., 1997). Absence of cardiomyopathy was also associated with shorter alleles.

Transcript and Protein

The FRDA transcript is comprised of six exons with expression in a wide range of tissues, with the highest levels of expression in the heart and spinal cord. Lower levels of expression were detected in the cerebellum, and no expression was demonstrated in the cerebral cortex. A yeast homolog has been identified for the FRDA gene (yeast frataxin homolog, *YFH1*), which encodes a mitochondrial protein involved in iron homeostasis and respiratory function (Babcock et al., 1997). Human frataxin was subsequently shown to be a mitochondrial protein. A deficient activity of the iron-sulfur (Fe-S) cluster-containing subunits of mitochondrial respiratory complexes I, II, and III in the endomyocardial biopsy of two unrelated FRDA patients has been reported (Rotig et al., 1997a, 1997b). It is possible that cellular degeneration in FRDA may be caused by mitochondrial dysfunction owing to abnormal iron accumulation, as observed in yeast cells deficient for a frataxin homolog. In RNase protection assays, the expanded GAA repeat was shown to interfere with in vitro transcription in a repeat length-dependent manner (Bidichandani et al., 1998).

REPEAT INSTABILITY IN CODING SEQUENCES

Spinal and Bulbar Muscular Atrophy

Clinical Features

Spinal and bulbar muscular atrophy (SBMA), also known as Kennedy's disease, is an X-linked recessive disorder characterized by degeneration of the spinal and bulbar motor neurons. The clinical features of SBMA consist of progressive muscle weakness and atrophy usually beginning in the third to fifth decade of life (Kennedy et al., 1968). The hallmark pathological feature is an atrophy of lower motor neurons throughout the entire spinal cord (Harding et al., 1982; Sobue et al., 1989). It affects 1 in 50,000 males and, in addition to the neurological features, also show gynecomastia, testicular atrophy, infertility, and androgen insensitivity.

Genetics

The SBMA locus was mapped to chromosome Xq11-12 in the same region where the androgen receptor gene (AR) was identified. Examination of the AR sequences in SBMA patients showed CAG repeat expansion within the first exon of the gene (Fischbeck et al., 1991; La Spada et al., 1991). In normal subjects the size of CAG repeat ranges from 11 to 34, and in affected individuals the expanded CAG repeats vary from 40 to 72 (La Spada et al., 1991).

Transcript and Protein

The androgen receptor (AR) is a member of the steroid/nuclear receptor superfamily, in which all members share basic structural and functional homology. These superfamily members have three distinct domains: the DNA binding domain, ligand binding domain, and amino terminal domain. However, despite the similarity in structure and function of the receptor superfamily, activation of different receptors leads to different cellular responses. The polyglutamine tract encoded by the tandem CAG repeat is in a portion of the AR protein that is not directly involved in hormone or DNA binding. There is no correlation between the size of the CAG repeat and the presence of altered in vitro androgen binding. However, there is a correlation between CAG repeat length and disease severity, with the mildest clinical manifestations being associated with the smallest CAG repeat (La Spada et al., 1991; Doyu et al., 1992).

The SBMA gene has varied expression in brain but is particularly prevalent in motor nuclei of cranial nerves and in motor neurons. Androgen influences the survival of facial and hypoglossal motor neurons and increases the rate of regeneration of their axons after axotomy. The expanded polyglutamine tract in the AR has been shown to result in a reduction in transcriptional activity in non-neuronal and neuronal systems (MacLean et al., 1997). In vitro studies by Butler et al. (1998) indicated the formation of cleavage products from mutant *AR* expression constructs in a repeat length-dependent manner when transfected into mouse neuroblastoma cells. A 74-kDa cleavage product was observed along with full length 110-kDa protein in cells expressing the mutant *AR* gene. Immunofluorescence microscopy on the transfected cells indicated that the mutant protein was translocated into the nucleus in the form of dense aggregates with very little of the normal protein in the nucleus.

Huntington's Disease

Clinical Features

Huntington's disease (HD) is an autosomal dominant inherited disorder characterized by chorea (involuntary movements of arms and legs), dystonia, intellectual impairment, and emotional disturbances (Folstein, 1989). Pathologically, HD is characterized by selective and premature neuronal death of striatal projection neurons early in the disease process and widespread, progressive neurodegeneration in other brain regions (cortex and hippocampus) in later stages. Mean age of onset is at 37 years of age, although 5% to 6% of cases present as juvenile HD with onset before age 20 (Folstein, 1989). Anticipation is observed, especially when the disease is inherited from the paternal line, often resulting in juvenile HD (Huntington's Disease Collaborative Research Group, 1993).

Genetics

HD was the first genetic disorder to be mapped through linkage analysis using a systematic screening of polymorphic genetic markers (Gusella et al., 1983). The initial mapping to the short arm of chromosome 4p16.3 was followed by a decade of intensive collaborative research effort leading to the identification of a novel gene (*IT15* for interesting transcript 15) of unknown function (Huntington's Disease Collaborative Research Group, 1993). The underlying mutation in HD is an expanded polyglutamine $(CAG)_n$ within exon 1 of the *IT15* gene. In the normal population, the number of CAG repeats in *IT15* ranges from 6 to 35, whereas in affected individuals, it ranges from 36 to 121 (Huntington's Disease Collaborative Research Group, 1993). Rubinzstein et al., (1996) examined a large number of borderline cases with 30 to 40 CAG repeats, and concluded that the disease can be manifested even with 35 repeats, while some individuals are not affected even though they carry 40 CAG repeat units in one or both alleles (Rubinzstein et al.,

1996). This finding indicates that though the expansion is the primary determinant for the disease, other modifying factors can affect the age of onset.

Transcript and Protein

The *IT15* gene consists of 67 exons spanning a genomic interval of about 180 kb and shows a high degree of sequence conservation with murine, rat, and pufferfish homologs (Strong et al., 1993; Barnes et al., 1994; Lin et al., 1994; Baxendale et al., 1995; Schmitt et al., 1995). The human and mouse sequences are 90% homologous in the coding region. The mouse *Hdh* sequence has seven CAG repeats that are nonpolymorphic likely owing to a CAA interruption after the first three CAG repeats. The gene encodes a 348-kDa protein, huntingtin, consisting of 3,144 amino acids. As in all CAG repeat expansion diseases, the CAG repeat is part of the coding region and encodes for polyglutamines (Gutekunst et al., 1995; Jou and Myers, 1995; Sharp et al., 1995). Although huntingtin is expressed ubiquitously (Li et al., 1993a) selective neurodegeneration is found in the brain, particularly in the striatum (caudate and putamen), and the cerebral cortex (Spargo et al., 1993; Vonsattel et al., 1985). The link between this selective neuropathology and abnormal huntingtin expression still remains unclear.

In the majority of cells, huntingtin was present in the cytoplasm, but it was also present in the nucleus of some tissues, including neurons. Wood and co-workers (1996) characterized murine huntingtin, and showed apparent variations in the subcellular localization of huntingtin in human, mouse, and rat brains. Subcellular fractionation revealed that in mouse brain, huntingtin localizes in the soluble S3 fraction; in rat brain huntingtin was localized in the soluble S3 fraction and also in the membrane p2 and p3 fractions; and in both affected and normal human brain huntingtin was membrane-bound, the distribution being essentially the same as that of synaptophysin. Trottier and co-workers (1995a, 1995b) have raised monoclonal antibodies against four different regions of the HD protein. Immunohistochemical analysis revealed that these monoclonal antibodies detect huntingtin in the perikarya and varicosities of some neurons as well as in neuropiles, and there was punctate staining likely to represent nerve endings. Immunocytofluorescence studies with COS cells transfected with normal CAGs (HD7 or HD15) or expanded CAGs (HD73) showed a cytoplasmic rather than nuclear localization of huntingtin in human brains. This observation agrees with other studies by DiFiglia and colleagues (1995) and Gutekunst and co-workers (1995). All these studies suggested that both normal and mutant huntingtin are localized in cytoplasm. However, a small proportion of mutant huntingtin is localized in the nucleus.

Li and co-workers (1995) identified HD-associating protein (HAP)-1 using the yeast two hybrid system. Cell extracts containing huntingtin with 82 repeats yielded stronger binding compared with 44 repeats, and both proteins showed increased binding to GST-HAP1 compared to the proteins from normal individuals (19, 22 repeats). HAP 1 is enriched in the brain, suggesting a possible basis for the selective brain pathology of HD. Kalchman and colleagues (1996) have identified a human ubiquitin-conjugating enzyme (hE2-25K) as a protein that interacts with huntingtin. It has complete amino acid identity with the bovine E2-25K protein and is strikingly similar to the UBC-1, -4, and -5 enzymes of *Saccharomyces cerevisiae*. This protein is highly expressed in brain, and a slightly larger protein recognized by an anti-E2-25K polyclonal antibody is selectively expressed in affected areas of the HD brain. The huntingtin E2-25K interaction is not influenced by CAG repeat length and huntingtin is ubiquinated. Burke and co-workers (1996) identified huntingtin and DRPLA as proteins which selectively interact with the enzyme glyceraldehyde 3 phosphate dehydrogenase (GAPDH). The interaction was initially demonstrated using an affinity column and occurred in a CAG repeat-dependent manner. Though such a CAG repeat-dependent alteration has not been found with other triplet disorders such as SCA1, androgen receptor, and SBMA (Koshy et al., 1996), the interaction of GAPDH is provocative, and has been implicated in cell death (Ikeda et al., 1996). Wanker et al. (1997) discovered huntingtin-interacting protein (HIP1), which binds specifically to the N-terminus of huntingtin. The predicted amino acid sequence of the HIP1 fragment exhibits significant similarity to cytoskeletal proteins, suggesting that HIP1 and huntingtin play a functional role in the cell filament networks. HIP1 is enriched in human brain regions and is expressed ubiquitously at low levels. DiFiglia and co-workers (1997) demonstrated that an amino-terminal fragment of mutant huntingtin localizes to neuronal intranuclear inclusions (NIIs) and dystrophic neurites (DNs) in the HD cortex and striatum and that polyglutamine length influences the extent of huntingtin accumulation in these structures. Co-labeling of the aggregates with antibodies toward ubiquitin suggests that abnormal huntingtin is targeted for proteolysis but is resistant to removal.

Animal Models

To investigate the normal function of the HD gene, three groups independently generated a knockout mouse model. Nasir and colleagues (1995) created a targeted disruption in exon 5 the murine *Hdh* gene; homozygotes initiate gastrulation but do not proceed to organogenesis and die before embryonic day 8.5. Mice heterozygous for the mutation display increased motor activity and cognitive deficits, and significant neuronal loss in the subthalamic nucleus. Duyao and co-workers (1995) and Zeitlin and colleagues (1995) also generated knockout models for HD and observed that this was lethal in homozygote embryos. They proposed that Hdh protein is functionally indispensable for neurogenesis, and concluded that in the null mutants regionalized apoptotic cell death in the embryonic ectoderm, a layer expressing the *Hdh* gene, is much higher than normal.

Hodgson and colleagues (1996) crossed YAC transgenic mice expressing human normal huntingtin with mice containing a targeted disruption of the murine gene. The YAC-transgene-based expression of huntingtin occurs prior to 7.5 days of gestation. Human huntingtin expression in YAC transgenic mice followed an identical tissue distribution and subcellular localization pattern to that of murine endogenous protein and expression levels of the transgene were two to three times endogenous levels.

A mouse model for HD using exon 1 with 115 to 150 CAG repeats under human endogenous HD promoter control showed ubiquitous transgene expression at both the mRNA and protein level, and transgenic mice showed a progressive neurological phenotype, including choreiform-like movements, involuntary stereotypic movements, tremors and epileptic seizures (Mangiarini et al., 1996, 1997). However, these mice did not show any neurodegeneration, which is the hall mark of Huntington's disease in humans (Mangiarini et al., 1996, 1997). Cha et al. (1998) studied neurotransmitters in these mice transgenic for exon 1. Analysis of metabotropic glutamate receptors in symptomatic 12-week-old mice revealed decreased levels of mGluR1, mGluR2, and mGluR3 (but not mGluR5) subunits, as determined by glutamate receptor binding, protein immunoblotting, and in situ hybridization. Ionotrophic AMPA and kainic receptors were also decreased, while ionotropic NMDA receptors were not different compared to controls. Other neurotransmitter receptors known to be affected in HD were also decreased, including dopamine and muscarinic cholinergic, but not GABA receptors in the cortex and striatum. Davies and co-workers (1997) observed that these mice develop pronounced neuronal intranuclear inclusions, which also contain the proteins huntingtin and ubiquitin. Related observations were made by Scherzinger and colleagues (1997), who used recombinant HD exon 1 to show that site-specific proteolysis of the recombinant protein with 51 polyglutamines results in the formation of high-molecular-weight protein aggregates with a fibrillar or ribbon-like morphology.

Recently Reddy and co-workers (1998) generated a mouse model using full-length cDNA with normal and expanded CAG (48 and 89) repeats under heterologous cytomegalovirus (hCMV) promoter control. Transgenic mice with a CAG repeat expansion exhibited feet clasping upon suspension by the tail. Feet clasping was observed in most of the transgenic animals but not in those with the 16 CAG repeat or in age-matched wild-type control animals. At about 5 to 7 months of age, 40% of animals showed stereotypic behavior consisting of circling, backflips, jumping, and excessive grooming. This behavior lasts for 3 to 4 months, and is followed by hypoactivity, then there is an akinetic stage and finally the animals die. Neurodegeneration in the striatum (Figure 3), cerebral cortex, and hippocampus resembling the changes in human HD was also observed. Degenerated neurons appeared dark, shrunken with pyknotic or small densely staining nuclei and eosinophilic cytoplasm in the hippocampus. This observation was further confirmed by TUNEL assay, and gliosis (reactive astrocytosis) observed after immunostaining with anti-GFAP antibody (Figure 4). Immunostaining with a ubiquitin antibody showed intranuclear and perinuclear inclusions and diffuse staining.

Ordway and colleagues (1997) developed a knockin mouse model with 146 CAG repeats targeted into the mouse hypoxanthine phosphribosyltransferase (*Hprt*) gene. Mutant transgenic mice show a late-onset neurological phenotype that progresses to premature death. Seizures were observed in mice older than 18 weeks,

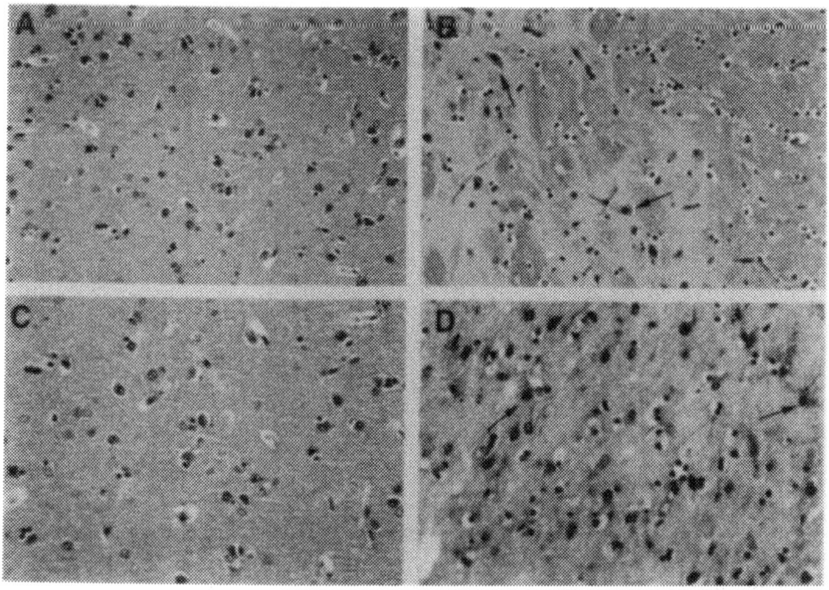

Figure 3. Comparative sections of human and mouse striatum. Microscopic sections were cut and stained with hematoxylin and eosin (H & E). Degenerating neurons were observed as scattered dark, shrunken cells with pykotic or small, densely staining nuclei and eosinophilic cytoplasm. Light micrographs at 630× original magnification from the striatum of wild-type mice (**A**) and normal human (**C**), and HD transgenic mice (**B**) and HD human (**D**). Arrows indicate representative degenerating neurons in **B** and **D**.

and feet clasping was noticed at 12 weeks of age. There was no difference in the weight of the brain in mutant transgenic mice compared to age-matched controls. Homozygous knockin mice with a targeted introduction of expanded 50 CAG repeats showed less of the characteristic behavior, aberrant brain development, and prenatal lethality. By contrast, mice with normal levels of mutant huntingtin did not display these anomalies, indicating that the expanded CAG repeat does not eliminate or detectably impair huntingtin's neurological function (White et al., 1997).

Dentatorubral-Pallidoluysian Atrophy

Clinical Features

Dentatorubral-pallidoluysian atrophy (DRPLA), also known as Haw-River syndrome (HRS), is an autosomal dominant disorder characterized by myoclonus epilepsy, dementia, ataxia, and choreoathetosis. The mean age of onset is 30 years of age, although onset of symptoms has been reported from less than 10 years to the seventh decade of life. Neuropathological changes involve degeneration of the

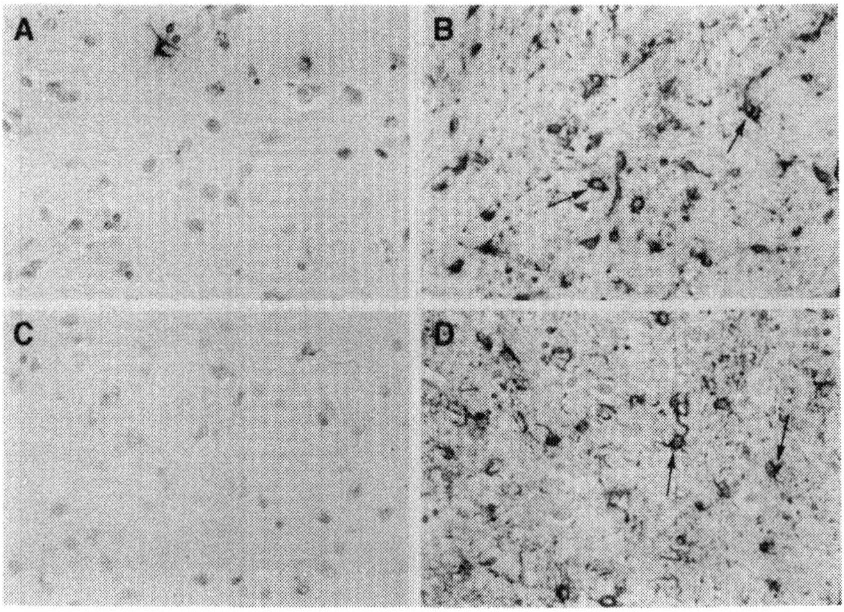

Figure 4. Comparative sections of human and mouse striatum. Microscopic sections were cut and stained with glial fibriallary acidic protein (GFAP). Degenerating neurons were observed by as reactive astrocytosis after GFAP immunostaining. Light micrographs at 630× original magnification from the striatum of wild-type (**A**) and normal human (**C**), and HD transgenic mice (**B**) and HD human (**D**). Arrows indicate representative reactive glial cells in **B** and **D**.

pallidofugal and dentatorubral pathways. In DRPLA, just as in HD, considerable variation in both clinical manifestations and age of onset have been observed even within the family. Though the disease is predominantly found in individuals of Japanese ancestry, the allelic disorder HRS has been described in African-American families (Burke et al., 1994). Phenotypic differences between HRS and DRPLA include the absence of myoclonic seizures in HRS as well as the presence of extensive demyelinization of the subcortical white matter, basal ganglia calcifications, and neuroaxonal dystrophy, which are not seen in DRPLA (Burke et al., 1994).

Genetics

The DRPLA locus was mapped on the short arm of chromosome 12p12 (Nagafuchi et al., 1994a, 1994b). Because of the overlapping clinical features with HD, DRPLA was suspected to be caused by a triplet repeat expansion. One of the brain cDNAs that contained polymorphic CAGs (Li et al., 1993) and mapped to the same region of chromosome 12 was examined and found to show expansion in affected individuals (Koide et al., 1994; Nagafuchi et al., 1994a, 1994b). The CAG repeat size varies from 7 to 23 in normal individuals and shows an expansion to approximately 49 to 75 in DRPLA patients. As with other disorders caused by expanded CAG repeats, expansion is usually associated with paternal transmission. The repeat size correlates inversely with age of onset of symptoms and with disease severity. Patients with progressive myoclonus epilepsy phenotype had larger expansions (62 to 79), had onset younger than 20 years of age (Ikeuchi et al., 1995) and inherited their expanded allele from their affected fathers.

Transcript and Protein

The atrophin protein is approximately 190-kDa in size (Yazawa et al., 1995) with no difference in distribution or expression level between normal and mutant atrophin in brain tissues and lymphoblastoid cells. Two-dimensional gel electrophoresis revealed that the DRPLA gene product consists of a number of isoelectric variants with atrophin from brain tissues having fewer isoelectric variants than atrophin from lymphoblastoid cells, suggesting that post-translational modification exists in brain tissues that is different from that in lymphoblastoid cells (Yazawa et al., 1997). Miyashita and colleagues (1997) identified a cleavage site near the N-terminal end of atrophin during apoptosis induced by VP-16, staurosporine or glucocorticoid. In vitro translated atrophin was cleaved by caspase 3, a member of cystine protease family that is important in the signaling for apoptosis. However, cleavage does not appear to be influenced by length of polyglutamine stretch (Miyashita et al., 1997).

Spinocerebellar Ataxia Type 1

Clinical Features

Spinocerebellar ataxia type 1 (SCA1) is an autosomal dominant neurological disorder characterized by gait ataxia, dysarthria, dysmetria, nystagmus, and variable degrees of muscle wasting. As the disease progresses hyper-reflexia may develop and the ataxia worsens. Extrapyramidal signs including dystonia and chorea may develop in the later stages of the disease. Onset often occurs during the third and fourth decades of life, with a mean age of onset at 30. Juvenile onset cases (onset before 15 years of age) have been described as occurring predominantly in offspring of affected males; disease progression is more rapid and ataxia is accompanied by mental retardation. The pathological hallmark in SCA1 is the degeneration of cerebellar Purkinje cells. Postmortem examination of affected brains revealed that disease affects the cerebellum, brain stem, and spinocerebellar tracts (see Zoghbi, 1995, for review). Cerebellum was reduced in size with loss of Purkinje cells and dentate nucleus cells.

Genetics

The *SCA1* gene was mapped to 6p22-p23 (Ranum et al., 1991; Zoghbi et al., 1991) and the gene shows expansion of an unstable CAG repeat in its coding region (Orr et al., 1993). The repeat has a perfect CAG configuration on expanded alleles, whereas it is interrupted by one to three CAT units on normal alleles. Normal alleles have CAG repeats in the order of 16 to 39 copies, whereas SCA1 alleles have CAG expansions in the range of 40 to 81 repeat units (Ranum et al., 1994a, 1994b; Matilla et al., 1993; Jodice et al., 1994). There appears to be no overlap in allele sizes between normal and SCA1 chromosomes. The murine and human ataxin 1 are highly homologous; however, CAG repeats are virtually absent in mouse, suggesting that the polyglutamine stretch is not essential for normal function of the gene (Banfi et al., 1996). In situ hybridization studies in mice indicate that during cerebellar development, there is a transient burst of *SCA1* expression at postnatal day 14 when the murine cerebellar cortex becomes physiologically functional. There is also marked expression of *SCA1* in mesenchymal cells of the vertebral discs during development of the spinal column. These results suggest that the normal *SCA1* gene has a role at specific stages of both cerebellar and vertebral column development (Banfi et al., 1996).

Transcript and Protein

Localization studies of SCA1 protein revealed that wild-type human ataxin 1 and transgenic mice for SCA1 had nuclear localization in Purkinje cells. Using a yeast two hybrid system, an interaction of GAPDH with SCA1 was detected (Koshy et al., 1996). In addition, these investigators showed that wild-type and mutant ataxin 1 are capable of forming homo- and heterodimers. More recently, Matilla and co-workers (1998) identified another SCA1-interacting protein known as leucine-rich acidic nuclear protein (LANP). LANP is expressed predominantly in Purkinje cells and the interaction between LANP and ataxin 1 becomes stronger with increasing number of CAG repeats. Immunofluorescence studies demonstrate that both LANP and ataxin 1 co-localize in nuclear matrix-associated structures (Matilla et al., 1997). Cummings and colleagues (1998) observed a single nuclear inclusion body in affected neurons of SCA1 patients and in brain sections from mutant transgenic mice for SCA1 (see later discussion) using antibodies to both ataxin 1 and ubiquitin. The inclusions stain positively for the 20S proteosome and molecular chaperone. Similarly, HeLa cells transfected with mutant ataxin 1 develop nuclear aggregates, which co-localize with the 20S proteosome.

Animal Models

Burright and co-workers (1996) generated a transgenic mouse model for SCA1 with expanded CAG repeats which developed ataxia and Purkinje cell neurodegeneration. Behavioral analysis of SCA1 transgenic mice revealed that at about 5 weeks of age, the mutant transgenic mice have an impaired performance in rotorod, indicating neurological deficits in balance and coordination (Clark et al., 1997). The mutant mice showed an increased initial exploratory behavior but with increasing age, these animals develop incoordination with concomitant progressive neuronal dendritic and somatic atrophy but relatively little Purkinje cell loss. In Purkinje cells of transgenic mice expressing a wild-type SCA1 allele, both normal and mutant ataxin 1 localized throughout the nucleus and to multiple nuclear structures that were 0.5 μm in size, but mutant ataxin 1 localized to a single large 2-μm nuclear structure (Skinner et al., 1997). The fraction of Purkinje cells containing large inclusions of mutant ataxin 1 increased with age from 25% at 6 weeks of age to 90% at 12 weeks. In addition, monkey kidney COS cells (which do not express endogenous SCA1 gene) when transfected with human SCA1 cDNA constructs also showed localization of the protein to the nucleus. Seventy-five percent of the transfected cells with normal ataxin 1 showed a diffuse staining accompanied by many small structures of less than 1 μm in diameter, whereas 60% of transfected cells with mutant ataxin had nuclear structures that were fewer in number and larger in size. Intergenerational CAG repeat instability was observed in these transgenic mice only when the transgene was maternally transmitted (Kaytor et al., 1997). To understand the pathogenesis of SCA1, Matilla and colleagues (1998) have generated a mouse model with a deletion in the SCA1 gene. The mice lacking ataxin 1 were viable, fertile, and did not show any evidence of ataxia or neurodegeneration.

Spinocerebellar Ataxia Type 2

Clinical Features

Spinocerebellar ataxia 2 (SCA2) accounts for 13% of patients with autosomal dominant cerebellar ataxia without retinal degeneration (Geschwind et al., 1997). A wide spectrum of clinical phenotypes was observed among SCA2 patients, including typical mild dominant ataxia, facial fasiculations and lid retraction, and early-onset ataxia with a rapid course, chorea, and dementia. Unlike SCA1, where the saccade eye movement amplitude was significantly increased, resulting in hypermetria, in SCA2 the saccade velocity is markedly decreased (Rivaud-Pechoux et al., 1998). In addition, in SCA2 progressive neuronal death extends beyond the cerebellar (Purkinje cells, some granule cells) and the spinal/brain stem systems (motoneurons, dorsal root ganglion cells, inferior olive, pons nuclei, locus coeruleus, saccade circuitry) to include neurodegeneration in the striatum, globus pallidus, substantia nigra, and sometimes in the cerebral cortex (Georg Auburger, personal communication). Age of onset varies from 2 to 65 years, with 40% of patients presenting before 25 years of age in a large Cuban kindred (Auburger et al., 1990).

Genetics

The *SCA2* gene was mapped to human chromosome 12q24.1 (Gispert et al., 1993), and the gene was subsequently identified by three groups independently using three different methodologies (Imbert et al., 1996; Pulst et al., 1996; Sanpei et al., 1996). Pulst and colleagues (1996) identified the entire *SCA2* gene on chromosome 12q24.1 using a combination of positional cloning and candidate gene approaches. Sanpei and co-workers (1996) isolated a series of cDNA clones using primer sequences that flanked a CAG repeat in patients with SCA2. Imbert and colleagues (1996) screened cDNA expression libraries using an antibody specific for polyglutamine repeats (Trottier et al., 1995a, 1995b) and identified six novel genes containing $(CAG)_n$ stretches. Each of these approaches was successful in identifying the ataxin 2 gene. The SCA2 repeat is unusual in that 94% of the normal alleles possess 22 repeats despite having a range of 15 to 24 CAGs (Imbert et al., 1996; Pulst et al., 1996; Sanpei et al., 1996) and is interrupted by one to three CAA repeats. In contrast, SCA2 patient chromosomes contained pure CAG expanded repeats ranging in size from 35 to 64 units (Imbert et al., 1996; Pulst et al., 1996; Sanpei et al., 1996; Lorenzetti et al., 1997). There was a strong inverse correlation between the size of the repeat and the age of onset of symptoms. No overlap in ataxin 2 allele sizes was detected between normal and disease chromosomes.

Transcript and Protein

The ataxin 2 protein product consists of about 1,313 amino acids (Sanpei et al., 1996) . Northern blot analysis revealed a 4.5-kb band in different human tissues, including brain. Sequence and amino acid analyses of human and mouse *SCA2* sequences revealed 89% and 91% identity at the nucleotide and amino acid level, respectively (Nechiporuk et al., 1998). However, there was no extended polyglutamine tract in the mouse *SCA2* cDNA, suggesting that the normal function of SCA2 is not dependent on this domain. The SCA2 mutant protein is cytoplasmic in location and has an apparent molecular weight of 150 kD (Imbert et al., 1996). Immunohistochemical staining using affinity-purified antibodies to ataxin 2 demonstrated that the protein is expressed in the cytoplasm of Purkinje cells as well as in other neurons of the CNS (Nechiporuk et al., 1998).

Spinocerebellar Ataxia Type 3

Clinical Features

Spinocerebellar ataxia type 3 (SCA3) and Machado-Joseph disease (MJD) are allelic disorders wherein MJD has been commonly associated in affected individuals of Japanese or Portuguese Azorean descent. The disease is inherited in an autosomal dominant manner and is characterized by cerebellar ataxia, a variable combination of pyramidal and extrapyramidal signs, peripheral palsy, amyotrophy facial and lingual fasciculations, ophthalmoplegia, and exophthalmos. The most common pathological findings include degeneration within the deep basal ganglia, the brain stem, the spinal cord and, to a lesser extent, the cerebellum (Takiyama et al., 1994). Age on onset is typically after the fourth decade of life.

Genetics

Kawaguchi and colleagues (1994) screened a human brain cDNA library using an oligonucleotide probe with 13 CTG repeats on the assumption that SCA3/MJD is also caused by triplet repeat expansions. One of the cDNA clones that contained a CAG repeat was mapped to human chromosome 14q32.1, the location to which the locus for SCA3 and MJD was previously determined (Takiyama et al., 1993; Kawaguchi et al., 1994; Stevanin et al., 1994, 1995). In normal individuals, the ataxin 3 gene was found to contain between 12 and 37 CAG repeats, whereas affected individuals showed expansion of the repeat number in the range of 61 to 84 (Maruyama et al., 1995). Unlike other diseases caused by CAG repeat expansions, the polyglutamine repeat in ataxin 3 lies near the carboxyl terminus.

Transcript and Protein

Transfection of COS cells with a portion of the ataxin 3 cDNA expressing expanded CAG repeats resulted in the induction of apoptosis where the protein products precipitated into large covalently modified forms (Ikeda et al., 1996). Paulson et al. (1997) studied intranuclear inclusions in SCA3 patients' brain tissues. Ubiquitin immunostaining revealed that ataxin 3 protein accumulates selectively in ubiquitinated intranuclear inclusions in neurons of affected brain. It was observed that an expanded polyglutamine-containing fragment can recruit full-length ataxin 3 into insoluble aggregates. The normal ataxin 3 is primarily localized in the cytoplasm.

Animal Models

Transgenic mice expressing a truncated version of the cDNA expressing expanded polyglutamine repeat showed phenotypic anomalies, including ataxic postures and gait disturbance at 4 weeks of age (Ikeda et al., 1996). By contrast, none of the mice transgenic for the C-terminal end of ataxin 3 expressing 35 polyglutamines or full-length cDNA with expanded repeats showed any phenotype. Strong ataxic phenotypes were more prominent with higher copy number of truncated cDNA with expanded repeats. Significant atrophy in the cerebellum of transgenic animals was observed compared to control brains.

Warrack and co-workers (1998) generated a *Drosophila* model for SCA3 and observed that flies expressing ataxin 3 with polyglutamine expansions form neuronal inclusions and show late-onset degeneration. Differential sensitivity to the mutant transgene was observed among the different cell types, with neurons being particularly susceptible.

Spinocerebellar Ataxia Type 6

Clinical Features

Spinocerebellar ataxia type 6 (SCA6) is an autosomal dominant disorder characterized predominantly by mild but slowly progressive cerebellar ataxia of the limbs, dysarthria, nystagmus, and mild vibratory and proprioceptive sensory loss. Onset of the disease is usually after 55 years of age (Schols et al., 1997), although the range can be from 24 to 63 years of age. The disease is insidious and usually progresses over a period of 20 to 30 years, after initial mild episodes of a sense of momentary imbalance and incoordination, which eventually lead to impairment of gait. Toward the end stage of the disease, patients are severely impaired and are often wheelchair-bound. In a few older patients, difficulty in swallowing has been observed, suggesting involvement of the brain stem. The pathological features of SCA6 include isolated cerebellar atrophy and extensive loss of Purkinje cells in the cerebellar cortex with proliferation of Bergmann glia (Gomez et al., 1997; Takahashi et al., 1998) and moderate loss of granule cells and dentate nucleus (Zhuchenko et al., 1997). Neuronal loss in the inferior olivary complex has also been described, which is likely to be secondary to the cerebellar cortical lesion (Takahashi et al., 1998).

Genetics

Zhuchenko and co-workers (1997) tested DNA samples from a collection of patients with late-onset neurogenic diseases with polymorphic CAG repeats. An expansion of the CAG repeat in the human alpha(1A)-voltage-dependent Ca^{2+} channel (*CACNL1A4*) gene, which maps to 19p13, was detected in SCA6 samples (Matsuyama et al., 1997). Point mutations in the same gene have been previously described for two distinct disorders, hemiplegic migraine and episodic ataxia types 2 (Ophoff et al., 1996). The CAG repeat length in the *CACNL1A4* gene is inversely correlated with age of onset. SCA6 chromosomes contained 21 to 30 CAG repeats, whereas the normal alleles showed 6 to 18 repeats (Matsuyama et al., 1997). However, in a family initially classified as autosomal dominant cerebellar ataxia of unknown type, Jodice et al. (1997) found an intergenerational allele size change in the *CACNL1A4* gene, showing that a $(CAG)_{20}$ allele was associated with the phenotype of episodic ataxia type 2 and a $(CAG)_{25}$ allele with progressive cerebellar ataxia. These results suggested that EA2 and SCA6 may be the same disorder with a high phenotypic variability.

Transcript and Protein

The *CACNL1A4* gene has transcript size of about 8.5 kb, which includes 47 exons spanning a genomic interval of 300 kb. In some alternatively spliced variants, of which six isoforms have been identified, the CAG repeat is not translated and is found in the 3′ UTR of the gene (Zhuchenko et al., 1997).

Animal Models

It is interesting that naturally occurring mouse mutants with point mutations in the homologous mouse gene *Cacnl1a4* demonstrate ataxic phenotype, as seen in tottering and leaner (Doyle et al., 1997). However, in these mice the mutation is inherited as an autosomal recessive trait.

Spinocerebellar Ataxia Type 7

Clinical Features

Spinocerebellar ataxia type 7 (SCA7) is an autosomal dominant neurodegenerative disorder characterized by cerebellar ataxia that may be associated with opthalmoplegia, loss of vision, dysarthria, pyramidal and extrapyramidal signs, deep sensory loss, or dementia. The retinopathy, which is variable, initially starts as macular degeneration and progresses to the periphery (Anttinen et al., 1986; Enevoldson et al., 1994). Degeneration in the regions of cerebellar cortex, basis pontis, inferior olive, and retinal ganglion cells constitutes the pathologicical features in SCA7 (Konigsmark and Weiner, 1970; Berciano, 1982; Gouw et al., 1994). Anticipation is observed and is more pronounced in paternal rather than in maternal transmissions (Benomar et al., 1994; Gouw et al., 1995).

Genetics

The locus for SCA7 was mapped to chromosome 3p21.1-p12 (Benomar et al., 1995). That a CAG repeat expansion underlies the mutation in SCA7 was shown in two unrelated SCA 7 patients in whom a protein product of 130 kDa was detected by a monoclonal antibody that immunoreacts specifically with proteins with long polyglutamine tracts (Trottier et al., 1995a, 1995b; Stevanin et al., 1996). Further confirmation was obtained with a technique called repeat expansion detection (RED), a method in which a thermolabile ligase is used to detect repeat expansions directly from genomic DNA (Lindblad et al., 1996). RED products of 150 to 240 bp were found in all affected individuals analyzed from eight SCA7 families, and these fragments were found to co-segregate with the disease, suggesting strongly that a $(CAG)_n$ expansion is the cause of SCA7 (Linblad et al., 1996). A gene of unknown function containing a CAG repeat that is expanded in SCA7 patients was subsequently identified from the candidate interval (David et al., 1997; Koob et al., 1998). On mutated alleles, CAG repeat size ranges from 38 to 200 repeats, whereas on normal alleles it varied from 7 to 35 repeats (David et al., 1997; Johansson et al., 1998). Patients with greater than 59 CAG repeats were found to have visual impairment as the initial presentation of symptoms with ataxia as a later feature (Johansson et al., 1998). As with other diseases caused by polyglutamine expansions, larger repeat lengths are associated with earlier onset, and more severe and rapid clinical course. In SCA7, larger expansions also correlate with a higher frequency of decreased vision opthalmoplegia, extensor plantar responses, and scoliosis (David et al., 1998). Furthermore, SCA7 also shows a high degree of instability during meiotic transmission, resulting mostly in expansions whose magnitude (mean increase of 10 to 16 CAG repeats) is greater than the other neurodegenerative disorders caused by polyglutamine expansions (David et al., 1998).

Transcript and Protein

The SCA7 gene on Northern blot analysis shows a transcript size of 7.5. Sequence analysis indicates a 2,727-bp open reading frame encoding the ataxin 7 protein consisting of 892 amino acids with a nuclear localization signal. Abundant expression of the SCA7 transcript is found in heart, placenta, skeletal muscle, and pancreas, with fainter signals in brain, lung, liver, and kidney (David et al., 1997). Expression is ubiquitous in the central nervous system, with particularly high expression in cerebellum.

Recently, neuronal intranuclear inclusions have been described in SCA7 patients' brain sections. Nuclear inclusion bodies were most commonly found in the inferior olivary complex, a site of severe neuronal loss in SCA7, but also in other regions including the cerebral cortex, which is not affected in SCA7 (Holmberg et al., 1998).

POSSIBLE TRIPLET REPEAT EXPANSION IN OTHER DISEASES

Most of the diseases caused by $(CAG)_n$ repeat expansions share certain clinical features, such as neuronal degeneration, dominant inheritance patterns, and genetic anticipation. Anticipation, resulting in earlier age of onset in successive generations, is largely a result of to the tendency of the affected parent's allele to expand in the germ line, particularly when transmitted from the paternal line (Albin and Tagle, 1995; Hummerich and Lehrach, 1995; Warren, 1996; Zoghbi, 1996) and is a strong determinant for these diseases. Several strategies have been employed in the past to identify genes or cDNAs with triple repeats that may be associated with disorders of the brain. These include screening of brain cDNA libraries (Riggins et al., 1992; Li et al., 1993a, 1993b; Jiang et al., 1995; Bulle et al., 1997; Margolis et al., 1997; Reddy et al., 1997), screening of genomic DNA from normal (Gastier et al., 1996) or patient DNA (Schalling et al., 1993; Sanpei et al., 1996; Koob et al., 1998), detection of polyglutamine- expanded proteins from patient samples (Trottier et al., 1995a, 1995b; Imbert et al., 1996; Stevanin et al., 1996), searches in the EST database (Néri et al., 1996; Bulle et al., 1997), and FISH (Haaf et al., 1996). These approaches have led to the identification of the genes for SBMA, DRPLA, SCA2, and SCA6 (La Spada et al., 1991; Koide et al., 1994; Nagafuchi et al., 1994a, 1994b; Imbert et al., 1996; Sanpei et al., 1996; Zhuchenko et al., 1997; Koob et al., 1998). There are an estimated 200 to 300 genes in the human genome that contain polymorphic trinucleotide repeats. Conceivably, there may be other clinical disorders, known or yet to be described, that may be due to triplet repeat expansions.

These include diseases where CAG repeat expansions have been indirectly implicated after analysis of genomic DNA from patients with schizophrenia and bipolar disorder (Lindblad et al., 1995; Morris et al., 1995; O'Donovan et al., 1995), for glutamine expansions in the protein lysates from patients with autosomal dominant cerebellar ataxias (Stevanin et al., 1996), and for autosomal dominant pure spastic paraplegia (Nielsen et al., 1997).

The genetic basis for some forms of schizophrenia and bipolar disorder is widely recognized. Twin studies have indicated a 46% concordance for schizophrenia in monozygotic or genetically identical twin pairs compared to a concordance of 14% amongst dizygotic or nonidentical twin pairs. These diseases are likely the result of the interaction of multiple genes (termed polygenic or multifactorial inheritance) wherein some of the genes involved have a major effect, while others play a relatively minor role. Anticipation has been previously documented in both schizophrenia (Bassett and Honer, 1994; Petronis and Kennedy, 1995) and bipolar disorder (McInnis et al., 1993), which was the basis for evaluating the presence of CAG repeat expansions in genomic DNA from patients (Lindblad et al., 1995; Morris et al., 1995; O'Donovan et al., 1995).

Among the autosomal dominant cerebellar ataxias, the genes for spinocerebellar ataxia 4 and 5 have yet to be determined. SCA4 is an autosomal dominant, late-onset spinocerebellar ataxia characterized by the presence of a prominent axonal sensory neuropathy (Flanigan et al., 1996). The SCA4 locus has been assigned to human chromosome 16q22.1 (Flanigan et al., 1996). SCA5 represents a class of mild, autosomal dominant cerebellar ataxias that, in general, is not life threatening (Ranum et al., 1994a, 1994b) and includes families descended from the paternal grandparents of President Abraham Lincoln. Disease onset varied from 10 to 68 years of age and anticipation was evident. The SCA5 locus has been mapped to the centromeric region of chromosome 11p11-q11 (Ranum et al., 1994a, 1994b).

An association between CAG repeat expansion and autosomal dominant pure spastic paraplegia (ADPSP) has recently been established (Nielsen et al., 1997). ADPSP is a neurodegenerative disorder characterized by slowly progressive spasticity of the legs, hyperreflexia, and Babinski sign. Multiple forms of the disease exist and are associated with locus heterogeneity. One form, mapped to chromosome 2p21-p24, showed CAG repeat expansions of at least 60 CAG repeats (Nielsen et al., 1997) using the RED method. Furthermore, the CAG repeat expansion for this locus is likely to be translated as indicated by the immuno-specific detection of an expanded polyglutamine-containing protein in patient cell lysates.

Other neurodegenerative and neuropsychiatric diseases are likely to be caused by triplet repeat expansions. A second locus for myotonic dystrophy (DM2) has been identified in human chromosome 3q (Ranum et al., 1998). DM2 shows clinical overlap with DM, including myotonia, proximal and distal limb weakness, frontal balding, cataracts, and cardiac arrhythmias. It would be interesting if the DM2 gene, when identified, also shows a trinucleotide repeat expansion.

MECHANISMS OF TRIPLET REPEAT INSTABILITY

It is currently unknown how the genetic instability observed in trinucleotide repeat sequences arises. Somatic and gonadal mosaicism for alleles carrying expanded trinucleotide repeats (Gourdon et al., 1997; Montermini et al., 1997; Machkhas et al., 1998; see Bates and Lehrach, 1994, for review) indicates that the instability may originate during the process of replication of DNA. This is to be distinguished from nonreplication-based mechanisms involving primarily recombination (either homologous or nonhomologous chromosomal rearrangements) of tandem arrays of repeated sequences. The absence of reciprocal alterations in the sequences flanking the repeat region also argues that recombination is an unlikely mechanism in triplet repeat instability. The role, if any, of recombination events resulting in trinucleotide repeat expansion is not clear at present.

Polymerase or DNA strand slippage is a widely recognized mechanism that can lead to instability of simple repeat sequences during replication. (Sinden and Wells, 1992; Wells, 1996). During replication, transient dissociation of the nascent and template strands may lead to misaligned reassociation resulting in additional synthesis of the repeat sequences. But such a slippage would not account for the larger gain in repeat units in the course of expansions seen in triplet repeat diseases nor does it explain the preponderance of expansions rather than contractions in these diseases.

Secondary structures, such as hairpin formation of repeat sequences, formed by hydrogen bond interaction between mismatched base pairs can allow the formation of large-scale expansions in a modified version of the strand-slippage model (McMurray, 1995; Mitas et al., 1995). In the hairpin model, these structures can form ahead of the replication complex preventing the progression of the replication fork and causing a stutter in DNA synthesis. This lag in the replication fork may result in reinitiation of synthesis in the lagging strand, resolution of the hairpin loop, and its subsequent ligation to the newly synthesized strand. This process can occur either during meiosis to produce eggs and sperm, or during somatic cell divisions in early embryogenesis.

PATHOGENIC MECHANISMS OF TRIPLET REPEATS

Alterations in Transcription in Noncoding Triplet Expansions

The identification of trinucleotide repeat expansions as the underlying cause of several neurodegenerative disorders raises the important question of how repeat instability can lead to disease. There is a growing consensus that trinucleotide expansions in the noncoding region of the gene may disrupt the transcriptional machinery of the gene. In the *FMR1* gene, the site of expansion is part of a GC-rich region, called CpG island, that is methylated to inactivate one of the normal X chromosomes. Expansion of the CGG repeat in the X chromosome results in the persistent methylation status of the *FMR1* gene, resulting in loss of *FMR1* gene expression (Feng et al., 1995). Thus in FRAXA, the CGG repeat instability results in a loss of function mutation.

The dominant inheritance of myotonic dystrophy is difficult to rationalize with the fact that the expansion mutation lies outside the translated part of the gene and is not part of the protein product. Because nuclear histones are known to mediate general transcriptional repression along chromosomes, Wang and co-workers (1994) used electron microscopy to examine in vitro the nucleosome assembly of DNA containing repeating CTG triplets. The expanded triplet repeats have been shown to repress transcription through the creation of hyperstable nucleosomes (Wang and Griffith 1995). This can possibly result in alteration in the local chromatin structure, inhibition of the passage of transcription complexes, or prevention of the opening of the DNA before replication. They may also result in DNA polymerase slippage, pausing, or idling, leading to expansion of the triplet block (Wang et al., 1994). Alternatively, the mutant DM molecule has been postulated to act in a dominant negative fashion owing to the hyperstabilization of mRNA resulting in the accumulation of the expanded DM kinase transcripts. Another mode of action proposes that the expanded CTG repeat in the 3′ UTR of the DMPK gene can affect the transcription of neighboring genes. The DM region of chromosome 19 is gene rich, and it is possible that the repeat expansion may lead to dysfunction of a number of transcriptional units (Harris et al., 1996) including the DM locus-associated homeodomain protein (DMAHP) (Boucher et al., 1995), perhaps as a consequence of chromatin disruption. DMAHP is expressed in a number of human tissues, including skeletal muscle, heart, and brain (Boucher et al., 1995). A nuclease-hypersensitive site is found adjacent to the DM CTG repeat in wild-type alleles but is eliminated in expanded alleles converting the region surrounding the repeats to a more condensed chromatin structure (Otten and Tapscott, 1995) and disrupting an enhancer element found in this region (Klesert et al., 1997). The loss of the hypersensitive site can be correlated with a two- to fourfold reduction in steady state DMAHP transcript levels relative to wild-type controls.

The expansion of the GAA repeat in the FRDA gene has been shown recently to repress the expression of the gene (Cossee et al., 1997) with compound heterozygotes (those with point mutations on one allele and a GAA expansion on the other allele) showing intermediate levels of message (Bidichandani et al., 1998). This loss in function mutation has been seen as well in reduced levels of the protein product (Campuzano et al., 1996) where the mRNA and protein levels were found to be inversely proportional to the size of the expansion. It is likely that the GAA expansion can cause inhibition of transcription of the FRDA gene, and that it can cause abnormal processing or excision of the intron with the GAA repeat in heteronuclear RNA.

Gain in Function Mutation in Polyglutamine Expansions

The CAG repeats are translated into polyglutamine tracts in each of the respective proteins, raising the possibility that the pathogenic mechanism involves aberrant function or interaction of the mutated protein. The genes encode novel proteins with the exception of SBMA and SCA6, where the normal function of the genes is known. The mere loss of gene function in these genes does not lead to severe neurological phenotype but gives rise to clinically distinct features, such as androgen insensitivity in the case of the androgen receptor gene or familial hemiplegic migraine and episodic ataxia in the case of the *CACNL1A4* gene. In addition, the embryonic lethality of the *Hdh* nullizygous mutation in mice suggests that, while huntingtin is important for development, the expanded polyglutamine in HD does not act by inhibiting normal protein function. These findings argue for a gain in function mutation caused by polyglutamine expansions leading to neurodegeneration. The nature of this gain in function for these protein remains unclear at present.

Recently, ubiquitinated neuronal intranuclear inclusions have been found in the affected brain regions of patients with HD (DiFiglia et al., 1997) and have also been described in SCA1 and SCA3 cases (Paulson et al., 1997; Skinner et al., 1997). Neuronal intranuclear inclusions have also been found in mice expressing long tracts of polyglutamines in the neurons of these mice (Davies et al., 1997; Ordway et al., 1997). Ubiquitination is a common feature for neuropathological structures, such as the Lewy body in Parkinson's disease. Its presence in neuropathological structures is thought to reflect aberrant protein folding or degradation. Many ubiquitin positive pathological structures contain antigen determinants from cytoskeletal proteins or cytoskeletal-associated proteins and may thus act by altering cellular architecture. It is also possible that the polyglutamine aggregation to form these bodies may prove toxic to the cells. While the formation of neuronal intranuclear inclusions poses an intriguing possibility for pathogenesis, questions remain as to whether they are coincidental to the disease process or may even act as a protective response of the cells (Ordway et al., 1997) against polyglutamine toxicity via sequestration of the toxic products.

It has also been suggested that a gain-in-function effect can arise by the promotion of aberrant protein-protein interactions (Reddy and Housman, 1997) that can lead to impairment of neuronal function or cell death. Although expanded polyglutamines are causative factors in all CAG associated disorders, the target regions of neurodegeneration are different despite the ubiquitous expression of the genes. The mechanism by which neurodegeneration occurs may have a common toxic effect from expanded polyglutamines and their interaction with the rest of the protein (of the same gene) or interactions with other proteins involved in specific neurons leading to specific neuronal degeneration, or both. Transgenic mice expressing HD exon 1 with expanded polyglutamines exhibited neurological features, but failed to show neurodegeneration (Mangiarini et al., 1996). This indicates that the rest of the huntingtin protein has a crucial role in promoting neurodegeneration. A recent transgenic mouse model expressing full-length huntingtin (Reddy et al., 1998) exhibited a progressive neurological phenotype and showed selective neurodegeneration in the striatum, cerebral cortex, and hippocampus. However, it remains unclear how neurodegeneration occurs selectively in brain regions with different CAG-repeat-associated diseases. It is possible that the expressions of the interacting proteins are cell specific and are essential for proper functioning and maintenance of the susceptible neurons for each of the diseases. It is also possible that a cell specific protease (such as caspase 3) (Goldberg et al., 1996) can release or activate the effects of the polyglutamine tract.

CONCLUDING REMARKS

A pattern has been emerging since dynamic mutations were first described less than a decade ago in the kind and mode of action for trinucleotide repeat expansions. In general, the expansions found in noncoding sequences are able to undergo massive expansions from a normal range of 6 to 40 repeats to a mutated range of over 1,000 repeat units. This type of expansion can lead to transcriptional alterations by causing suppression of mRNA production, abnormal RNA processing, or even affecting the transcription of adjacent genes. Thus far, only CGG, CTG, and GAA repeat sequences have been found to be involved in these types of expansions. In contrast, expansions in the coding sequences are much more modest, ranging from 5 to 35 repeat units in the normal allele to an abnormal range of 40 to over 100 repeats. The CAG repeats that make up this type of expansions all code for polyglutamines and the proteins are ubiquitously expressed. In these polyglutamine expansions, the actual pathogenic effect of the abnormal protein is still unclear, although it appears to reflect a gain of function, eventually leading to neurodegeneration.

Without doubt, the number of diseases identified as being caused by triplet repeat expansions will continue to grow. It remains to be seen whether the common thread of neurological involvement, genetic anticipation, cellular mosaicism, and marked variability in phenotype will continue to define this class of diseases caused by dynamic mutations. Many questions remain to be answered. What is the mechanism of expansion? Will this mechanism hold true for all the types of triplet repeat sequences? What other trinucleotide, or for that matter, di-, tetra-, and other repeat combinations can cause human disease? Why is triplet repeat instability, and thus trinucleotide expanded diseases, unique to the human species? Will individuals with repeat lengths at the high end of the normal repeat spectrum eventually develop a mild form of the disease if they live long enough? Does repeat variability in the normal population influence the rate or degree of cognitive and neurological decline associated with the aging process? What causes the selective neuronal loss in the polyglutamine disorders? Would a shift in the CAG translation (i.e., polyserine or polyalanine) be capable of producing disease? Undoubtedly, many new questions will emerge as we gleam more insight into the dynamic nature of triplet repeats.

SUMMARY

In recent years, a new type of genetic mutation (i.e., the instability of trinucleotide repeats or dynamic mutation) has been identified as a mechanism by which expansion of short tandem repeat sequences can lead to neurological disorders. Triplet repeat expansion explains the unique inheritance pattern of genetic anticipation and phenotypic variability that is characteristic of these diseases. Since the initial description in 1991 of unstable, highly polymorphic trinucleotide repeats associated with fragile X syndrome and spinobulbar muscular atrophy, the list of genes affected by dynamic triplet repeats has grown to twelve. This chapter summarizes the data that have emerged on the genetics and neurobiology of these disorders. Several other diseases, including neuropsychiatric disorders such as schizophrenia, that share features of triplet repeat expansion diseases are presented. The current data on the mechanisms of instability and pathogenesis involved in these diseases are also discussed.

REFERENCES

Abitbol, M., Menini, C., Delezoide, A.L., Rhyner, T., Vekemans, M. & Mallet, J. (1993). Nucleus basalis magnocellularis and hippocampus are the major sites of FMR-1 expression in the human fetal brain. Nat. Genet. 4, 147–153.

Abrams, M., Doheny, K., Mazzacco, M., Knight, S., Baumgardner, T., Freund, L., Davies, K. & Reiss, A. (1997). Cognitive, behavioral, and neuroanatomical assessment of two unrelated male children expressing FRAXE. Am. J. Hum. Genet. 74, 73–81.

Albin, R.L. & Tagle, D.A. (1995). Genetics and molecular biology of Huntington's disease. Trends Neurosci. 18, 11–14.

Anttinen, A., Nikoskelainen, E., Martilla, R.J., Grenman, R., Falck, B., Aarnisalo, E. & Kalimo, H. (1986). Familial olivopontocerebellar atrophy with macular degeneration: a separate entity among the olivopontocerebellar atrophies. Acta. Neurol. Scand. 73, 180–190.

Ashley, C.T. Jr., Wilkinson, K.D., Reines, D. & Warren S.T. (1993). FMR 1 protein, conserved RNP family domains and selective RNA binding. Science 262, 563–566.

Aslanidis, C., Jansen, G., Amemiya, C., Shutler, G., Mahadevan, T., Tsilfidis, C., Chen, C., Alleman, J., Wormskamp, N.G.M., Vooijs, M., Buxton, J., Johnson, K., Smeets, H.J.M., Lennon, G.G., Carrano, A.V., Korneluk, R.G., Wieringa, B. & de Jong, P.J. (1992). Cloning of the essential myotonic dystrophy region and mapping of the putative defect. Nature 355, 548–551.

Auburger, G., Diaz, G.O., Capte, R.F., Sanchez, S.G., Perez, M.P., del Cueto, M.P., Menesses, M.G., Farrall, M., Williamson, R., Chamberlain, S. et al. (1990). Autosomal dominant ataxia: genetic evidence for locus heterogeneity from a Cuban founder effect population. Am. J. Hum. Genet. 46, 1163–1177.

Babcock, M., de Silva, D., Oaks, R., Davis-Kaplan, S., Jiralerspong, S., Montermini, L., Pandolfo, M. & Kaplan, J. (1997). Regulation of mitochondrial iron accumulation by Yfh1p, a putative homolog of frataxin. Science 276, 1709–1712.

Banfi, S., Sevadio, A., Chung, M., Capozzoli, F., Duvick, L., Elde, R., Zoghbi, H.Y. & Orr, H. T. (1996). Cloning and developmental expression analysis of the murine homolog of the spinocerebellar ataxia type 1 gene (SCA 1). Hum. Mol. Genet. 5, 33–40.

Barbeau, A., Roy, M., Sadibelouiz, M. & Wilensky, M.A. (1984). Recessive ataxia in Acadians and 'Cajuns'. Canad. J. Neurol. Sci. 11, 526–544.

Barnes, G.T., Duyao, M.P., Ambrose, C.M., McNeil, S., Persichetti, F., Srinidhi, J., Gusella, J.F. & MacDonald, M.E. (1994). Mouse Huntinton's disease gene homolog (*Hdh*). Somat. Cell Mol. Genet. 20, 87–97.

Barnicoat, A., Wang, Q., Turk, J., Green, E., Mathew, C., Flynn, G., Buckle, V., Hirst, M., Davies, K. & Bobrow, M. (1997). Clinical, cytogenetic and molecular analysis of three families with FRAXE. J. Med. Genet. 34, 13–17.

Bassett, A.S. & Honer, W.G. (1994). Evidence for anticipation in schizophrenia. Am. J. Hum. Genet. 54, 864–870.

Bates, G. & Lehrach, H. (1994). Trinucleotide repeat expansions and human genetic disease. Bioessays 16, 277–283.

Baxendale, S., Abdulla, S., Elgar, G., Buck, D., Berks, M., Micklem, G., Durbin, R., Bates, G., Brenner, S., Beck, S. & Leharch, H. (1995). Comparative sequence analysis of the human and the pufferfish Huntington's disease genes. Nat. Genet. 10, 67–76.

Benders, A.A.G., Groenen, P.J.T.A., Oerlmans, F.T.J.J., Veerkamp, J.H. & Wieringa, B. (1997). Myotonic dystrophy protein kinase is involved in the modulation of the Ca(2+) homeostasis in skeletal muscle cells. J. Clin. Invest. 100, 1440–1447.

Benomar, A., Le Guern, E., Durr, A., Ouhabi, H., Stevanin, G., Yahyaoni, M., Chkili, T., Agid, Y. & Brice, A. (1994). Autosomal dominant cerebellar ataxia with retinal degeneration (ADCA type II) is genetically different from ADCA type 1. Ann. Neurol. 35, 439–444.

Benomar, A., Krols, L., Stevanin, G., Cancel, G., Le Guern, E., David, G., Ouhabi, H., Martin, J.J., Durr, A., Zaim, A., Ravise, N., Busque, C., Penet, C., Van Regemorter. N., Weissenbach, J., Yahyaoui, M., Chkili, T., Agid, Y., Van Broeckhoven, C. & Brice, A. (1995). The gene for autosomal dominant cerebellar ataxia with pigmentary muscular dystrophy maps to chromosome 3p12-p21.1. Nat. Genet. 10, 84–88.

Berciano, J., Amado, J.A., Frejanes, J., Rebollo, M. & Vaquero, A. (1982). Familiam cerebellar ataxia and hypogonadotrophic hypogonadism: evidence for hypothalamic LHRH deficiency. J. Neurol. Neurosurg. Psychiatry 45, 747–751.

Bidichandani, S.I., Ashizawa, T. & Patel, P.I. (1998). The GAA triplet-repeat expansion in Friedreich ataxia interferes with transcription and may be associated with an unusual DNA structure. Am. J. Hum. Genet. 62, 111–121.

Boucher, C.A., King, S.K., Carey, N., Krahe, R., Winchester, C.L., Rahman, S., Creavin, T., Meghji, P., Bailey, M.E.S., Chartier, F.L., Brown, S.D., Siciliano, M.J. & Johnson, K.J. (1995). A novel homeodomain-encoding gene is associated with a large CpG island interrupted by the myotonic dystrophy unstable (CTG)n repeat. Hum. Mol. Genet. 4, 1919–1925.

Brook, J.D., McCurrach, M.E., Harley, H.G., Buckler, A.J., Church, D., Aburatani, H., Hunter, K., Stanton, V.P., Thirion, J.P., Hudson, T., Sohn, R., Zemelman, B., Snell, R. G., Rundle, S.A., Crow, S., Davies, J., Shelbourn, P., Buxton, J., Jones, C., Juvonen, V., Johnson, K., Harper, P. S., Shaw, D.J. & Housman, D.E. (1992). Molecular basis of myotonic dystrophy, expansion of a trinucleotide (CTG) repeat at the 3′ end of a transcript encoding a protein kinase family member. Cell 68, 799–808.

Bulle, F., Chiannikulchai, N., Pawlak, A., Weissenbach, J., Gyapay, G. & Guellaen, G. (1997). Identification and chromosomal location of human genes containing CAG/CTG repeats expressed in testis and brain. Genome Res. 7, 705–715.

Burke, J.R., Ikeuchi, T., Koide, R., Tsuiji, S., Yamada, M., Pericak-Vance, M.A. & Vance, J.M. (1994). Dentatorubral-pallidoluysian atrophy and Haw River syndrome. Lancet 344, 1711–1712.

Burke, J.R., England, J.J., Martin, M.E., Jou, Y.S., Myers, R.M., Roses, A.D., Vance, J.M., Strittmatter, W.J. (1996). Huntingtin and DRPLA proteins selectively interact with the enzyme GAPDH. Nat. Med. 2, 347–350.

Burright, E.N., Clark, H.B., Sevadio, A., Matilla, T., Fedderson, R.M., Yunis, W.S., Duvick, L.A., Zoghbi, H.Y. & Orr, H.T. (1996). SCA1 transgenic mice: a model for neurodegeneration caused by an expanded CAG trinucleotide repeat. Cell 82, 937–948.

Butler, R., Leigh, P.N., McPhal, J. & Gallo, J.M. (1998). Truncated forms of the androgen receptor are associated with polyglutamine expansion in X-linked spinal and bulbar muscular atrophy. Hum. Mol. Genet. 7, 121–127.

Buxton, J., Shelbourne, P., Davies, J., Jones, C., Van Tongeren, T., Aslanidis, C., de Jong, P., Jansen, G., Anvret, M., Riley, B., Williamson, R. & Johnson, K. (1992). Detection of an unstable fragment of DNA specific to individuals with myotonic dystrophy. Nature 355, 547–548.

Campuzano, V., Montermini, L., Molto, M.D., Pianese, L., Cossee, M., Cavalcanti, F., Monros, E., Rodius, F., Duclos, F., Monticelli, A., et al. (1996). Friedreich's ataxia: autosomal recessive disease caused by an intronic GAA triplet repeat expansion. Science 271, 1423–1427.

Carango, P., Noble, J.E., Marks, H.G. & Funanage, V.L. (1993). Absence of myotonic dystrophy protein kinase (DMPK) mRNA as a result of a triplet repeat expansion in myotonic dystrophy. Genomics 18, 304–348.

Caughey, J.E. & Myrianthopoulos, N.C. (1963). Dystrophia Myotonica and Related Disorders. : Charles C. Thomas, Springfield, IL.

Cha, J.H.J., Kosinki, C.M., Kerner, J.A., Alsdorf, S.A., Mangiarini, L., Davies, S.W., Penny, J.B., Bates, G.P. & Ypung, A. (1998). Altered brain neurotransmitter receptors in transgenic mice expressing a portion of an abnormal human huntingtonís disease gene. PNAS. 95, 6480–6485.

Chakrabarti, L., Knight, S.J.L., Flannery, A.V. & Davies, K.E. (1996). A candidate gene for mild mental handicap at the FRAXE fragile site. Hum. Mol. Genet. 5, 275–282.

Clark, H.B., Burright, E., Yunis, W.S., Larson, S., Wilcox, C., Hartman, B., Matilla, A., Zoghbi, H.Y. & Orr, H.T. (1997). Purkinje cell expression of a mutant allele of SCA1 in transgenic mice leads to disparate effects on motor behaviors, followed by a progressive cerebellar dysfunction and histological alterations. J. Neurosci. 17, 7385–7395.

Cossee, M., Campuzano, V., Koutnikova, H., Fischbeck, K., Mandel, J.-L., Koenig, M., Bidichandani, S.I., Patel, P.I., Molte, M.D., Canizares, J., De Frutos, R., Pianese, L., Cavalcanti, F., Onticelli, A., Cocozza, S., Montermini, L. & Pandolfo, M. (1997). Frataxin fracas [letter]. Nat. Genet. 15, 337–338.

Cummings, C.J., Mancini, M.A., Antalffy, B., DeFranco, D.B., Orr, H.T. & Zoghbi, H.Y. (1998). Chaperone suppression of aggregation and altered subcellular proteasome localization imply protein misfolding in SCA1. Nat. Genet. 19, 148–154.

David, G., Abbas, N., Stevanin, G., Durr, A., Yvert, G., Cancel, G., Weber, C., Imbert, G., Saudou, F., Antoniou, E., Drabkin, H., Gemmill, R., Giunti, P., Benomar, A., Wood, N., Ruberg, M., Agid, Y., Mandel, J-L. & Brice, A. (1997). Cloning of the SCA7 gene reveals a highly unstable CAG repeat expansion. Nat. Genet. 17, 65–70.

David, G., Stevanin, G., Cancel, G., Abbas, N., Benomar, A., Belal, S., Lebre, A.S., Abada Bendib, M., Grid, D., Holmberg, M., Yahyaoui, M., Hentati, F., Chkili, T., Agid, Y. & Brice, A. (1998). Molecular and clinical correlationms in autosomal dominant cerebellar ataxia with progressive macuilar dystrophy (SCA7). Hum. Mol. Genet. 7, 165–172.

Davies, S.W., Turmaine, M., Cozens, B.A., DiFiglia, M., Sharp, A.H., Ross, C.A., Scherzinger, E., Wanker, E.E., Mangiarini, L. & Bates, G.P. (1997). Formation of neuronal intranuclear inclusions underlies the neurological dysfunction in mice transgenic for the HD mutation. Cell 90, 537–548.

De Boulle, K., Vervek, A.J.M.H., Reyners, E., Vits, L., Hendrix, J., Van Roy, B., Van Den Bos, F., de Graaff, E., Oostra, B.A. & Willems, P.J. (1993). A point mutation in the FMR1 gene associated with fragile X mental retardation. Nat. Genet. 3, 31–35.

De Michele, G., Filla, A., Cavalcanti, F., De Maio, L., Pianese, L., Castaldo, I., Calabrese, O., Monticelli, A., Varrone, S., Campanella, G., Leone, M., Pandolfo, M. & Cocozza, S. (1994). Late onset Friedreich's disease: clinical features and mapping of mutation to the FRDA locus. J. Neurol. Neurosurg. Psychiat. 57, 977–979.

Devys, D., Lutz, Y., Rouyer, N., Bellocq, J.P. & Mandel, J. (1993). The FMR1 protein is cytoplasmic, most abundent in neurons and appears normal in carriers of a fragile X premutation. Nat. Genet. 4, 335–340.

DiFiglia, M., Sapp, E., Chase, K., Schwarz, C., Meloni, A., Young, C., Martin, E., Vonsattel, J.P., Carraway, R., Reeves, S., Boyce, F.M. & Aronin, N. (1995). Huntingtin is a cytoplasmic protein associated with vesicles in human and rat brain neurons. Neuron 14, 1075–1081.

DiFiglia, M., Sapp, E., Chase, K.O., Davies, S.W., Bates, G.P., Vonsattel, J.P. & Aronin, N. (1997). Aggregation of huntingtin in neuronal intranuclear inclusions and dystrophic neurites in brain. Science 277, 1990–1993.

Doyle, J., Lennon, X., & Stubbs, L. (1997). Mutations in the Cacnl1a4 calcium channel gene are associated with seizures, cerebellar degeneration, and ataxia in tottering and leaner mutant mice. Mamm. Genome 8, 113–120.

Doyu, M., Sobue, G., Mukai, E., Kachi, T., Yasuda, T., Mitsuma, T. & Takahashi, A. (1992). Severity of X-linked recessive bulbospinal neuronopathy correlates with size of the tandem CAG repeat in androgen receptor gene. Ann. Neurol. 32, 707–710.

Durr, A., Cossee, M., Agid, Y., Campuzano, V., Mignard, C., Penet, C., Mandel, J.-L., Brice, A. & Koenig, M. (1996). Clinical and genetic abnormalities in patients with Friedreich's ataxia. N. Eng. J. Med. 335, 1169–1175.

Dutch-Belgian Fragile X Consortium (1994). Fmrl knockout mice: a model to study fragile X mental retardation. Cell 78, 23–33.

Duyao, M.P., Auerbach, A.B., Ryan, A., Persichetti, F., Barnes, G.T., McNeil, S.M., Ge, P., Vonsattel, J-P., Gusella, J.F., Joyner, A.L. & MacDonald, M.E. (1995). Inactivation of the mouse huntington's disease gene homolog Hdh. Science 269, 408–410.

Eichler, E.E., Richards, S., Gibbs, R.A. & Nelson, D.L. (1993). Fine structure of the human FMR1 gene. Hum. Mol. Genet. 2, 1147–1153.

Enevoldson, T.P., Sanders, M.D. & Harding, A.E. (1994). Autosomal dominant cerebellar ataxia with pigmentary macular dystrophy. A clinical and genetic study of eight families. Brain 117, 445–460.

Feng, Y., Zhang, F., Lokey, L.K., Chastain, J.L., Lakkis, L., Eberhart, D. & Warren, S.T. (1995). Translational suppression by trinucleotide repeat expansion at FMR1. Science 268, 731–734.

Filla, A., De Michele, G., Cavalcanti, F., Pianese, L., Monticelli, A., Campanella, G. & Cocozza, S. (1996). The relationship between trinucleotide (GAA) repeat length and clinical features in Friedreich ataxia. Am. J. Hum. Genet. 59, 554–560.

Fischbek, K.H., Souders, D. & La Spada, A. (1991). A candidate gene for linked spinal muscular atrophy. Adv. Neurol. 56, 209–213.

Flanigan, K., Gardner, K., Alderson, K., Galster, B., Otterud, B., Leppert, M.F., Kaplan, C. & Ptacek, L.J. (1996). Autosomal dominant spinocerebellar ataxia with sensory axonal neuropathy (SCA4): clinical description and genetic localization to chromosome 16q22.1. Am. J. Hum. Genet. 59, 392–399.

Folstein, S.E. (1989). In: Huntington's Disease: A Disorder of Families. Johns Hopkins University Press, Baltimore.

Fu, Y.H., Kuhl, D.P.A., Pizzuti, A., Pieretti, M., Sutcliffe, J.S., Richards, S., Verkerk, A.J.M.H., Holden, J.J.A., Fenwick, R.G. Jr., Warren, S.T., Oostra, B.A., Nelson, D.L. & Caskey C.T. (1991). Variation of the CGG repeat at the fragile X site results in genetic instability, resolution of the Sherman paradox. Cell 67, 1047–1058.

Fu, Y.-H., Pizzuti, A., Fenwick, R.G., Jr., King, J., Rajnarayan, S., Dunne, P.W., Dubel, J., Nasser, G., Ashizawa, T., de Jong, P., Wieringa, B., Korneluk, R., Perryman, M.B., Epstein, H.F. & Caskey, C.T. (1992). An unstable triplet repeat in a gene related to myotonic muscular dystrophy. Science 255, 1256–1258.

Fu, Y.-H., Friedman, D.L., Richards, S., Pearlman, J.A., Gibbs, R.A., Pizzuti, A., Ashizawa, T., Perryman, M.B., Scarlato, G., Fenwick, R.G., Jr. & Caskey, C.T. (1993). Decreased expression of myotonin-protein kinase messenger RNA and protein in adult form of myotonic dystrophy. Science 260, 235–238.

Fujita, R., Agid, Y., Trouillas, P., Seck, A., Tommasi-Davenas, C., Driesel, A.J., Olek, K., Grzeschik, K.-H., Nakamura, Y., Mandel, J.L. & Hanauer, A. (1989). Confirmation of linkage of Friedreich ataxia to chromosome 9 and identification of a new closely linked marker. Genomics 4, 110–111.

Gastier, J.M., Brody, T., Pulido, J.C., Businga, T., Sunden, S., Hu, X., Maitra, S., Buetow, K.H., Murray, J.C., Sheffield, V.C., Boguski, M., Duyk, G. & Hudson, T.J. (1996). Development of a screening set for new (CAG/CTG)n dynamic mutations. Genomics 32, 75–85.

Gecz, J., Oostra, B., Hockey, A., Carbonell, P., Turner, G., Haan, E., Sutherland, G. & Mulley, J. (1997). FMR2 expression in families with FRAXE mental retardation. Hum. Mol. Genet. 6, 435–441.

Gerald, P.S. (1980). X-linked mental retardation and an X-chromosome marker [editorial]. N. Eng. J. Med. 303, 696–697.

Geschwind, D.H., Perlman, S., Figueroa, F., Treiman, L.J. & Pulst, S. (1997). The prevalence and wide clinical spectrum of the spinocerebellar atxia type 2 trinucleotide repeat in patients with autosomal dominant cerebellar ataxia. Am. J. Hum. Genet. 60, 842–850.

Giraud, F., Ayme, S., Mattei, J.F. & Mattei, M.G. (1976). Constitutional chromosomal breakage. Hum. Genet. 34, 125–136.

Gispert, S., Twells, R., Orozco, G., Brice, A., Weber, J., Heredero, L., Scheufler, K., Riley, B., Allotey, R. & Nothers, C. (1993). Chromosomal assignment of the second locus for autosomal dominant cerebellar ataxia (SCA2 to chromosome 12q23-24. Nat. Genet. 4, 295–299.

Goldberg, Y.P., Nicholson, D., Rasper, D.M., Kalchman, M.A., Koide, H.B., Graham, R.K., Bromm, M., Kazemi-Esfarjani, P., Thornberry, N., Vaillancourt, J.P. & Hayden, M.R. (1996). Cleavage of huntingtin by apopain, a proapoptotic cysteine protease, is modulated by the polyglutamine tract. Nat. Genet. 13, 442–449.

Gomez, C.M., Thomson, R.M., Gammack, J., Perlman, S.L., Dobyns, W.B., Truwit, C.L., Zee, D.S., Clarck, H.B. & Anderson, J.K. (1997). Spinocerebellar ataxia type 6: gaze-evoked and vertical nystagmus, Purkinje cell degeneration, and variable age of onset. Ann. Neurol. 42, 933–950.

Gourdon, G., Radvanyi, F., Lia, A.-S., Duros, C., Blanche, M., Abitbol, M., Junien, C. & Hofmann-Radvanyi, H. (1997). Moderate intergenerational and somatic instability of a 55-CTG repeat in transgenic mice. Nat. Genet. 15, 190–192.

Gouw, L.G., Digre, K.B., Harris, C.P., Haines, J.H. & Ptacek, L.J. (1994). Autosomal dominant cerebellar atxaia with retinal degeneration: clinical, neuropathologic, and genetic analysis of amlarge kindred. Neurology 44, 1441–1447.

Gouw, L.G., Kaplan, C.D., Haines, J.H., Digre, K.B., Rutledge, S.L., Matilla, A., Leppert, M., Zoghbi H.Y. & Ptacek, L.J. (1995). Retinal degeneration characterizes a spinocerebellar ataxia mapping to chromosome 3p. Nat. Genet. 10, 89–93.

Gu, Y., Shen, Y., Gibbs, R.A. & Nelson, D.L. (1996). Identification of FMR2, a novel gene associated with the FRAXE CGG repeat and CpG island. Nat. Genet. 13, 109–113.

Gusella, J.F., Wexler, N.S., Conneally, P.M., Naylor, S.L., Anderson, M.A., Tanzi, R.E., Atkins, P.C., Ottina, K., Wallace, M.R., Sakaguchi, A.Y., Youing, A.B., Shoulson, I., Bonilla, E. & Martin, J.B. (1983). A polymorphic DNA marker genetically linked Huntington's disease. Nature 306, 234–238.

Gutekunst, C.A., Levey, A.I., Heilman, C.J., Whaley, W.L., Yi, H., Nash, N.R., Rees, H.D., Madden, J.J. & Hersch, S.M. (1995). Identification and localization of huntingtin in brain and human lymphoblastoid cell lines with anti-fusion protein antibodies. Proc. Natl. Acad. Sci. U. S. A. 92, 8710–8714.

Haaf, T., Sirugo, G., Kidd, K.K. & Ward, D.C. (1996). Chromosomal localization of long trinucleotide repeats in the human genome by fluorescence in situ hybridization. Nat. Genet. 12, 183–185.

Hanauer, A., Chery, M., Fujita, R., Driesel, A.J., Gilgenkrantz, S. & Mandel, J.L. (1990). The Friedreich ataxia gene is assigned to chromosome 9q13-q21 by mapping of tightly linked markers and shows linkage disequilibrium with D9S15. Am. J. Hum. Genet. 46, 133–137.

Harding, A.E., Thomas, P.K., Baraitser, M., Bradbury, P.G., Morgan-Hughes, J.A. & Pongsford, J.R. (1982). X linked recessive bulbospinal neuronopathy: a report of ten cases. J. Neurol. Neurosurg. Psychiatry 45, 1012–1019.

Harley, H.G., Brook, J.D., Rundle, S.A., Crow, S., Reardon, W., Buckler, A.J., Harper, P.S., Housman, D.E. & Shaw, D.J. (1992) Expansion of an unstable DNA region and phenotypic variation in myotonic dystrophy. Nature 355, 545–546.

Harper, P.S. (1975). Congenital myotonic dystrophy in Britain. I. Clinical aspects. Arch. Dis. Child. 50, 505–513.

Harper, P.S. (1989). Myotonic Dystrophy, 2nd ed. W.B. Saunders, Philadelphia.

Harper, P.S. (1991). In: Huntington's Disease. W.B. Saunders, Philadelphia.

Harper, P.S. & Dyken, P.R. (1972). Early-onset dystrophia myotonica: evidence supporting a maternal environmental factor. Lancet II, 53–55.

Harris, S., Moncrieff, C. & Johnson, K. (1996). Myotonic dystrophy: will the real gene please step forward! Hum. Mol. Genet. 5, 1417–1423.

Harvey, J., Judge, C. & Wiener, S. (1977). Familial X-linked mental retardation with an X chromosome abnormality. J. Med. Genet. 14, 46–50.

Hinds, H.L., Ashley, C.T., Sutcliffe, J.S., Nelson, D.L., Warren, S.T., Housman, D.E. & Schalling, M. (1993). Tissue specific expression of FMR1 provides evidence for a functional role in fragile X syndrome. Nat. Genet. 3, 36–43.

Hodgson, J.G., Smith, D.J., McCutcheon, K., Koide, H.B., Nishiyama, K., Dinulos, M.B., Stevens, M.E., Bissada, N., Nasir J, Kanazawa, I., Distechi, C.M., Rubin, E.M. & Hayden, M.R. (1996). Human huntingtin derived from YAC transgenes compensates for loss of murine huntingtin by rescue of the embryonic lethal phenotype. Hum. Mol. Genet. 5, 1875–1885.

Holmberg, M., Duyckaerts, C., Durr, A., Cancel, G., Gourfinkel, I., Damier, P., Faucheux, B., Trottier, Y., Hirsch, E.C., Agid, Y. & Brice, A. (1998). Spinocerebella ataxia (SCA7): a neurodegenerative disorder with neuronal intranuclear inclusions. Hum. Mol. Genet. 7, 913–918.

Hummerich, H. & Lehrach, H. (1995). Trinucleotide repeat expansion and human disease. Electrophoresis 16, 1698–1704.

Huntington's Disease Collaborative Research Group (1993). A novel gene containing a trinucleotide repeat that is expanded and unstable on Huntington's disease chromosomes. Cell 72, 971–983.

Ikeda, H., Yamaguchi, M., Sugai, S., Aze, Y., Narumiya, S. & Kakizuka, A. (1996). Expanded polyglutamine in the Machado-Joseph disease protein induces cell death in vitro and in vivo. Nat. Genet. 8, 221–228.

Ikeuchi, T., Koide, R., Onodera, O., Tanaka, H., Oyake, M., Takano, H. & Tsuiji, S. (1995). Detatorubral pallidorubrallusian atrophy (DRPLA). Molecular basis for wide clinical features of DRPLA. Clin. Neurosci. 3, 23–27.

Imbert, G., Saudou, F., Yvert, G., Devys, D., Trottier, Y., Garnier, J.-M., Weber, C., Mandel, J.L., Cancel, G., Abbas, N., Dürr, A., Didierjean, O., Stevanin, G., Agid, Y. & Brice, A. (1996). Cloning of the gene for spinocerebellar ataxia 2 reveals a locus with high sensitivity to expanded CAG/glutamine repeats. Nat. Genet. 14, 285–291.

Ionov, Y., Peinado, M.A., Malkhosyan, S., Shibata, D. & Perucho, M. (1993). Ubiquitous somatic mutations in simple repeated sequences reveal a new mechanism for colorectal carcinogenesis. Nature 363, 558–561.

Jansen, G., Mahadevan, M., Amemiya, C., Wormskamp, N., Segers, B., Hendriks, W., O'Hoy, K., Baird, S., Sabourin, L., Lennon, G., Jap, P.L., Iles, D., Coerwinkel, M., Hofker, M., Carrano, A.V., de Jong, P.J., Korneluk, R. & Wieringa, B. (1992). Characterization of the myotonic dystrophy region predicts multiple protein isoform-encoding mRNAs. Nat. Genet.1, 261–266.

Jansen, G., Groenen, P.J.T.A., Bachner, D., Jap, P.H.K., Coerwinkel, M., Oerlemans, F., van den Broeck, W., Gohlsch, B., Pette, D., Plomp, J.J., Molenaar, P.C., Nederhoff, M.G., van Echteld, C.J., Dekker, M., Berns, A., Hameister, H. & Wieringa, B. (1996). Abnormal myotonic dystrophy protein kinase levels produce only mild myopathy in mice. Nat. Genet. 13, 316–324.

Jiang, J.X., Deprez, R.H., Zwarthoff, E.C. & Riegman, P.H. (1995). Characterization of four novel CAG repeat-containing cDNAs. Genomics 30, 91–93.

Jodice, C., Matuano, E., Veneziano, L., Trehel, F., Sabbadini, G., Galandriello, L., Francia, A., Spadaro, M., Pierelli, F., Salvi, F., Ophoff, R.A., Frants, R.R. & Frontali, M. (1993). Episodic ataxia type 6 (SCA6) due to CAG repeat expansion in the CACNAIA gene on chromosome 19p. Hum. Mol. Genet. 6, 1973–1978.

Jodice, C., Malaspina, P., Persichetti, F., Novello, A., Spadaro, M., Giunti, P., Morocutti, C., Terrenato, L., Harding, A.E. & Frontali, M. (1994). Effects of trinucleotide repeat length and parental sex on phenotypic variation in spinocerebellar ataxia 1. Am. J. Hum. Genet. 54, 959–965.

Johansson, J., Forsgren, L., Sandgren, O., Brice, A., Holmgren, G. & Holmberg, M. (1998). Expanded CAG repeats in Swedish spinocerebellar ataxia type 7 (SCA7) patients: effect of CAG repeat length on the clinical manifestation. Hum. Mol. Genet. 7, 171–176.

Jones, C., Slijepcevic, P., Marsh, S., Baker, E., Langdon, W.Y., Richards, R.I. & Tunnacliffe, A. (1994). Physical linkage of the fragile site FRA11B and a Jacobsen syndrome chromosome deletion breakpoint in 11q23.3. Hum. Mol. Genet. 3, 2123–2130.

Jou, Y.S. & Myers, R.M. (1995). Evidence from antibody studies that the CAG repeat in the Huntington disease gene is expressed in the protein. Hum. Mol. Genet. 4, 465–469.

Kalchman, M.A., Graham, R.K., Xi, G., Koide, H.B., Hodgeson, J.G., Graham, K.C., Goldberg, Y.P., Gietz, R.D., Pickart, C.M. & Hayden, M.R. (1996). Huntingtin is ubiquitinated and interacts with a specific ubiquitin conjugating enzyme. J. Biol. Chem. 271, 19385–19394.

Kawaguchi, Y., Okamoto, T., Taniwaki, M., Aizawa, M., Inoue, M., Katayama, S., Kawakami, H., Nakamura, S., Nishimura, M., Akiguchi, I., et al. (1994). CAG expansions in a novel gene for Machado-Joseph disease at chromosome 14q32.1. Nat. Genet. 8, 221–228.

Kaytor, M.D., Burright, E.N., Duvick, L.A., Zoghbi, H.Y. & Orr, H.T. (1997). Increased trinucleotide repeat instability with advanced maternal age. Hum. Mol. Genet. 6, 2135–2139.

Kennedy, W.R., Alter, M. & Sung, J.H. (1968). Progressive proximal spinal and bulbar muscular atrophy of late onset: a sex-linked recessive trait. Neurology 18, 671–680.

Klesert, T.R., Otten, A.D., Bird, T.D. & Tapscott, S.J. (1997). Trinucleotide repeat expansion at the myotonic dystrophy locus reduces expression of DMAHP. Nat. Genet. 16, 402–406.

Knight, S.J., Flannery, A.V., Hirst, M.C., Campbell, L., Christodoulou, Z., Phelps, S.R., Pointon, J., Middleton-Price, H.R., Barnicoat, A., Pembrey, M.E., Holland, J., Oostra, B.A., Bobrow, M. & Davies, K.E. (1993). Trinucleotide repeat amplification and hypermethylation of a CpG island in FRAXE mental retardation. Cell 74, 127–134.

Knight, S.J.L., Voelckel, M.A., Hirst, M.C., Flannery, A.V., Moncla, A. & Davies K.E. (1994). Triplet repeat expansion at the fragile X locus and X linked mild mental handicap. Am. J. Hum. Genet. 55, 81–86.

Knight, S.J.L., Ritchie, R.J., Chakrabarti, L., Cross, G., Taylor, G.R., Mueller, R.F., Hurst, J., Peterson, J., Yates, J.R.W., Dow, D.J. & Davies, K.E. (1996). A study of FRAXE in mentally retarded individuals referred for fragile X syndrome (FRAXE) testing in the United Kingdom. Am. J. Hum. Genet. 58, 906–913.

Koide, R., Ikeuchi, T., Onodera, O., Tanaka, H., Igarashi, S., Endo, K., Takahashi, H., Kondo, R., Ishikawa, A., Hayashi, T., et al. (1994). Unstable expansion of CAG repeat in hereditary dentatorubral-pallidoluysian atrophy (DRPLA). Nat. Genet. 6, 9–13.

Konigsmark, B.W. & Weiner, P.P. (1970). The olivopontocerebellar atrophies. Medicine 49, 227–241.

Koob, M.D., Benzow, K.A., Bird, T.D., Day, J.W., Moseley, M.L. & Ranum, L.P. (1998). Rapid cloning of expanded trinucleotide repeat sequences from genomic DNA. Nat. Genet. 18, 72–75.

Koshy, B., Matilla, T., Burright, E.N., Merry, D.E., Fischbeck, K.H., Orr, H.T., Zoghbi, H.Y. (1996). Spinocerebellar ataxi a type 1 and spinobulbar muscular atrophy gene products interact with glyceraldehyde 3 phosphate dehydrogenase. Hum. Mol. Genet. 5, 1311–1318.

La Spada, A.R., Wilson, E.M., Lubahn, D.B., Harding, A.E. & Fischbeck, K.H. (1991). Androgen receptor gene mutations in X-linked spinal and bulbar muscular atrophy. Nature 352, 77–79.

Li, S.H., McInnis, M.G., Margolis, R.L., Antonarakis, S.E. & Ross, C.A. (1993a). Novel triplet repeat containing genes in human brain: cloning, expression, and length polymorphisms. Genomics 16, 572–579.

Li, S.H., Schilling, G., Young, W.S.D., Li, X.J., Margolis, R.L., Stine, O.C., Wagster, M.V., Abbott, M.H., Franz, M.L., Ranen, N.G., et al. (1993b). Huntington's disease gene (IT15) is widely expressed in human and rat tissues. Neuron 11, 985–993.

Li, X.J., Li, S.H., Sharp, A.H., Nuclfora, F.C., Schilling, G., Lanahan, A., Worley, P., Snyder, S.H. & Ross, C.A. (1995). A huntingtin associated protein enriched with implications for pathology. Nature 378, 398–402.

Lin, B., Nasir, J., MacDonald, H., Hutchinson, G., Graham, R.K., Rommens, J.M. & Hayden, M.R. (1994). Sequence of murine Huntington's disease gene: evidence for conservation, alternate splicing and polymorphism in a triplet (CCG) repeat. Hum. Mol. Genet. 3, 85–92.

Lindblad, K., Nylander, P.O., De bruyn, A., Sourey, D., Zander, C., Engstrom, C., Holmgren, G., Hudson, T., Chotai, J., Mendlewicz, J., et al. (1995). Detection of expanded CAG repeats in bipolar affective disorder using the repeat expansion detection (RED) method. Neurobiol. Dis. 2, 55–62.

Lindblad, K., Savontaus, M.L., Stevanin, G., Holmberg, M., Digre, K., Zander, C., Ehrsson, H., David, G., Benomar, A., Nikoskelainen, E., Trottier, Y., Holmberg, G., Ptacek, L.J., Anttinen, A., Brice, A. & Schalling, M. (1996). An expanded CAG repeat sequence in spinocerebellar ataxia type 7. Gen. Res. 10, 965–971.

Lorenzetti D., Bohlega, S. & Zoghbi, H.Y. (1997). The expansion of the CAG repeat in ataxin-2 is a frequent cause of autosomal dominant spinocerebellar ataxia. Neurology 49, 1009–1013.

Lubs, H.A. Jr. (1969). A marker X chromosome. Am. J. Hum. Genet. 21, 231–244.

Machkhas, H., Bidichandani, S.I., Patel, P.I. & Harati, Y. (1998). A mild case of Friedreich ataxia: lymphocyte and sural nerve analysis for GAA repeat length reveals somatic mosaicism. Muscle Nerve 21, 390–393.

MacLean, H.E., Warne, G.L. & Zajac, J.D. (1997). Localization of functional domains in the androgen receptor. J. Steroid. Biochem. Molec. Biol. 62, 233–242.

Mahadevan, M., Tsilfidis, C., Sabourin, L., Shutler, G., Amemiya, C., Jansen, G., Neville, C., Narang, M., Barcelo, J., O'Hoy, K., Leblond, S., Earle-Macdonald, J., de Jong, P.J., Wieringa, B. & Korneluk, R.G. (1992). Myotonic dystrophy mutation: an unstable CTG repeat in the 3′ untranslated region of the gene. Science 255, 1253–1255.

Mahadevan, M.S., Amemiya, C., Jansen, G., Sabourin, L., Baird, S., Neville, C.E., Wormskamp, N., Segers, B., Batzer, M., Lamerdin, J., de Jong, P., Wieringa, B. & Korneluk, R.G. (1993). Structure and genomic sequence of the myotonic dystrophy (DM kinase) gene. Hum. Molec. Genet. 2, 299–304.

Mangiarini, L., Sathasivam, K., Seller, M., Cozens, B., Harper, A., Hetherington, C., Lawton, M., Trottier, Y., Lehrach, H., Davies, S.W. & Bates, G.P. (1996). Exon 1 of the HD gene with an expanded CAG repeat is sufficient to cause a progressive neurological phenotype in transgenic mice. Cell 87, 493–506.

Mangiarini, L., Sathasivam, K., Mahal, A., Mott, R., Seller, M. & Bates, G.P. (1997). Instability of highly expanded CAG repeats in mice transgenic for the Huntington's disease mutation. Nat. Genet. 15, 197–200.

Margolis, R.L., Abraham, M.R., Gatchel, S.B., Li, S.H., Kidwai, A.S., Breschel, T.S., Stne, O.C., Callahan, C., McInnins, M.G. & Ross, C.A. (1997). cDNAs with long CAG trinucleotide repeats from human brain. Hum. Genet. 100, 114–122.

Maruyama, H., Nakamura, S., Matsuyama, Z., Sakai, T., Doyu, M., Sobue, G., Seto, M., Tsuiihata, M., Oh-I, T., Nishio, T., et al. (1995). Molecular features of the CAG repeats and clinical manifestation of Machado-Joseph disease. Hum. Mol. Genet. 4, 807–812.

Matilla, T., Volpini, V., Genis, D., Rosell, J., Corral, J., Davalos, A., Molins, A. & Estivill, X. (1993). Presymptomatic analysis of spinocerebellar ataxia type 1 (SCA 1) via the expansion of the SCA

1 CAG repeat in a large pedigree displaying anticipation and parental male bias. Hum. Mol. Genet. 2, 2123–2128.

Matilla, A., Koshy, B.T., Cummings, C.J., Isobe, T., Orr, H.T. & Zoghbi, H.Y. (1997). The cerebellar leucin-rich acidic nuclear protein interacts with ataxin 1. Nature 389, 974–978.

Matilla, A., Roberson, E.D., Banfi, S., Morales, J., Armstrong, D.L., Burright, E.N., Orr, H.T., Seeatt, J.D., Zoghbi, H.Y. & Matzuk, M.M. (1998). Mice lacking ataxin 1 display learning deficits and decreased hippocampal paired-pulse facilitation. J. Neurosci. 18, 5508–5516.

Matsuyama, Z., Kawakami, H., Maruyama, H., Izumi, Y., Komure, O., Udaka, F., Kameyama, M., Nisio, T., Kuroda, Y., Nishimura, M. & Nakamura, S. (1997). Molecular features of the CAG repeats of spinocerebellar ataxia 6 (SCA6). Hum. Mol. Genet. 6, 1283–1287.

McInnis, M.G., McMahon, F.J., Chase, G.A., Simpson, S.G., Ross, C.A. & DePaolo, J.R., Jr. (1993). Anticipation in bipolar affective disorder. Am. J. Hum. Genet. 53, 385–390.

McMurray, C.T. (1995). Mechanisms of DNA expansion. Chromosoma 104, 2–13.

Meijer, H., de Graaff, E., Merckx, D.M., Jongbloed, R.J., de Die-Smulders, C.E., Englen, J.J., Fryns, J.P., Curfs, P.M. & Oostra, B.A. (1994). A deletion of 1.6 kb proximal to the CGG repeat of the FMR1 gene causes the clinical phenotype of the fragile X syndrome. Hum. Mol. Genet. 3, 615–620.

Mitas, M., Yu, A., Dill, J. & Haworth, I.S. (1995). The trinucleotide repeat sequence d(CGG)15 forms a heat stable hairpin containing Gsyn. Ganti base pairs. Biochemistry 34, 12803–12811.

Miyashita, T., Okamura-Ohi, Y., Mito, Y., Nagafuchi, S. & Yamada, M. (1997). Dentatorubral pallidoluysian atrophy (DRPLA) protein is cleaved by caspase 3 during apoptosis J. Biol. Chem. 272, 29238–29242.

Modrich, P. (1991). Mechanisms and biological effects of mismatch repair. Ann. Rev. Genet. 55, 659–665.

Monckton, D.G., Coolbaugh, M.I., Ashizawa, K., Siciliano, M.J. & Caskey, C.T. (1997). Hypermutable myotonic dystrophy CTG repeats in transgenic mice. Nat. Genet. 15, 193–196.

Monroe, E., Molto, M.D., Martinez, F., Canizares, J., Blanca, J., Vilchez, J.J., Prieto, F., de Frutos, R. & Palau, F. (1997). Phenotype correlation and intergenerational dynamics of the Friedreich ataxia GAA trinucleotide repeat. Am. J. Hum. Genet. 61, 101–110.

Montermini, L., Rodius, F., Pianese, L., Molto, M.D., Cossee, M., Campuzano, V., Cavalcanti, F., Monticelli, A., Palau, F., Gyapay, G., Wenhert, M., Zara, F., Patel, P.I., Cocozza, S., Koenig, M. & Pandolfo, M. (1995). The Friedreich ataxia critical region spans a 150-kb interval on chromosome 9q13. Am. J. Hum. Genet. 57, 1061–1067.

Montermini, L., Kish, S.J., Jiralerspong, S., Lamarche, J.B. & Pandolfo, M. (1997). Somatic mosaicism for Friedreich's ataxia GAA triplet repeat expansions in the central nervous system. Neurology 49, 606–610.

Morris, A.G., Gaitonde, E., McKenna, P.J., Mollon, J.D. & Hunt, D.M. (1995). CAG repeat expansions and schizophrenia: association with disease in females and with early age-at-onset. Hum. Mol. Genet. 4, 1957–1961.

Nagafuchi, S., Yanagisawa, H., Ohsaki, E., Shirayama, T., Tadokora, K., Inonue, T. & Yamada, M. (1994a). Structure and expression of the gene responsible for the triplet repeat disorder, dentatorubral and pallidoluysian atrophy (DRPLA). Nat. Genet. 8, 177–182.

Nagafuchi, S., Yanagisawa, H., Sato, K., Shirayama, T., Ohsaki, E., Bundo, M., Takeda, T., Tadokoro, K., Kondo, I., Murayama, N., et al. (1994b). Dentatorubral and pallidoluysian atrophy expansion of an unstable CAG trinucleotide on chromosome 12p. Nat. Genet. 6, 14–18.

Nancarrow, J.K., Kremer, E., Holman, K., Eyre, H., Doggett, N.A., Le Paslier, D., Callen, D.F., Sutherland, G.R. & Richards, R.I. (1994). Implications of FRA16A structure for the mechanism of chromosomal fragile site genesis. Science 264, 1938–1941.

Nasir, J., Floresco, S.B., Diewert, V.M., Richman, J.M., Zeiser, J., Borowski, A., Marth, J.D., Philips, A.G. & Hayden, M.R. (1995). Targeted disruption of the Huntington's disease gene results in embryonic lethality and behavioral and morphological changes in heterozygotes. Cell 81, 811–823.

Nechiporuk, T., Huynh, D.P., Figueroa, K., Sahba, S., Nechiporuk, A. & Pulst, S.M. (1998). The mouse SCA2 gene: cDNA sequence, alternative splicing and protein expression. Hum. Mol Genet. 7, 1301–1309.

Néri, C., Albanese, V., Lebre, A.-S., Holbert, S., Saada, C., Bougueleret, L., Meier-Ewert, S., Le Gall, I., Millasseau, P., Bui, H., Giudicelli, C., Massart, C., Guillou, S., Gervy, P., Poullier, E., Rigault, P., Weissenbach, J., Lennon, G., Chumakov, I., Dausset, J., Lehrach, H., Cohen, D. & Cann, H.M. (1996). Survey of CAG/CTG repeats in human cDNAs representing new genes: candidates for inherited neurological disorders. Hum. Mol. Genet. 5, 1001–1009.

Nielsen, J.E., Koefoed, P., Abell, K., Hasholt, L., Eiberg, H., Fenger, K., Neibuhr, E. & Sorensen, S.A. (1997). CAG repeat expansion is autosomal dominant pure spastic paraplegia linked to chromosome 2p21-24. Hum. Mol. Genet. 6, 1811–1816.

O'Donovan, M.C., Guy, C., Craddock, N., Murphy, K.C., Cardno, A.G., Jones, L.A., Owen, M.J. & McGuffin, P. (1995). Expanded CAG repeats in schizophrenia and bipolar disorder. Nat. Genet. 10, 380–381.

Ophoff, R.A., Terwindt, G.M., Vergouwe, M.N., Eijk, R.V., Oefner, P.J., Hoffman, S.M.G, Lamerdin, J.E., Mohrenweiser, H.M., Bulman, D.E., Ferrari, M., Haan, J., Lindhout, D., Ommen, G.J.B.V., Hofker, M. H., Ferrari, M.D. & Frants, R.R. (1996). Familial hemiplegic migraine and episodic ataxia type 2 are caused by mutations in the Ca2+ channel gene CACNL1A4. Cell 87, 543–552.

Ordway, J.M., Tallaksen-Greene, S., Gutekunst, C.-A., Bernstein, E.M., Cearley, J.A., Wiener, H.W., Dure, L.S., IV, Lindsey, R., Hersch, S.M., Jope, R.S., Albin, R.L. & Detloff, P.J. (1997). Ectopically expressed CAG repeats cause intranuclear inclusions and a progressive late onset neurological phenotype in the mouse. Cell 91, 753–764.

Orr, H.T., Chung, M.Y., Banfi, S., Kwiatkowski, T.J., Jr., Servadio, A., Beaudet, A.L., McCall, A.E., Duvick, L.A., Ranum, L.P. & Zoghbi, H.Y. (1993). Expansion of an unstable trinucleotide CAG repeat in spinocerebellar ataxia type 1. Nat. Genet. 4, 221–226.

Otten, A.D. & Tapscott, S.J. (1995). Triple-repeat expansion in myotonic dystrophy alters the adjacent chromatin structure. Proc. Nat. Acad. Sci. U. S. A. 92, 5465–5469.

Parrish, J.E., Oostra, B.A., Verkerk, A.J., Richards, C.S., Reynolds, J., Spikes, A.S., Shaffer, L.G. & Nelson, D.L. (1994). Isolation of a GCC repeat showing expansion in FRAXF, a fragile site distal to FRAXA and FRAXE. Nat. Genet. 8, 229–235.

Paulson, H.L., Perez, M.K., Trottier, Y., Trojanowski, J.Q., Subramony, S.H., Das, S.S., Vig, P., Mandel, J.L., Fischbeck, K.H. & Pittman, R.N. (1997). Intranuclear inclusions of expanded polyglutamine protein in spinocerebellar ataxia type 3. Neuron 19, 333–344.

Penrose, L.S. (1948). The problems of anticipation in pedigrees of dystrophia myotonica. Ann. Eugen. 14, 125–132.

Petronis, A. & Kennedy, J.L. (1995). Unstable genes—unstable minds? Am. J. Psychiatry 152, 164–172.

Pianese, L., Cavalcanti, F., De Michele, G., Filla, A., Campanella, G., Calabrese, O., Castaldo, I., Monticelli, A. & Cocozza, S. (1997). The effect of parental gender on the GAA dynamic mutation in the FRDA gene [letter]. Am. J. Hum. Genet. 60, 460–463.

Potter, N.T., Yanagisawa, H. & Yamada, M. (1996). Different origins of expanded repeats for Haw River syndrome and dentatorubral-pallidoluysian atrophy. Lancet 347, 1271.

Pulst, S.M., Nechipork, A., Nechiporuk, T., Gispert, S., Chen, X.N., Lopes-Cendes, I., Pearlman, S., Starkman, S., Orozco-Diaz, G., Lunkes, A., DeJong, P., Rouleau, G.A., Auburger, G., Korenberg, J.R., Carla, F. & Sahba, S. (1996). Moderate expansion of a normally biallelic trinucleotide repeat in spinocerebellar ataxia type 2. Nat. Genet. 14, 269–276.

Raimondi, E., Antonelli, A., Driesel, A.J. & Pandolfo, M. (1990). Regional localization by in situ *hybridization of a human* chromosome 9 marker tightly linked to the Friedreich's ataxia locus. Hum. Genet. 85, 125–126.

Ranum, L.P., Duvick, L.A., Rich, S.S., Schut, L.J., Litt, M. & Orr, H.T. (1991). Localization of the ausomal dominant HLA linked spinocerebellar ataxia (SCA1) locus, in two kindreds, within an 8-cM subregion of chromosome 6p. Am. J. Hum. Genet. 49, 31–41.

Ranum, L.P.W., Chung, M.Y., Banfi, S., Bryer, A., Schut, L.J., Ramesar, R., Duvick, L.A., McCall, A.E., Subromony, S.H., Goldfrab, L., Gomez, C., Sandkuij, L.A., Orr, H.T., Zoghbi, H.Y. (1994a). Molecular and clinical correlations in spinocerebellar ataxi a type 1 (SCA1). Evidence for familial effects on the age of onset. Am. J. Hum. Genet. 55, 244–252.

Ranum, L.P.W., Schut, L.J., Lundgren, J.K., Orr, H.T. & Livingston, D.M. (1994b). Spinocerebellar ataxia type 5 in a family descended from the grandparents of President Lincoln maps to chromosome 11. Nat. Genet. 8, 280–284.

Ranum, L.P., Rasmussen, P.F., Benzow, K.A., Koob, M.D. & Day, J.W. (1998). Genetic mapping of a second myotonic dystrophy locus. Nat. Genet. 19, 196–198.

Reddy, P.S. & Housman, D.E. (1997). The complex pathology of trinucleotide repeats. Curr. Opin. Cell Biol. 9, 364–372.

Reddy, S., Smith, D.B., Rich, M.M., Leferovich, J.M., Reilly, P., Davis, B.M., Tran, K., Rayburn, H., Bronson, R., Cros, D., Balice-Gordon, R.J. & Housman, D. (1996). Mice lacking the myotonic dystrophy protein kinase develop a late onset progressive myopathy. Nat. Genet. 13, 325–335.

Reddy, P.H., Stockburger, E., Gillevet, P. & Tagle, D.A. (1997). Mapping and characterization of nevel (CAG)n repeat cDNAs from adult human brain derived by the oligo capture method. Genomics 46, 174–182.

Reddy, P.H., Williams, M., Charles, V., Garrett, L., Buchanan, L.P., Whetsell, W.O., Jr., Miller, G. & Tagle, D.A. (1998). Behavioural abnormalities and selective neuronal loss in HD transgenic mice expressing mutated full-length HD cDNA. Nat. Genet. 20, 198–202.

Reiss, A.L., Kazazian, H.H. Jr., Krebs, C.M., McAugham, A., Boem, C.D., Abrams, M.T. & Nelson D.L. (1994). Frequency and stability of the fragile X premutation. Hum. Mol. Genet. 3, 393–398.

Richards, R.I. & Sutherland, G.R. (1992). Dynamic mutations: a new class of mutations causing human disease. Cell 70, 709–712.

Riggins, G.J., Lokey, L., Chastain, J.L., Leiner, H.A., Sherman, S.L., Wilkinson, K.D. & Warren, S.T. (1992). Human genes containing polymorphic trinucleotide repeats [published erratum appears in Nat. Genet. (1993) 3, 273]. Nat. Genet. 2, 186–191.

Rivaud-Pechpox, S., Durr, A., Gaymard, B., Cancel, G., Ploner, C.J., Agid, Y., Brice, A. & Pierrot-Deseilligny, C. (1998). Eye movement abnormalities correlate with genotype in autosomal dominant cerebellar ataxia type I. Ann. Neurol. 43, 297–302.

Rotig, A., de Lonlay, P., Chretien, D., Foury, F., Koenig, M., Sidi, D., Munnich, A. & Rustin, P. (1997a). Aconitase and mitochondrial iron-sulphur protein deficiency in Friedreich ataxia. Nat. Genet. 17, 215–217.

Rotig, A., deLonlay, P., Chretien, D., Foury, F., Koenig, M., Sidi, D., Munnich, A. & Rustin, P. (1997b). Frataxin gene expansion causes aconitase and mitochondrial iron-sulfur protein deficiency in Fredreich ataxia. Nat. Genet. 17, 215–217.

Rubinsztein, D.C., Leggo, J., Coles, R., Almqvist, E., Biancalana, V., Cassiman, J.J., Chotai, K., Connarty, M., Crauford, D., Curtis, A., Curtis, D., Davidson, M.J., Differ, A.M., Dode, C., Dodge, A., Frontali, M., Ranen, N., Stine, O.C., Sherr, M., Abbott, M.H., Franz, M.L., Graham, C.A., Harper, P.S., Hadreen, J.C., Hayden, M.R. & Ross C.A. (1996). Phenotype characterization of individuals with 30-40 CAG repeats in the Huntington's disease (HD) gene reveals HD cases with 36 repeats and apparently normal elderly individuals with 36-39. Am. J. Hum. Genet. 59, 16–22.

Sanpei, K., Takano, H., Igarashi, S., Sato, T., Oyake, M., Sasaki, H., Wakisaka, A., Tashiro, K., Ishida, Y., Ikeuchi, T., Koide, R., Saito, M., Sato, A., Tanaka, T., Hanyu, S., Takiyama, Y., Nishizawa, M., Shimizu, N., Nomura, Y., Segawa, M., Iwabuchi, K., Eguchi, I., Tanaka, H., Takahashi, H. & Tsuji, S. (1996). Identification of the spinocerebellar ataxia type 2 gene using a direct identification of repeat expansion and cloning technique, DIRECT. Nat. Genet. 14, 277–284.

Schalling, M., Hudson, T.J., Buetow, K.H. & Housman, D.E. (1993). Direct detection of novel expanded trinucleotide repeats in the human genome. Nat. Genet. 4, 135–139.

Scherzinger, E., Lurz, R., Turmaine, M., Mangiarini, L., Hollenbach, B., Hasenbank, R., Bates, G.P., Davies, S.W., Lehrach, H. & Wanker, E.E. (1997). Huntingtin-encoded polyglutamine expansions form amyloid-like protein aggregates in vitro and in vivo. Cell 90, 549–558.

Schmitt, I., Bachner, D., Megow, D., Henklein, P., Hameister, H., Epplen, J.T. & Reiss, O. (1995). Expression of the Huntington's disease in rodents: cloning the rat homologue and evidence for downregulation in nonneuronal tissues during development. Hum. Mol. Genet. 4, 1173–1182.

Schols, L., Reiss, O., Amoiridis, G., Reiss, A., Przuntek, H. & Epplen J.T. (1997). Genetic diagnosis, classification and clinical hereditary ataxia disease entities. Fortchr. Neurol. Psychiatry. 65, 79–89.

Schols, L., Kruger, R., Amoiridis, G., Przuntek, H., Epplen, J.T. & Riess, O. (1998). Spinocerebellar ataxia type 6: genotype and phenotype in German kindreds. J. Neurol. Neurosurg. Psychiatry 64, 67–73.

Sharp, A.H., Loev, S.J., Schilling, G., Li, S.H., Li, X.J., Bao, J., Wagster, M.V., Kotzuk, J.A., Steiner, J.P., Lo, A., et al. (1995). Widespread expression of Huntington's disease gene (IT15) protein product. Neuron 14, 1065–1074.

Shaw, D.J., McCurrach, M., Rundle, S.A., Harley, H.G., Crow, S.R., Sohn, R., Thirion, J.-P., Hamshere, M.G., Buckler, A.J., Harper, P.S., Housman, D.E. & Brook, J.D. (1993). Genomic organization and transcriptional units at the myotonic dystrophy locus. Genomics 18, 673–679.

Shaw, J., Lichter, P., Driesel, A.J., Williamson, R. & Chamberlain, S. (1990). Regional localization of the Friedreich ataxia locus to human chromosome 9q13-q21.1. Cytogenet. Cell Genet. 53, 221–224.

Sherman, S.L., Jacobs, P.A., Morton, N.E., Froster-Iskenius, U., Howard-Peebles, P.N., Nielsen, K.B., Partington, M.W., Sutherland, G.R., Turner, G. & Watson, M. (1985). Further segregation analysis of the fragile X syndrome with special reference to transmitting males. Hum. Genet. 69, 289–299.

Sinden, R.R. & Wells, R.D. (1992). DNA structure, mutations, and human genetic disease. Curr. Opin. Biotechnol. 3, 612–622.

Siomi, H., Siomi, M.C., Nussbaum, R.L. & Dreyfuss, G. (1993). The protein product of the fragile X gene, FMR1, has characteristic of an RNA-binding protein. Cell 74, 291–298.

Siomi, H., Choi, M., Siomi, M., Nussbaum, R. & Dreyfuss (1994). Essential role for KH domains in RNA binding by a mutation in the KH domains of FMR1 that causes fragile X syndrome. Cell 77, 33–39.

Skinner, P.J., Koshy, B.T., Cummings, C.J., Klement, I.A., Helin, K., Servadio, A., Zoghbi, H.Y. & Orr, H.T. (1997). Ataxin-1 with an expanded glutamine tract alters nuclear matrix-associated structures. Nature 389, 971–974.

Sobue, G., Hashizume, Y., Mukai, E., Hirayama, M., Mitsuma, T. & Takahashi, A. (1989). X-linked recessive bulbospinal neuropathy. A clinico pathological study. Brain 112, 209–232.

Spargo, E., Everall, I.P. & Lantos, P.L. (1993). Neuronal loss in the hippocampus in Huntington's disease: a comparison with HIV infection. J Neurol. Neurosurg. Psychiatry 56, 487–491.

Stevanin, G., Sousa, P.S., Cancel, G., Durr, A., Dubourg, O., Nicholson, G.A., Weissenbach, J., Jardim, E., Agid, Y., Cassa, E., et al. (1994). The gene for Machado-Joseph disease maps to the same 3-cM intervals as the spinal cerebellar ataxia 3 gene on chromosome 14q. Neurobiol. Dis. 1, 79–82.

Stevanin, G., Cancel, G., Didierjean, O., Durr, A., Cassa, E., Feingold, J., Agid, Y. & Brice, A. (1995). Linkage disequilibrium at the Machado-Joseph disease/spinal cerebellar ataxia 3 locus: evidence for a common founder effect in French and portuguese-Brazilian families as well as a second ancestral Portuguese-Azorean mutation. Am. J. Hum. Genet. 57, 1247–1250.

Stevanin, G., Trottier, Y., Cancel, G., Durr, A., David, G., Didierjean, O., Burk, K., Imbert, G., Saudou, F., Abada-Bendib, M., Gourfinkel-An, I., Benomar, A., Abbas, N., Klockgether, T., Grid, D., Agid, Y., Mandel, J.L. & Brice, A. (1996). Screening for proteins with polyglutamine expansions in autosomal dominant cerebellar ataxias. Hum. Mol. Genet. 5, 1887–1892.

Strong, T.V., Tagle, D.A., Valdes, J.M., Elmer, L.W., Boehm, K., Swaroop, M., Kaatz, K.W., Collins, F.S. & Albin, R.L. (1993). Widespread expression of the human and rat Huntington's disease gene in brain and nonneural tissues. Nat. Genet. 5, 259–265.

Sutherland, G.R. (1977). Fragile sites on human chromosomes, demonstration of their dependence on the type of tissue culture medium. Science 197, 265–266.

Sutherland, G.R. & Baker, E. (1992). Characterization of a new rare fragile site easily confused with the fragile X. Hum. Mol. Genet. 1, 111–113.

Sutherland, G.R. & Richards, R.I. (1993). Dynamic mutations on the move. J. Med. Genet. 30, 978.

Takahashi, H., Ikeuchi, T., Honma, Y., Hayashi, S. & Tsuiji S. (1998). Autosomal dominant cerebellar ataxia (SCA6): clinical, genetic and neuropathological study in a family. Acta Neuropathol. 95, 333–337.

Takiyama, Y., Oyanagi, S., Kawashima, S., Sakamoto, S., Karube, S., Shimazaki, S., Soutome, M., Endo, K., Ohta, S., Kagawa, Y., Kanazawa, I., Mizuno, Y., Yoshida, M., Yuasa, T., Horikawa, Y., Oyanagi, K., Nagai, H., Kondo, T., Inizuka, T., Onodera, O. & Tsuiji. (1993). The gene for Machado-Joseph disease maps to human chromosome 14q. Nat. Genet. 4, 300–304.

Takiyama, Y., Oyanagi, S., Kawashima, S., Sakamoto, H., Saito, K., Yoshida, M., Tsuji, S., Mizuno, Y. & Nishizawa, M. (1994). A clinical and pathologic study of a large Japanese family with Machado-Joseph disease tightly linked to the DNAS markers on chromosome 14q. Neurology 44, 1302–1308.

Tanaka, K., Takeshita, K. & Takita, M. (1981). Deoxycholic acid, a candidate for the maternal intrauterine factor in early-onset myotonic dystrophy [letter]. Lancet I, 1046–1047.

Thibodeau, S.N., French, A.J., Roche, P.C., Cunningham, J.M., Tester, D.J., Lindor, N.M., Moslein, G., Baker, S.M., Liskay, R.M., Burgart, L.J., Honchel, R., Halling, K.C. (1996). Altered expression of hMSH2 and hMLH1 in tumors with microsatellite instability and genetic alterations in mismatch repair genes. Cancer Res. 56, 4836–4840.

Thibodeau, S.N., French, A.J., Cunningham, J.M., Tester, D., Burgart, L.J., Roche, P.C., McDonnell, S.K., Schaid, D.J., Vockley, C.W., Michels, V.V., Farr, G.H., Jr. & O'Connell, M.J. (1998). Microsatellite instability in colorectal cancer: different mutator phenotypes and the principal involvement of hMLH1. Cancer Res. 58, 1713–1718.

Trottier Y., Devys, D., Imbert, G., Saudou, F., An, I., Lutz, Y., Weber, C., Agid, Y., Hirsch, E.C. & Mandel J.-L. (1995a). Cellular localization of the Huntington's disease protein and discrimination of the normal and mutated form. Nat. Genet. 10, 104–110.

Trottier, Y., Lutz, Y., Stevanin, G., Imbert, G., Devys, D., Cancel, G., Saudou, F., Weber, C., David, G., Tora, L., et al. (1995b). Polyglutamine expansion as a pathological epitope in Huntington's disease and four dominant cerebellar ataxias. Nature 378, 403–406.

Tsilfidis, C., MacKenzie, A.E., Mettler, G., Barcelo, J. & Korneluk, R.G. (1992). Correlation between CTG trinucleotide repeat length and frequency of severe congenital myotonic dystrophy. Nat. Genet. 1, 192–195.

Verheij, C., Bakker, C.E., de Graaff, E., Keulemans, J., Willemsen, R., Verkerk, A.J.M.H., GalJaard, H., Reuser, A.J.J., Hoogeveen, A.T. & Oostra, B.A. (1993). Characterization and localization of FMR1 gene product associated with the fragile X syndrome. Nature 363, 722–724.

Verkerk, A.J.M.H., Pieretti, M., Sutcliffe, J.S., Fu, Y.H., Kuhl, D.P., Pizzuti, A., Reiner, O., Richards, S., Victoria, M.F., Zhang, F.P., Eussen, B.E., Ommen, G.-J.B., Blonden, L.A.J., Riggins, G.J., Chastain, J.L., Kunst, C.B., Galjaard, H., Caskey, C.T., Nelson, D.L., Oostra, B.A. & Warren, S.T. (1991). Identification of a gene (FMR-1) containing a CGG repeat coincident with a breakpoint cluster region exhibiting length variation in fragile X syndrome. Cell 65, 905–914.

Vonsattel, J.P., Myers, R.H., Stevens, T.J., Ferrante, R.J., Bird, E.D. & Richardson, E.P., Jr. (1985). Neuropathological classification of Huntington's disease. J. Neuropathol. Exp. Neurol. 44, 559–577.

Wang, Y.-H. & Griffith, J. (1995). Expanded CTG triplet blocks from the myotonic dystrophy gene create the strongest known natural nucleosome positioning elements. Genomics 25, 570–573.

Wang, Y.-H., Amirhaeri, S., Kang, S., Wells, R.D. & Griffith, J.D. (1994). Preferential nucleosome assembly at DNA triplet repeats from the myotonic dystrophy gene. Science 265, 669–671.

Wanker, E.E., Rovira, C., Scherzinger, E., Hasenbank, R., Walter, S., Tait, D., Colicelli, J. & Lehrach, H. (1997). HIP 1: a huntingtin interacting protein isolated by the yeast two hybrid system. Hum. Mol. Genet. 6, 487–495.

Warrack, J.M., Paulson, H.L., Gray-Board, G.L., Bui, Q.T., Fischbeck, K.H., Pittman, R.N. & Bonini, N.M. (1998). Expanded polyglutamine protein forms nuclear inclusions and causes neural degeneration in *Drosophila*. Cell 93, 939–949.

Warren, S.T. (1996). The expanding world of trinucleotide repeats. Science 271, 1374–1375.

Wells, R.D. (1996). Molecular basis of genetic instability of triplet repeats. Biol. Chem. 271, 2875–2878.

White, J.K., Auerbach, W., Duyao, M.P., Vansattel, J.-P., Gusella, J.F., Joyner, A.L. & Macdonald, M.E. (1997). Huntingtin is required for neurogenesis and is not impaired by the Huntington's disease CAG expansion. Nat. Genet. 17, 404–410.

Williamson, R., Bowcock, A., Kidd, K., Pearson, D., Schmidtke, J., Chan, H.S., Chipperfield, M., Cooper, D.N., Hewitt, J., Lewitter, F., Maidek, B.Q., Riciuti, F. & Treck, R. (1990). Report of the DNA committee and catalogues of cloned and mapped genes and polymorphisms. Cyto. Cell Genet. 55, 457.

Wood, J.D., MacMillan, J.C., Harper, P.S., Lowenstein, P.R. & Jones, A.L. (1996). Partial characterization of murine huntingtin and apparent variations in the subcellular localization of huntingtin in human, mouse and rat brain. Hum. Mol. Genet. 5, 481–487.

Yazawa, I., Nukina, N., Hashida, H., Goto, J., Yamada, M. & Kanazawa, I. (1995). Abnormal gene product identified in hereditary dentatorubral-pallidoluysian atrophy (DRPLA) brain. Nat. Genet. 10, 99–103.

Yazawa, I., Nukina, N., Goto, J., Kurisaki, H., Hebisawa, A. & Kanazawa, I. (1997). Expression of dentatorubral-pallidoluysian atrophy (DRPLA) proteins in patients. Neurosci. Lett. 225, 53–56.

Zeitlin, S.Z., Liu, J.-P., Chapman, D.L., Papaioannou, V.E. & Efstratiadis, A. (1995). Increased apoptosis and early embryonic lethality in mice nullizygous for the Huntington's disease gene homologue. Nat. Genet. 11, 155–163.

Zhuchenko, O., Bailey, J., Bonnen, P., Ashizawa, T., Stockton, D.W., Amos, C., Dobyns, W.B., Subramony, S.H., Zoghbi, H.Y. & Lee, C.C. (1997). Autosomal dominant cerebellar ataxia (SCA6) associated with small polyglutamine expansions in the alpha 1A-voltage-dependent calcium channel. Nat. Genet. 15, 62–69.

Zoghbi, H.Y. (1995). Spinocerebellar ataxia type 1. Clin. Neurosci. 3, 5–11.

Zoghbi, H.Y. (1996). The expanding world of ataxins. Nat. Genet. 14, 237–238.

Chapter 3

The Genetic Basis and Molecular Pathogenesis of Huntington's Disease

NEIL W. KOWALL, STEPHAN KUEMMERLE, and ROBERT J. FERRANTE

INTRODUCTION

Huntington's disease (HD) is an autosomal dominant neurodegenerative disorder characterized clinically by progressive cognitive decline and chorea. Onset is typically in adulthood but a more fulminant juvenile form is well recognized. Initial clinical manifestations may be behavioral so the diagnosis may be missed for years until the classical movement disorder develops. Psychosis, obsessive thought disorder, and dementia are common. Mood disorders, personality changes, irritable and explosive behavior, schizophrenia-like behavior, suicidal behavior, sexuality changes, and specific cognitive deficits can occur. The choreiform movement disorder is progressive, affects the extremities and face, and is associated with

Advances in Cell Aging and Gerontology
Volume 3, pages 81–92

ISBN: 0-7623-0405-7

slowness and clumsiness of fine movements and postural instability (Thompson et al., 1988). Reflexes are typically unaffected and tone is reduced in early stages but is increased later (Homberg and Huttunen, 1994). There are no sensory abnormalities. In late stages patients are cachectic, dystonic, rigid, and bedridden. The disease typically runs its course over several years. Studies show that the likelihood of HD in a patient with the typical clinical features of this disorder but no history of affected relatives is at least 75%. The most plausible explanations for seemingly sporadic patients with HD are nonpaternity and mild, late-onset disease that is overlooked by other family members (Bateman et al., 1992), but new mutations may also occur (Andrew and Hayden, 1995; Alford et al., 1996).

GENETIC BASIS OF HUNTINGTON'S DISEASE

The discovery of the HD gene on chromosome 4 in 1993 was a major step forward (Huntington's Disease Collaborative Research Group, 1993; Gusella and MacDonald, 1995). The gene was initially called IT 15 (interesting transcript 15) or the huntingtin gene. The genetic abnormality was unexpected; rather than a simple missense mutation or deletion, the HD gene is expanded with an excess number of CAG trinucleotide repeats that result in a long stretch of polyglutamine in the expressed protein. Expansion of trinucleotide repeats is now recognized as a major cause of neurological disease (Plassart and Fontaine, 1994). At least eight disorders result from trinucleotide repeat expansion: X-linked spinal and bulbar muscular atrophy (Kennedy's syndrome, SBMA), two fragile X syndromes of mental retardation (FRAXA and FRAXE), myotonic dystrophy, HD, spinocerebellar ataxia type 1 (SCA1, 6p), spinocerebellar ataxia type 3/Machado-Joseph disease (SCA3/MJD, 14q) and dentatorubral-pallidoluysian atrophy (DRPLA, 12p). The expanded trinucleotide repeats are unstable, and the phenomenon of anticipation (i.e., worsening of disease phenotype over successive generations) correlates with increasing expansion size. These disorders may be subdivided into two classes: Fragile X and myotonic dystrophy are multisystem disorders usually associated with large expansions of untranslated repeats, whereas the neurodegenerative disorders—SBMA, Huntington's disease, SCA1, SCA3, and DRPLA—are caused by smaller expansions of CAG repeats within the protein coding portion of the gene. Polyglutamine expansion appears to be a common element that may lead to a toxic gain of function effect (La Spada et al., 1994). SBMA, or Kennedy's syndrome, is an X-linked, adult-onset motor neuronopathy caused by expansion of a trinucleotide (CAG) repeat in the androgen receptor gene. The length of this repeat varies as it is passed down through SBMA families, and correlates inversely with the age of onset of the disease. The motor neuron degeneration that occurs in this disease is probably caused by a toxic gain of function in the androgen receptor protein (Brooks and Fischbeck, 1995).

NEURONAL DEGENERATION IN HUNTINGTON'S DISEASE

Even though the genetic abnormality in HD has been defined, the underlying mechanism of neuronal degeneration has not been discovered. Neuropathological studies of human brain identified selective patterns of neuronal degeneration in HD and have driven the creation of animal models (Kowall et al., 1987; Beal, 1992). Initially it was hypothesized that excitotoxic injury caused neuronal degeneration in HD (Coyle and Schwarcz, 1976; McGeer and McGeer, 1976). Animal studies showed that NMDA-type excitotoxins reproduce the patterns of neuronal loss found in HD (Beal et al., 1991; Ferrante et al., 1993). An abnormality affecting the NMDA receptor, reduced levels of an endogenous NMDA antagonist or increased production of an endogenous excitotoxin were thought to be potential causes of HD but no abnormality related to glutamate receptors was identified (Beal et al., 1990, 1992; Bruyn and Stoof, 1990; Pearson and Reynolds, 1992). More recently it has been thought that HD may result from impaired mitochondrial oxidative phosphorylation because an identical pattern of differential neuronal loss can be produced in rodents and primates by mitochondrial electron transport chain inhibitors (Beal, 1992; Brouillet et al., 1994, 1995). These inhibitors cause partial energy failure that triggers NMDA-receptor mediated excitotoxic injury, decreased free radical scavenging and increased production of free radicals (Beal, 1994). Other indirect lines of evidence suggest a possible role for mitochondrial abnormalities in the pathogenesis of HD as well (Blass et al., 1988; Parker et al., 1990; Gu et al., 1996).

There is direct evidence of abnormal electron transport chain activity in HD. Decreased activity of complex II/III of the electron transport chain has been found in the caudate nucleus, but not in other brain areas in HD (Mann et al., 1990; Gu et al., 1996). Cytochrome oxidase (complex IV) abnormalities have also been reported in the caudate nucleus (Brennan et al., 1985; Gu et al., 1996). Studies of platelets from HD patients suggest that complex I activity may be selectively decreased in HD patients although it is normal in at-risk family members (Parker et al., 1990). Complex I is composed of over 30 subunits, the majority of which are encoded by nuclear DNA (Hatefi, 1985). Other electron transport chain complexes, including complex II, complex III (ubiquinol cytochrome c reductase), and complex IV (cytochrome oxidase) are normal in blood platelets (Parker et al., 1990). Depletion of complex IV activity parallels the pattern of neuronal loss in HD striatum (Ferrante et al., 1988).

Other indirect lines of evidence suggest a possible role for mitochondria in the pathogenesis of HD. Tellez-Nagel et al. (1973) performed ultrastructural studies on brain biopsies of four HD patients and found evidence of mitochondrial abnormalities and increased lipofuscin, a pigment that accumulates as a consequence of free radical–mediated membrane damage. The delayed onset of HD may be a result of the contribution of age-related mutations of mitochondrial DNA that cause decreased mitochondrial function, as shown by Trounce and co-workers in skeletal muscle (Trounce et al., 1989). Even though the neuron is postmitotic, the mitochondria in neurons continue to proliferate (Linnane et al., 1989). Focal putaminal degeneration has been reported in a family with Leber's disease caused by a point mutation in NADH-dehydrogenase (Larsson et al., 1991). Encephalopathy also occurs in patients with MELAS, another disease characterized by a deficiency in complex I (Wallace, 1991). Jenkins and co-workers have found increased anaerobic metabolism in the cerebral cortex of patients with HD using magnetic resonance spectroscopy, further suggesting that an abnormality of oxidative phosphorylation occurs in this disease (Jenkins et al., 1993).

Systemic administration of electron transport chain inhibitors to animals causes focal basal ganglia pathology. Intravascular administration of rotenone, an electron transport chain complex 1 inhibitor, causes degeneration of the striatum and pallidum in rats (Ferrante et al., 1997b). Oral or intraperitoneal administration of 3-nitropropionic acid (3NP), an electron transport chain complex II inhibitor, produces striking focal pathology that resembles excitotoxin lesions (Borlongan et al., 1995; Brouillet et al., 1995). Lesions can be blocked with the NMDA antagonist MK801 or free radical spin trapping agents (Schulz et al., 1995). Interestingly, chronic food restriction, a manipulation that extends life span in rodents and monkeys (Finch and Morgan, 1997), greatly reduces the vulnerability of striatal neurons to 3NP toxicity and improves behavioral outcome (Bruce-Keller et al., 1998). The latter findings support a role for age-related increases in free radical production or neuronal susceptibility to free radical damage, or both, in HD.

How do these experimental findings relate to the underlying genetic cause of HD? The answer lies with a better understanding of the physiological role and pathological effects of the huntingtin gene product.

NORMAL DISTRIBUTION AND FUNCTION OF HUNTINGTON IN THE BRAIN

Sharp et al. (1995) found widespread expression of huntingtin, most prominently in neurons with no enrichment in the striatum. They found that it was localized to the cytoplasm, especially in nerve terminals, and was loosely associated with membranes and the cytoskeleton. Hersh and colleagues used immunocytochemical methods with antipeptide antibodies and found that huntingtin is present throughout the brain enriched in large neurons and striatal patches associated with microtubules, and to a lesser degree synaptic vesicles (Gutekunst et al., 1995). Trottier et al. (1995) used a series of monoclonal antibodies raised against different parts of the protein and found huntingtin in perikarya of some neurons, neuropil, varicosities, and as punctate staining likely to be synapses. DiFiglia et al. (1995) found huntingtin associated with synaptic vesicles, especially in the somatodendritic compartment. Hoogeveen et al. (1993) found a cytoplasmic localization in various cell types including neurons. In most neurons, huntingtin was present in the nucleus. No difference in intracellular localization was found between normal and mutant cells (Hoogeveen et al., 1993). Most recently, Ferrante et al. (1997a) have identified that huntingtin is heterogeneously distributed and enriched in neurons that degenerate in HD. Double staining studies show that it is co-localized with calbindin and is not present in NADPH diaphorase neurons. The preferential distribution of huntingtin in neurons that are vulnerable to degeneration in HD is consistent with the hypothesis that mutant huntingtin manifests a toxin "gain-of-function" that directly causes preferential degeneration of neurons expressing high levels of the protein. This observation has been confirmed by others (Kosinski et al., 1997), but some investigators dispute this finding (Gourfinkel-An et al., 1997).

The physiological role of huntingtin remains a mystery. Targeted disruption of the HD gene leads to fetal death, possibly owing to increased apoptosis (Duyao et al., 1995; Zeitlin et al., 1995). Recent work suggests that huntingtin may be involved in the intracellular trafficking of nutrients in early embryonic stages (Dragatsis and Zeitlin, 1998). The localization of huntingtin with microtubules suggests it could play a role in intracellular or axonal transport (Tukamoto et al., 1997). Excitotoxic lesions of the striatum lead to increased expression of neuronal huntingtin (Tatter et al., 1995). It is likely that huntingtin plays a role in cellular survival or response to injury, or both, but little more can be stated at the present time.

Huntingtin-associated proteins have been described. Huntingtin-associated protein-1 (HAP-1) is enriched in brain and binds with increased affinity to mutant huntingtin (Li et al., 1995). However, HAP-1 is not enriched in areas of HD pathology (Bertaux et al., 1998). A second protein called huntingtin-interacting protein (HIP-1) interacts with huntingtin in a similar manner (Kalchman et al., 1997). HIP-1 is associated with the cytoskeleton and abnormal binding may contribute to a cytoskeletal disruption and cell death in HD.

TRINUCLEOTIDE REPEAT EXPANSION AND THE MOLECULAR PATHOGENESIS OF HUNTINGTON'S DISEASE

Nakayabu et al. (1998) recently found that mismatching of nucleotide pairs makes double-stranded DNA unstable and triggers the slippage of DNA polymerase, thereby leading to expansion of trinucleotide repeats. Kang et al. (1995) found that DNA triplets beyond a certain length interfere with the progression of DNA polymerase, giving rise to expanded sequences. Surprisingly, CTG repeats are expanded at least eight times more frequently than other triplet permutations in *Escherichia coli*. The structure of CTG repeats, or unique aspects of their interaction with DNA polymerase, may explain this effect.

The relationship between expanded polyglutamine tracts in huntingtin and the well-established defects in cellular energetics in HD has not been explained. Burke and associates (1996) reported that huntingtin binds to the glycolytic enzyme glyceraldehyde-3-phosphate dehydrogenase. There is a suggestion, which remains to be reconfirmed, that long stretches of polyglutamine may inhibit activity of this enzyme. Because neurons are highly dependent on glucose as a source of energy, impaired glycolysis could lead to neuronal cell death. Recently, huntingtin has been shown to specifically bind to cystathionine β-synthase (Boutell et al., 1998). Inhibition of this enzyme would lead to increased homocysteine levels that could potentially lead to excitotoxicity. In either case, the pattern of degeneration in HD may reflect differential sensitivity to energy impairment and the distribution pattern of huntingtin in specific neuronal populations.

DISTRIBUTION OF HUNTINGTIN IN HUNTINGTON'S DISEASE: NUCLEAR INCLUSIONS

When the distribution of huntingtin was studied in HD brain, an unexpected finding was made. Intense huntingtin immunoreactivity was found in neuronal intranuclear inclusions and in dystrophic neurites in the striatum and cerebral cortex (DiFiglia et al., 1997). These aggregates also contain ubiquitin immunoreactivity. Tissue transglutaminase may catalyze cross-linking reactions between cellular proteins and the expanded polyglutamine domain in huntingtin, leading to the deposition of high molecular weight protein aggregates (Gentile et al., 1998). Transglutaminase has also been localized to the nucleus and may contribute to nuclear inclusion formation (Lesort et al., 1998). It is possible that these inclusions are neurotoxic and lead to the neuronal degeneration in HD. Bates and colleagues (Davies et al., 1997) found that transgenic mice expressing mutant huntingtin develop similar neuronal intranuclear inclusions prior to developing neurological signs. Transfection of neuroblastoma cells with mutant huntingtin constructs results in cytoplasmic and cellular aggregates (Cooper et al., 1998). Truncated huntingtin forms perinuclear aggregates more readily than full-length huntingtin and increases the susceptibility of cells to death following apoptotic stimuli (Hackam et al., 1998; Martindale et al., 1998). Caspase-3 appears to be the enzyme responsible for proteolytic cleavage of huntingtin and other mutant proteins with expanded polyglutamine tracts that accumulate in trinucleotide repeat disorders (Wellington et al., 1998). Indeed, all of the neurological diseases associated with expanded trinucleotide repeats may be characterized by neuronal nuclear inclusions (Ross, 1997; Davies et al., 1998). Huntingtin has a propensity to associate with other intracellular inclusions and been reported to be present in both neurofibrillary tangles and Pick bodies (Singhrao et al., 1998).

The hypothesis that intracellular aggregates are toxic and directly lead to neuronal cell death has been challenged. Intraneuronal inclusions are prominent in spared subsets of neurons in both HD brain (Figure 1) and in transgenic animal models of HD (Figure 2). Indeed, in spinocerebellar ataxia type 7, inclusion distribution is very widespread and does not correspond to the restricted distribution of cell loss (Holmberg et al., 1998).

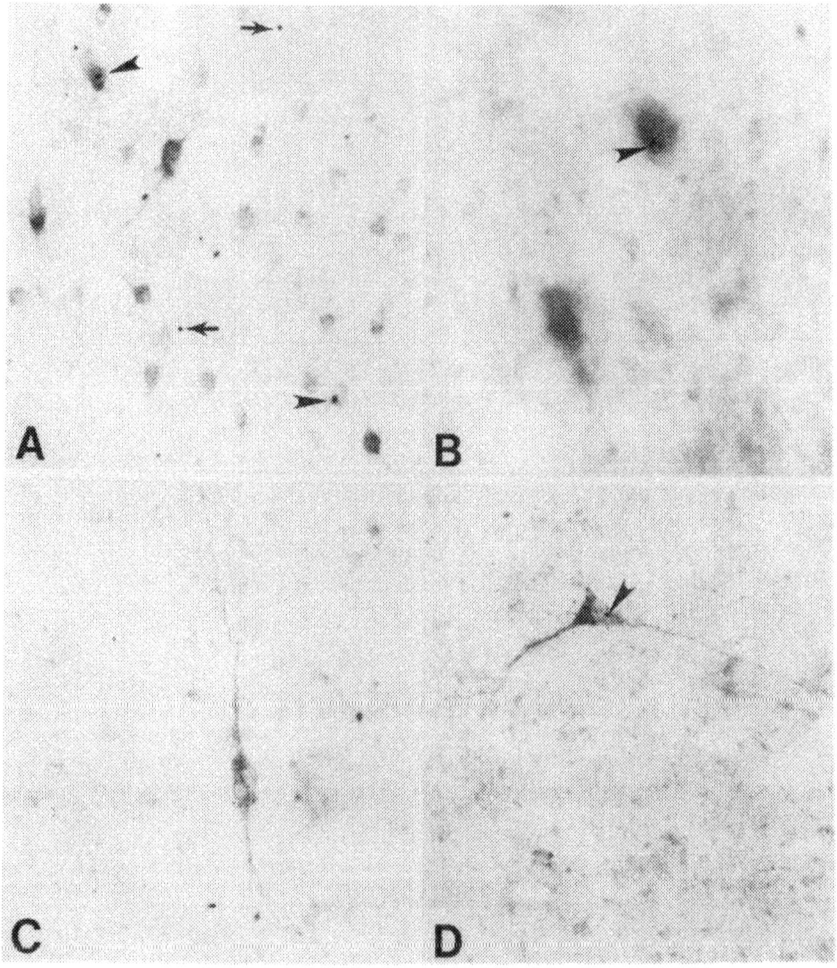

Figure 1. Double immuno- and enzyme histochemical staining for N-terminal huntingtin (EM-48, courtesy of Dr. Xiao Li, Emory University) immunostaining in striatal neuronal populations from Huntington's disease (HD) tissue specimens and those containing calbindin (**A**), acetycholinesterase (**B**), and NADPH-diaphorase (**C** and **D**). Intranuclear neuronal aggregates can be observed in neurons (*arrowheads*) in vulnerable calbindin neurons (**A**). These inclusions are also present in unstained neurons (*arrows*) in this tissue preparation. Cytosolic aggregation of EM-48 is present in many NADPH-diaphorase neurons (**C**). Of great interest is the observation that nuclear aggregates are also seen in spared acetycholinesterase and NADPH-diaphorase striatal neurons in HD (**B** and **D**), suggesting a dissociation between huntingtin aggregation and neuronal death in this disease.

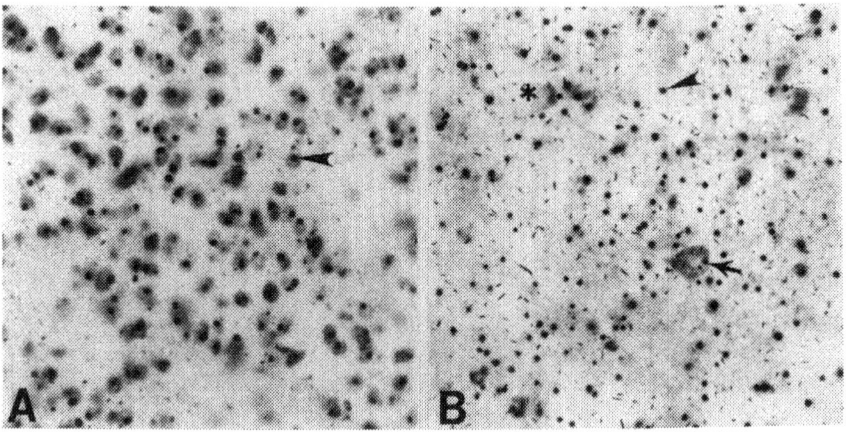

Figure 2. Double immunostaining for N-terminal huntingtin (EM-48) and NADPH-diaphorase enzyme histochemical staining in a transgenic mouse model for Huntington's disease (Gill Bates; R6/2). Large numbers of intranuclear neuronal inclusion bodies are present within the striatum of these animals (**A**, *arrowhead*). Intranuclear neuronal inclusions are also observed within NADPH-diaphorase striatal neurons as in the human HD.

ACKNOWLEDGMENTS

This work was supported by NIH grants AG 13846 (NWK), NS 35255 (RJF), and the Department of Veterans Affairs.

REFERENCES

Alford, R.L., Ashizawa T., Jankovic J., Caskey, C.T. & Richards, C.S. (1996). Molecular detection of new mutations, resolution of ambiguous results and complex genetic counseling issues in Huntington disease. Am. J. Med. Genet. 66, 281–286.

Andrew, S.E. & Hayden, M.R. (1995). Origins and evolution of Huntington disease chromosomes. Neurodegeneration 4, 239–244.

Bateman, D., Boughey, A.M., Scaravilli, F., Marsden, C.D. & Harding, A.E. (1992). A follow-up study of isolated cases of suspected Huntington's disease. Ann. Neurol. 31, 293–298.

Beal, M.F. (1992). Does impairment of energy metabolism result in excitotoxic neuronal death in neurodegenerative illnesses? Ann. Neurol. 31, 119–130.

Beal, M.F. (1994). Huntington's disease, energy, and excitotoxicity. Neurobiol. Aging 15, 275–276.

Beal, M.F., Matson, W.R., Swartz, K.J., Gamache, P.H. & Bird, E.D. (1990). Kynurenine pathway measurements in Huntington's disease striatum: evidence for reduced formation of kynurenic acid. J. Neurochem. 55, 1327–1339.

Beal, M.F., Swartz, K.J., Ferrante, R.J. & Kowall, N.W. (1991). Chronic quinolinic acid lesions in rats closely resemble Huntington's disease. J. Neurosci. 11, 1649–1659.

Beal, M.F., Matson, W.R., Storey, E., Milbury, P., Ryan. E.A., Ogawa, T. & Bird, E.D. (1992). Kynurenic acid concentrations are reduced in Huntington's disease cerebral cortex [see comments]. Neurol. Sci. 108, 80–87.

Bertaux, F., Sharp, A.H., Ross, C.A., Lehrach, H., Bates, G.P. & Wanker, E. (1998). HAP1-huntingtin interactions do not contribute to the molecular pathology in Huntington's disease transgenic mice. FEBS Lett. 426, 229–232.

Blass, J.P., Sheu, R.K. & Cedarbaum, J.M. (1988). Energy metabolism in disorders of the nervous system [review]. Rev. Neurol. 144, 543–563.

Borlongan, C.V., Koutouzis, T.K., Randall, T.S., Freeman, T.B., Cahill, D.W. & Sanberg, P.R. (1995). Systemic 3-nitropropionic acid: behavioral deficits and striatal damage in adult rats. Brain Res. Bull. 36, 549–556.

Boutell, J.M., Wood, J.D., Harper, P.S. & Jones, A.L. (1998). Huntingtin interacts with cystathionine beta-synthase. Hum. Mol. Genet. 7, 371–378.

Brennan, W.A.J., Bird, E.D. & Aprille, J.R. (1985). Regional mitochondrial respiratory activity in Huntington's disease brain. J. Neurochem. 44, 1948–1950.

Brooks, B.P. & Fischbeck, K.H. (1995). Spinal and bulbar muscular atrophy: a trinucleotide-repeat expansion neurodegenerative disease. Trends Neurosci. 18, 459–461.

Brouillet, E., Hyman, B.T., Jenkins, B.G., Henshaw, D.R., Schulz, J.B., Sodhi, P., Rosen, B.R. & Beal, M.F. (1994). Systemic or local administration of azide produces striatal lesions by an energy impairment-induced excitotoxic mechanism. Exp. Neurol. 129, 175–182.

Brouillet, E., Hantraye, P., Ferrante, R.J., Dolan, R., Leroy, W.A., Kowall, N.W. & Beal, M.F. (1995). Chronic mitochondrial energy impairment produces selective striatal degeneration and abnormal choreiform movements in primates. Proc. Natl. Acad. Sci. U. S. A. 92, 7105–7109.

Bruce-Keller, A.J., Umberger, G., McFall, R. & Mattson, M.P. (1999). Food restriction reduces brain damage and improves behavioral outcome following excitotoxic and metabolic insults. Ann. Neurol. 45, 8–15.

Bruyn, R.P. & Stoof, J.C. (1990). The quinolinic acid hypothesis in Huntington's chorea. J. Neurol. Sci. 95, 29–38.

Burke, J.R., Enghild, J.J., Martin, M.E., Jou, Y-S., Myers, R.M., Roses, A.D., Vance, J.M. & Strittmatter, W.J. (1996). Huntingtin and DRPLA proteins selectively interact with the enzyme GADPH. Nat. Med. 2, 347–350.

Cooper, J.K., Schilling, G., Peters, M.F., Herring, W.J., Sharp, A.H., Kaminsky, Z., Masone, J., Khan, F.A., Delanoy, M., Borchelt, D.R., Dawson, V.L., Dawson, T.M. & Ross, C.A. (1998). Truncated N-terminal fragments of huntingtin with expanded glutamine repeats form nuclear and cytoplasmic aggregates in cell culture. Hum. Mol. Genet. 7, 791–800.

Coyle, J.T. & Schwarcz, R. (1976). Lesions of striatal neurons with kainic acid provides an animal model for Huntington's chorea. Nature 263, 244–246.

Davies, S.W., Turmaine, M., Cozens, B.A., DiFiglia, M., Sharp, A.H., Ross, C.A., Scherzinger, E., Wanker, E.E., Mangiarini, L. & Bates, G.P. (1997). Formation of neuronal intranuclear inclusions underlies the neurological dysfunction in mice transgenic for the HD mutation. Cell 90, 537–548.

Davies, S.W., Beardsall, K., Turmaine, M., DiFiglia, M., Aronin, N. & Bates, G.P. (1998). Are neuronal intranuclear inclusions the common neuropathology of triplet-repeat disorders with polyglutamine-repeat expansions? Lancet 351, 131–133.

DiFiglia, M., Sapp, E., Chase, K., Schwarz, C., Meloni, A., Young, C., Martin, E., Vonsattel, J.P., Carraway, R., Reeves, S.A. et al. (1995). Huntingtin is a cytoplasmic protein associated with vesicles in human and rat brain neurons. Neuron 14, 1075–1081.

DiFiglia, M., Sapp, E., Chase, K.O., Davies, S.W., Bates, G.P., Vonsattel, J.P. & Aronin, N. (1997). Aggregation of huntingtin in neuronal intranuclear inclusions and dystrophic neurites in brain. Science 277, 1990–1993.

Dragatsis, I. & Zeitlin, A. (1998). Mouse mutant embryos lacking huntingtin are rescued from lethality by wild-type extraembryonic tissues. Development 125, 1529–1539.

Duyao, M.P., Auerbach, A.B., Ryan, A., Persichetti, F., Barnes, G.T., McNeil, S.M., Ge, P., Vonsattel, J.P., Gusella, J.F. & Joyner, A.L. (1995). Inactivation of the mouse Huntington's disease gene homolog Hdh. Science 269, 407–410.

Ferrante, R.J., Kowall, N.W. & Richardson, E.P. (1988). Patch-matrix distribution of cholecystokinin and cytochrome oxidase activity in normal and Huntington's disease striatum. Soc. Neurosci. Abs. 14, 1046.

Ferrante, R.J., Kowall, N.W., Cipolloni, P.B., Storey, E. & Beal, M.F. (1993). Excitotoxin lesions in primates as a model for Huntington's disease: histopathologic and neurochemical characterization. Exp. Neurol. 119, 46–71.

Ferrante, R.J., Gutekunst, C.A., Persichetti, F., McNeil, S.M., Kowall, N.W., Gusella, J.F., MacDonald, M.E., Beal, M.F. & Hersch, S.M. (1997a). Heterogeneous topographic and cellular distribution of huntingtin expression in the normal human neostriatum. J. Neurosci. 17, 3052–3063.

Ferrante, R.J., Shulz, J.B., Kowall, N.W. & Beal, M.F. (1997b). Systemic administration of rotenone produces selective damage in the striatum and globus pallidus, but not the substsantia nigra. Brain Res. 753, 157–162.

Finch, C.E. & Morgan, T.E. (1997). Food restriction and brain aging. Adv. Cell Aging Gerontol. 2, 279–297.

Gentile, V., Sepe, C., Calvani, M., Melone, M.A., Cotrufo, R., Cooper, A.J., Blass, J.P. & Peluso, G. (1998). Tissue transglutaminase-catalyzed formation of high-molecular-weight aggregates in vitro is favored with long polyglutamine domains: a possible mechanism contributing to CAG-triplet diseases. Arch. Biochem. Biophys. 352, 314–321.

Gourfinkel-An, I., Cancel, G., Trottier, Y., Devys, D., Tora, L., Lutz, Y., Imbert, G., Saudou, F., Stevanin, G., Agid, Y., Brice, A., Mandel, J.L. & Hirsch, E.C. (1997). Differential distribution of the normal and mutated forms of huntingtin in the human brain. Ann. Neurol. 42, 712–719.

Gu, M., Gash, M.T., Mann, V.M., Javoy, A.F., Cooper, J.M. & Schapira, A.H. (1996). Mitochondrial defect in Huntington's disease caudate nucleus. Ann. Neurol. 39, 385–389.

Gusella, J.F. & MacDonald, M.E. (1995). Huntington's disease [review]. Sem. Cell Biol. 6, 21–28.

Gutekunst, C.A., Levey, A.I., Heilman, C.J., Whaley, W.L., Yi, H., Nash, N.R., Rees, H.D., Madden, J.J. & Hersch, S.M. (1995). Identification and localization of huntingtin in brain and human lymphoblastoid cell lines with anti-fusion protein antibodies. Proc. Natl. Acad. Sci. U. S. A. 92, 8710–8714.

Hackam, A.S., Singaraja, R., Wellington, C.L., Metzler, M., McCutcheon, K., Zhang, T., Kalchman, M. & Hayden M.R. (1998). The influence of huntingtin protein size on nuclear localization and cellular toxicity. J. Cell Biol. 141, 1097–1105.

Hatefi, Y. (1985). The mitochondrial electron transport and oxidative phosphorylation system. Annu. Rev. Biochem. 54, 1015–1069.

Holmberg, M., Duyckaerts, C., Durr, A., Cancel, G., Gourfinkel-An, I., Damier, P., Faucheux, B., Trottier, Y., Hirsch, E.C., Agid, Y. & Brice, A. (1998). Spinocerebellar ataxia type 7 (SCA7): a neurodegenerative disorder with neuronal intranuclear inclusions. Hum. Mol. Genet. 7, 913–918.

Homberg, V. & Huttunen, J. (1994). Muscle tone in Huntington's disease. J. Neurol. Sci. 121, 147–154.

Hoogeveen, A.T., Willemsen, R., Meyer, N., de Rooij, K.E., Roos, R.A., Van Ommen, O.G., & Galjaard, H. (1993). Characterization and localization of the Huntington disease gene product. Hum. Mol. Genet. 2, 2069–2073.

Huntington's Disease Collaborative Research Group. (1993). A novel gene containing a trinucleotide repeat that is expanded and unstable on Huntington's disease chromosomes. Cell 72, 971–983.

Jenkins, B.G., Koroshetz, W.J., Beal, M.F. & Rosen, B.R. (1993). Evidence for impairment of energy metabolism in vivo in Huntington's disease using localized 1H NMR spectroscopy. Neurology 43, 2689–2695.

Kalchman, M.A., Koide, H.B., McCutcheon, K., Graham, R.K., Nichol, K., Nishiyama, K., Kazemi-Esfarjani, P., Lynn, F.C., Wellington, C., Metzler, M., Goldberg, Y.P., Kanazawa, I., Gietz, R.D. &

Hayden, M.R. (1997). HIP1, a human homologue of *S. cerevisiae* Sla2p, interacts with membrane-associated huntingtin in the brain. Nat. Genet. 16, 44–53.

Kang, S., Ohshima, K., Shimizu, M., Amirhaeri, S. & Wells, R.D. (1995). Pausing of DNA synthesis in vitro at specific loci in CTG and CGG triplet repeats from human hereditary disease genes. J. Biol. Chem. 270, 27014–27021.

Kosinski, C.M., Cha, J.H., Young, A.B., Persichetti, F., MacDonald, M., Gusella, J.F., Penney, J.B., Jr. & Standaert, D.G. (1997). Huntingtin immunoreactivity in the rat neostriatum: differential accumulation in projection and interneurons. Exp. Neurol. 144, 239–247.

Kowall, N.W., Ferrante, R.J. & Martin, J.B. (1987). Patterns of cell loss in Huntington's disease. TINS 10, 24–29.

La Spada, A.R., Paulson, H.L. & Fischbeck, K.H. (1994). Trinucleotide repeat expansion in neurological disease. Ann. Neurol. 36, 814–822.

Larsson, N.-G., Andersen, O., Home, E., Oldfors, A. & Wahlstrom, J. (1991). Leber's hereditary optic neuropathy and Complex I deficiency in muscle. Ann. Neurol 30, 701–708.

Lesort, M., Attanavanich, K., Zhang, J. & Johnson, G.V. (1998). Distinct nuclear localization and activity of tissue transglutaminase. J. Biol. Chem. 273, 11991–11994.

Li, X.J., Li, S.H., Sharp, A.H., Nucifora, F.C., Jr., Schilling, G., Lanahan, A., Worley, P., Snyder, S.H. & Ross, C.A. (1995). A huntingtin-associated protein enriched in brain with implications for pathology. Nature 378, 398–402.

Linnane, A.W., Marzuki, S., Ozawa, T. & Tanaka, M. (1989). Mitochondrial DNA mutations as an important contributor to ageing and degenerative diseases. Lancet 1, 642–645.

Mann, V.M., Cooper, J.M., Javoy-Agid, F., Agid, F., Jenner, P. & Schapira, A.H.V. (1990). Mitochondrial function and parental sex effect in Huntington's disease. Lancet 336, 749.

Martindale, D., Hackam, A., Wieczorek, A., Ellerby, L., Wellington, C., McCutcheon, K., Singaraja, R., Kazemi-Esfarjani, P., Devon, R., Kim, S.U., Bredesen, D.E., Tufaro, F. & Hayden, M.R. (1998). Length of huntingtin and its polyglutamine tract influences localization and frequency of intracellular aggregates. Nat. Genet. 18, 150–4.

McGeer, E.G. & McGeer, P.L. (1976). Duplication of biochemical changes of Huntington's chorea by intrastriatal injections of glutamic and kainic acid. Nature 263, 517–519.

Nakayabu, M., Miwa, S., Suzuki, M., Izuta, S., Sobue, G. & Yoshida, S. (1998). Mismatched nucleotides may facilitate expansion of trinucleotide repeats in genetic diseases. Nucleic Acids Res. 26, 1980–1984.

Parker, W.D.J., Boyson, S.J., Luder, A.S. & Parks, J.K. (1990). Evidence for a defect in NADH:ubiquinone oxidoreductase (complex I) in Huntington's diseases. Neurology 40, 1231–1234.

Pearson, S.J. & Reynolds, G.P. (1992). Increased brain concentrations of a neurotoxin, 3-hydroxykynurenine, in Huntington's disease. Neurosci. Lett. 144, 199–201.

Plassart, E. & Fontaine, B. (1994). Genes with triplet repeats: a new class of mutations causing neurological diseases. Biomed. Pharmacother. 48, 191–197.

Ross, C.A. (1997). Intranuclear neuronal inclusions: a common pathogenic mechanism for glutamine-repeat neurodegenerative diseases? Neuron 19, 1147–1150.

Schulz, J., Henshaw, D., Siwek, D., MacGarvey, U., Jenkins, B., Ferrante, R., Cipolloni, P., Kowall, N., Rosen, B. & Beal, M. (1995). Involvement of free radicals in excitotoxicity in vivo. Neurochem 64, 2239–2247.

Sharp, A.H., Loev, S.J., Schilling, G., Li ,S.H., Li, X.J., Bao, J., Wagster, M.V., Kotzuk, J.A., Steiner, J.P., Lo, A. et al. (1995). Widespread expression of Huntington's disease gene (IT15) protein product. Neuron 14, 1065–1074.

Singhrao, S.K., Thomas, P., Wood, J.D., MacMillan, J.C., Neal, J.W., Harper, P.S. & Jones, A.L. (1998). Huntingtin protein colocalizes with lesions of neurodegenerative diseases: an investigation in Huntington's, Alzheimer's, and Pick's diseases. Exp. Neurol. 150, 213–222.

Tatter, S.B., Galpern, W.R., Hoogeveen, A.T. & Isacson, O. (1995). Effects of striatal excitotoxicity on huntingtin-like immunoreactivity. Neuroreport 6, 1125–1129.

Tellez-Nagel, I., Johnson, A.B. & Terry, R.D. (1973). Ultrastructural and histochemical study of cerebral biopsies in Huntington's chorea. Adv. Neurol. 1, 387–398.

Thompson, P.D., Berardelli, A,. Rothwell, J.C., Day, B.L., Dick, J.P., Benecke, R. & Marsden, C.D. (1988). The coexistence of bradykinesia and chorea in Huntington's disease and its implications for theories of basal ganglia control of movement. Brain 111, 223–244.

Trottier, Y,, Devysm D., Imbert, G., Saudou, F., An, I., Lutz, Y., Weber, C., Agid, Y., Hirsch, E.C. & Mandel, J.L. (1995). Cellular localization of the Huntington's disease protein and discrimination of the normal and mutated form. Nat. Genet. 10, 104–110.

Trounce, I., Byrne, E. & Marzuki, S. (1989). Decline in skeletal muscle mitochondrial respiratory chain function: possible factor in ageing. Lancet 1 (8639), 637–639.

Tukamoto, T., Nukina, N., Ide, K. & Kanazawa, I. (1997). Huntington's disease gene product, huntingtin, associates with microtubules in vitro. Arch. Med. Res. 28, 513–516.

Wallace, D.C. (1991). Mitochondrial genes and neuromuscular disease. Res. Publ. Assoc. Res. Nerv. Ment. Dis. 69, 101–120.

Wellington, C.L., Ellerby, L.M., Hackam, A.S., Margolis, R.L., Trifiro, M.A., Singaraja, R., McCutcheon, K., Salvesen, G.S., Propp, S.S., Bromm, M., Rowland, K.J., Zhang, T., Rasper, D., Roy, S., Thornberry, N., Pinsky, L., Kakizuka, A., Ross, C.A., Nicholson, D.W., Bredesen, D.E. & Hayden, M.R. (1998). Caspase cleavage of gene products associated with triplet expansion disorders generates truncated fragments containing the polyglutamine tract. J. Biol. Chem. 273, 9158–9167.

Zeitlin, S., Liu, J.P., Chapman, D.L., Papaioannou, V.E. & Efstratiadis, A. (1995). Increased apoptosis and early embryonic lethality in mice nullizygous for the Huntington's disease gene homologue. Nat. Genet. 11, 155–163.

Chapter 4

Genetic Abnormalities in Amyotrophic Lateral Sclerosis

EDWARD J. KASARSKIS and DARET K. ST. CLAIR

INTRODUCTION

The past decade has witnessed major conceptual advances in understanding the pathophysiology of amyotrophic lateral sclerosis (ALS, known as Lou Gehrig's Disease in the United States [Kasarskis and Winslow, 1989; Reider and Paulson, 1997]). Identification of mutations in the Cu/Zn superoxide dismutase (SOD [SOD1]) gene in some patients with the familial, autosomal dominant form of ALS has focused attention on free radicals as integral to the process of neurodegeneration in this disorder. Transgenic mice bearing the mutant human Cu/Zn SOD develop a progressive, fatal neurodegenerative disease which replicates the pathological findings seen in human ALS in many respects. These mice should prove to be increasingly valuable as a test model for new therapeutic agents.

Advances in Cell Aging and Gerontology
Volume 3, pages 93–133

ISBN: 0-7623-0405-7

In parallel, other studies have placed glutamate excitotoxicity and apoptosis in the pathophysiological chain leading to motor neuron death in ALS. Despite these advances, the initiating event(s) in ALS remain unidentified. It is a common clinical observation that, except for developing a progressive fatal neurodegenerative disease, ALS patients are in general, healthy. They do not have a plethora of other medical problems and many are taking no medications at the onset of their illness. The factor(s) that initiate the neurodegenerative process and the failure of the normal compensatory or protective mechanisms that permit the spread of neurodegeneration along the neuraxis remain speculative at the present time.

The decade following the Decade of the Brain will undoubtedly chronicle continued advances in understanding this disease. Patients, families, and their physicians eagerly await the fruits of molecular neuroscience for effective treatment to arrest the relentless progression of this fatal disease.

CLINICAL FEATURES OF AMYOTROPHIC LATERAL SCLEROSIS

Amyotrophic lateral sclerosis is a chronic, age-associated, neurodegenerative disease that is recognized primarily by progressive atrophy and weakness of most voluntary skeletal muscles. The illness was first characterized as a distinct clinical entity by Charcot in 1874 although others (e.g., Aran, Duchenne) had reported patients with progressive muscular atrophy as early as 1850. The core clinical features of ALS are attributed to so-called lower motor neuron deficits (atrophy, fasciculations, flaccid weakness) and also to upper motor neuron deficits (weakness, spasticity, and hyperactive tendon reflexes). The evolution of weakness commences insidiously, usually in a single region (e.g., in a single lower limb), and spreads over time to involve other bodily regions while regions previously affected experience continued worsening. The balance of upper motor neuron spasticity and lower motor neuron flaccidity will dictate the overall net muscle tone at any given time. However, as the disease progresses, profound flaccid paralysis obscures the evidence of upper motor involvement in most patients. During the entire course of a patient's illness, sensation, ocular motility, bladder/bowel/sexual function, and cognition remain clinically unaffected. It is this latter feature which makes ALS particularly challenging for the patient, family, and health care providers.

The terminology used to describe the clinical syndromes is, at times, confusing. The generic term, motor neuron disease (MND), is preferred in Europe and in this schema, ALS is a specific subtype encompassing the clinical features as described above (Charot-type ALS). Classic ALS is considered to include patients who exhibit either upper motor neuron, limb lower motor neuron, or bulbar-onset disease with spread over time to involve all regions. In some patients, however, the involvement is restricted to only lower motor neurons innervating the limbs (termed, progressive muscular atrophy, or PMA), to the bulbar region (progressive bulbar palsy, or PBP), or to the upper motor neurons (primary lateral sclerosis, or PLS). The situation is even more complex when one considers atypical patients who exhibit the core symptoms of progressive weakness but also experience cognitive impairment, parkinsonian features, or disturbances in ocular motility to varying degrees. The prevalence and prognosis of these more restricted or atypical variants differs from the more common classic form of ALS.

The diagnosis of ALS is based on the clinical features of the illness and exclusion of conditions that may share aspects of ALS to some degree. There is no "ALS test" that will assure the clinician and family of the diagnosis. The clinical features required for the diagnosis of ALS have been standardized as the World Federation of Neurology "El Escorial" criteria, which have gained wide acceptance for clinical management of patients and their recruitment into clinical trials (CNTF, BDNF, gabapentin, IGF-1). Briefly, the diagnosis of ALS requires the insidious onset of weakness in a bodily region with spread over time, evidence of upper and lower motor signs, normal sensation and sphincter function, and exclusion of rare disorders which at times can mimic ALS. When the Escorial criteria are met, the diagnostic certainty approaches 95% in the hands of experienced neurologists (Chaudhuri et al., 1995).

The clinical impression of ALS is supported by electrophysiological (Tandan and Bradley, 1985b) and imaging studies. Electromyography (EMG) provides evidence for denervation (fibrillations, fasciculations, positive waves) and reinervation (long duration, high-amplitude motor unit potentials) in muscles of at least three limbs and/or tongue. In this context, the EMG is an extension of the neurological examination and can demonstrate changes consistent with ongoing denervation and chronic reinervation, which may not be clinically evident on examination. Importantly, the EMG can exclude myopathic conditions that may at times emulate ALS in the severity of muscle weakness. Nerve conduction studies examine the electrophysiological integrity of motor and sensory axons and their myelin sheaths. Extensions of these electrophysiological tests are helpful to exclude proximal multifocal conduction block of the action potential or defects in synaptic transmission at the cholinergic neuromuscular junction. Magnetic resonance imaging of the cervical spinal cord is frequently indicated to rule out structural lesions (which can cause hyperreflexia in the lower extremities) or of the proximal nerve roots (which can cause flaccid weakness in the upper extremities). In selected patients, a muscle biopsy is indicated to search for neurogenic atrophy and to eliminate myopathic conditions, although this is infrequently performed.

The majority of patients do not have a family history of ALS and are deemed to have "sporadic" ALS. However, about 5% to 10% of patients have another affected family member and are considered to have familial ALS (FALS). Typically, an autosomal dominant pattern of inheritance is apparent. Genetic study of FALS has provided important insights into the mechanism of neurodegeneration.

Both sporadic and familial ALS are relentlessly progressive until death, most often from respiratory insufficiency. The course of an individual patient may be quantified by measuring isometric power, pulmonary functions, or a variety of clinimetric scales (Andres et al., 1986; Brooks, 1996; Brooks et al., 1996). In general, the course of ALS is smoothly and linearly progressive. Examination of Lou Gehrig's performance as a baseball player offers graphic evidence of the deterioration in motor function due to ALS in a single individual (Kasarskis and Winslow, 1989). It is a common clinical observation that once the pace of a patient's ALS is established, it remains relatively constant during the course of the disease. Therefore, some patients progress rapidly whereas others progress slowly. Although the molecular basis for the differences in the rate of evolution of ALS is not understood, a poorer prognosis is associated with older age and with "bulbar"-onset of weakness (i.e., weakness of oropharyngeal muscles).

Population studies afford a complimentary perspective of ALS. The worldwide incidence of ALS is approximately 0.5 to 2 patients per 100,000 population per year and the prevalence is 2 to 8 per 100,000. The historical exceptions have been the Western Pacific foci of ALS in the Kii peninsula of Japan, among the Chamorros of Guam, and among groups in West New Guinea. ALS is more common in males and the incidence is clearly age-related, with the peak incidence occurring between

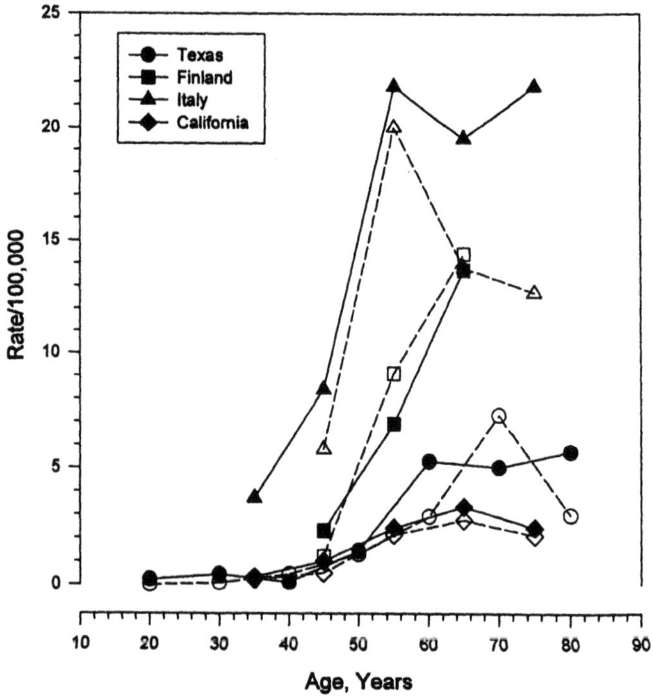

Figure 1. Age-specific incidence of ALS in males and females in four studies of ALS epidemiology (Bracco et al., 1979; Murros & Fogelholm, 1983; Annegers et al., 1991; Norris et al., 1993). Males: closed symbols; Females: open symbols.

55 and 75 years in both sexes (Figure 1) (Bracco et al., 1979; Murros and Fogelholm, 1983; Annegers et al., 1991; Norris et al., 1993). Despite numerous attempts, no environmental risk factors have been conclusively identified, although farming and potential exposure to certain toxins are considered to be possibly related.

Population studies reveal 50% of an initial cohort will survive 3 to 4 years following the onset of weakness. Importantly, about 20% to 30% of ALS patients survive beyond 5 years and 10% beyond 10 years. Risk factors for early death include older age, respiratory or bulbar onset of weakness, and rapid evolution of weakness. Although there is a male preponderance, the rate of progression and prognosis for survival do not differ between males and females. A temporal trend toward improved survival is apparent (Bracco et al., 1979; Murros and Fogelholm, 1983; Annegers et al., 1991; Norris et al., 1993; Mitsumoto et al., 1998a). The cause of this apparent improvement in survival is unclear but may be related to improved diagnosis, earlier referral to centers specializing in ALS, or better symptomatic care.

NEUROPATHOLOGY OF AMYOTROPHIC LATERAL SCLEROSIS

Defining the neuropathological features of ALS is a requisite initial step in understanding the clinical features and conceptualizing the mechanism(s) of neurodegeneration in this disease. Knowledge of the pathology is derived almost exclusively from examination of tissue taken from patients who were symptomatic for 2 to 5 years prior to death. Very little is known about the pathology of human ALS at its earliest clinical stages.

The primary pathological features of advanced, end-stage ALS have been understood for years and consist of a loss of motor neurons in the ventral spinal cord and in the primary motor cortex (Lawyer and Netsky, 1953; Brownell et al., 1970; Hirano et al., 1984; Tandan and Bradley, 1985b; Chou, 1992). Secondary axonal loss in the dorsal and ventral corticospinal tracts and in the ventral nerve roots is present, as would be anticipated. In addition, neurons and their axons originating in the motor nuclei of cranial nerves V, VII, IX, X, and XII undergo degeneration as well. Posterior nerve roots, the motor neurons of cranial nerves III, IV, and VI, and the motor neurons of Onufrowicz (Onuf's nucleus) in the sacral spinal cord appear to be intact. Reactive astrogliosis in the cortex and spinal cord and fragmentation of the Golgi apparatus are present (Gonatas et al., 1992; Schiffer et al., 1996). Taken together, these pathological features serve to explain the clinical manifestations well and justify the concept of ALS as a motor neuron disorder.

The core neuropathology of ALS has been expanded, however, by evaluating patients who have been artificially ventilated, thereby extending their survival beyond the typical terminal event of respiratory failure. In these cases, most of which have originated from Japan, more widespread neuronal loss and gliosis have been observed. In addition to the spinal and cortical motor neurons, involved areas include cranial nerve III, Clarke's column, the red nucleus, substantia nigra, globus pallidus, subthalamus, thalamus, dentate nucleus, and the pontine reticular substance and tegmentum (Hayashi and Kato, 1989). Myelin loss was observed in the superior cerebellar peduncle, the central tegmental tract, and the medial longitudinal fasciculus. These cases were unusual inasmuch as the patients experienced rapid progression of weakness to a "locked-in" state with mean duration of only 12.6 months from the onset of weakness to tracheostomy and continuous assisted ventilation (Hayashi & Kato, 1989). However, others have observed involvement of alpha and gamma motor neurons, Clarke's column and spinocerebellar pathways, thalamus, corpus callosum, and superior colliculus in non-Japanese, nonventilated cases (Swash et al., 1988; Swash and Schwartz, 1992; Lowe, 1994).

A number of neuronal inclusion bodies are observed in ALS spinal motor neurons including Bunina bodies, Lewy body-like hyaline inclusions, basophilic inclusions, and skein-like inclusions (Nakamura et al., 1997). In addition, proximal axonal spheroids containing neurofilaments are frequent in ALS motor neurons. Inclusion bodies are ubiquitin positive to varying degrees. Intraneuronal, ubiquitin-positive inclusions are observed in ALS cases in a large number of other anatomical regions, including hippocampal granule cells and pyramidal neurons, dorsal root ganglia, Clarke's column, the intermediolateral column of the thoracic cord, reticular formation, nonmotor cerebral cortex, and Onuf's nucleus (Lowe, 1994; Kakita et al., 1997).

Neurofilaments accumulate in axonal spheroids of spinal motor neurons in ALS (Carpenter, 1968; Hirano et al., 1984; Leigh and Swash, 1991; Lowe, 1994). The neurofilament aggregations resemble those seen in experimental models of axonal toxicity (Griffin et al., 1978; Troncoso et al., 1982), motor neuron diseases in nontransgenic animals (Cork et al., 1988), and in transgenic animals overexpressing light or heavy neurofilament proteins (Lee et al., 1994b; Tu et al., 1996, 1997). Neurofilaments in human sporadic ALS are abnormally phosphorylated (Manetto et al., 1988), ubiquitinated (Lowe, 1994; Migheli et al., 1994), and co-localize with Cu/Zn SOD/n-NOS/calmodulin/citrulline/cGMP/nitrotyrosine (Brown, 1954; Chou et al., 1996). Cerebrospinal fluid from ALS patients facilitates the phosphorylation of neurofilaments in neuronal or spinal cord cultures (Nagaraja et al., 1994; Rao et al., 1995). The presumed functional consequence of these changes is an alteration in the rate of fast axonal conduction (Sasaki and Iwata, 1996).

These observations raise important questions regarding the apparent restriction of symptoms to the motor system in ALS when observed clinically *vis à vis* the wider distribution of pathological involvement, much of which is subclinical. One possibility is that many, and perhaps all, neurons are susceptible to neurodegeneration in ALS patients but the degenerative process occurs most rapidly in spinal, bulbar, and cortical motor neurons. In this view, artificial ventilation simply allows the patient to survive long enough to allow the complete neuropathological expression of the disease. Alternatively, the expanded pathological changes reported primarily from Japan may define an ALS variant that is prevalent in that population or conceivably are related to bouts of recurrent nocturnal hypoxia prior to permanent ventilatory support. Several authors have conceptualized ALS as a "multisystem atrophy" whose clinical signature is overwhelmingly dominated by progressive weakness (Brownell et al., 1970; Hayashi and Kato, 1989; Hayashi et al., 1991; Swash and Schwartz, 1992; Lowe, 1994). The totality of the neuropathological picture calls into question the notion of an exquisitely focused, selective degeneration of motor neurons, which is the logical inference from the clinical picture.

The literature regarding the neuropathology of FALS will need to be re-evaluated in the future in the context of identified genotypes inasmuch as the unique features of FALS were defined prior to the description of mutations in the Cu/Zn SOD gene. In FALS (genotype unknown), the neurodegeneration in Clarke's column and its efferent spinocerebellar tract are reported and considered to be more prevalent than in sporadic ALS. In addition, axonal degeneration in the posterior columns and Lewy body-like inclusions in spinal motor neurons are frequently seen.

THEORIES OF CAUSATION

Nongenetic Factors

It is clear that the motor neuron disorders, including Charcot-type classic ALS, are heterogeneous in terms of etiology. From a logical point of view, there are at a minimum three etiologies: mutations in Cu/Zn SOD in some FALS patients, unidentified mutations in other FALS patients, and sporadic ALS of unknown etiology. It is likely that there are many causes of motor neuron degeneration that result in the MND syndrome. In this context, many theories of causation have arisen in an attempt to explain the apparently selective degeneration of spinal motor neurons. The proposed etiologies have been thoughtfully advanced based on established mechanisms of neural damage seen in other diseases or experimental models, many of which have been well-summarized in several reviews (Conradi et al., 1982b; Tandan and Bradley, 1985b; Heiman-Patterson et al., 1986; Festoff, 1987; Mitchell, 1987; Williams and Windebank, 1991; Mitsumoto et al., 1998a). The major proposals have encompassed the following mechanisms: Nongenetic (immune, viral or infectious, paraneoplastic, toxic metals) and genetic mutations which lead to an alteration in cytoskeletal neurofilaments or predispose the neuron to excitotoxic or oxidative damage. The oxidative damage hypothesis is discussed subsequently in the context of mutations in Cu/Zn SOD.

Nongenetic factors may, in fact, be very important in the initiation of the ALS neurodegenerative process, but any proposed mechanism is extremely difficult to prove. It is generally agreed that approximately 50% of motor neurons must be lost before a person recognizes the onset of clinical weakness which initiates the process of neurological evaluation (Sharrard, 1955). This implies that the process of neural injury leading to neurodegeneration by an exogenous toxic factor could have occurred many years before the recognition of weakness. A corollary that flows from this formulation is that any putative exogenous agent may not be detectable in body fluids at the time of diagnosis or in tissues at the time of autopsy. Therefore, the potential involvement of exogenous agents in causing ALS cannot be categorically dismissed. Indeed, it is conceivable that exogenous environmental factors could initiate the process of neurodegeneration and, in consort with genetic polymorphisms or mutations in critical genes that normally serve to limit neurotoxicity, could lead to the development of clinical ALS.

Immune Factors

The attraction of immune-based theories is the high degree of target specificity which could easily account for selective destruction of motor neurons. The evidence supporting an immune basis for ALS is, however, not compelling. Initial studies reported a toxic effect of ALS serum on neurons in culture or on erythrocytes (Wolfgram and Myers, 1973; Ronnevi et al., 1987). Periodically, studies have reported the presence of paraproteins or monoclonal gamma globulins in patients with ALS or syndromes of progressive muscular atrophy (Rowland et al., 1982; Freddo et al., 1986; Shy et al., 1986; Fishman et al., 1991; Smith et al., 1992). Experimental models of immune motor neuron disease have been developed in guinea pigs (Smith et al., 1993), providing evidence to support the potential of immune-based attack on the motor system. Immunotherapy of human motor neuron syndromes has provided mixed results. Some patients with lower motor neuron signs in the context of multifocal conduction block on EMG, lymphoproliferative disease, or anti-GM1 ganglioside antibodies are potentially treatable with immunosuppression, but these appear to be a small minority of patients with syndromes of progressive weakness (Dalakas et al., 1994; Pestronk et al., 1994; Tan et al., 1994). Treatment with intravenous immunoglobulin, plasmapharesis with or without azathioprine, cyclosporine, intravenous cyclophosphamide and corticosteroids, intrathecal corticosteroids, or total lymphoid irradiation have been ineffective in slowing the progression of weakness in the more classical ALS (Pieper & Fields, 1957; Norris et al., 1978; Olarte et al., 1980; Kelemen et al., 1983; Brown et al., 1986; Appel et al., 1988; Dalakas et al., 1994; Drachman et al., 1994; Tan et al., 1994).

Viral or Infectious Factors

The evidence supporting a viral/infectious/transmissible basis for ALS has been generally negative (see Salazar-Grueso and Roos, 1992; Jubelt and Drucker, 1993; Jubelt, 1998, for review). Searches for evidence of viral infection in postmortem tissue from ALS patients using antiviral antibodies have been negative (Kascsak et al., 1982). The syndrome of post-polio progressive muscular atrophy ("post-polio syndrome"), although resembling ALS by sharing the features of progressive lower motor neuron-type weakness, is distinct from ALS on clinical and pathological grounds (Dalakas et al. 1986; Cwik and Mitsumoto, 1992; Jubelt and Drucker, 1993; Ito and Hirano, 1994). Moreover, epidemiological studies have not demonstrated a reduction in the incidence of ALS in populations vaccinated against polio (Swingler et al., 1992). The current status of studies searching for the presence of viral genomic material in ALS tissues has been recently reviewed (Mitsumoto et al., 1998). Empirical trials of antiviral therapy in ALS have been negative to date (Norris et al., 1974; Farkkila et al., 1984; Mora et al., 1986; Smith and Norris, 1988; Westarp et al., 1992, 1993).

Paraneoplastic Factors

The prevalence of cancer does not appear to be increased in ALS (Barron and Rodichok, 1982; Posner, 1995). Extremely rare patients with an ALS-like syndrome or lower motor neuron syndrome have been reported in patients with cancers or lymphoma (see Mitsumoto et al., 1998a, for review).

Toxic Metals

Rare patients with ALS-like syndromes resulting from toxic metal exposure have been reported (Kantarjian, 1953; Brown, 1954; Currier and Hearer, 1968; Boothby et al., 1974; Petkau et al., 1974; Adams et al., 1983). Analysis of tissues and bodily fluids from ALS patients reveal increased concentrations of potentially toxic metals, including mercury, manganese, and lead (Kasarskis, 1992). Aluminum appears to be increased in the central nervous system of patients from the Western Pacific foci of ALS but not in the more common, sporadic ALS (reviewed in Kasarskis, 1992). Although aluminum can induce experimental motor neuron degeneration (Strong and Garruto, 1991), it appears not to be elevated in spinal motor neurons from patients with sporadic ALS (Kasarskis et al., 1995). However, intraneuronal iron has been reported to be elevated in these same tissues (Kasarskis et al., 1995), a finding which needs to be replicated by others. Chelation therapy of ALS patients appears to be ineffective in uncontrolled clinical trials (Campbell and Williams, 1968; Conradi et al., 1982a, 1982b; Kurlander and Patten, 1979) and can possibly accelerate the rate of deterioration (Conradi et al., 1982a, 1982b).

Genetic Abnormalities

In an attempt to reconcile and integrate the various theories of causation, several authors have conceptualized neurodegeneration in ALS as a multistep process (Eisen, 1995; Mitsumoto et al., 1998a). At least two steps are envisaged: first, a mechanism, possibly environmental in nature, which serves to initiate the process of neurodegeneration; second, other factors that facilitate the spread of neurodegeneration from one region of the neuraxis to another. Examples of potential initiating factors are toxic metals, a neurotrophic virus, or agents active at the neuromuscular junction. Mutations or polymorphisms in one of several genes could serve to modify the nervous system's response to environmental factors and perpetuate or extend the initial injury. Areas under active investigation include alterations in neurofilaments and cytoskeleton, glutamate excitotoxicity, and enzymes involved in the detoxification of reactive oxygen and nitrogen species.

Neurofilaments

Because abnormal accumulations of neurofilaments are a conspicuous pathological feature of ALS, genes encoding cytoskeletal and neurofilament proteins are an obvious area to examine for mutations or polymorphisms. The major structural components of the cytoskeleton are microtubules, neurofilaments, microfilaments, and various microtubule-associated proteins.

Work to date has focused entirely on the neurofilaments as summarized in Table 1. Figlewicz et al. (1994) reported deletions in the region of the NF-H gene encoding the KSP (Lys-Ser-Pro) repeat domain five ALS patients. Rooke et al. (1996) evaluated 117 unrelated individuals derived from families with autosomal dominant, non-SOD1 FALS patients for mutations in the KSP region of the NF-H gene and failed to find either mutations or polymorphisms in affected individuals. A comprehensive analysis of the neurofilament genes in 100 FALS patients, not linked to the SOD1 locus, and an additional 75 sporadic ALS patients has been performed by Vechio et al. (1996). They identified a series of polymorphisms in each gene that were distributed in both affected and control individuals, suggesting that the polymorphic variants in the NF genes are not likely to be important in the pathogenesis of motor neuron degeneration. In addition, the expression of the NF-L and NF-M genes was reported to be normal in ALS (and Parkinson's disease) patients but decreased in brain tissue in Alzheimer's disease (Kittur et al., 1994).

Studies in transgenic animals overexpressing neurofilament proteins (Table 2) provide convincing animal models for ALS in many ways and focus attention on the neurobiology of long axons, which are characteristically affected in ALS. Côté et al. (1993) and Xu et al. (1993) created transgenic mice overexpressing the human NF-H and murine NF-L genes, respectively, at three to four times endogenous levels. In both models, denervation atrophy and accumulations of neurofilaments in motor and sensory neurons were observed without evidence of motor neuron death. Moreover, axonal transport was affected, causing impairment of movement of actin, tubulin, and neurofilament proteins (Collard et al., 1995). These studies indicate that inducing an imbalance in the proportion of normal cytoskeletal proteins can disrupt the axonal cytoskeleton, causing retraction of the distal axonal branches from their synaptic contact with skeletal muscle. M.K. Lee et al. (1994) used site-directed mutagenesis of the murine NF-L gene to create a substitution of Leu→Pro at codon 394. Transgenic animals overexpressing this construct exhibited neurofilament aggregation, denervation atrophy, and selective death of spinal motor neurons, pathological features that emulate the human disease quite well. Although the initial studies searching for mutations in neurofilament genes in ALS have been negative, animal studies emphasize the critical importance of cytoskeletal homeostasis in the survival of spinal motor neurons (Williamson et al., 1996).

Table 1. Mutations or Polymorphisms in Sporadic and Familial Amyotrophic Lateral Sclerosis

Gene	*Patient Population*	*Mutation/Polymorphisms*	*Author, Year*
Neurofilament Genes			
Neurofilament heavy subunit, KSP repeat region	ALS (sporadic?)	Deletions in 5 patients	Figlewicz et al., 1994
Neurofilament heavy subunit, KSP repeat region	Autosomal dominant FALS, not linked to the SOD1 locus	No mutations in 117 unrelated individuals	Rooke et al., 1996
Neurofilament light subunit	FALS, controls	Polymorphism Leu222Leu (C→T) Asp469Ans (G→A)	Vechio et al., 1996
Neurofilament medium subunit	FALS, sporadic ALS, controls	Polymorphism Ile129Met (T→G) Asn368Asn (T→C) Ala474Thr (G→A)	Vechio et al., 1996
Neurofilament heavy subunit	FALS, sporadic ALS, controls	Polymorphism Leu158Leu (T→C) Ala401Ala (T→C) Glu460Lys (G→A Pro615Leu (C→T) Ala805Glu (C→A)	Vechio et al., 1996

(continued)

Table 1. Continued

Gene	*Patient Population*	*Mutation/Polymorphisms*	*Author, Year*
Glutamate Homeostasis			
EAAT2	Sporadic and familial ALS	No genomic mutations detected	Aoki et al., 1998
EAAT2	Sporadic and familial ALS	Aberrant mRNA species Intron 7 retention Exon 9 skipping	Lin et al., 1998
Antioxidant Cellular Defense			
Cu/Zn SOD (SOD1)	See Table 3		
Mn SOD (SOD2), exons 2,3,4	Sporadic ALS	No mutations	Parboosingh et al., 1995
Mn SOD (SOD2)	Sporadic ALS	Polymorphism Ala(-9)Val Ile58Thr (Not observed)	Tomblyn et al., 1998
Catalase	Sporadic ALS	No mutations	Parboosingh et al., 1995
Other Candidate Genes			
X-linked	Sporadic ALS	Unknown	Siddique, 1998
SMN	Sporadic and familial ALS	No deletions of the telomeric copy; normal distribution of centromeric deletions	Moulard et al., 1998
2q33-q35	Recessive FALS	Unknown	Hentati et al., 1994

Note: FALS, familial amyototripic lateral sclerosis.

Table 2. Transgenic Animal Models of Motor Neuron Disease

Gene	*Mutation*	*Author, Year*
NF-L	Leu394Pro [NF-L(Pro)]	Lee et al., 1994
NF-H (human)	Overexpression	Côté et al., 1993
NF-L (murine)	Overexpression	Xu et al., 1993
Cu/Zn SOD (SOD1, human)	Ala4Val	Gurney et al.,1994
	Gly37Arg	Wong et al.,1995
	Gly85Arg	Bruijn et al., 1997
	Gly86Arg	Ripps et al., 1995
	Gly93Ala	Gurney et al., 1994

Glutamate Excitotoxicity

The neurotoxic effects of exogenous glutamate are widely recognized and the cellular transport mechanisms to regulate extracellular glutamate after presynaptic release are well-understood. Four separate transporter proteins, termed EAAT (excitatory amino acid transporter), are known and have the following cellular localizations: EAAT1, astrocytes and Bergman glia in cerebellum; EAAT2, astrocytes; EAAT3, neurons; and EAAT4, Purkinje cells (Kanai et al., 1993; Furuta et al., 1997; Lin et al., 1998). The potential association of glutamate neurotoxicity stems from initial observations of elevated fasting plasma glutamate in ALS patients (Plaitakis and Caroscio, 1987). Subsequent studies demonstrated abnormal clearance of glutamate after an oral load, elevated cerebrospinal fluid (CSF) levels of aspartate and glutamate, and elevated CSF levels of NAAG (N-acetyl aspartyl glutamate) and NAA (N-acetylaspartate) in ALS (Coyle et al., 1989; Rothstein et al., 1991; Iwasaki et al., 1992; Rothstein et al., 1998). The concentration of glutamate was reduced in central nervous system tissues, supporting the notion that glutamate metabolism might be altered in ALS (Perry et al., 1987; Plaitakis et al., 1988).

In a series of studies, Rothstein and colleagues demonstrated a specific reduction in the amount of EAAT2 protein in ALS motor cortex and spinal cord (Rothstein et al., 1995) after initially showing impaired sodium-dependent glutamate uptake in these same tissues (Rothstein et al., 1992). More recently, this same group examined the EAAT2 gene and failed to find mutations in exonic sequences but did demonstrate a mutation in intron 7 and a silent (G→A) mutation in exon 5 (Aoki et al., 1998). Further investigation revealed the presence of multiple abnormal EAAT2 mRNA species in brain and CSF of ALS patients encompassing partial retention of intron 7 and skipping of exon 9 (Lin et al., 1998). Using transient expression in COS cells, Lin et al. (1998) showed that the aberrant mRNAs caused a loss of normal EAAT2 protein and activity by a dominant negative effect on the normal EAAT2 gene or caused a rapid degradation of the dimeric EAAT2 protein containing an abnormal protein product (Lin et al., 1998). Loss of EAAT2 protein would account for the observations of impaired glutamate uptake by astrocytes, thereby facilitating neuronal excitotoxicity.

In vitro and animal models of EAAT inhibition have supported the concept that glutamate excitotoxicity is functionally important in the pathogenesis of ALS. Using organotypic cultures, Rothstein et al. (1993) induced motor neuron death in spinal cord organotypic cultures by chronically inhibiting glutamate uptake. In parallel studies, decreasing EAAT1 and EAAT2 protein by inhibition of translation using antisense oligonucleotides replicated the motor neuron loss (Rothstein et al., 1995). Finally, chronic administration of EAAT1 antisense oligonucleotides intrathecally caused motor neuron degeneration and paralysis in rats (Rothstein et al., 1995).

Clinical trials of presumed glutamate-modulatory drugs have produced mixed results in ALS. Riluzole administration, now approved for clinical use by the Food and Drug Administration in the United States, prolonged survival in patients but the effect was not marked (Bensimon et al., 1994). Recently gabapentin, a widely available anticonvulsant drug, slowed the rate of progression of weakness in a pilot trial (Miller et al., 1996). Other strategies using lamotrigine (another anticonvulsant), dextromethorphan, or branched chain amino acids were without effect on altering either the progression of weakness or survival (Testa et al., 1989; Asmark et al., 1993; Eisen et al., 1993; Tandan et al., 1996). More focused pharmacological approaches based on the recent information about EAAT2 are clearly needed.

Copper Zinc Superoxide Dismutase (SOD1)

Undoubtedly the most significant discovery in understanding the pathogenesis of ALS has been the identifications of mutations in the Cu/Zn SOD gene by Rosen et al. (1993) and Deng et al. (1993). These groups first demonstrated that some patients with FALS (approximately 20% to 30%) harbor missense mutations in the gene encoding Cu/Zn SOD, a finding widely confirmed by others (Ogasawara et al., 1993; Rosen et al., 1993; Aoki et al., 1994; Elshafey et al., 1994; Esteban et al., 1994; Hirano et al., 1994; Kawamata et al., 1994; Nakano et al., 1994; Rosen et al., 1994; Pramatarova et al., 1995). Over 50 mutations in the Cu/Zn SOD gene have been identified to date (summarized in Table 3). The majority of mutations are missense although several nonsense mutations, insertions, and deletions have been documented. Initially it was believed that the mutations were restricted to exons 1,2,4, and 5 but recently mutations in exon 3 have been identified as well (Andersen et al., 1997; Shaw et al., 1998).

The phenotypic expression of SOD1 mutations varies considerably (Andersen et al., 1997; Cudkowicz et al., 1997). In general, ALS patients with SOD1 mutations have an earlier onset of weakness compared to patients with sporadic disease. However, the rates of progression differ markedly. For example, the most common mutation (Ala4Val) is associated with a rapid evolution of disease with a mean survival of approximately 12 to 15 months. Other mutations (e.g., Gly37Arg) are associated with long survival of 18.7 years. Most mutations have been identified in

Table 3. Mutations in the Cu/Zn SOD Gene in Patients with Amyotrophic Lateral Sclerosis

Exon	*Codon*	*Amino Acid Substitution*	*Nucleotide Substitution*	*Mean Age at Onset*	*Mean Survival (yrs)*	*Author, Year*
Missense Mutations						
1	4	Ala→Val	GCC→GTC	47.8; 47.0	1.4; 1.0	Deng et al., 1993; Cudkowicz et al., 1997; Juneja et al., 1997
1	4	Ala→Thr	GCC→ACC	40.0	0.75	Takahashi et al., 1994; Nakano et al., 1994
1	6	Cys→Phe	TGC→TTT			Morita et al., 1996
1	7	Val→Glu	GTG→GAG	36.0	4.0	Hirano et al., 1994
1	8	Leu→Gln	CTG→CAG			Bereznai et al., 1997
1	14	Val→Gly	GTG→GGG			Andersen et al., 1997
1	14	Val→Met	GTG→ATG			Deng et al., 1995
1	16	Gly→Ser	GGC→AGC			Kawamata et al., 1996
1	21	Glu→Lys	GAG→AAG			Jones et al., 1994
1	21	Glu→Gly	GAG→GGG			Moulard et al., 1995
2	37	Gly→Arg	GGA→AGA	40.0	18.7	Rosen et al., 1993; Cudkowicz et al., 1997
2	38	Leu→Val	CTG→GTG	41.5; 44.9	2.8	Rosen et al., 1993; Robberecht et al., 1994; Cudkowicz et al., 1997
2	41	Gly→Ser	GGC→AGC	50.8; 46.8	0.9; 1.0	Rosen et al., 1993; Rainero et al., 1994; Cudkowicz et al., 1997
2	41	Gly→Asp	GGC→GAC	46.0	17.0	Rosen et al., 1993; Cudkowicz et al., 1997
2	43	His→Arg	CAT→CGT	42.8; 49.8	2.8	Rosen et al., 1993; Deng et al., 1993; Cudkowicz et al., 1997

(continued)

Table 3. Continued

Exon	Codon	Amino Acid Substitution	Nucleotide Substitution	Mean Age at Onset	Mean Survival (yrs)	Author, Year
2	46	His→Arg	CAT→CGT	48.6	17.0	Aoki et al., 1994
2	48	His→Gln	CAT→CAG	54	0.75	Orrell et al., 1997; Shaw et al., 1997; Enayat et al., 1995
3	72	Gly→Ser	GGT→AGT	29	1.3	Shaw et al., 1998
3	76	Asp→Tyr	GAT→TAT			Andersen et al., 1997
4	84	Leu→Val	TTG→GTG	53.8	1.8	Aoki et al., 1995; Deng et al., 1995
4	84	Leu→Phe	TGG→TCG	45	2.1+	Shaw et al., 1998
4	85	Gly→Arg	GGC→CGC			Rosen et al., 1993
4	86	Asn→Ser				Maeda et al., 1997
4	90	Asp→Ala	GAC→GCC	29.7	6+	Andersen et al., 1995, 1996; Själander et al., 1995; Robberecht et al., 1996; Jackson et al., 1997
4	93	Gly→Ala	GGT→GCT	47.9	2.2	Rosen et al., 1993; Cudkowicz et al., 1997
4	93	Gly→Cys	GGT→TGT	47.4	10.1	Rosen et al., 1993; Cudkowicz et al., 1997
4	93	Gly→Arg	GGT→CGT			Elshafey et al., 1994
4	93	Gly→Asp	GGT→GAT	35.8; 48.3	5.7; 10.5	Esteban et al., 1994; Orrell et al., 1995; Cudkowicz et al., 1997
4	93	Gly→Ser	GGT→AGT			Kawamata, 1994
4	93	Gly→Val	GGT→GTT			Hosler et al., 1996; Orrell et al., 1997
4	100	Glu→Gly	GAA→GGA	46.9; 46.0	4.0; 4.8; 5.1	Rosen et al., 1993; Calder et al., 1995; Cudkowicz et al., 1997; Juneja et al., 1997
4	100	Glu→Lys	GAA→AAA			Deng, unpubl: cited in Siddique, 1996
4	101	Asp→Asn	GAT→AAT			Jones et al., 1994
4	101	Asp→Gly	GAT→GGT			Orrell et al., 1997
4	104	Ile→Phe	ATC→TTC			Ikeda et al., 1996

4	106	Leu→Val	CTC→GTC	40.0; 35.5	2.3	Rosen et al., 1993; Kawamata et al., 1994; Cudkowicz et al., 1997
4	108	Gly→Val	GGA→GTA			Orrell et al., 1997
4	112	Ile→Thr	ATC→ACC	44.0	0.9	Esteban et al., 1994; Cudkowicz et al., 1997; Enayat et al., 1995
4	113	Ile→Thr	ATT→ACT	42.0; 46.0; 58.9	2.0; 3.5	Rosen et al., 1993; Suthers et al., 1994; Orrell et al., 1995; Cudkowicz et al., 1997; Jackson et al., 1997; Jones et al., 1994; Enayat et al., 1995
4	115	Arg→Gly	CGC→GGC	59	2.5	Kostrzewa et al., 1994
5	124	Asp→Val	GAT→GTT			Hosler, 1996
5	125	Asp→His	GAC→CAC			Orrell et al., 1997; Enayat et al., 1995
5	126	Leu→STOP	TTG→TAG			Zu, cited in Siddique, 1996
5	134	Ser→Asn	AGT→AAT			Watanabe et al., 1996
5	139	Asn→Lys	AAC→AAA			Pramatarova et al., 1995
5	144	Leu→Phe	TTG→TTC			Deng et al., 1993
5	144	Leu→Ser	TTG-TCG	42.5	12.3	Sapp et al., 1995; Cudkowicz et al., 1997
5	145	Ala→Thr	GCT→ACT	48.0	1.6	Sapp et al., 1995; Cudkowicz et al., 1997
5	146	Cys→Arg	TGT→CGT			Kawamata et al., 1996
5	148	Val→Gly	GTA→GGA	43.5	2.3	Deng et al., 1993; Cudkowicz et al., 1997
5	148	Val→Ile	GTA→ATA		1.5	Ikeda et al., 1995
5	149	Ile→Thr	ATT→ACT			Pramatarova et al., 1995; Enayat et al., 1995
5	151	Ile→Thr	ATT→ACT			Kostrzewa et al., 1996
Deletions						
5	126; STOP at 131		TTG→__G	42	2	Pramatarova et al., 1994, 1995; Kadekawa et al., 1997; Kato et al., 1996
5	126; STOP at 130		GAC T_T_GGC (126) GAC GGC			Nakashima et al., 1995
5			3 bp deletion			Hosler et al., 1996

(continued)

Table 3. Continued

Exon	Codon	Amino Acid Substitution	Nucleotide Substitution	Mean Age at Onset	Mean Survival (yrs)	Author, Year
Insertions						
4	118	Val→ Lys-Thr-Gly-Pro-STOP	G→AAAAC	35.0	1.3	Jackson et al., 1997 (SALS)
5	133	Glu→		44.0	2.4	Cudkowicz et al., 1997
5	STOP at 133		AAT GAT* (132)TAG			Orrell et al., 1997
5	STOP at 133		TTG(126)G*G*G* G*GG			Hansen et al., 1996
Intronic Mutations						
	Intron	+Phe-Leu-Gln	T→G	54.0	2.8	Cudkowicz et al., 1997
	Intron	+Phe-Leu-Glu	T→G, 10 bp before exon 5			Sapp et al., 1995
	Intron	+Phe-Phe-Thr-Gly-Pro-STOP	A→G, 11 bp before exon 5			Zu, cited in Siddique, 1996

patients from families with a clear autosomal dominant pattern of inheritance, although SOD1 mutations have been found in patients with apparently sporadic disease (e.g., Gly72Ser; Asp90Ala; Ile113Thr).

The phenotypic features of the SOD1 mutations differ among themselves and from sporadic ALS in several regards. It appears that spasticity and bulbar onset disease are uncommon features of FALS patients with SOD1 mutations (Andersen et al., 1996; Cudkowicz et al., 1997). However, signs and symptoms of sensory disturbance, cerebellar, or autonomic dysfunction can be elicited, indicating involvement of neural systems other than the voluntary motor system. The number of published autopsies of FALS patients with defined genotypes is limited, but they do confirm the involvement of posterior columns and spinocerebellar tracts in addition to the anticipated death of spinal motor neurons.

Active dimeric Cu/Zn SOD is distributed in the cytosol of neurons and other non-neural tissues. Copper ion, by cycling between its oxidized and reduced states, is required for enzymatic activity. The majority of the SOD1 mutations in FALS are distributed in parts of the protein forming a β-barrel structure or in the loops guarding the copper-containing active site. Few mutations, most notably His46Arg and His48Gln, affect the metal-binding sites.

Initially it was hypothesized that a reduction in SOD1 activity was the critical factor in the development of the ALS phenotype. Indeed, SOD1 activity is reduced between 20% and 50% in FALS tissues and for some mutations, the biological half-life of SOD1 dimers containing mutant protein is reduced (Bowling et al., 1993; Deng et al., 1993; Borchelt et al., 1994; Robberecht et al., 1994). However, SOD1 knockout mice do not develop an ALS-like syndrome although motor neurons are more sensitive to axotomy-induced degeneration (Reaume et al., 1996). Finally, transgenic mice expressing mutant human SOD1 genes develop an ALS-like motor neuron disease in the presence of normal levels of murine SOD1, whereas transgenics expressing the wild-type human SOD1 gene remain free of disease (Gurney et al., 1994; Dal Canto and Gurney, 1995). Evidence has emerged indicating that mutant SOD1 acquires a new, toxic property consisting of increased reactivity toward peroxynitrite or with hydrogen peroxide, or both. Structural studies of mutant SOD1 protein indicate that the active site copper ion may be spatially more accessible to potential substrates such as hydrogen peroxide or peroxynitrite, which are normally excluded by the normal tertiary conformation of the wild-type protein. The generation of reactive nitrogen species and hydroxyl free radicals by mutant SOD1 have some experimental support (Beckman et al., 1993; Wiedau-Pazos et al., 1996).

Transgenic mice have been created that express five different mutant human Cu/Zn SOD genes, as indicated in Table 2. The mice expressing the Gly37Arg and Gly93Ala mutations in high copy number are similar. Both develop progressive limb paralysis commencing at 3 months of age leading to death by 5 months (Gurney et al., 1994). Ultimately spinal motor neurons degenerate in these transgenic mice (Wong et al., 1995; Gurney et al., 1994). Although Cu/Zn SOD is localized to the cytoplasm, the earliest pathological changes in presymptomatic transgenic mice consist of swollen axons and dendrites containing large, dilated mitochondria (Dal Canto & Gurney, 1994; Gurney et al., 1994; Chiu et al., 1995; Wong et al., 1995; Tu et al., 1996). As in human ALS, the terminal stage is characterized by atrophic motor neurons with neurofilamentous cytoplasmic inclusions (Dal Canto and Gurney, 1995). Treatment of these transgenic mice from weaning with antioxidants such as vitamin E and selenium (a component of glutathione peroxidase) delays the onset and progression of paralysis without affecting survival (Gurney et al., 1996). Riluzole and gabapentin prolonged survival to a slight degree but did not affect the appearance of weakness (Gurney et al., 1996). The Glu86Arg transgenic mice develop paralysis at 3 months of age with progression as in the Gly37Arg and Gly93Ala mice (Ripps et al., 1995). However, in contrast, the former do not develop cytoplasmic vacuoles (Morrison et al., 1996). Neuronal number is diminished and surviving neurons are pyknotic with phosphorylated neurofilaments. The Gly85Arg mice develop the onset of weakness after 8 to 14 months of age, depending on copy number. Despite the delayed appearance of weakness, the evolution of paralysis, once it is initiated, is fulminant (Bruijn et al., 1997). Lewy body-like neuronal and astrocytic inclusions are seen, which are SOD1-positive, but vacuolated mitochondria are not observed. Although the Ala4Val mutation is associated with rapid progression of weakness in human FALS, the initial studies with transgenic mice bearing this mutation did not develop weakness and these mice have received little further attention (Gurney et al., 1994).

The mutant Cu/Zn SOD transgenic mice are important from several perspectives. Firstly, they provide an animal model of human FALS that recapitulates many features of the human disease. Secondly, the mice afford an opportunity to study the earliest pathological changes that occur in the presymptomatic state, which is not possible in humans. Studies with the Gly37Arg and Gly93Ala mice have identified mitochondrial vacuolation as an early pathological change, supporting the notion that mitochondrial failure of adenosine triphosphate (ATP) production may be a critical initial step in the process of neurodegeneration. Thirdly, these mice provide test systems of potential drug therapy that might lead to more efficient drug development in ALS.

Manganese Superoxide Disumutase (SOD2)

Parboosingh et al. (1995) were the first to examine the Mn SOD genome for mutations in 73 patients with FALS not linked to the SOD1 locus using SSCP analysis. They found no mutations in exons 3, 4, or 5 in FALS patients or controls but were unable to successfully amplify exons 1 and 2 for evaluation. Using a similar approach, they did not find mutations in any exon of the catalase gene.

Tomblyn et al. (1998) evaluated the Mn SOD genome in sporadic ALS patients by direct sequencing. The polymorphic variant Ile58Thr previously described by Borgstahl et al. (1996) was not found in either ALS patients or controls. A second polymorphic variant in the mitochondrial targeting sequence, Ala(-9)Val (Shimoda-

Matsubayashi et al., 1996), was detected in both ALS and controls with the Val/Val genotype overrepresented in the ALS population. St. Clair and colleagues also detected a previously-unrecognized polymorphic variant in the 5′-flanking region of the Mn SOD gene consisting of a loss of a G from a sequence consisting of 11 Gs (Xu et al., unpublished observations, 1998). The functional significance of these findings have not been established, although the polymorphism in the mitochondrial targeting sequence may alter the mitochondrial endowment of Mn SOD.

Other Candidate Genes

Moulard et al. (1998) failed to find deletions of the telomeric copy of the SMN (survival motor neuron) gene in sporadic and familial ALS (see Table 1). Such deletions are found in more than 90% of children with infantile MND and in some adult spinal muscular atrophies (AMS type IV). Moreover, the centromeric copy of SMN, which can be deleted in up to 5% of the general population, was distributed similarly in ALS and controls. Siddique et al. (unpublished abstract, 1998) have reported, in a preliminary communication, a family with apparently sporadic ALS with genetic linkage to a locus on the X chromosome. Hentati et al. (1994) have found another FALS family with a recessive pattern of inheritance with linkage to 2q33-q35. The genes remain unidentified and the subject of ongoing investigation.

INTEGRATED APPROACH TO UNDERSTANDING MOTOR NEURON DEGENERATION IN AMYOTROPHIC LATERAL SCLEROSIS

Recent advances in understanding ALS have indicated three interrelated areas of importance in the process of neurodegeneration: excitotoxicity, oxidative stress, and the cytoskeleton. Much of the evidence to support these mechanisms of neuronal injury derives from animal models and tissue culture studies. However, investigations of postmortem specimens from ALS patients offer support for these mechanisms as well. As outlined earlier, abnormalities of neurofilament accumulation in ALS spinal motor neurons are apparent (Rouleau et al., 1996), which also are a pathological characteristic of transgenic mice expressing mutant Cu/Zn SOD (Tu et al., 1997). Evidence is accumulating that oxidant damage to cellular constituents occurs in sporadic ALS (Bergeron, 1995), including increased protein carbonyl concentration in ALS cortex (Bowling et al., 1993) and spinal cord (Shaw et al., 1995) and elevated nitrotyrosine immunoreactivity in ALS spinal motor neurons (Abe et al., 1995). In addition, there is evidence that ALS fibroblasts have increased sensitivity to oxidative stress (Kidson et al., 1983; Aguirre et al., 1998). Recent studies have shown that levels of 4-hydroxynonenal, a toxic aldehydic product of membrane lipid peroxidation, are increased in spinal cord tissue and cerebrospinal fluid from ALS patients (Pedersen et al., 1998b; Smith et al., 1998). Exposure of cultured motor neuron-like cells to Fe^{2+} and 4-hydroxynonenal results in impairment of glutamate and glutamate transport (Pedersen et al., 1998a), suggesting a central role for membrane lipid peroxidation in rendering motor neurons vulnerable to excitotoxicity in ALS.

A model described by Beal (1995) and others (Coyle & Puttfarcken, 1993) attempts to integrate the mechanisms of excitotoxicity, oxidative damage, and neurofilament accumulation in the pathogenesis of age-associated neurodegenerative diseases. Cognizant of the critical role of mitochondrial energy production for neuronal survival, we have recast Beal's hypothesis to emphasize the importance of intramitochondrial antioxidant defense in this process (Figure 2), as will be described subsequently.

In tissues with a high oxygen consumption, free radicals such as the superoxide anion ($O_2^-\bullet$) are formed by the incomplete reduction of molecular oxygen (Fridovich, 1978; Chance et al., 1979). Free radicals are constantly generated *in vivo* by many physiological reactions such as mitochondrial respiration, oxygen transport by hemoglobin (reviewed in Halliwell and Gutteridge [1989]), and activation of N-methyl-D-aspartate (NMDA) receptors (Bondy and Lee, 1993; Lafon-Cazal et al., 1993). The superoxide radical itself is known to directly damage many biomolecules (Kono and Fridovich, 1982; Kim et al., 1986; Kuo et al., 1987; McCord and Russell, 1988). The toxicity of the superoxide radical is further

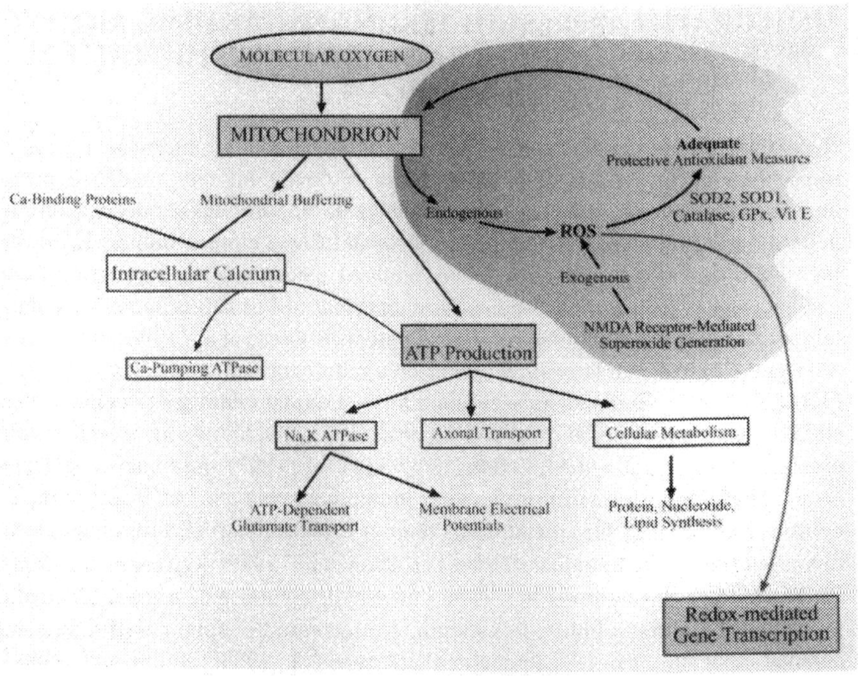

Figure 2. The central role of mitochondria in neuronal energy production. Adequate antioxidant protection of mitochondria is essential for their continued functioning.

enhanced by the metal-catalyzed generation of the highly reactive hydroxyl free radical (OH•), as shown in the equations in Figure 3 (Hibbs et al., 1988).

Transition metals such as Fe are potentially toxic to cells because of their ability to facilitate the generation of OH• in vivo. The central nervous system is rich in Fe, most of which is protein bound (Halliwell and Gutteridge, 1989). Under normal circumstances protein-bound Fe reacts slowly, if at all, to form hydroxyl radicals. However when Fe is released from its protein ligands, it becomes available for the metal-catalyzed reaction within its immediate locale. Thus, conditions that cause an increase in free Fe are potentially cytopathic. Our work (Kasarskis et al., 1995) and that of others (Ince et al., 1994) indicates that Fe is increased in ALS spinal cord and in motor neurons. This finding suggests that ALS patients may be more susceptible to Fe-catalyzed OH• radical damage (Liu et al., 1994).

Of all organelles, the mitochondrion is particularly vulnerable to oxidant damage. This occurs because the majority of intracellular free radicals are generated locally within the mitochondrion by the incomplete reduction of molecular oxygen by the electron transport chain (Richter et al., 1988; Linnane et al., 1989). Some components of the electron transport chain, such as the NADH-coenzyme Q reductase complex and the reduced form of coenzyme Q itself, leak electrons onto oxygen which produce a univalent reduction to form superoxide radicals (Boveris and Cadenas, 1982; Halliwell and Gutteridge, 1989). Generation of the OH• radical and formation of the free radical–induced adduct, 8-hydroxydeoxyguanosine, in mitochondrial DNA during mitrochondrial electron transfer have been demonstrated (Giulivi et al., 1995).

Humans must also cope with reactive nitrogen species (RNS) such as nitric oxide (NO•). Nitric oxide is produced in vivo by many cell types such as neurons, endothelial cells, fibroblasts, muscle cells, and phagocytes (Palmer et al., 1987; Hibbs et al., 1988; Billiar et al., 1989; Schmidt et al., 1989; Stuehr and Nathan, 1989; Radomski et al., 1990; Snyder and Bredt, 1991). Within the central and peripheral nervous systems, NO• has been shown to be a neurotransmitter. Nitric oxide can react with O_2• to form peroxynitrite ($ONOO^-$), a strong oxidant with reactivity similar to that of OH• (Radi et al., 1991). It has recently been shown that NO•-mediated neurotoxicity is generated, at least in part, by its reaction with O_2•

$$\text{Oxidized metal complex} + O_2\bullet \rightarrow O_2 + \text{Reduced metal complex}$$

$$\text{Reduced metal complex} + H_2O_2 \rightarrow OH\bullet + OH^- + \text{Oxidized metal complex}$$

$$\text{Net: } O_2\bullet + H_2O_2 \xrightarrow[\text{catalyst}]{\text{metal}} O_2 + OH\bullet + OH^-$$

Figure 3. Generation of hydroxyl free radicals from superoxide and hydrogen peroxide in the presence of a metal catalyst.

leading to the formation of $ONOO^-$ (Lipton et al., 1993). The superoxide radical thus emerges as the key progenitor for the two most reactive oxygen species in biological systems ($ONOO^-$, OH•), which must be effectively detoxified to ensure neuronal survival.

In most tissues, the removal of superoxide radicals and H_2O_2 is accomplished by the sequential action of SOD and catalase or glutathione peroxidase (GPx). In the brain, it appears that GPx and glutathione may be present in high concentration in astrocytes (Silvka et al., 1987; Raps et al., 1989; Damier et al., 1993), although this distribution might be developmentally regulated (Makar et al., 1994). Effective detoxification of the superoxide radical and related reactive oxygen species may require the coordinated expression of many genes in both neurons and glia. The family of SODs are metalloenzymes which catalyze the dismutation of superoxide radicals. Three distinct SODs are found in humans: a homodimeric Cu/Zn SOD, found mainly in the cytosol (McCord and Fridovich, 1969); a homotetrameric Mn SOD in the mitochondrial matrix (Weisiger and Fridovich, 1973); and a homotetrameric glycosylated Cu/Zn SOD in extracellular space (ECSOD) (Marklund, 1982). Thus, all three enzymes catalyze the identical reaction but in different cellular compartments.

The biological importance of Mn SOD has been demonstrated in many biological systems. For example:

1. Inactivation of Mn SOD genes in *Escherichia coli* increases the mutation frequency under aerobic conditions (Carlioz and Touati, 1986; Farr et al., 1986).
2. Elimination of the Mn SOD gene in *Saccharomyces cerevisiae* increases its sensitivity to oxygen (van Loon et al., 1986).
3. Induction of Mn SOD by interleukin-1 and tumor necrosis factor (TNF) in tumor cells increases their resistance to subsequent killing by TNF (Wong et al., 1989).
4. Induction of SODs in mammalian cells and tissues is accompanied by an increase in tolerance to toxic agents that induce oxidative stress (Crapo and Tierney, 1974; Stevens and Autor, 1977a, 1997b; Frank et al., 1981; Kasemset and Oberley, 1984; Oberley et al., 1987; Schiavorne and Hassan, 1987; Weiss and Kumar, 1988; Mimnaugh et al., 1989; Sinha and Mimnaugh, 1990; Spitz et al., 1992; Wan and St.Clair, 1993a, 1993b).
5. Transfection of Mn SOD or Cu/Zn SOD cDNA into plant or mammalian cells renders them resistant to paraquat (Krall et al., 1988; Gruber et al., 1990; Bowler et al., 1991; St.Clair et al., 1991), TNF (Wong et al., 1989), adriamycin, and radiation-induced cytotoxicity (Hirose et al., 1993).

Indirect evidence argues that Mn SOD may be of importance in neural protection against ALS-type degeneration based on its immunocytochemical distribution among susceptible and resistant motor neurons (Wakai et al., 1994). Other indications for the primacy of Mn SOD over Cu/Zn SOD in neuronal antioxidant defense comes from the study of knockout mice. SOD1 null mice develop normally, exhibit normal levels of lipid peroxidation products and protein carbonyls (Reaume et al., 1996), but have increased sensitivity to neural death following axonal injury (Reaume et al., 1996). In contrast, null mutations of Mn SOD in mice results in neonatal death from a cardiomyopathy (Li et al., 1995; Lebovitz et al., 1996). Treating SOD2 null mice with Mn SOD mimics, which do not cross the blood-brain-barrier, extends survival and permits the development of spongiform degeneration of the cortex and brain stem (Melov et al., 1998). In other models, Mn SOD deficiency promotes cerebral infarction after ischemia (Murakami et al., 1998). Mn SOD expression in motor neurons increased following transection of their axons (Yoneda et al., 1992) or in experimental optic neuritis (Qi et al., 1997), whereas Cu/Zn SOD expression did not change. Finally, Mn SOD confers resistance to NMDA and nitric oxide–induced neurotoxicity in nNOS neurons (Gonzalez-Zulueta et al., 1998), prevents methylmercury toxicity in HeLa cells (Naganuma et al., 1998), and protects hippocampal neurons against oxidative-stress induced apoptosis (Mattson et al., 1997; Keller et al., 1998). Therefore, the expression of Mn SOD is essential for normal growth and longevity in an aerobic environment and for the development of cellular resistance to oxygen radical–mediated toxicity. Why is the mitochondrial form of SOD so critical?

The mitochondrion occupies a critical and vulnerable position in neurons (Bowling and Beal, 1995). Because the mitochondrion is at the epicenter of free radical formation, it might be anticipated that mitochondrial DNA, proteins, and lipids (Halliwell, 1992) may bear the initial brunt of oxidative assault on the neuron. Mitochondrial DNA is highly susceptible to damage because mitochondrial DNA is not protected by histone and some of the DNA repair systems are ineffective (Pettepher et al., 1991). Oxidative damage to DNA increases with age and is more prevalent in mitrochondrial DNA than in nuclear DNA (Richter et al., 1988; Cortopassi and Arnheim, 1990; Corral-Debrinski et al., 1992; Hayakawa et al., 1992; Simonetti et al., 1992; Mecocci et al., 1993; Baffoli et al., 1994; Lee H. et al., 1994; Yen et al., 1994). Mutations in any of the mitochondrial genes which result in abnormal subunits of either cytochrome oxidase, cytochrome b-c_1, NADH dehydrogenase, or ATPase complexes may lead to a defective function of these enzymes. Additionally, free radicals can directly inactivate mitochondrial enzymes, causing mitochondrial dysfunction and increased generation of reactive oxygen species.

Our view postulates that motor neuron degeneration in ALS derives primarily from the age-associated failure of mitochondrial ATP production owing to the accumulation of oxidative mitochondrial damage (Figure 4). In neurons, mitochondria are distributed along the axons to furnish the energy required to maintain the structure and function of the distal extensions of the neuron. Thus, the mitochondria can be viewed as a mobile system to decentralize the capability for energy generation throughout the neuron, even to its distal extremities. In this context, the

ATP produced is utilized to maintain ionic gradients and membrane potential and to power axonal transport at sites physically far removed from the soma where ATP is important for cellular metabolism and calcium homeostasis (see Figure 2) (Bostock et al., 1995). Inadequate mitochondrial ATP production would result in impaired protein synthesis for neuronal repair, increased susceptibility to glutamate excitotoxicity, and facilitation of calcium-dependent neurotoxic processes in the soma. In the periphery, decreased ATP production could impair axonal transport and affect the structure and function of neurons in this manner. Thus, the deleterious consequences of poorly endowed mitochondria might be predicted to be most evident in neurons with long axonal projections such as the cortical and spinal motor neurons, which degenerate in ALS. Such neurons appear to be doubly vulnerable, susceptible to processes that primarily target the neuronal soma, such as glutamate excitotoxicity, or the axon itself leading to neurodeneration. Therefore, we propose that the effectiveness of mitochondrial antioxidant defense is the most critical factor in the survival of such neurons.

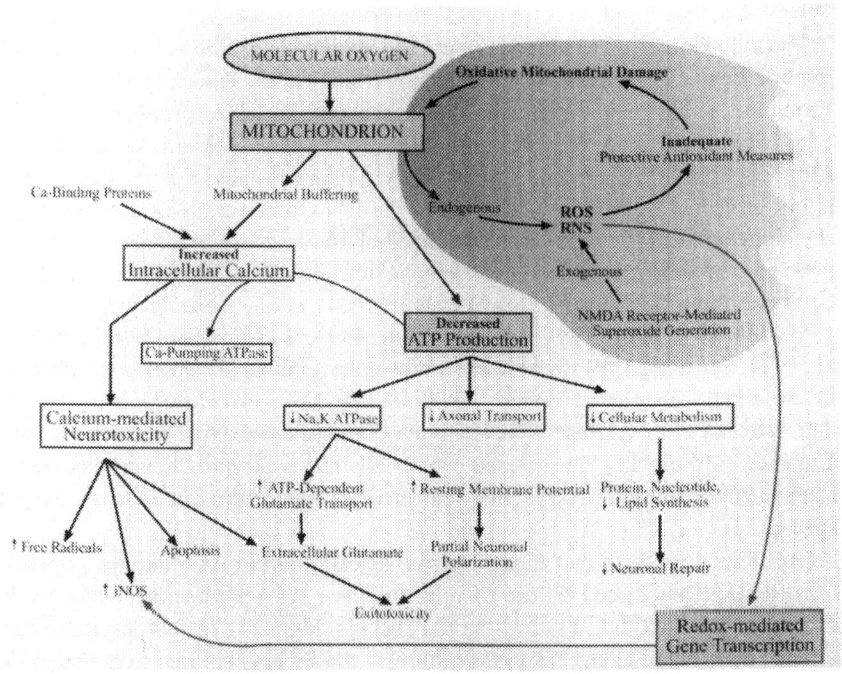

Figure 4. Progressive failure of mitochondrial ATP production as a cause of motor neuron death in ALS. The consequences of decreased ATP production are illustrated which may jeopardize the survival of spinal and cortical motor neurons.

ACKNOWLEDGMENTS

Original research by the authors is supported by NIH grants CA49797, CA59835, HL 03544, NIEHS Training Grant ES 07266, the Veterans Affairs Research Service, and the ALS Association.

REFERENCES

Abe, K., Pan, L.-H., Watanabe, M., Kato, T. & Itoyama, Y. (1995). Induction of nitrotyrosine-like immunoreactivity in the lower motor neuron of amyotrophic lateral sclerosis. Neurosci. Lett. 199, 152–154.

Adams, C.R., Ziegler, D.K. & Lin, J.T. (1983). Mercury intoxication simulating amyotrophic lateral sclerosis. JAMA. 250, 642–643.

Aguirre, T., Van Den Bosch, L., Goetschalckx, K., Tilkin, P., Mathijs, G., Cassiman, J.J. & Robberecht, W. (1998). Increased sensitivity of fibroblasts from amyotrophic lateral sclerosis patients to oxidative stress. Ann. Neurol. 43, 452–457.

Andersen, P.M., Forsgren, L., Binzer, M., Nilsson, P., AlaHurula, V., Keranen, M.L., Bergmark, L., Saarinen, A., Haltia, T., Tarvainen, I., Kinnunen, E., Udd, B. & Marklund, S.L. (1996). Autosomal recessive adult-onset amyotrophic lateral sclerosis associated with homozygosity for Asp90Ala CuZn-superoxide dismutase mutation—a clinical and genealogical study of 36 patients. Brain 119, 1153–1172.

Andersen, P.M., Nilsson, P., Ala-Hurula, V., Keranen, M., Tarvainen, I. & Haltia, T. (1995). Amyotrophic lateral sclerosis associated with homozygosity for an Asp90Ala mutation in CuZn-superoxide dismutase. Nat. Genet. 10, 61–66.

Andersen, P.M., Nilsson, P., Keranen, M., Forsgren, L., Hagglund, J., Karlsborg, M., Ronnevi, L.-O., Gredal, O. & Marklund, S.L. (1997). Phenotypic heterogeneity in motor neuron disease patients with CuZn-superoxide dismutase mutations in Scandinavia. Brain 120, 1723–1737.

Andres, P.L., Hedlund, W., Finison, L.J., Conlon, T., Felnus, M. & Munsat, T.L. (1986). Quantitative motor assessment in amyotrophic lateral sclerosis. Neurology 36, 937–941.

Annegers, J.F., Appel, S., Lee, J.R.J. & Perkins, P. (1991). Incidence and prevalence of amyotrophic lateral sclerosis in Harris County, Texas, 1985-1988. Arch. Neurol. 48, 589–593.

Aoki, M., Ogasawara, M., Matsubara, Y., Narisawa, K., Nakamura, S. & Itoyama, Y. (1994). Familial amyotrophic lateral sclerosis (ALS) in Japan associated with a H46R mutation in Cu/Zn superoxide dismutase gene: a possible new subtype of familial ALS. J. Neurol. Sci. 126, 77–83.

Aoki, M., Abe, K., Houi, K., Ogasawara, M., Matsubara, Y., Kobayashi, T., Mochio, S., Narisawa, K. & Itoyama, Y. (1995). Variance of age at onset in a Japanese family with amyotrophic lateral sclerosis associated with a novel Cu/Zn superoxide dismutase mutation. Ann. Neurol. 37, 676–679.

Aoki, M., Lin, C.G., Rothstein, J.D., Geller, B., Hosler, B., Munsat, T.L., Horvitz, H.R. & Brown, R.H., Jr. (1998). Mutations in the glutamate transporter EAAT2 gene do not cause abnormal EAAT2 transcripts in amyotrophic lateral sclerosis. Ann. Neurol. 43, 645–653.

Appel, S.H., Stewart, S.S. & Appel, V. (1988). A double-blind study of cyclosporine in amyotrophic lateral sclerosis. Arch. Neurol. 43, 234–238.

Asmark, H., Aquilonius, S.M. & Gillberg, P.G. (1993). A pilot trial of dextromethorphan in amyotrophic lateral sclerosis. J. Neurol. Neurosurg. Psychiat. 56, 197–200.

Baffoli, D., Scacco, S.C. & Vergari, R. (1994). Decline with age of the respiratory chain activity in human skeletal muscle. Biochim. Biophys. Acta 1226, 73–82.

Barron, K.D. & Rodichok, L.D. (1982). Cancer and disorders of motor neurons. In: Human Motor Neuron Diseases (Rowland, L.P., ed.), pp. 267–272. New York: Raven Press, New York.

Beal, M.F. (1995). Aging, energy, and oxidative stress in neurodegenerative diseases. Ann. Neurol. 38, 357–366.

Beckman, J.S., Carson, M., Smith, C.D. & Koppenol, W.H. (1993). ALS, SOD and peroxynitrite. Nature 364, 584.

Bensimon, G., Lacomblez, L. & Meininger, V. (1994). A controlled trial of riluzole in amyotrophic lateral sclerosis. N. Engl. J. Med. 330, 585–591.

Bereznai, B., Winkler, A., Borasio, G.D. & Gasser, T. (1997). A novel SOD1 mutation in an Austrian family with amyotrophic lateral sclerosis. Neuromusc. Disord. 7, 113–116.

Bergeron, C. (1995). Oxidative stress: its role in the pathogenesis of amyotrophic lateral sclerosis. J. Neurol. Sci. 129 (suppl.), 81–84.

Billiar, T.R., Curran, R.D., Stuehr, D.J., West, M.A., Bentz, B.G. & Simmons, R.L. (1989). An L-arginine-dependent mechanism mediates Kupffer cell inhibition of hepatocyte protein synthesis in vitro. J. Expt. Med. 169, 1467–1472.

Bondy, S.C. & Lee, D.K. (1993). Oxidative stress induced by glutamate receptor agonists. Brain Res. 61, 229–233.

Boothby, J.A., DeJesus, P.V. & Rowland, L.P. (1974). Reversible forms of motor neuron disease: Lead "neuritis." Arch. Neurol. 31, 18–23.

Borchelt, D.R., Lee, M.K., Slunt, H.S., Guarnieri, M., Xu, Z.S. & Wong, P.C. (1994). Superoxide dismutase 1 with mutations linked to familial amyotrophic lateral sclerosis possesses significant activity. Proc. Natl. Acad. Sci. U. S. A. 91, 8292–8296.

Borgstahl, G.E.O., Parge, H.E., Hickey, M.J., Johnson, M., Boissinot, M., Hallewell, R.A., Lepock, J.R., Cabelli, D.E. & Tainer, J.A. (1996). Human mitochondrial manganese superoxide dismutase polymorphic variant Ile58Thr reduces activity by destabilizing the tetrameric interface. Biochemistry 35, 4287–4297.

Bostock, H., Sharief, M.K., Reid, G. & Murray, N.M.F. (1995). Axonal ion channel dysfunction in amyotrophic lateral sclerosis. Brain 118, 217–225.

Boveris, A. & Cadenas, E. (1982). Production of superoxide radicals and hydrogen peroxide in mitochondria. In: Superoxide Dismutase (Oberley, L.W., ed.), pp. 15–30. CRC Press, Boca Raton FL.

Bowler, C., Sloonten, L., Vandenbranden, S., De Rycke, R., Botterman, J. & Sybesma, C. (1991). Manganese superoxide dismutase can reduce cellular damage mediated by oxygen radicals in transgenic plants. EMBO J. 10, 1723–1732.

Bowling, A.C. & Beal, M.F. (1995). Bioenergetic and oxidative stress in neurodegenerative diseases. Life Sci. 56, 1151–1171.

Bowling, A.C., Schulz, J.B., Brown, R.H. & Beal, M.F. (1993). Superoxide dismutase activity, oxidative damage, and mitochondrial energy metabolism in familial and sporadic amyotrophic lateral sclerosis. J. Neurochem. 61, 2322–2325.

Bracco, L., Antuono, P. & Amaducci, L. (1979). Study of epidemiological and etiological factors of amyotrophic lateral sclerosis in the province of Florence, Italy. Acta Neurol. Scand. 60, 112–124.

Brooks, B.R. (1996). Natural history of ALS: symptoms, strength, pulmonary function, and disability. Neurology 47, S71–S81.

Brooks, B.R., Sanjak, M., Ringel, S., England, J., Brinkmann, J., Pestronk, A., Florence, J., Mitsumoto, H., Szirony, K., Wittes, J., Charatan, M., Stambler, N. & Cedarbaum, J.M. (1996). The amyotrophic lateral sclerosis functional rating scale: assessment of activities of daily living in patients with amyotrophic lateral sclerosis. Arch. Neurol. 53, 141–147.

Brown, I.A. (1954). Chronic mercurialism. A cause of the clinical syndrome of amyotrophic lateral sclerosis. Arch. Neurol. Psychiatry 72, 647–681.

Brown, R.H., Hauser, S.L. & Harrington, H. (1986). Failure of immunosuppression with a 10- to 14-day course of high-dose intravenous cyclophosphamide to alter the progression of amyotrophic lateral sclerosis. Arch. Neurol. 43, 383–384.

Brownell, B., Oppenheimer, D.R. & Hughes, J.T. (1970). The central nervous system in motor neurone disease. J. Neurol. Neurosurg. Psychiatry 33, 338–357.

Bruijn, L.I., Becher, M.W., Lee, M.K., Anderson, K.L., Jenkins, N.A., Copeland, N.G., Sisodia, S.S., Rothstein, J.D., Borchelt, D.R., Price, D.L. & Cleveland, D.W. (1997). ALS-linked SOD1 mutant G85R mediates damage to astrocytes and promotes rapidly progressive disease with SOD1-containing inclusions. Neuron 18, 327–338.

Calder, V.L., Domigan, N.M. & George, P.M. (1995). Superoxide dismutase (glu100gly) in a family with inherited motor neuron disease: detection of mutant superoxide dismutase activity and the presence of heterodimers. Neurosci. Lett. 189, 143–146.

Campbell, A.M.G. & Williams, E.R. (1968). Chronic lead intoxication mimicking motor neurone disease. Br. Med. J. 4, 582

Carlioz, A. & Touati, D. (1986). Is superoxide dismutase necessary for aerobic life? EMBO J. 5, 623–630.

Carpenter, S. (1968). Proximal axonal enlargement in motor neuron disease. Neurology 18, 841–851.

Chance, B., Sies, H. & Boveris, A. (1979). Hydroperoxide metabolism in mammalian organs. Physiol. Rev. 59, 527–605.

Chaudhuri, K.R., Crump, S., Al-Sarraj, S., Anderson, V., Cavanagh, J. & Leigh, P.N. (1995). The validation of El Escorial criteria for the diagnosis of amyotrophic lateral sclerosis: A clinicopathological study. J. Neurol. Sci. 129 (suppl), 11–12.

Chiu, A.Y., Zhai, P., Dal Canto, M.C., Peters, T.M., Kwon, Y.W., Prattis, S.M.& Gurney, M.E. (1995). Age-dependent penetrance of disease in a transgenic mouse model of familial amyotrophic lateral sclerosis. Mol. Cell. Neurosci. 6, 349–362.

Chou, S.M. (1992). Pathology-Light microscopy of amyotrophic lateral sclerosis. In: Handbook of Amyotrophic Lateral Sclerosis (Smith, R.A., ed.), pp. 133–181. Marcel Dekker, New York.

Chou, S.M., Wang, H.S. & Taniguchi, A. (1996). Role of SOD-1 and nitric oxide/cyclic GMP cascade on neurofilament aggregation in ALS/MND. J. Neurol. Sci. 139 (suppl), 16–26.

Collard, J.-F., Côté, F. & Julien, J.-P. (1995). Defective axonal transport in a transgenic mouse model of amyotrophic lateral sclerosis. Nature 375, 61–64.

Conradi, S., Ronnevi, L.-O., Nise, G. & Vesterberg, O. (1982a). Long term penicillamine treatment in amyotrophic lateral sclerosis with parallel determination of lead in blood, plasma and urine. Acta Neurol. Scand. 65, 203–211.

Conradi, S., Ronnevi, L.-O. & Norris, F.H. (1982b). Motor neuron disease and toxic metals. Adv. Neurol. 36, 201–231.

Cork, L.C., Troncoso, J.C. & Klavano, G.G. (1988). Neurofilamentous abnormalities in motor neurons in spontaneously occurring animal disorders. J. Neuropath. Expt . Neurol. 47, 420–431.

Corral-Debrinski, M., Horton, T. & Lott, M.T. (1992). Mitochondrial DNA deletions in human brain: regional variability and increase with advanced age. Nat. Genet. 2, 324–329.

Cortopassi, G.A. & Arnheim, N. (1990). Detection of a specific mitochondrial DNA deletion in tissues of older humans. Nucleic Acids Res. 18, 6927–6933.

Côté, F., Collard, J.-F. & Julien, J.-P. (1993). Progressive neuronopathy in transgenic mice expressing the human neurofilament heavy gene: A mouse model of amyotrophic lateral sclerosis. Cell 73, 35–46.

Coyle, J.T. & Puttfarcken, P. (1993). Oxidative stress, glutamate, and neurodegenerative disorders. Science 262, 689–695.

Coyle, J.T., Robinson, M.B. & Blakely, R.D. (1989). The neurobiology of N-acetyl-aspartyl glutamate. In: Allosteric Modulation of Amino Acid Receptors: Therapeutic Implications (Barnard, E.A. & Cossta, E., eds.), pp. 319–333. Raven Press, New York.

Crapo, J.D. & Tierney, D.F. (1974). Superoxide dismutase and pulmonary oxygen toxicity. Am. J. Physiol. 226, 1401–1407.

Cudkowicz, M.E., McKenna-Yasek, D., Sapp, P.E., Chin, W., Geller, B., Hayden, D.L., Schoenfeld, D.A., Hosler, B.A., Horvitz, H.R. & Brown, R.H. (1997). Epidemiology of mutations in superoxide dismutase in amyotrophic lateral sclerosis. Ann. Neurol. 41, 210–221.

Currier, R.D. & Hearer, A.F. (1968). Amyotrophic lateral sclerosis and metallic toxins. Arch. Environ. Health 17, 712–719.

Cwik, V.A. & Mitsumoto, H. (1992). Postpoliomyelitis syndrome. In: Handbook of Amyotrophic Lateral Sclerosis (Smith, R.A., ed.), pp. 77–91. Marcel Dekker, New York.

Dal Canto, M.C. & Gurney, M.E. (1994). Development of central nervous system pathology in a murine transgenic model of human amyotrophic lateral sclerosis. Am. J. Pathol. 145, 1271–1279.

Dal Canto, M.C. & Gurney, M.E. (1995). Neuropathological changes in two lines of mice carrying a transgene for mutant human Cu,Zn SOD, and in mice overexpressing wild type human SOD: a model of familial amyotrophic lateral sclerosis (FALS). Brain Res. 676, 25–40.

Dalakas, M.C., Elder, G. & Hallett, M. (1986). A long-term follow-up study of patients with post-poliomyelitis neuromuscular symptoms. N. Engl. J. Med. 314, 959–963.

Dalakas, M.C., Stein, D.P. & Otero, C. (1994). Effect of high-dose intravenous immunoglobulin on amyotrophic lateral sclerosis and multifocal motor neuropathy. Arch. Neurol. 51, 861–864.

Damier, P., Hirsch, E., Zhang, P., Agid, Y. & Javoy-Agid, F. (1993). Glutatione peroxidase, glial cells and Parkinson's disease. Neurosci. 52, 1–6.

Deng, H.X., Hentati, A., Tainer, J.A., Iqbal, Z., Cayabyab, A. & Hung, W.Y. (1993). Amyotrophic lateral sclerosis and structural defects in Cu,Zn superoxide dismutase. Science, 261, 1047–1051.

Deng, H.X., Tainer, J.A., Mitsumoto, H., Ohnishi, A., He, X., Hung, W.Y., Zhao, Y., Juneja, T., Hentati, A. & Siddique, T. (1995). Two novel SOD1 mutations in patients with familial amyotrophic lateral sclerosis. Hum. Mol. Genet. 4, 1113–1116.

Drachman, D.B., Chaudhry, V. & Cornblath, D. (1994). Trial of immunosuppression in amyotrophic lateral sclerosis using total lymphoid irradiation. Ann. Neurol. 35, 142–150.

Eisen, A. (1995). Amyotrophic lateral sclerosis is a multifactorial disease. Muscle Nerve 18, 741–752.

Eisen, A., Stewart, H., Schulzer, M. & Cameron, D. (1993). Anti-glutamate therapy in amyotrophic lateral sclerosis - a trial using lamotrigine. Can. J. Neurol. Sci. 20, 297–301.

Elshafey, A., Lanyon, W.G. & Connor, J.M. (1994). Identification of a new missense point mutation in exon 4 of the Cu/Zn superoxide dismutase (SOD-1) gene in a family with amyotrophic lateral sclerosis. Hum. Mol. Genet. 3, 363–364.

Enayat, Z.E., Orrell, R.W., Claus, A., Ludolph, A., Bachus, R., Brockmüller, J. Ray-Chaudhuri, K., Radunovic, A., Shaw, C., Wilkinson, J., King, A., Swash, M., Leight, P.N., de Belleroche, J. & Powell, J. (1995). Two novel mutations in the gene for copper zinc superoxide dismutase in UK families with amyotrophic lateral sclerosis. Hum. Mol. Genet. 4, 1239–1240.

Esteban, J., Rosen, D.R., Bowling, A.C., Sapp, P., McKenna-Yasek, D. & Oregan, J.P. (1994). Identification of two novel mutations and a new polymorphism in the gene for Cu/Zn superoxide dismutase in patients with amyotrophic lateral sclerosis. Hum. Mol. Genet. 3, 997–998.

Farkkila, M., Iivanainen, M. & Roine, R. (1984). Neurotoxic and other side effects of high-dose interferon in amyotrophic lateral sclerosis. Acta Neurol. Scand. 69, 42–46.

Farr, S.B., D'Ari, R. & Touati, D. (1986). Oxygen-dependent mutagenesis in *Escherichia coli* lacking superoxide dismutase. Proc. Natl. Acad. Sci. U. S. A. 83, 8268–8272.

Festoff, B. (1987). Amyotrophic lateral sclerosis. In: Handbook of the Spinal Cord (Davidoff, R.A., ed.), pp. 607–664. Marcel Dekker, New York.

Figlewicz, D.A., Krizus, A., Martinoli, M.G., Meininger, V., Dib, M. & Rouleau, G.A. (1994). Variants of the heavy neurofilament subunit are associated with the development of amyotrophic lateral sclerosis. Hum. Mol. Genet. 3, 1757–1761.

Fishman, P.S., Shy, M.E., Hart, D.E., Thompson, P.E. & Cashman, N.R. (1991). Antibodies to the ganglioside GD1b in a patient with motor neuron disease and thyroid adenoma. Arch. Neurol. 48, 1188–1190.

Frank, L., Lapidus, R. & Ahmad, K. (1981). Protection against radiation pneumonitis in rats: Role of lung antioxidant enzymes. Fed. Proc. 40, 425.

Freddo, L., Yu, R.K. & Latov, N. (1986). Gangliosides GM1 and DG1b are antigens for IgM-protein in a patient with motor neuron disease. Neurology 36, 454–458.

Fridovich, I. (1978). The biology of oxygen radicals. The superoxide radical is an agent of oxygen toxicity: superoxide dismutases provide an important defense. Science 201, 875–880.

Furuta, A., Martin, L.J., Lin, C.L., Dykes-Hoberg, M. & Rothstein, J.D. (1997). Cellular and synaptic localization of the neuronal glutamate transporters. Excitatory amino acid transport 3 and 4. Neuroscience 81, 1031–1042.

Giulivi, C., Boveris, A. & Cadenas, E. (1995). Hydroxyl radical generation during mitochondrial electron transfer and the formation of 8-hydroxyguanosine in mitochondrial DNA. Arch. Biochem. Biophys. 316, 909–916.

Gonatas, N.K., Stieber, A., Mourelatos, Z., Chen, Y., Gonatas, J.O. & Appel, S.H. (1992). Fragmentation of the Golgi apparatus of motor neurons in amyotrophic lateral sclerosis. Am. J. Pathol. 140, 731–737.

Gonzalez-Zulueta, M., Ensz, L.M., Mukhina, G., Lebovitz, R.M., Zwacka, R.M., Engelhardt, J.F., Oberley, L.W., Dawson, V.L. & Dawson, T.M. (1998). Manganese superoxide dismutase protects nNOS neurons from NMDA and nitric oxide-mediated neurotoxicity. J. Neurosci. 18, 2040–2055.

Griffin, J.W., Hoffman, P.N. & Clark, A.W. (1978). Slow axonal transport of neurofilament protiens: impairment by beta, beta′-iminodipropionitrile administration. Science 202, 633–635.

Gruber, M.Y., Glick, B.R. & Thompson, J.E. (1990). Cloned manganese superoxide dismutase reduces oxidative stress in *Eshcerichia coli* and *Anacystis nidulans*. Proc. Natl. Acad. Sci. U. S. A. 87, 2608–2612.

Gurney, M.E., Pu, H., Chiu, A.Y., Dal Canto, M.C., Polchow, C.Y. & Alexander, D.D. (1994). Motor neuron degeneration in mice that express a human Cu,Zn superoxide dismutase mutation. Science 264, 1772–1775.

Gurney, M.E., Cutting, F.B., Zhai, P., Doble, A., Taylor, C.P. & Andrus, P.K. (1996). Benefit of vitamin E, riluzole, and gabapentin in a transgenic model of familial amyotrophic lateral sclerosis. Ann. Neurol. 39, 147–157.

Halliwell, B. (1992). Reactive oxygen species and the central nervous system. J. Neurochem. 59, 1609–1623.

Halliwell, B. & Gutteridge, J.M.C. (1989). Free Radicals in Biology and Medicine. Clarendon Press, Oxford.

Hansen, C., Gredal, O., Karlsborg, M., Lerdelin, L., Andersen, P. M., Nilson, P., Marklund, S.L. & Nielsen, K.B. (1996). Two new SOD1 mutations in ALS: a four base pair insertion in exon5 and a missense mutation in exon 3. 7th International Symposium on ALS/MND, Chicago, IL.

Hayakawa, M., Hattori, K., Sugiyama, S. & Ozawa, T. (1992). Age-associated oxygen damage and mutations in mitochondrial DNA in human hearts. Biochem. Biophys. Res. Commun. 189, 979–985.

Hayashi, H. & Kato, S. (1989). Total manifestations of amyotrophic lateral sclerosis. ALS in the totally locked-in state. J. Neurol. Sci. 93, 19–35.

Hayashi, H., Kato, S. & Kawada, A. (1991). Amyotrophic lateral sclerosis patients living beyond respiratory failure. J. Neurol. Sci. 105, 73–78.

Heiman-Patterson, T.D., Gudesblatt, M.J. & Tahmoush, A.J. (1986). Etiologic considerations and research trends in amyotrophic lateral sclerosis. In: ALS—A Guide to Patient Care (Caroscio, J.T., ed.), pp. 85–102. Thieme, New York.

Hentati, A., Bejaoui, K., Pericak-Vance, M.A., Hentati, F., Speer, M.C. & Hung, W.Y. (1994). Linkage of recessive familial amyotrophic lateral sclerosis to chromosome 2q33-q35. Nat. Genet. 7, 425–428.

Hibbs, J.B., Taintor, R.R., Vavrin, Z. & Rachlin, E.M. (1988). Nitric oxide: a cytotoxic activated macrophage effector molecule. Biochem. Biophys. Res. Commun. 157, 87–94.

Hirano, A., Donnefeld, H., Sasaki, S. & Nakano, I. (1984). Fine structural observations of neurofilamentous changes in amyotrophic lateral sclerosis. J. Neuropath. Expt. Neurol. 43, 461–470.

Hirano, M., Fujii, J., Nagai, Y., Sonobe, M., Okamoto, K. & Araki, H. (1994). A new variant Cu/Zn superoxide dismutase (val(7)-glu) deduced from lymphocyte mRNA sequences from Japanese patients with familial amyotrophic lateral sclerosis. Biochem. Biophys. Res. Commun. 204, 572–577.

Hirose, K., Longo, D.L., Oppenheim, J.J. & Matsushima, K. (1993). Overexpression of mitochondrial manganese superoxide dismutase promotes the survival of tumor cells exposed to interleukin-1, tumor necrosis factor, selected anticancer drugs, and ionizing radiation. FASEB J. 7, 361–368.

Hosler, B.A., Nicholson, G.A., Sapp, P.C., Chin, W., Orrell, R.W., Debelleroche, J.S., Esteban, J., Hayward, L.J., Mckennayasek, D., Yeung, L., Cherryson, A.K., Dench, J.E., Wilton, S.D., Laing, N.G., Horvitz, H.R. & Brown, R.H. (1996). Three novel mutations and two variants in the gene for Cu/Zn superoxide dismutase in familial amyotrophic lateral sclerosis. Neuromusc. Disord. 6, 361–366.

Ikeda, M., Abe, K., Aoki, M., Ogasawara, M., Kameya, T., Watanabe, M., Shoji, M., Hirai, S. & Itoyama, Y. (1995). A novel point mutation in the Cu/Zn superoxide dismutase gene in a patient with familial amyotrophic lateral sclerosis. Hum. Mol. Genet. 4, 491–492.

Ikeda, M., Abe, K., Aoki, M., Sahara, M., Watanabe, M., Shoji, M., St. George-Hyslop, P.H., Hirai, S. & Itoyama, Y. (1996). Variable clinical symptoms in the pedigrees of familial amyotrophic lateral sclerosis with novel missense point mutations in the Cu/Zn superoxide dismutase gene. In: Amyotrophic Lateral Sclerosis: Progress and Perspectives in Basic Research and Clinical Application (Nakano, I. & Hirano, A., eds.), p. 285. Elsevier, The Netherlands.

Ince, P.G., Shaw, P.J., Candy, J.M., Mantle, D., Tandon, L. & Ehmann, W.D. (1994). Iron, selenium and glutathione peroxidase activity are elevated in sporadic motor neuron disease. Neurosci. Lett. 182, 87–90.

Ito, H. & Hirano, A. (1994). Comparative study of spinal cord ubiquitin expression in post-poliomyelitis and sporadic amyotrophic lateral sclerosis. Acta Neuropath. 87, 425–429.

Iwasaki, Y., Ikeda, K. & Kinoshita, M. (1992). Plasma amino acid levels in patients with amyotrophic lateral sclerosis. J. Neurol. Sci. 107, 219–222.

Jackson, M., Al-Chalabi, A., Enayat, Z.E., Chioza, B., Leigh, P.N., & Morrison, K.E. (1997). Copper/zinc superoxide dismutase 1 and sporadic amyotrophic lateral sclerosis: analysis of 155 cases and identification of a novel insertion mutation. Ann. Neurol. 42, 803–807.

Jones, C.T., Shaw, P.J., Chari, G. & Brock, D.J.H. (1994a). Identification of a novel exon 4 SOD1 mutation in a sporadic amyotrophic lateral sclerosis patient. Mol. Cell Probe. 8, 329–330.

Jones, C.T., Swingler, R.J. & Brock, D.J. (1994b). Identification of a novel SOD1 mutation in an apparently sporadic amyotrophic lateral sclerosis patient and the detection of Ile113Thr in three others. Hum. Mol. Genet. 3, 649–650.

Jubelt, B. (1998). Viruses and motor neuron diseases. Adv. Neurol. 56, 463–472.

Jubelt, B. & Drucker, J. (1993). Post-polio syndrome: an update. Semin. Neurol. 13, 283–289.

Juneja, T., Pericak-Vance, M.A., Laing, N.G., Dave, S. & Siddique, T. (1997). Prognosis in familial amyotrophic lateral sclerosis: progression and survival in patients with glu100gly and ala4val mutations in Cu,Zn superoxide dismutase. Neurology 48, 55–57.

Kadekawa, J., Fujimura, H., Ogawa, Y., Hattori, N., Kaido, M., Nishimura, T., Yoshikawa, H., Shirahata, N., Sakoda, S. & Yanagihara, T. (1997). A clinicopathological study of a patient with familial amyotrophic lateral sclerosis associated with a two base pair deletion in the copper/zinc superoxide dismutase (SOD1) gene. Acta Neuropath. 94, 617–622.

Kakita, A., Oyanagi, K., Nagai, H. & Takahashi, H. (1997). Eosinophilic intranuclear inclusions in the hippocampal pyramidal neurons of a patient with amyotrophic lateral sclerosis. Acta Neuropath. 93, 532–536.

Kanai, Y., Smith, C.P. & Hediger, M.A. (1993). A new family of neurotransmitter transporters - the high-affinity glutamate transporters. FASEB J. 7, 1450–1459.

Kantarjian, A.D. (1953). A syndrome clinically resembling amyotrophic lateral sclerosis following chronic mercurialism. Neurology 11, 639.

Kasarskis, E.J. (1992). Neurotoxicology: heavy metals. In: Handbook of Amyotrophic Lateral Sclerosis (Smith, R.A., ed.), pp. 559–574. Marcel Dekker, New York.

Kasarskis, E.J. & Winslow, M. (1989). When did Lou Gehrig's personal illness begin? Neurology 39, 1243–1245.

Kasarskis, E.J., Tandon, L., Lovell, M.A. & Ehmann, W.D. (1995). Aluminum, calcium, and iron in the spinal cord of patients with sporadic amyotrophic lateral sclerosis using laser microprobe mass spectroscopy: a preliminary study. J. Neurol. Sci. 130, 203–208.

Kascsak, R.J., Carp, R.I. & Vilcek, J.T. (1982). Virological studies in amyotrophic lateral sclerosis. Muscle Nerve 5, 93–101.

Kasemset, D. & Oberley, L.W. (1984). Regulation of Mn-SOD activity in the mouse heart: Glucose effect. Biochem. Biophys. Res. Commun. 122, 682–686.

Kato, S., Shimoda, M., Watanabe, Y., Nakashima, K., Takahashi, K. & Ohama, E. (1996). Familial amyotrophic lateral sclerosis with a two base pair deletion in superoxide dismutase 1 gene: multisystem degeneration with intracytoplasmic hyaline inclusions in astrocytes. J. Neuropath. Exp. Neurol. 55, 1089–1101.

Kawamata, J., Hasegawa, H., Shimohama, S., Kimura, J., Tanaka, S. & Ueda, K. (1994). Leu106Val (CTC GTC) mutation of superoxide dismutase-1 gene in patient with familial amyotrophic lateral sclerosis in Japan. Lancet 343, 1501.

Kawamata, J., Shimohama, S., Hasegawa, H., Imura, T., Kimura, J. & Ueda, K. (1996). Deletion and point mutations in superoxide dismutase-1 gene in amyotrophic lateral sclerosis. In: Amyotrophic Lateral Sclerosis: Progress and Perspectives in Basic Research and Clinical Application (Nakano, I. & Hirano, A., eds.), p. 276. Elsevier, The Netherlands.

Kawamata, J., Shimohama, S., Takano, S., Harada, K., Ueda, K. & Kimura, J. (1997). Novel G16S (GGC-AGC) mutation in the SOD-1 gene in a patient with apparently sporadic young-onset amyotrophic lateral sclerosis. Hum. Mutat. 9, 356–358.

Kelemen, J., Hedlund, W. & Orlin, J.B. (1983). Plasmapharesis with immunosuppression in amyotrophic lateral sclerosis. Arch. Neurol. 40, 752–753.

Keller, J.N., Kindy, M.S., Holtsberg, F.W., St. Clair, D.K., Yen, H., Germeyer, A., Steiner, S.M., Bruce-Keller, A.J., Hutchins, J.B. & Mattson, M.P. (1998). Mitochondrial manganese superoxide dismutase prevents neural apoptosis and reduces ischemic brain injury: suppression of peroxynitrite production, lipid peroxidation, and mitochondrial dysfunction. J. Neurosci. 18, 687–697.

Kidson, C., Chen, P., Imray, P. & Gipps, E. (1983). Nervous system disease associated with dominant cellular radiosensitivity. J. Cell. Biol. Suppl 7B, 209.

Kim, H., Minard, P., Legoy, M.D. & Thomas, D. (1986). Inactivation of 3-alpha-hydroxysteroid dehydrogenase by superoxide redicals. Modification of histidine and cysteine residues causes the conformational change. Biochem. J. 233, 493.

Kittur, S., Hoh, J., Endo, H., Tourtellotte, W., Weeks, B.S., Markesbery, W. & Alder, W. (1994). Cytoskeletal neurofilament gene expresison in brain tissue from Alzheimer's disease patients. I. Decrease in NF-L and NF-M message. J. Geriat. Psychiat. Neurol. 7, 153–158.

Kono, Y., & Fridovich, I. (1982). Superoxide radical inhibits catalase. J. Biol. Chem. 262, 5751.

Kostrzewa, M., Burcklehmann, U. & Muller, U. (1994). Autosomal dominant amyotrophic lateral sclerosis: a novel mutation in the Cu/Zn superoxide dismutase-1 gene. Hum. Mol. Genet. 3, 2261–2262.

Kostrzewa, M., Damian, M.S. & Muller, U. (1996). Superoxide dismutase 1: identification of a novel mutation in a case of familial amyotrophic lateral sclerosis. Hum. Genet. 98, 48–50.

Krall, J., Bagley, A.C., Mullenbach, G.T., Hallewell, R.A. & Lynch, R.E. (1988). Superoxide mediates the toxicity of paraquat for cultured mammalian cells. J. Biol. Chem. 263, 1910–1914.

Kuo, C.F., Mashino, T. & Fridovich, I. (1987). Alpha, beta-dihydroxyisovalerate dehydratase: a superoxide-sensitive enzyme. J. Biol. Chem. 262, 4724–4747.

Kurlander, H.M. & Patten, B.M. (1979). Metals in spinal cord tissue of patients dying of motor neuron disease. Ann. Neurol. 6, 21–24.

Lafon-Cazal, M., Pietri, C., Culcasi, M. & Brockaert, J. (1993). NMDA-dependent superoxide production and neurotoxicity. Nature 364, 535–537.

Lawyer, T. & Netsky, M.G. (1953). Amyotrophic lateral sclerosis. A clinicoanatomic study of fifty-three cases. Arch. Neurol. Psychiat. 69, 171–192.

Lebovitz, R.M., Zhang, H., Vogel, H., Cartwright, J., Dionne, L. & Lu, N. (1996). Neurodegeneration, myocardial injury, and perinatal death in mitochondrial superoxide dismutase-deficient mice. Proc. Natl. Acad. Sci. U. S. A. 93, 9782–9787.

Lee, H., Pang, C. & Hsu, H. (1994). Differential accumulations of 4,977 bp deletion in mitochondrial DNA of various tissues in human aging. Biochim. Biophys. Acta 1226, 37–43.

Lee, M.K., Marszalek, J.R. & Cleveland, D.W. (1994). A mutant neurofilament subunit causes massive, selective motor neuron death: implications for the pathogenesis of human motor neuron disease. Neuron 13, 975–988.

Leigh, P.N. & Swash, M. (1991). Cytoskeletal pathology in motor neuron diseases. Adv. Neurol. 56, 115–124.

Li, Y., Huang, T., Carlson, E.J., Melov, S., Ursell, P.C. & Olson, J.L. (1995). Dilated cardiomyopathy and neonatal lethality in mutant mice lacking manganese superoxide dismutase. Nat. Genet. 11, 376–381.

Lin, C.G., Bristol, L.A., Jin, L., Dykes-Hoberg, M., Crawford, T., Clawson, L. & Rothstein, J.D. (1998). Aberrant RNA processing in a neurodegenerative disease: the cause for absent EAAT2, a glutamate transporter, in amyotrophic lateral sclerosis. Neuron 20, 589–602.

Linnane, A.W., Marzuki, S., Ozawa, T. & Tanaka, M. (1989). Mitochondrial DNA mutations as an important contribution to aging and degenerative diseases. Lancet 1, 642–645.

Lipton, S.A., Choi, Y., Pan, Z., Lei, S., Chen, H. & Sucher, J.J. (1993). A redox-based mechanism for the neuroprotective and neurodestructive effects of nitric oxide and related nitroso-compounds. Nature 364, 626–632.

Liu, D.X., Yang, R.M., Yan, X.X. & Mcadoo, D.J. (1994). Hydroxyl radicals generated in vivo kill neurons in the rat spinal cord - electrophysiological, histological, and neurochemical results. J. Neurochem. 62, 37–44.

Lowe, J. (1994). New pathological findings in amyotrophic lateral sclerosis. J. Neurol. Sci. 124, 38–51.

Maeda, T., Kurahashi, K., Matsunaga, M., Inoue, K. & Inoue, M. (1997). On intra-familial clinical diversities of a familial amyotrophic lateral sclerosis with a point mutation of Cu/Zn superoxide dismutase (Asn 86-Ser). Brain Nerve (Japan) 49, 847–851.

Makar, T.K., Nedergaard, M., Preuss, A., Gelbard, A.S., Perumal, A.S. & Cooper, A.J.L. (1994). Vitamin E, ascorbate, glutathione, glutathione disulfide, and enzymes of glutathione metabolism in cultures of chick astrocytes and neurons: evidence that astrocytes play an important role in antioxidative processes in the brain. J. Neurochem. 62, 45–53.

Manetto, V., Sternberger, N.H. & Perry, G. (1988). Phosphorylation of neurofilaments is altered in amyotrophic lateral sclerosis. J. Neuropath. Expt. Neurol. 47, 642–653.

Marklund, S.L. (1982). Human copper-containing superoxide dismutase of high molecular weight. Proc. Natl. Acad. Sci. U. S. A. 79, 7634–7638.

Mattson, M.P., Goodman, Y., Luo, H., Fu, W. & Furukawa, K. (1997). Activation of NF-κB protects hippocampal neurons against oxidatinve stress-induced apoptosis: evidence for induction of manganese superoxide dismutase and suppression of peroxynitirite production and protein tyrosine nitration. J. Neurosci. Res. 49, 681–697.

McCord, J.M. & Fridovich, I. (1969). Superoxide dismutase: an enzymic function for erythrocuprein (hemocuprein). J. Biol. Chem. 244, 6049–6055.

McCord, J.M. & Russell, W.J. (1988). Superoxide inactivates creatine phosphokinase during reperfusion of ischemic heart. In: Oxy-radicals in Molecular Biology and Pathology (Ceruti, P.A., Fridovich, I. & McCord, J.M., eds.), pp. 27–35. Alan R. Liss, New York.

Mecocci, P., Macgarvey, U., Kaufman, A.E., Koontz, D., Shoffner, J.M. & Wallace, D.C. (1993). Oxidative damage to mitochondrial DNA shows marked age-dependent increases in human brain. Ann. Neurol. 34, 609–616.

Melov, S., Schneider, J.A., Day, B.J., Hinerfeld, D., Coskun, P., Mirra, S.S., Crapo, J.D. & Wallace, D.C. (1998). A novel neurological phenotype in mice lacking mitochondrial manganese superoxide dismutase. Nat. Genet. 18, 159–163.

Migheli, A., Attanasio, A. & Schiffer, D. (1994). Ubiquitin and neurofilament expression in anterior horn cells in amyotrophic lateral sclerosis: possible clues to the pathogenesis. Neuropathol. Appl. Neurobiol. 20, 282–289.

Miller, R.G., Moore, D., Young, L.A., Armon, C., Barohn, R.J., Bromberg, M.B., Bryan, W.W., Gelinas, D.F., Mendoza, M.C., Neville, H.E., Parry, G.J., Petajan, J.H., Ravits, J.M., Ringel, S.P., Ross, M.A., Quien, A., Murphy, J., Brinkman, J., Burns, D., Parsons, J., Braiman, J., Kelkar, P., Hartsock, L., Heffner, M., Moses, D., Riddle, K., Furtado, M., Doyle, J., Graves, J., Aulenbacher, D., Herbelin, L., Ivy, N., Forshew, D., Mason, V., Trafton, L., Barbee, M., Garofalo, E., Hayes, L., Leiderman, D. & Pierce, M. (1996). Placebo-controlled trial of gabapentin in patients with amyotrophic lateral sclerosis. Neurology 47, 1383–1388.

Mimnaugh, E.G., Dusre, L., Atwell, J. & Myers, C.E. (1989). Differential oxygen radical susceptibility of adriamycin-sensitive and -resistant MCF-7 human breast tumor cells. Cancer Res. 49, 8–15.

Mitchell, J.D. (1987). Heavy metals and trace elements in amyotrophic lateral sclerosis. Neurol. Clin. North Am. 5, 43–60.

Mitsumoto, H., Chad, D.A. & Pioro, E.P. (1998a). Amyotrophic Lateral Sclerosis. F.A. Davis, Philadelphia.

Mitsumoto, H., Chad, D.A. & Pioro, E.P. (1998b). Hypotheses for viral and other transmissible agents in ALS. In: Amyotrophic Lateral Sclerosis (Mitsumoto, H., Chad, D.A., & Pioro, E.P, eds.), pp. 239–247. F.A. Davis, Philadelphia.

Mora, J.S., Munsat, T.L. & Kao, K. (1986). Intrathecal administration of natural human interferon alpha in amyotrophic lateral sclerosis. Neurology 36, 1137–1140.

Morita, M., Aoki, M., Abe, K., Hasegawa, T., Sakuma, R., Onodera, Y., Ichikawa, N., Nishizawa, M. & Itoyama, Y. (1996). A novel two-base mutation in the Cu/Zn superoxide dismutase gene associated with familial amyotrophic lateral sclerosis in Japan. Neurosci. Lett. 205, 79–82.

Morrison, B.M., Gordon, J.W., Ripps, M.E. & Morrison, J.H. (1996). Quantitative immunocytochemical analysis of the spinal cord in G86R superoxide dismutase transgenic mice: neurochemical correlates of selective vulnerability. J. Comp. Neurol. 373, 619–631.

Moulard, B., Camu, W., Brice, A., Salachas, F., Meininger, V. & Malafosse, A. (1995). A previously undescribed mutation in the SOD1 gene in a French family with atypical ALS. International Symposium on ALS/MND, Dublin, Ireland.

Moulard, B., Salachas, F., Chassande, B., Briolotti, V., Meininger, V., Malafosse, A. & Camu, W. (1998). Association between centromeric deletions of the SMN gene and sporadic adult-onset lower motor neuron disease. Ann. Neurol. 43, 640–644.

Murakami, K., Kondo, T., Kawase, M., Li, Y., Sato, S., Chen, S.F. & Chan, P.H. (1998). Mitochondrial susceptibility to oxidative stress exacerbates cerebral infarction that follows permanent focal cerebral ischemia in mutant mice with manganese superoxide dismutase deficiency. J. Neurosci. 18, 205–213.

Murros, K. & Fogelholm, R. (1983). Amyotrophic lateral sclerosis in Middle-Finland: an epidemiological study. Acta Neurol. Scand. 67, 41–47.

Naganuma, A., Miura, K., Tanaka-Kagawa, T., Kitahara, J., Seko, Y., Toyoda, H. & Imura, N. (1998). Overexpression of manganese-superoxide dismutase prevents methylmercury toxicity in HeLa cells. Life Sci. 62, PL157–PL161.

Nagaraja, T.N., Gourie-Devi, M., Nalini, A. & Raju, T.R. (1994). Neurofilament phosphorylation is enhanced in cultured chick spinal cord neurons exposed to cerebrospinal fluid from amyotrophic lateral sclerosis patients. Acta Neuropath. 88, 349–352.

Nakamura, S., Kawamoto, Y., Nakano, S., Ikemoto, A., Akiguchi, I. & Kimura, J. (1997). Cyclin-dependent kinase 5 in Lewy body-like inclusions in anterior horn cells of a patient with sporadic amyotrophic lateral sclerosis. Neurology 48, 267–270.

Nakano, R., Sato, S., Inuzuka, T., Sakimura, K., Mishina, M. & Takahashi, H. (1994). A novel mutation in Cu/Zn superoxide dismutase gene in Japanese familial amyotrophic lateral sclerosis. Biochem. Biophys. Res. Commun. 200, 695–703.

Nakashima, K., Watanabe, Y., Kuno, N., Nanba, E. & Takahashi, K. (1995). Abnormality of Cu/Zn superoxide dismutase (SOD1) activity in Japanese familial amyotrophic lateral sclerosis with two base pair deletion in the SOD1 gene. Neurology 45, 1019–1020.

Norris, F., Shepherd, R., Denys, E., U, K., Mukai, E., Elias, L., Holden, D. & Norris, H. (1993). Onset, natural history and outcome in idiopathic adult motor neuron disease. J. Neurol. Sci. 118, 48–55.

Norris, F.H., Calanchini, P.R., Fallat, R.J., Panchari, S. & Jewett, B. (1974). The administration of guanidine in amyotrophic lateral sclerosis. Neurology 24, 721–728.

Norris, F.H., Denys, E.H. & Mielke, C.H. (1978). Plasmapharesis in amyotrophic lateral sclerosis. Muscle Nerve 1, 342.

Oberley, L.W., St. Clair, D.K., Autor, A.P. & Oberley, T.D. (1987). Increase in manganese superoxide dismutase activitiy in the mouse heart after x-irradiation. Arch. Biochem. Biophys. 254, 69–80.

Ogasawara, M., Matsubara, Y., Narisawa, K., Aoki, M., Nakamura, S. & Itoyama, Y. (1993). Mild ALS in Japan associated with novel SOD mutation. Nat. Genet. 5, 323–324.

Olarte, M.R., Schoenfeldt, R.B. & McKiernan, G. (1980). Plasmapharesis in amyotrophic lateral sclerosis. Ann. Neurol. 8, 644–645.

Orrell, R., Debelleroche, J., Marklund, S., Bowe, F. & Hallewell, R. (1995a). A novel SOD mutant and ALS. Nature 374, 504–505.

Orrell, R.W., King, A.W. & Hilton, D.A. (1995b). Familial amyotrophic lateral sclerosis with a point mutation of SOD1: intrafamilial heterogeneity of disease duration associated with neurofibrillary tangles. J. Neurol. Neurosurg. Psychiatry 59, 266–270.

Orrell, R.W., Habgood, J.J., Gardiner, I., King, A.W., Bowe, F.A., Hallewell, R.A., Marklund, S., Greenwood, J. & de Belleroche, J. (1997). Clinical and functional investigation of 10 missense mutations and a novel frameshift insertion mutation of the gene for copper-zinc superoxide dismutase in UK families with amyotrophic lateral sclerosis. Neurology 48, 746–751.

Palmer, R.M.J., Ferrige, A.G. & Moncada, S. (1987). Nitric oxide release accounts for the biological activity of endothelium-derived relaxing factor. Nature 327, 524–526.

Parboosingh, J.S., Rouleau, G.A., Meninger, V., McKenna-Yasek, D., Brown, R.H. & Figlewicz, D.A. (1995). Absence of mutations in the Mn superoxide dismutase or catalase genes in familial amyotrophic lateral sclerosis. Neuromusc. Disord. 5, 7–10.

Pedersen, W.A., Cashman, N. & Mattson, M.P. (1998a). The lipid peroxidation product 4-hydroxynonenal impairs glutamate and glucose transport and choline acetyltransferase activity in NSC-19 motor neuron cells. Exp. Neurol. In press.

Pedersen, W.A., Fu, W., Markesbery, W.R., Appel, S.H., Smith, R.G., Keller, J.N. & Mattson, M.P. (1998b). Protein modification by the lipid peroxidation product 4-hydroxynonenal in spinal cord tissue of ALS patients. Ann. Neurol. In press.

Perry, T.L., Hansen, S. & Jones, K. (1987). Brain glutamate deficiency in amyotrophic lateral sclerosis. Neurology 37, 1845–1848.

Pestronk, A., Lopate, G., Kornberg, A.J., Elliott, J.L., Blume, G. & Yee, W.C. (1994). Distal lower motor neuron syndrome with high-titer serum IgM anti-GM1 antibodies: improvement following immunotherapy with monthly plasma exchange and intravenous cyclophosphamide. Neurology 44, 2027–2031.

Petkau, A., Sawatzky, A., Hillier, C.R. & Hoogstraten, J. (1974). Lead content of neuromuscular tissues in amyotrophic lateral sclerosis: Case report and other considerations. Br. J. Indust. Med. 31, 275–287.

Pettepher, C.C., Ledoux, S.P., Bohr, V.A. & Wilson, G.L. (1991). Repair of alkali-labile sites within the mitochondrial DNA of RINr 38 cells after exposure to the nitrosourea streptozotocin. J. Biol. Chem. 266, 3113–3117.

Pieper, S.J.L. & Fields, W.S. (1957). Failure of ALS to respond to intrathecal steroid and vitamin B_{12}. Arch. Neurol. 19, 522–526.

Plaitakis, A. & Caroscio, J.T. (1987). Abnormal glutamate metabolism in amyotrophic lateral sclerosis. Ann. Neurol. 22, 575–579.

Plaitakis, A., Constantakakis, E. & Smith, J. (1988). The neuro-excitotoxic amino acids glutamate and aspartate are altered in the spinal cord and brain in amyotrophic lateral sclerosis. Ann. Neurol. 24, 446–449.

Posner, J.B. (1995). Complications of Cancer. F.A. Davis, Philadelphia.

Pramatarova, A., Goto, J., Nanba, E., Nakashima, K., Takahashi, K., Takagi, A. & Kanazawa, I. (1994). A two base pair deletion in the SOD1 gene causes familial amyotrophic lateral sclerosis. Hum. Mol. Genet. 3, 2061–2062.

Pramatarova, A., Figlewicz, D.A., Krizus, A., Han, F.Y., Ceballos-Picot, I. & Nicole, A. (1995). Identification of new mutations in the Cu/Zn superoxide dismutase gene of patients with familial amyotrophic lateral sclerosis. Am. J. Hum. Genet. 56, 592–596.

Qi, X., Guy, J., Nick, H., Valentine, J. & Rao, N. (1997). Increase of manganese superoxide dismutase, but not of Cu/Zn-SOD, in experimental optic neuritis. Invest. Ophthal. Vis. Sci. 38, 1203–1212.

Radi, R., Beckman, J.S., Bush, K.M. & Freeman, B.A. (1991). Peroxynitrite-induced membrane lipid peroxidation: The cytotoxic potential of superoxide and nitric oxide. Arch. Biochem. Biophys. 288, 481–487.

Radomski, M.W., Palmer, R.M.J. & Moncada, S. (1990). Glucocorticoids inhibit the expression of an inducible, but not the constitutive, nitric oxide synthase in vascular endothelial cells. Proc. Natl. Acad. Sci. U. S. A. 87, 10043–10047.

Rainero, I., Pinessi, L. & Tsuda, T. (1994). SOD1 missense mutation in an Italian family with ALS. Neurology 44, 347–349.

Rao, M.S., Devi, M.G., Nalini, A., Shahani, N. & Raju, T.R. (1995). Neurofilament phosphorylation is increased in ventral horn neurons of neonatal rat spinal cord exposed to cerebrospinal fluid from patients with amyotrophic lateral sclerosis. Neurodegen. 4, 397–401.

Raps, S.P., Lai, J.C., Hertz, L. & Cooper, A.J. (1989). Glutathione is present in high concentrations in cultured astrocytes but not in cultured neurons. Brain Res. 493, 398–401.

Reaume, A.G., Elliott, J.L., Hoffman, E.K., Kowall, N.W., Ferrante, R.J., Siwek, D.F., Wilcox, H.M., Flood, D.G., Beal, M.F., Brown, R.H., Jr., Scott, R.W. & Snider, W.D. (1996). Motor neurons in Cu/Zn superoxide dismutase-deficient mice develop normally but exhibit enhanced cell death after axonal injury. Nat. Genet. 13, 43–47.

Reider, C.R. & Paulson, G.W. (1997). Lou Gehrig and amyotrophic lateral sclerosis. Is Vitamin E to be revisited? Arch. Neurol. 54, 527–528.

Richter, C., Park, J.-W. & Ames, B.N. (1988). Normal oxidative damage to mitochondrial and nuclear DNA is extensive. Proc. Natl. Acad. Sci. U. S. A. 85, 6465–6467.

Ripps, M.E., Huntley, G.W., Hof, P.R., Morrison, J.H. & Gordon, J.W. (1995). Transgenic mice expressing an altered murine superoxide dismutase gene provide an animal model of amyotrophic lateral sclerosis. Proc. Natl. Acad. Sci. U. S. A. 92, 689–693.

Robberecht, W., Sapp, P., Viaene, M.K., Rosen, D., McKenna-Yasek, D. & Haines, J. (1994). Cu/Zn superoxide dismutase activity in familial and sporadic amyotrophic lateral sclerosis. J Neurochem. 62, 384–387.

Robberecht, W., Aguirre, T., VandenBosch, L., Tilkin, P., Cassiman, J.J. & Matthijs, G. (1996). D90A heterozygosity in the SOD1 gene is associated with familial and apparently sporadic amyotrophic lateral sclerosis. Neurology 47, 1336–1339.

Ronnevi, L.-O., Conradi, S. & Karlsson, E. (1987). Nature and properties of cytotoxic plasma activity in amyotrophic lateral sclerosis. Muscle Nerve 10, 734–743.

Rooke, K., Figlewicz, D.A., Han, F.Y. & Rouleau, G.A. (1996). Analysis of the KSP repeat of the neurofilament heavy subunit in familial amyotrophic lateral sclerosis. Neurology 46, 789–790.

Rosen, D.R., Siddique, T., Patterson, D., Figlewicz, D.A., Sapp, P. & Hentati, A. (1993). Mutations in Cu/Zn superoxide dismutase gene are associated with familial amyotrophic lateral sclerosis. Nature 362, 59–62.

Rosen, D.R., Bowling, A.C., Patterson, D., Usdin, T.B., Sapp, P. & Mezey, E. (1994). A frequent ala 4 to val superoxide dismutase-1 mutation is associated with a rapidly progressive familial amyotrophic lateral sclerosis. Hum. Mol. Genet. 3, 981–987.

Rothstein, J.D., Kuncl, R., Chaudhry, V., Clawson, L., Cornblath, D.R. & Coyle, J.T. (1991). Excitatory amino acids in amyotrophic lateral sclerosis—an update. Ann. Neurol. 30, 224–225.

Rothstein, J.D., Martin, L.J. & Kuncl, R.W. (1992). Decreased glutamate transport by the brain and spinal cord in amyotrophic lateral sclerosis. N. Engl. J. Med. 326, 1464–1468.

Rothstein, J.D., Jin, L., Dykes-Hoberg, M. & Kuncl, R.W. (1993). Chronic inhibition of glutamate uptake produces a model of slow neurotoxicity. Proc. Natl. Acad. Sci. USA 90, 6591–6595.

Rothstein, J.D., Serratrice, G. & Munsat, T. (1995a). Excitotoxic mechanisms in the pathogenesis of amyotrophic lateral sclerosis. Adv. Neurol. 68, 7–20.

Rothstein, J.D., Van Kammen, M., Levey, A.I., Martin, L.J. & Kuncl, R.W. (1995b). Selective loss of glial glutamate transporter GLT-1 in amyotrophic lateral sclerosis. Ann. Neurol. 38, 73–84.

Rothstein, J.D., Tsai, G., Kuncl, R.W., Clawson, L., Cornblath, D.R., Drachman, D.B., Pestronk, A., Stauch, B.L. & Coyle, J.T. (1998). Abnormal excitatory amino acid metabolism in amyotrophic lateral sclerosis. Ann. Neurol. 28, 18–25.

Rouleau, G.A., Clark, A.W., Rooke, K., Pramatarova, A., Krizus, A. & Suchowersky, O. (1996). SOD1 mutation is associated with accumulation of neurofilaments in amyotrophic lateral sclerosis. Ann. Neurol. 39, 128–131.

Rowland, L.P., Defendini, R. & Sherman, W. (1982). Macroglobulinemia with peripheral neuropathy simulating motor neuron disease. Ann. Neurol. 11, 532–536.

Salazar-Grueso, E.F. & Roos, R.P. (1992). Viruses and amyotrophic lateral sclerosis. In: Handbook of Amyotrophic Lateral Sclerosis (Smith, R.A., ed.), pp. 453–477. Marcel Dekker, New York.

Sapp, P.C., Rose, D.R., Hosler, B.A., Esteban, J., McKenna-Yasek, D., O'Regan, J.P., Horvitz, H.R. & Brown, R.H.J. (1995). Identification of three novel mutations in the gene for Cu/Zn superoxide dismutase in patients with familial amyotrophic lateral sclerosis. Neuromusc. Dis. 5, 353–357.

Sasaki, S. & Iwata, M. (1996). Impairment of fast axonal transport in the proximal axons of anterior horn neurons in amyotrophic lateral sclerosis. Neurology 47, 535–540.

Schiavorne, J.R. & Hassan, H.M. (1987). Biosynthesis of superoxide dismutase in eight prokaryotes: effects of oxygen, paraquat, and iron chelator. FEMS Microbiol. Lett. 42, 33–38.

Schiffer, D., Cordera, S., Cavalla, P. & Migheli, A. (1996). Reactive astrogliosis of the spinal cord in amyotrophic lateral sclerosis. J. Neurol. Sci. 139, 27–33.

Schmidt, H.H.W., Seifert, R. & Bohme, E. (1989). Formation and release of nitric oxide from human neutrophils and HL-60 cells induced by a chemotactic peptide, platelet activating factor and leukotriene B4. FEBS Lett. 244, 357–360.

Sharrard, W.J.W. (1955). The distribution of the permanent paralysis in the lower limb in poliomyelitis: a clinical and pathological study. J. Bone Joint Surg. 37B, 540–548.

Shaw, C.E., Enayat, Z.E., Powell, J.F., Anderson, V.E.R., Radunovic, A., Al-Sarraj, S. & Leigh, P.N. (1997). Familial amyotrophic lateral sclerosis. Molecular pathology of a patient with a SOD1 mutation. Neurology 49, 1612–1616.

Shaw, C.E., Enayat, Z.E., Chioza, B.A., Al-Chalabi, A., Radunovic, A., Powell, J.F. & Leigh, P.N. (1998). Mutations in all five exons of SOD-1 may cause ALS. Ann. Neurol. 43, 390–394.

Shaw, P.J., Ince, P.G., Falkous, G. & Mantle, D. (1995). Oxidative damage to protein in sporadic motor neuron disease spinal cord. Ann. Neurol. 38, 691–695.

Shimoda-Matsubayashi, S., Matsumine, H., Kobayashi, T., Nakagawa-Hattori, Y., Shimizu, Y. & Mizuno, Y. (1996). Structural dimorphism in the mitochondrial targeting sequence in the human manganese superoxide dismutase gene—a predictive evidence for conformational change to influence mitochondrial transport and a study of allelic association in Parkinson's disease. Biochem. Biophys. Res. Commun. 226, 561–565.

Shy, M.E., Rowland, L.P. & Smith, T. (1986). Motor neuron disease and plasma cell dyscrasia. Neurology 36, 1429–1436.

Siddique, T. (1996). ALS: molecular clues to the jigsaw puzzle of neuronal degeneration. Cold Spring Harbor Symp. Quant. Biol. 61, 699–708.

Silvka, A., Mytilineou, C. & Cohen, G. (1987). Histochemical evaluation of glutathione in brain. Brain Res. 409, 275–284.

Simonetti, S., Chen, X. & DiMauro, S. (1992). Accumulation of deletions in human mitochondrial DNA during normal aging: analysis by quantitative PCR. Biochim. Biophys. Acta 1180, 113–122.

Sinha, B.K. & Mimnaugh, E.G. (1990). Free radicals and anticancer drug resistance: oxygen free radicals in the mechanisms of drug cytotoxicity and resistance by certain tumors. Free Radical. Biol. Med. 8, 567–581.

Själänder, A., Beckman, G., Deng, H.X., Iqbal, Z., Tainer, J.A. & Siddique, T. (1995). The D90A mutations results in a polymorphism of Cu,Zn superoxide dismutase that is prevalent in northern Sweden and Finland. Hum. Mol. Genet. 4, 1105–1108.

Smith, R.A. & Norris, F.H. (1988). Treatment of amyotrophic lateral sclerosis with interferon. In: Interferon Treatment of Neurological Disorders (Smith, R.A., ed.), pp. 265–275. Marcel Dekker, New York.

Smith, R.G., Hamilton, S., Hofmann, F., Schneider, T., Nastainczyk, W. & Birnbaumer, L. (1992). Serum antibodies to L-type calcium channels in patients with amyotrophic lateral sclerosis. N. Engl. J. Med. 327, 1721–1728.

Smith, R.G., Engelhardt, J.I., Tajti, J. & Appel, S.H. (1993). Experimental immune-mediated motor neuron diseases—models for human ALS. Brain Res. Bull. 30, 373–380.

Smith, R.G., Henry, Y.K., Mattson, M.P. & Appel, S.H. (1998). Presence of 4-hydroxynonenal in CSF of patients with sporadic ALS. Ann. Neurol. In press.

Snyder, S.H. & Bredt, D.S. (1991). Nitric oxide as a neuronal messenger. Trends Pharmacol. Sci. 12, 125–128.

Spitz, D.R., Adams, D.T., Sherman, C.M. & Roberts, R.J. (1992). Mechanisms of cellular resistance to hydrogen peroxide, hyperoxia, and 4-hydroxy-2-nonenal toxicity: the significance of increased catalase activity in H_2O_2-resistant fibroblasts. Arch. Biochem. Biophys. 292, 221–227.

St. Clair, D.K., Oberley, T.D. & Ho, Y. (1991). Overproduction of human Mn-superoxide dismutase modulates paraquat-mediated toxicity in mammalian cells. FEBS Lett. 293, 199–202.

Stevens, J.B. & Autor, A.P. (1977a). Induction of superoxide dismutase by oxygen in neonatal rat lung. J. Biol. Chem. 252, 3509–3514.

Stevens, J.B. & Autor, A.P. (1977b). Oxygen-induced synthesis of superoxide dismutase and catalase in pulmonary macrophages of neonatal rats. Lab. Invest. 37, 470–478.

Strong, M.J. & Garruto, R.M. (1991). Chronic aluminum-induced motor neuron degeneration—clinical, neuropathological and molecular biological aspects. Can. J. Neurol. Sci. 18, 428–431.

Stuehr, D.J. & Nathan, C.F. (1989). Nitric oxide: a macrophage product responsible for cytostasis and respiratory inhibition in tumor target cells. J. Expt. Med. 169, 1543–1555.

Suthers, G., Laing, N., Wilton, S., Dorosz, S. & Waddy, H. (1994). "Sporadic" motoneuron disease due to familial SOD1 mutation with low penetrance. Lancet 344, 1773.

Swash, M., Scholtz, C.L. & Vowles, G. (1988). Selective and asymmetric vulnerability of corticospinal and spinocerebellar tract in motor neuron disease. J. Neurol. Neurosurg. Psychiatry 51, 785–789.

Swash, M. & Schwartz, M.S. (1992). What do we really know about amyotrophic lateral sclerosis? J. Neurol. Sci. 113, 4–16.

Swingler, R.J., Fraser, H. & Warlow, C.P. (1992). Motor neuron disease and polio in Scotland. J. Neurol. Neurosurg. Psychiatry 55, 1116–1120.

Takahashi, H., Makifuchi, T., Nakano, R., Sato, S., Inuzuka, T. & Sakimura, K. (1994). Familial amyotrophic lateral sclerosis with a mutation in the Cu/Zn superoxide dismutase gene. Acta Neuropathol. 88, 185–188.

Tan, E., Lynn, J. & Amato, A.A. (1994). Immunosuppressive treatment in motor neuron syndromes. Attempts to distinguish a treatable disorder. Arch. Neurol. 51, 194–200.

Tandan, R. & Bradley, W.G. (1985a). Amyotrophic lateral sclerosis: Part 2. Etiopathogenesis. Ann. Neurol. 18, 419–431.

Tandan, R. & Bradley, W.G. (1985b). Amyotrophic lateral sclerosis: Part 1. Clinical features, pathology, and ethical issues in management. Ann. Neurol. 18, 271–280.

Tandan, R., Bromberg, M.B., Forshew, D., Fries, T.J., Badger, G.J., Carpenter, J., Krusinski, P.B., Betts, E.F., Arciero, K. & Nau, K. (1996). A controlled trial of amino acid therapy in amyotrophic lateral sclerosis: 1. Clinical, functional, and maximum isometric torque data. Neurology 47, 1220–1226.

Testa, D., Caraceni, T. & Fetoni, V. (1989). Branched-chain amino acids in the treatment of amyotrophic lateral sclerosis. J. Neurol. 236, 445–447.

Tomblyn, M., Kasarskis, E.J., Xu, Y. & St. Clair, D.K. (1998). Distribution of MnSOD polymorphisms in sporadic ALS patients. J. Molec. Neurosci. 10, 65–66.

Troncoso, J.C., Price, D.L., & Griffin, J.W. (1982). Neurofibrillary axonal pathology in aluminum intoxication. Ann. Neurol. 12, 278–283.

Tu, P., Raju, P., Robinson, K.A., Gurney, M.E., Trojanowski, J.Q. & Lee, V.M.Y. (1996). Transgenic mice carrying a human mutant superoxide dismutase transgene develop neuronal cytoskeletal pathology resembling human amyotrophic lateral sclerosis lesions. Proc. Natl. Acad. Sci. U. S. A. 93, 3155–3160.

Tu, P., Gurney, M.E., Julien, J.P., Lee, V.M. & Trojanowski, J.Q. (1997). Oxidative stress, mutant SOD1, and neurofilament pathology in transgenic mouse models of human motor neuron disease. Lab. Invest. 76, 441–456.

van Loon, A.P.G.M., Pesold-Hurt, B. & Schatz, G. (1986). A yeast mutant lacking mitochondrial manganese-superoxide dismutase is hypersensitive to oxygen. Proc. Natl. Acad. Sci. U. S. A. 83, 3820–3824.

Vechio, J.D., Bruijn, L.I., Xu, Z.S., Brown, R.H. & Cleveland, D.W. (1996). Sequence variants in human neurofilament proteins: absence of linkage to familial amyotrophic lateral sclerosis. Ann. Neurol. 40, 603–610.

Wakai, M., Mokuno, K., Hashizume, Y. & Kato, K. (1994). An immunohistochemical study of the neuronal expression of manganese superoxide dismutase in sporadic amyotrophic lateral sclerosis. Acta Neuropathol. 88, 151–158.

Wan, X.S. & St .Clair, D.K. (1993a). Differential cytotoxicity of buthionine sulfoximine to normal and transformed human lung fibroblast cells. Cancer Chemother. Pharmacol. 33, 210–214.

Wan, X.S. & St. Clair, D.K. (1993b). Thiol-modulating agents increase manganese superoxide dismutase activity in human lung fibroblasts. Arch. Biochem. Biophys. 304, 89–93.

Watanabe, Y., Kono, Y., Nanba, E., Nakashima, K. & Takahashi, K. (1996). Abnormal Cu,Zn-superoxide dismutase (SOD1) in familial amyotrophic lateral sclerosis with two base-pair deletion in SOD1 gene. In: Free Radicals in Brain Physiology and Disorders (Packer, L., Hiramatsu, M. & Yoshikawa, T., Eds.), pp. 371–376. Academic Press, San Diego.

Weisiger, R.A. & Fridovich, I. (1973). Superoxide dismutase. Organelle specificity. J. Biol. Chem. 248, 4793–4796.

Weiss, J.F. & Kumar, K.S. (1988). Antioxidant Mechanisms in Radiation Unjury and Radioprotection. In: Cellular Antioxidant Defense Mechanism. (Chow, C.K., Ed.), pp. 163–189. CRC Press, Boca Raton, FL.

Westarp, M.E., Westphal, K.P., Kolde, G., Wollinsky, K.H., Westarp, M.P. & Dickob, M. (1992). Dermal, serological and CSF shanges in amyotrophic lateral sclerosis with and without intrathecal interferon-beta treatment. Int. J Clin. Pharm. Ther. Toxicol. 30, 81–93.

Westarp, M.E., Bartmann, P., Rossler, J., Geiger, E., Westphal, K.P. & Schreiber, H. (1993). Antiretroviral therapy in sporadic adult amyotrophic lateral sclerosis. Neuroreport 4, 819–822.

Wiedau-Pazos, M., Goto, J.J., Rabizadeh, S., Gralla, E.B., Roe, J.A. & Lee, M.K. (1996). Altered reactivity of superoxide dismutase in familial amyotrophic lateral sclerosis. Science 271, 515–518.

Williams, D.B. & Windebank, A.J. (1991). Motor neuron disease (amyotrophic lateral sclerosis). Mayo Clin. Proc. 66, 54–82.

Williamson, T.L., Marszalek, J.R., Vechio, J.D., Bruijn, L.I., Lee, M.K., Xu, Z., Brown, R.H. & Cleveland, D.W. (1996). Neurofilaments, radial growth of axons, and mechanisms of motor neuron disease. Cold Spring Harbor Symp. Quant. Biol. 61, 709–723.

Wolfgram, F. & Myers, L. (1973). Amyotrophic lateral sclerosis: effect of serum on anterior horn cells in tissue culture. Science 179, 579–580.

Wong, G.H.W., Elwell, J.H., Oberley, L.W. & Goeddel, D.V. (1989). Manganous superoxide dismutase is essential for cellular resistance to cytotoxicity of tumor necrosis factor. Cell 58, 923–931.

Wong, P.C., Pardo, C.A., Borchelt, D.R., Lee, M.K., Copeland, N.G. & Jenkins, N.A. (1995). An adverse property of a familial ALS-linked SOD1 mutation causes motor neuron disease characterized by vacuolar degeneration of mitochondria. Neuron 14, 1105–1116.

Xu, Z., Cork, L.C. & Griffin, J.W. (1993). Increased expression of neurofilament subunit NF-L produces morphological alterations that resemble the pathology of human motor neuron disease. Cell 73, 23–33.

Yen, T., King, K. & Lee, H. (1994). Age-dependent increase of mitochondrial DNA deletions together with lipid peroxides and superoxide dismutase in human liver mitochondria. Free Rad. Biol. Med. 16, 207–214.

Yoneda, T., Inagaki, S., Hayashi, Y., Nomura, T. & Takagi, H. (1992). Differential regulation of manganese and copper/zinc superoxide dismutases by the facial nerve transection. Brain Res. 582, 342–345.

Chapter 5

Human Prion Diseases

BERNARDINO GHETTI and PIERLUIGI GAMBETTI

Advances in Cell Aging and Gerontology
Volume 3, pages 135–187

ISBN: 0-7623-0405-7

INTRODUCTION

Prion diseases are fatal neurodegenerative disorders that affect humans and animals. They represent a heterogeneous group that is unique among neurological diseases as it includes genetic, transmitted, and sporadic forms, and displays a wide spectrum of clinical phenotypes and histopathological patterns. It is now evident that the prion protein (PrP) is central to the pathogenesis of these disorders whether inherited, infectious, or idiopathic (referred to as sporadic in the literature). In humans, prion diseases include Creutzfeldt-Jakob disease (CJD), Gerstmann-Sträussler-Scheinker syndrome (GSS), kuru, fatal familial insomnia (FFI), and prion protein cerebral amyloid angiopathy (PrP-CAA). In animals, the most notable forms are scrapie in sheep and bovine spongiform encephalopathy in cattle. The sporadic, the infectious, and many of the genetically determined forms of human prion diseases as well as the animal diseases can be transmitted across and within animal species. Pathogenetically they all appear to share a mechanism previously unrecognized, which differs from the established mechanisms of infectivity. The central principle involves changes in conformation of the normal PrP that generate a pathogenic PrP conformer, which is capable of converting other normal PrP molecules into the abnormal form. This conversion results in degenerative cascade, and it has the potential to transmit the disease.

The first degenerative disorder reported in the scientific literature and later recognized to belong to the prion disease group is that described in the "H" family in 1912, 1928, and 1936. The authors detailed a hereditary, chronic (slowly progressive) disorder observed in multiple generations and characterized clinically by a cerebellar syndrome, pyramidal signs, and dementia. Neuropathologically, there was cerebral and cerebellar atrophy and deposition of amorphous material later recognized as amyloid (Seitelberger, 1981). The disease was subsequently named Gerstmann-Sträussler syndrome, and the relationship to prion diseases was clarified in 1981 (Masters et al., 1981).

CJD was described as a rapidly progressive sporadic disease. The original descriptions of Creutzfeldt (1920, 1921) and Jakob (1921a, 1921b, 1921c; 1923) have until now presented difficulties of interpretation and reconciliation with our present understanding of the disease. According to Richardson (1977), Creutzfeldt's case can most likely be excluded from the classification as a spongiform encephalopathy on the basis of his own clinical and pathological description. From Jakob's cases, only two, it would appear, had the clinical and pathological features accepted today as diagnostic of CJD (Masters et al., 1981; DeArmond and Prusiner, 1997). Clinically, CJD is characterized by a prodromal syndrome including fatigue, sleep disturbance, and general malaise; and by deficits in higher cortical function, developing into a state of profound dementia. Visual impairment or cerebellar dysfunction may dominate the picture. Extrapyramidal signs, rigidity, and other neurological signs may also be evident. Myoclonus appears in a large proportion of patients at varied times throughout the illness. Pathologically, nerve cell loss, vacuolation of neuropil, and severe astrocytic gliosis are the hallmarks of the disease.

Kuru afflicted the Fore Tribe of New Guinea and was acquired by ingestion during cannibalistic rituals (Gajdusek and Zigas, 1957). Kuru has a duration of approximately 1 year and is clinically characterized by progressive ataxia, tremor, and dysarthria often associated with emotional lability. The pathological hallmark is the Kuru plaques—punched out PrP amyloid deposits—are present in about half of the cases, and occur more often in the granular cell layer, but are also in the molecular layer of the cerebellum, cerebral cortex, and basal ganglia. Other types of plaques such as diffuse and "florid" are only occasionally present. The other consistent changes are neuronal loss, gliosis, and spongiosis (Parchi et al., 1998b). The lesions are present throughout the nervous system, but they are frequently more severe in the basal ganglia, thalamus, and brain stem. The similarity between the histopathologies of kuru and scrapie, a disease of sheep known to be transmissible, was noted by William Hadlow, a veterinary pathologist, who recommended experiments to test the transmissibility of kuru (Hadlow, 1959). Gajdusek, Gibbs, and their collaborators succeeded in transmitting a spongiform encephalopathy to primates by inoculating brain tissue of kuru, CJD, and GSS patients (Gajdusek et al., 1966; Gibbs et al., 1968; Masters et al., 1981).

FFI is a hereditary prion disease characterized by a unique phenotype linked to a mutation at codon 178 of the prion protein gene cosegregating with the methionine polymorphism at codon 129 of the mutated allele. It is characterized by disturbances of the wake-sleep cycle, dysautonomia, and somatomotor manifestations (myoclonus, ataxia, dysarthria, spasticity.)

PrP-CAA is a rare variant characterized by dementia and vascular PrP amyloidosis. It is caused by a stop codon mutation at *PRNP* codon 145 (Ghetti et al., 1996b).

In 1981, macromolecular structures named scrapie-associated fibrils (SAFs) were identified by electron microscopy in scrapie-infected brain tissue, revealing the first molecular link to the disease (Merz et al., 1981). However, the contemporary era of prion diseases was initiated by the work of Prusiner and his colleagues who, in 1982, discovered that a single protease-resistant protein with a molecular weight of 27 to 30 kD was consistently associated with purified fractions of brain extracts enriched for scrapie infectivity (Bolton et al., 1982). Although Prusiner tentatively identified the abnormal protein with the infectious agent and designated it as "prion" (Prusiner, 1982), it was discovered that the prion was an abnormal, protease-resistant isoform of a protein that is normally expressed in the brain (Chesebro et al., 1985; Oesch et al., 1985). The normal protein, which was named prion protein, is encoded by the *PRNP* gene.

GENE STRUCTURE OF *PRNP*

PRNP is located on the short arm of chromosome 20 (Sparkes et al., 1986). The 15-kb gene is comprised of two short, noncoding exons and a 2.4-kb exon 3, which contains the coding region (Kretzschmar et al., 1986). The coding region is 759-bp long and is followed by a 1.64-kb nontranslated 3′ region (Kretzschmar et al., 1986; Puckett et al., 1991). The promoter is a GC-rich housekeeping gene promoter, which lacks a TATA box. The entire gene has been sequenced (Gen Bank accession number U29185).

THE NORMAL PRION PROTEIN

The normal prion protein (PrP^C) is a membrane glycoprotein. After being synthesized in the endoplasmic reticulum (ER), PrP^C undergoes a series of co-translational and post-translational modifications that include the attachment of a glycosylphosphotidyl inositol (GPI) anchor (Stahl et al., 1987) and the nonobligatory addition of one or two N-linked oligosaccharide chains (Caughey et al., 1989; Endo et al., 1989). The oligosaccharide chains are then modified in the ER and Golgi to become either high-mannose complex or hybrid glycans (Caughey et al., 1989; Endo et al., 1989). The differential glycosylation leads to the formation of three PrP^C glycoforms with distinct electrophoretic mobility: (1) the fully glycosylated form that is thought to contain two glycan chains, one complex and one hybrid; (2) an intermediate form with only one complex chain; and (3) a unglycosylated form (Caughey et al., 1989; Chen et al., 1995; Petersen et al., 1996). Ultimately, most of the PrP^C is anchored to the cell surface by the GPI anchor. Cycling of PrP^C between the plasma membrane and the endosomal compartment has been observed in culture cells (Shyng et al., 1993). During this process, PrP^C undergoes endoproteolytic cleavage and N-terminally truncated forms are produced. In human brain, a cleavage at amino acid residue 111 or 112 generates the major product of PrP^C proteolytic processing (Chen et al., 1995).

The physiological function of PrP^C in the central nervous system has not been determined. Available evidence indicates that PrP^C (1) is transported by fast axonal transport (Borchelt et al., 1994), (2) plays a role in GABA-receptor-mediated synaptic inhibition and long-term potentiation (Collinge et al., 1994; Whittington et al., 1995), and (3) is involved in circadian rhythms and sleep regulation (Tobler et al., 1996).

CONVERSION OF PrP^C INTO PROTEASE-RESISTANT PrP (PrP^{res}): THE PATHOGENESIS OF PRION DISEASES

There is unanimous consent that PrP is crucial for the pathogenesis of infectious and sporadic prion diseases; however, the exact nature of the infectious agent causing these disorders is still the subject of significant controversy. In addition to the virus theory that follows a more conventional way of thinking, the protein-only hypothesis, or prion hypothesis, supports the theory that a protein may act as the transmissible agent (Griffith, 1967; Pattison and Jones, 1967). Evidence in support of this hypothesis reveals that the central event in the pathogenesis of prion diseases is a change in conformation resulting in the conversion of PrP^C into a conformer (PrP^{res}) that has the identical amino acid sequence and post-translational modifications of PrP^C but differs from PrP^C in the secondary structure, resistance to digestion with proteases, and pathogenicity (Hope et al., 1986; Turk et al., 1988; Caughey et al., 1991a; Gasset et al., 1993; Pan et al., 1993; Safar et al., 1993).

According to the prion hypothesis (Prusiner et al., 1991), the change in conformation of PrP^C into PrP^{res} would be (1) induced by the exposure to exogenous PrPres in the prion diseases transmitted by infection, (2) an almost invariable consequence of the instability of PrP in the presence of a mutation in the familial or inherited form, and (3) the result of a spontaneous, random event in the sporadic form.

Two primary models have been proposed to account for the mechanism of PrP^{res}-induced PrP^c conversion (Prusiner et al., 1991; Jarrett and Lansbury, 1993). According to one model (Prusiner et al., 1991), a PrP^C and PrP^{res} molecules would combine to form a heterodimer, which would be followed by the conversion of the PrP^C to the PrP^{res} form. Dissociation of the complex would release previously formed and newly formed PrP^{res} molecules, which would continue the conversion process in an autocatalytic chain reaction. The other model asserts that the conversion takes place following the interaction of PrP^C with an ordered aggregate of PrP^{res} called "nucleus" or "seed" (Jarrett and Lansbury, 1993).

PrP^C-PrP^{res} dimerization and PrP^C conversion into PrP^{res} is believed to occur at the cell membrane or in the endosome shortly after internalization, or in both, whereas storage of the newly formed PrP^{res} takes place in lysosomes (Borchelt et al., 1992; Caughey et al., 1991b; Taraboulos et al., 1992; McKinley et al., 1991).

CHARACTERISTICS OF THE PROTEASE-RESISTANT PRION PROTEIN (PrP^{res})

When PrP isolated from the brain of prion disease–affected subjects is treated with a protease, such as proteinase K (PK), the N-terminal portion up to residue ~90 is digested while the remaining C-terminal portion remains intact (McKinley et al., 1983). Thus, the PK-resistant fragment migrates faster on gel than the full-length PrP. This mechanism applies to each of the three PrP glycoforms, which include (1) a highly glycosylated form that contains two complex glycan chains, (2) an intermediate form with only one complex chain, and (3) a form that is unglycosylated (Prusiner 1982; Chen et al., 1995; Petersen et al., 1996). The size of the PrP^{res} fragments and the ratio of the PrP^{res} forms that are differently glycosylated, also referred to as glycoforms, are other features that aid in characterizing the various inherited prion diseases. The first observation, that the PrP^{res} isolated from different human prion diseases, could be determined on the basis of the size of the fragment generated by the PK digestion and the ratio of the glycoforms, was made in 1992 (Medori et al., 1992). Since then it has been demonstrated that following deglycosylation to eliminate the heterogeneity generated by the presence of the glycoforms, the major PrP^{res} core fragments associated with human prion diseases exhibit only two principal sizes, which migrate on gel at 20 to 21 kDa and 18 to 19 kDa, respectively. These two PrP^{res} fragments are respectively identified as PrP^{res} types 1 and 2 (Parchi et al., 1996c). PrP^{res} types 1 and 2 may be further subdivided according to the pattern determined by the ratio of the three glycoforms. For example, the PrP^{res} associated with the insertion mutations is characterized by the predominance of the intermediate glycoform, whereas in the CJD linked to the E200K mutation and in FFI, the glycoform ratio is consistently dominated by the highly glycosylated form while the unglycosylated form is markedly underrepresented (Monari et al., 1994; Parchi et al., 1996a). Thus, it seems appropriate to characterize the inherited prion diseases not only according to the haplotype but also taking into consideration the type of PrP^{res} as determined by the size of the PrP^{res} fragment and the ratio of the three glycoforms.

The precise role of the PrP^{res} type in the phenotypic expression of the inherited prion diseases remains to be determined. The different size of the PrP^{res} fragments following treatment with exogenous PK has been shown to result from cleavage at different sites of the full-length PrP^{res} (Monari et al., 1994). This, in turn, is probably due to different conformations of PrP^{res}, which expose distinct cleavage sites (Monari et al., 1994). The different ratio of the glycoforms may result from various causes such as different topographic origin of the PrP^{res} within the brain in different prion diseases, preferential conversion of PrP^C to PrP^{res} of one glycoform over the others, or underrepresentation of one of the glycoforms due to the altered metabolism caused by the mutation; the underrepresented form becoming thus less available for conversion to PrP^{res} (also see later discussion). Transmission of inherited and sporadic prion diseases to receptive animals has underlined the importance of the PrP^{res} typing (Telling et al., 1996). Transgenic mice expressing a chimeric wild-type human-mouse PrP^C develop a prion disease after the intracerebral inoculation with human brain homogenates carrying PrP^{res} of either type 1 or 2 (Telling et al., 1996). The size of the PrP^{res} fragment expressed by the inoculated mice corresponds exactly to the PrP^{res} type present in the inoculum. Since the recipient mice did not carry any of the donor's *PRNP* mutations and were isogenic, this remarkable finding indicates that the distinct conformations associated with PrP^{res} types 1 and 2 can be reproduced independently of the genetic information. However, the ratio of the glycoforms present in the recipient's PrP^{res} was different from that of the donor's PrP^{res}, indicating that conformation and glycoform ratio are independently determined (Telling et al., 1996). The information necessary for the specification of the cleavage site determining the size of the PrP^{res} fragment is likely to be contained in the conformation of the donor's PrP^{res}, whereas the glycosylation is most likely controlled by the host according to the cell population involved and glycoform convertibility as mentioned earlier.

HUMAN PRION DISEASES

The cloning of the *PRNP* and the detection of the PrP^{res} in the brain of affected subjects have provided two powerful molecular tests for the diagnosis and characterization of prion diseases. This, in turn, has led not only to a clearer understanding of known prion diseases, but also to the identification of new variants, widening the array of these disorders. Regardless of their etiology, prion diseases, whether inherited, acquired by infection, or idiopathic, express three distinct, major phenotypes: CJD, FFI, and GSS. A minority of cases of the inherited form, generally associated with insertional mutations and a stop codon mutation, cannot be accommodated in any of these phenotypes and form an heterogeneous group. The classification of the human prion diseases, according to form and phenotype, is given in Table 1. The disease phenotype is determined by clinical, pathological, and protein molecular features. Relevant clinical features are the duration of symptoms; the time of presentation of certain clinical signs, such as cognitive impairment or ataxia; and the presence or absence of pseudoperiodic sharp waves (PSW) on the electroencephalogram (EEG). The pathological phenotypes of prion diseases are defined by the individual relative amount and topography of five basic lesions: (1) spongiform degeneration, (2) astrogliosis, (3) loss of neurons, (4) deposits of PrP^{res}, and (5) neurofibrillary tangles. The PrP deposits manifest two distinct patterns: they can be associated with structural lesions that are visible also following routine stains, such as the punched out or kuru plaques, the multicentric plaques of GSS, and the "florid" plaques of the new variant CJD; or they can be free of histological lesions and are detectable only following immunostaining.

Table 1. Genotype and Phenotype of Inherited Prion Diseases

Form	*Clinical Syndromes*	*Etiology*
Inherited	GSS Familial CJD FFI Atypical dementias	Germline PRNP mutation
Idiopathic	CJD Atypical CJD Sporadic form of FFI	(?)
Acquired	Kuru Iatrogenic CJD New variant CJD	Cannibalism Accidental inoculatin with infectious human tissue Exposure to BSE(?)

Notes: BSE, bovine spongiform encephalopathy; CJD, Creutzfeldt-Jakob disease: FFI, fatal familial insomnia; GSS, Gerstmann-Sträussler-Scheinker syndrome.

INHERITED PRION DISEASES

General Features

At present, there are 23 known *PRNP* mutations and three polymorphisms. The mutations are of three types: (1) *point mutations,* which include 14 mutations distributed predominantly in the carboxyl two-thirds of the *PRNP* coding regions; (2) *insertional mutations,* made of 24 additional base pair (bp) repeats located between codon 51 and 91; and (3) *stop codon* mutations, of which one is known to result in the premature termination of synthesis and to yield a truncated PrP. The three polymorphisms include the Met/Val at codon 129, the Glu/Lys at codon 219, and the deletion of one 24 bp repeat. The polymorphism at codon 129 has a twofold effect:

1. On the mutant allele, the polymorphic codons 129 and 219 affect basic aspects of the disease phenotype. Therefore, it is more appropriate to identify each *PRNP* genotype associated with inherited prion diseases not only by the mutation, but also by the codon 129 (or other polymorphic codons) present on the mutant allele (i.e., with the haplotype). Currently, 25 disease-associated *PRNP* haplotypes are known that result in CJD, FFI, GSS, mixed CJD and GSS, PrP-CAA, and undefined phenotypes, the latter two phenotypes being observed with the insertion mutations.
2. On the normal allele, the 129 polymorphism may influence some phenotypic features such as age at onset and duration of the disease. Genotypes and phenotypes of the *PRNP* pathogenic mutations known at this time are reported in Table 2.

Gerstmann-Sträussler-Scheinker Disease Phenotype

Evolution of the Concept

GSS is a chronic hereditary autosomal dominant disease characterized by cerebellar and pyramidal signs and by cognitive decline, which may evolve into severe dementia (Ghetti et al., 1995). Amyotrophy and parkinsonian signs may be present early or late in the course of the disease. Pathologically, the phenotype is characterized by the presence of PrP-amyloid plaques in the cerebral and cerebellar cortex and by pyramidal tract degeneration. Specific features such as neurofibrillary tangles (NFTs), spongiform changes, Lewy bodies, and a combination of these may be differentiating the various GSS haplotypes (variants) (Ghetti et al., 1996a; Mirra et al., 1997). The *PRNP* mutations associated with GSS are shown in Table 2. In view of the consistent presence of PrP amyloid deposits in GSS and other hereditary prion disease variants, the term hereditary prion protein amyloidosis has also been introduced to include forms with extensive PrP amyloid deposition in disorders that have a phenotype different from GSS. This nomenclature would allow us to include in this group the forms designated as "inherited prion diseases with variable phenotypes" and the variant characterized clinically by severe dementia and

Table 2. Genotype and Phenotype of Inherited Prion Diseases

Genotype	*Onset (yr)*	*Duration*	*Clinical and Pathological Features*
Gerstmann-Sträussler-Scheinker Disease (GSS) Phenotype			
P102L-129M	30–62	1–10 yr	Slowly progressive cerebellar syndrome with late dementia, extrapyramidal and pyramidal signs. Some cases (shorter duration) overlap with CJD. PrP amyloid deposits in the cerebellum and, to a lesser extent, in the cerebrum. Variable degree of spongiosis, neuronal loss and astrogliosis. No NFTs.
P102L-129M-219K	31–34	4 yr	Differ from the above P102L form for the less prominent cerebellar signs. Few PrP plaques in the cerebral and cerebellar cortices. No spongiosis.
P102L-129V	33	12 yr	Seizures, numbness, gait difficulties, dysarthria, long tract signs. No dementia. Widespread PrP plaques with no spongiosis.
P105L-129V	40–50	6–12 yr	Spastic paraparesis progressing to quadriparesis; late dementia; no myoclonus; and only mild cerebellar signs. PrP amyloid deposits, neuronal loss, and gliosis in the cerebral cortex and, to a lesser extent, in the striatum and thalamus. No spongiform changes and NFTs.
A117V-129V	20–64	1–11 yr	Dementia, parkinsonism, pyramidal signs. Occasional cerebellar signs. Widespread PrP amyloid deposits in the cerebrum and, more rarely, in the cerebellum associated with variable degree of spongiform changes, neuronal loss, and astrogliosis. No NFTs.
F198S-129V	34–71	3–12 yr	Like P102L GSS subtype, but with a more chronic course (no overlap with CJD). Like P102L GSS subtype but with more extensive PrP amyloid deposits, NFTs in the cerebral cortex and subcortical nuclei and inconspicuous spongiosis.
D202N-129V	73	6 yr	Dementia, cerebellar symptoms, and cognitive deficits, but no myoclonus. No spongiform degeneration and abundant PrP-amyloid deposition in the cerebrum and cerebellum. NFTs in the cerebral cortex.
Q212P-129M	60	8 yr	Cerebellar symptoms but no dementia. No spongiform degeneration, moderate amount of PrP-immunopositive deposits, and mild amyloid deposition in the cerebrum and cerebellum. Lewy bodies in the neocortex and substantia nigra.
Q217R-129V	62–66	5–6 yr	Slowly progressive dementia; cerebellar and extrapyramidal signs. Like F1985 GSS subtype but with the most severe lesions in the cerebral cortex, thalamus, and amygdala.

(*continued*)

Table 2. Continued

Genotype	*Onset (yr)*	*Duration*	*Clinical and Pathological Features*
Cerebral Amyloid Angiopathy Phenotype			
Y145STOP-129M	38	21 yr	Slowly progressive dementia. PrP amyloid deposits in the cerebral and cerebellar cortices associated with NFTs in the neocortex, hippocampus, and subcortical nuclei. PrP amyloid angiopathy. No spongiosis.
Creutzfeldt-Jakob Disease (CJD) Phenotype			
D178N-129V	26–56	9–51 mo	Dementia, ataxia, myoclonus, extrapyramidal and pyramidal signs. Spongiosis, neuronal loss and astrogliosis in the cerebral cortex (most severe), striatum and thalamus (least severe), while the cerebellum is spared.
V180I-129M	66–85	1–2 yr	Similar to typical sCJD but with a slower progression. Pathological features, like typical sCJD.
T183A-129M	45	4 yr	Personality changes followed by dementia and parkinsonism. Atrophy with spongiform degeneration in the cerebral cortex and, to a lesser extent, in the basal ganglia.
E200K-129M	35–66	2–41 mo	Similar to typical sCJD. Atypical signs such as supranuclear palsy and peripheral neuropathy in some cases. Pathological features, like typical sCJD.
H208R-129M	60	7 mo	Like typical sCJD.
V210I-129M	49–70	3–5 mo	Like typical sCJD.
M232R-129M	55–70	4–24 mo	Like typical sCJD.
Fatal Familial Insomnia (FFI) Phenotype			
D178N-129M	20–71	6–33 mo	Reduction of total sleep time, enacted dreams, sympathetic hyperactivity, myoclonus, ataxia; late dementia, pyramidal and extrapyramidal signs in the cases with a relatively long duration (> 1 yr). Preferential thalamic and olivary atrophy. Spongiform changes in the cerebral cortex in the subjects with a duration of symptoms longer than 1 yr.
Variable Phenotype: Insertional Mutations			
Ins 24 bp-129M	73	4 mo	Like typical sCJD. (Pathological features NA.)
Ins 48 bp-129M	58	3 mo	Like typical sCJD.
Ins 96 bp-129M	56	2 mo	Like typical sCJD.
Ins 96 bp-129V	82	4 mo	Examined at terminal stage showed akinetic mutism, diffuse myoclonus, and pyramidal signs. (Pathological features NA.)
Ins 120 bp-129M	31, 45	5, 15 yr	Progressive dementia, myoclonus, cerebellar and extrapyramidal signs Spongiosis, gliosis and neuronal loss (no information on topography, severity and presence of PrP deposits). CJD phenotype.

(continued)

Table 2. Continued

Genotype	*Onset (yr)*	*Duration*	*Clinical and Pathological Features*
Ins 144 bp-129M	22–53	3 mo–18 yr	Similar to 120-bp insertion subtype. Most cases show a CJD phenotype with spongiosis, gliosis, and neuronal loss. One case had kurulike plaques in the cerebellum. Some cases show only mild aspecific gliosis and neuronal loss. PrP patches in the cerebellum.
Ins 168 bp-129M	23–35	7–13 yr	Similar to 120-bp insertion subtype. Mild gliosis and neuronal loss, and no spongiosis in one case, CJD phenotype in another.
Ins 192 bp-129V	21–54	5 mo–6 yr	Similar to 120-bp insertion subtype. Spongiosis, gliosis, and neuronal loss. PrP multicentric amyloid plaques with widespread distribution. GSS-like phenotype.
Ins 216 bp-129M	32–55	2.5–4+ yr	Similar to 120-bp insertion subtype PrP amyloid plaques in the cerebellum, cerebral cortex, and striatum. No obvious neuronal loss, gliosis, or spongiosis. GSS-like phenotype.

Notes: NA, not available; NFT, neurofibrillary tangles; PrP, prion protein; sCJD, sporadic Creutzfeldt-Jakob disease.

pathologically by the presence of a prion protein cerebral amyloid angiopathy and NFT.

In GSS the symptomatology occurs in the third to seventh decades; the mean duration of illness is 5 years. The incidence of GSS is believed to be less than 2 per 100 million; however, it is probably underestimated since GSS may present as a syndrome mimicking spinocerebellar degeneration, olivopontocerebellar atrophy, spastic paraparesis, parkinsonism, and dementia.

Although the neuropathological diagnosis of GSS is based on the presence of PrP-amyloid deposits, their distribution and extent vary substantially between families. Amyloid is accompanied by glial proliferation and by loss of neuronal processes and perikarya, leading to variable degrees of atrophy of the affected regions. The clinical phenotypes are associated with mutations of *PRNP*, allelic polymorphism, PrP^{res} characteristics, and possibly with environmental and tissue-specific factors.

GSS with the P102L mutation coupled with the 129M codon (GSSP102L, 129M)

Epidemiology and genetics. This mutation is the most common in GSS. The P102L mutation has been found in Japan (Doh-ura et al., 1989), Britain (Hsiao et al., 1989), Austria (Hainfellner et al., 1995), Germany (Goldgaber et al., 1989), Italy (Kretzschmar et al., 1992), Israel (Ashkenazi Jew) (Goldhammer et al., 1993), France (Laplanche et al., 1994), Mexico (E. Alonso et al., unpublished), Canada, and United States (Young et al., 1995). The P102L mutation was the first point mutation described in a prion disease, and genetic linkage of the mutation to the disease has been shown (Hsiao et al., 1989; Speer et al., 1991).

A CCG to CTG mutation at codon 102 causes a leucine (L) for proline (P) substitution. The mutation has high penetrance; in fact, individuals carrying the mutant allele develop the disease between the fourth and the seventh decade. The P102L *PRNP* mutation is also shared by two recently described rare GSS phenotypes. The genotypic differences between the three diseases resides in the codon 129 and in the codon 219 located on the mutant allele. The former encodes methionine or valine, the latter encodes glutamic acid or lysine. The P102L mutation has most likely occurred more than once since it has been observed in families of different ethnic groups.

Clinical manifestations of GSS P102L 129M. The phenotype is characterized by a progressive cerebellar syndrome with ataxia, dysarthria, and incoordination of saccadic movements and also by pyramidal and pseudobulbar signs. In the advanced stages, mental and behavioral deterioration leading to dementia or akinetic mutism occur. The onset develops clinically in the fourth to sixth decades of life, and the duration of the disease ranges from a few months to 6 years. In some instances, the disease presents a rapid course of 5 to 9 months with a clinical picture indistinguishable from that of CJD (Barbanti et al., 1994). Considerable intrafamilial phenotypic variability may be observed (Adam et al., 1982; Hainfellner et al., 1995; Young et al., 1995). Myoclonus and pseudoperiodic sharp wave discharges (PSDs) in the EEG, a finding of diagnostic relevance in CJD, occurs in some of the GSS P102L patients.

Neuropathology. Deposits of fibrillar and nonfibrillar PrP in the cerebral and cerebellar parenchyma against a background of variable spongiform changes are consistently found (Masters et al., 1981; Adam et al., 1982; Hainfellner et al., 1995; Piccardo et al., 1995). Spongiform changes, neuronal loss, and astrocytosis vary in severity even among patients of the same kindred, and are most severe when the course of the illness is rapid (Adam et al., 1982; Barbanti et al., 1994; Hainfellner et al., 1995).

Characteristics and allele origin of PrP^{res}. Recent studies have indicated the presence of two major PrP^{res} fragments in brain of GSS P102L 129M (Piccardo et al., 1997; Parchi et al., 1998a). One has a Mr of approximately 21 kDa, the other of approximately 8 kDa; both are also present in intact tissue. The 21-kDa fragment is similar to the PrP^{res} type 1 described in sporadic CJD (Parchi et al., 1996c; Parchi et al., 1998a). However, the ratio of the three principle glycoforms of PrP^{res} is significantly different and shows, in the GSS P102P 129M, a D pattern in which the diglycosylated form is the most, and the unglycosylated form the least, represented form. The 8-kDa fragment is similar to those described in other variants of

GSS (Piccardo et al., 1998a). Sequencing and mass spectrometry have demonstrated that the N-terminus of the 21-kDa fragment begins at residues 78 and 82, while the N-terminus of the 8-kDa fragment begins at residues 78, 80, and 82, and the C-terminus ends at a position spanning residues 147 through 150 (Parchi et al., 1998a). Thus, the 21-kDa and 8-kDa fragments differ exclusively in the C-terminal, since the former has an intact C-terminus of PrP^{res}, and the latter is truncated at both N- and C-terminal ends. Both fragments derive only from the mutant allele as shown by the continual presence of the mutated L residue at position 102. It was also shown that the presence of the 21-kDa form correlates with the presence of spongiform degeneration and "synaptic" pattern of PrP deposition, whereas the 8-kDa fragment is found in brain regions showing PrP positive multicentric amyloid deposits (Parchi et al., 1998a). These data further indicate that the neuropathology of prion diseases largely depends on the type of PrP^{res} fragment that forms in vivo. Because the formation of PrP^{res} fragments of 8 kDa with ragged N- and C-termini is not observed in CJD or FFI and appears to be shared by most GSS subtypes, it may represent a molecular marker for this disorder.

Transmissibility. Masters demonstrated that the inoculation into nonhuman primates of brain homogenates obtained from GSS patients induced a spongiform encephalopathy in the recipient animals (Masters et al., 1981). Some of the patients from whom transmission was obtained, developed the P102L mutation. One of the patients was the member of the "W" family reported by Rosenthal et al. (1976). Subsequently, intracerebral inoculation into marmosets from another patient of the "W" family induced a spongiform encephalopathy indistinguishable from that seen in marmosets inoculated with brain tissue from a case of CJD (Baker et al., 1990). Transmission experiments from P102L GSS patients to mice resulted similarly in the development of spongiform degeneration (Manuelidis et al., 1987; Tateishi et al., 1996). It is significant that the donor tissue is characterized by the presence of PrP amyloid and severe spongiform changes, whereas the recipient primates and rodents develop a rapidly progressing disease with severe spongiform degeneration but not PrP amyloid deposition. Transmission is not consistently obtained from P102L patients.

GSS with the P102L Mutation and the 129M and the 219K Codons on the Mutant Allele (GSSP102L, 129M, 219K)

Epidemiology and genetics. Only one family has been reported. The P102L mutation was detected in coupling with methionine at residue 129 and lysine at residue 219.

Clinical manifestations. Symptomatic subjects of this family had either dementia or cerebellar signs.

Neuropathology. Neuropathological studies demonstrated mild PrP deposition in the cerebral and cerebellar cortex and basal ganglia with no amyloid or spongiform changes (Furukawa et al., 1995).

Characteristics and allele origin of PrP^{res}. No biochemical data are available.

Transmissibility. No data are available.

GSS with the P102L Mutation and the 129V Codon on the Mutant Allele (GSSP102L, 129V)

Epidemiology and genetics. Although there are two known instances of P102L GSS disease with homozygosity for valine at codon 129, most likely from different families, only one patient has been well documented (Telling et al., 1995; Young et al., 1997). One subject, who was homozygous for valine (GTG) at codon 129, had the P102L mutation, CCG to CTG, on one allele. This patient, thus, had the *PRNP* P102L mutation in coupling with valine at residue 129.

Clinical manifestations. The clinical course and duration were significantly different from that observed in the most common form of GSS P102L. The documented patient had seizures, long tract signs, but no dementia. The duration of 12 years was unusually long. Typically, in GSS P102L-M129 the clinical onset is with cerebellar signs, seizures are generally not observed, dementia is frequently seen, and the duration is of approximately 5 years.

Neuropathology. The neuropathological findings in this patient differ from those frequently seen in P102L-M129 regarding the absence of spongiform changes. The involvement of the corticospinal, spinocerebellar, and gracile tracts, in addition to the presence of PrP deposits in the substantia gelatinosa may be correlated with the severe postural and sensory abnormalities observed in this patient (Young et al., 1997).

Characteristics and allele origin of PrP^{res}. No biochemical data are available.

Transmissibility. No data are available.

GSS with the P105L Mutation and the 129V Codon on the Mutant Allele (GSSP105L, 129V)

Epidemiology and genetics. Four families have been reported in Japan and recently, another patient has been reported. However, it is not clear whether he belongs to a previously reported family (Yamazaki et al., 1997). A proline (CCA) to leucine (CTA) substitution at codon 105 on a Val129 allele has been found in patients from four families (Nakazato et al., 1991; Amano et al., 1992; Terao et al.,

1992; Kitamoto et al., 1993a, 1993c; Yamada et al., 1993). It is not known whether the mutation has occurred more than once.

Clinical manifestations. Spastic gait, hyperreflexia, and the Babinski sign dominate the picture in the initial stages (Kitamoto et al., 1993a; Yamada et al., 1993). Extrapyramidal signs such as fine finger tremor and rigidity of limbs may be observed. Paraparesis progresses to tetraparesis and is accompanied by emotional incontinence and dementia. Myoclonus, PSDs in EEG or severe cerebellar signs have not been reported. The age at the onset of the clinical signs is in the fourth and fifth decades of life, the duration of the disease ranges from 6 to 12 years.

Neuropathology. PrP deposits in the neocortex, especially the motor area, striatum, and thalamus, are found. Multicentric PrP-amyloid plaques and diffuse deposits are present in superficial and deep layers of the neocortex respectively, in association with neuronal loss and astrocytosis. NFTs are occasionally observed, but not spongiform changes. In the case recently reported, a 57-year-old woman with dementia and gait disturbance, but not spastic paraparesis, numerous NFTs were seen in the cerebral cortex and several subcortical nuclei (Yamazaki et al., 1997). In most cases, amyloid plaques are rare in the cerebellum, while axonal losses occur in the pyramidal tracts (Nakazato et al., 1991; Amano et al., 1992; Terao et al., 1992; Kitamoto et al., 1993a; Yamada et al., 1993).

Characteristics and allele origin of PrP^{res}. No biochemical data are available.

Transmissibility. No data are available.

GSS with the A117V Mutation and the 129V Codon on the Mutant Allele (GSSA117V, 129V)

Epidemiology and genetics. The mutation has been evident in six families: two French (Alsatian) families, and a British family as well as three U.S. kindreds, including one recently identified (Doh-ura et al., 1989; Hsiao et al., 1991a; Mastrianni et al., 1995; Ghetti et al., 1998).

The alanine (A) to valine (V) substitution at residue 117 results from a cytosine to thymidine mutation in the second position of the codon. The mutant codon also contains the noncoding, or "silent," adenine to guanine mutation in the third position, so that the codon change from wild-type is GCA to GTG. These families may share a common founder, since the GCG codon at 117 is a rare polymorphism. However, the GCG variant contains a CpG dinucleotide so this mutation could also have arisen more than once.

Clinical manifestations. The clinical phenotypes are presenile dementia in the Alsatian family (Warter et al., 1981; Tranchant et al., 1991, 1992; Mohr et al., 1994), presenile dementia, pyramidal signs, and parkinsonism in the U.S.-German kindred (Nochlin et al., 1989; Hsiao et al., 1992), severe ataxia, dysarthria, mild parkinsonism, and dementia in a patient from the family reported by Mastrianni et al. (1995). In the Alsatian and in the U.S.-German families, the age at the onset of the clinical signs is 19 to 64 years and 23 to 58 years, respectively (Nochlin et al., 1989; Tranchant et al., 1992), while the duration of the disease ranges from 1 to 11 and 2 to 6 years, respectively. The clinical phenotype in the Alsatian family is variable (Mohr et al., 1994); although the three patients studied in the first and second generations of the published pedigree exhibited dementia as the main clinical symptom, affected subjects from subsequent generations showed the association of dementia, pyramidal and pseudobulbar signs with ataxia, extrapyramidal symptoms, amyotrophy, myoclonus and tonic-clonic seizures. EEGs did not show PSDs.

Neuropathology. There are PrP-amyloid deposits and PrP deposits without the tinctorial properties of amyloid that are widespread throughout the cerebrum, but rare or absent in the cerebellum of subjects with dementia alone (Warter et al., 1981; Nochlin et al., 1989; Tranchant et al., 1991; Mohr et al., 1994; Mastrianni et al., 1995). Numerous PrP-amyloid deposits in the cerebral cortex, basal ganglia, and thalamus, as well as in the cerebellum were found in three patients from the Alsatian family (Mohr et al., 1994), who had died at 24, 39, and 73 years. In the latter, who survived 9 years after the onset of the clinical signs, NFTs were numerous in the cerebral cortex, in contrast with the rare deposits of amyloid β-peptide (Aβ). Spongiform changes are variable in severity and extent. In patients of the U.S.-German family, PrP-amyloid deposits are prominent in the cerebral cortex and striatum, but not in the cerebellum; NFTs are occasionally found. In the family discovered by Mastrianni et al. (1995), there is a conspicuous deposition of PrP amyloid in the cerebellum.

Characteristics of PrP^{res} and allele of origin of PrP amyloid. Recently, it has been shown that distinctive PrP isoforms are observed in GSS A117V patients with and without NFTs (Lievens et al., 1998). In all cases, low molecular weight PrP peptides of approximately 7 to 8 kDa were detected using antibody 3F4 directed to the mid-region of PrP. The smallest amyloid subunit corresponds to peptides spanning residues 58 to 150 and 81 to 150. These peptides originated from the mutant allele (Tagliavini et al., 1995).

Transmissibility. Transmission from one case was tested in mice and was negative (Tateishi et al., 1996).

GSS with the F198S Mutation and the 129V Codon on the Mutant Allele (GSSF198S, 129V)

Epidemiology and genetics. Two families have been reported in the United States (Farlow et al., 1989; Ghetti et al., 1989; Hsiao et al., 1992; Ghetti et al., 1995; Mirra et al., 1997). A phenylalanine (TTC) to serine (TCC) substitution at codon 198 on a Val129 allele has been described in patients from these two U.S. families of Caucasian ethnic origin (Dlouhy et al., 1992; Hsiao et al., 1992; Mirra et al., 1997). This GSS variant has been extensively studied in a family from Indiana, the so-called Indiana kindred (IK), and its disorder has been referred to as GSS-IK.

Clinical manifestations. The clinical phenotype is characterized by cognitive cerebellar and pyramidal signs. The primary symptoms are gradual loss of short-term memory and progressive clumsiness in walking, bradykinesia, rigidity, dysarthria, and dementia. Signs of cognitive impairment and eye-movement abnormalities may be detected by specific tests before clinical onset of symptoms. Psychotic depression has been observed in several patients; tremor is mild or absent. Symptoms may progress slowly over 5 years or rapidly over as little as 1 year. The age at onset of clinical signs is 40 to 71 years; patients homozygous for Val at codon 129 have clinical signs more than 10 years earlier, on average, than heterozygous patients (Dlouhy et al., 1992). The duration of the disease ranges from 2 to 12 years (Farlow et al., 1989; Ghetti et al., 1992, 1995).

Neuropathology. The phenotype is characterized by presence of severe PrP deposition and amyloid formation in the cerebral and cerebellar parenchyma as well as neurofibrillary lesions in the cerebral gray matter (Ghetti et al., 1989; Giaccone et al., 1990; Ghetti et al., 1992; Giaccone et al., 1992; Tagliavini et al., 1993; Ghetti et al., 1995). In the 12 patients and the one nonsymptomatic carrier studied, unicentric and multicentric PrP-amyloid deposits are distributed throughout the gray structures of the cerebrum, cerebellum, and midbrain. Amyloid deposition is severe in frontal, insular, temporal and parietal cortex; the highest concentration of deposits is in layers one, four, five, and six. A moderate involvement is evident in the hippocampus, where plaques occur predominantly within the stratum lacunosum-moleculare of the CA1 sector and subiculum. PrP deposits are numerous in the claustrum; the caudate nucleus; putamen; the anterior, dorsomedial, ventrolateral, and lateral dorsal nuclei of the thalamus; the cerebellar molecular layer; the mesencephalic tegmentum; the substantia nigra; and periaqueductal grey matter. However, the degree of amyloid formation in these areas varies. Amyloid deposits are surrounded by astrocytes, astrocytic processes, and microglial cells. In the neocortex, many amyloid cores are associated with abnormal neurites, so that when these lesions are analyzed with classical stains, they are morphologically similar to neuritic plaques of Alzheimer's disease (Ghetti et al., 1989, 1992). The neurites immunoreact with antibodies to tau, ubiquitin, and to N- and C-terminal domains of the β-amyloid precursor protein (βPP). The accumulation of βPP in nerve cell processes is not associated with extracellular deposition of Aβ, except in older patients where Aβ immunoreactivity also may be observed around PrP-amyloid deposits (Bugiani et al., 1993; Ghetti et al., 1995).

NFTs and neuropil threads are found in large numbers in the neocortex and in the subcortical grey matter. Cortical regions especially affected are the frontal, cingulate, parietal, insular, and parahippocampal cortex. In the remaining cortical regions, NFTs are present but less numerous. Of the subcortical grey areas, the caudate nucleus, putamen, the nucleus basalis, the midbrain and pontine nuclei, the substantia nigra, the griseum centrale, and the locus coeruleus, show a variable degree of involvement.

Moderate to severe cerebral and cerebellar atrophy, nerve cell loss, and gliosis are found in the neocortex, striatum, red nucleus, substantia nigra, cerebellum, locus coeruleus, and inferior olivary nucleus, and iron deposition in the globus pallidus, striatum, red nucleus, and substantia nigra are observable. Spongiform changes are inconspicuous.

A neurologically asymptomatic subject, who had the F198S mutation and was heterozygous at codon 129, died at the age of 42. In this subject, PrP-amyloid deposits were already numerous in the cerebellar cortex but rare in the cerebrum cortex.

Characteristics of PrP^{res}. PK digestion of the brain extracts generated three prominent broad bands of approximately 27 to 29, 18 to 19, and 8 kDa, and a weaker but sharp band at 33 kDa, as detected with antibody 3F4 (Piccardo et al., 1996). The latter band (33 kDa) may be attributed to PrP or to cross reactivity of antibody 3F4 with residual PK, or both. The stoichiometry among the PrP species differed from that of the undigested peptides for a notable increase in the signal of the low molecular weight band. No PrP signal was observed in PK-treated and nontreated brain extracts when antibody 3F4 was absorbed against a PrP peptide spanning residues 102 to 114.

N-deglycosylation of non-PK–treated extracts with PNGase F resulted in disappearance of the 33- to 35-kDa band accompanied by an increased signal of the 28- to 30-kDa band (Piccardo et al., 1996). The 28- to 30-kDa band, seen with antibody 3F4, is consistent with the molecular weight of deglycosylated full-length PrP, as shown by a similar species detected with antibodies raised against synthetic peptides homologous to residues 23 to 40 and 220 to 231 of human PrP (PrP23-40 and PrP220-231) (Piccardo et al., 1996). In non-PK–treated brain extracts, the electrophoretic mobility of the 19 to 20 and 9 kDa bands was not modified by deglycosylation.

The combination of PK and enzymatic deglycosylation with PNGase F treatment generated a pattern similar to that of PK treatment alone, with prominent fragments at approximately 27 to 29, 18 to 19 and 8 kDa. These PrP fragments were immunoreactive with antibody 3F4 and with antisera AS 6800 (raised against a synthetic peptide homologous to residues 89 to 104 of human PrP), but not with PrP23-40 and PrP220-231.

To investigate the sensitivity of PrP to PK, samples from the cerebellum were exposed to PK under nondenaturing and denaturing conditions. Under nondenaturing conditions, the major PrP isoforms present in GSS-IK retained partial PK resistance, the intensity of the signal after 4 hours being similar to that observed after 1 hour of enzyme treatment. Conversely, denatured PrP was completely degraded after 30 min digestion with PK at 37°C in the presence of sodium dodecyl sulphate (Piccardo et al., 1996).

PK-resistant PrP fragments of similar electrophoretic mobility were seen in all brain regions examined (frontal cortex, caudate nucleus, and cerebellum) of two IK patients analyzed. In semiquantitative experiments (similar amounts of total protein loaded), comparable signals were observed in samples from the cerebellum and caudate nucleus, two areas that have high and low amount of amyloid respectively (Piccardo et al., 1996).

In immunoblot studies, the strongest signal was obtained from tissue corresponding to a patient who had the longest clinical course of the disease (12 years). To follow up on this observation, we also studied PK-resistant PrP obtained from the cerebella of five additional patients of the IK, who had a duration of clinical signs varying from 2 to 7 years. Similar electrophoretic patterns were observed, and high amounts of PrP were present in all cases regardless of the duration of disease. Nevertheless, in repeated experiments the patient with 12 years' duration of clinical signs always exhibited the most intense signal.

PrP was localized in the microsomal fraction. Sarkosyl-soluble and PK-sensitive PrP isoforms from this fraction were seen as prominent bands of approximately 33 to 35 kDa in both control (familial AD) and GSS-IK (Piccardo et al., 1996). In addition, Sarkosyl-insoluble PrP was present as four major PrP species of approximately 33 to 35, 28 to 30, 19 to 20 and 9 kDa (Piccardo et al., 1996). PK digestion of these samples generated three prominent bands of 27 to 29, 18 to 19, and 8 kDa in GSS-IK, comparable to the PK-resistant species present in brain homogenates. As expected, no Sarkosyl-insoluble PrP was present in control extracts (Piccardo et al., 1996).

Characteristics and allele origin of PrP amyloid. The biochemical composition of PrP amyloid was first determined in brain tissue samples obtained from patients of the Indiana kindred (Tagliavini et al., 1991, 1994) carrying the F198S mutation in coupling with 129V (Dlouhy et al., 1992; Hsiao et al., 1992). Amyloid cores were isolated by a procedure combining buffer extraction, sieving, collagenase digestion, and sucrose gradient centrifugation. Proteins were extracted from amyloid fibrils with formic acid, purified by gel filtration chromatography and reverse-phase HPLC, analyzed by SDS-PAGE and immunoblot, and sequenced. The amyloid preparations contained two major peptides of 11 and 7 kDa spanning residues 58 to 150 and 81 to 150 of PrP, respectively. The amyloid peptides had ragged N- and C-termini.

The finding that the amyloid protein was an N- and C-terminal truncated fragment of PrP was substantiated by immunostaining brain sections with antisera raised against synthetic peptides homologous to residues 23 to 40, 90 to 102, 127 to 147 and 220 to 231 of human PrP. The amyloid cores were strongly immunoreactive with antibodies that recognized epitopes located in the midregion of the molecule while only the periphery of the cores was immunostained by antibodies to N- or C-terminal domains. In addition, antisera to the N- and C-termini of PrP labeled large areas of the neuropil that did not possess the tinctorial and optical properties of amyloid. Immunogold electron microscopy showed that antibodies to the mid-region of PrP decorated fibrils of amyloid cores while antisera to N- and C-terminal epitopes labeled amorphous material at the periphery of the cores or dispersed in the neuropil. These data suggest that amyloid deposition in GSS is accompanied by accumulation of PrP peptides without amyloid characteristics (Giaccone et al., 1992).

In GSS-IK, the amyloid protein does not include the region containing the amino acid substitution. To establish whether amyloid peptides originate from mutant protein alone or both mutant and wild-type PrP, we analyzed patients heterozygous M/V at codon 129 and used 129V as a marker for protein from the mutant allele. Amino acid sequencing and electrospray mass spectrometry of peptides generated by digestion of the amyloid protein with endoproteinase Lys-C showed that the samples contained only peptides with 129V, suggesting that only mutant PrP was involved in amyloid formation (Tagliavini et al., 1994).

Characteristics of paired helical filaments. Paired helical filament-enriched fractions obtained from the neocortex of IK patients contained SDS-soluble isoforms with electrophoretic mobility and immunochemical profile corresponding to the isoforms extracted from the brain of patients with Alzheimer's disease (Tagliavini et al., 1993). These proteins migrate between 60 and 68 kDa, immunoreact with antibodies to the N- and C-termini of, and require dephosphorylation to be accessible to Tau-1. Thus, the immunocytochemical findings are consistent with those of the western blot analysis showing that significant similarity exists between GSS-IK and Alzheimer's disease regarding the Alz50, T46 and Tau-1 immunostaining of NFTs.

Transmissibility. In the case of the Indiana kindred, brain tissue and buffy coat from one affected individual were inoculated into hamsters in two experiments in the laboratory of Drs. E. and L. Manuelidis. No pathological changes were observed in the primary transmission attempt, nor in the second and third serial passage (Dr. L. Manuelidis, personal communication). Tissue homogenates from another F198S patient have been inoculated into hamsters and mice and no transmission has occurred (Ghetti et al., 1992; Hsiao et al., 1992). Amyloid enriched fractions and tissue homogenates from IK patients have been inoculated into marmosets and no transmission has occurred 30 months after inoculation (Baker, H.F., personal communication). Studies are in progress in Dr. Collinge's laboratory to determine whether transgenic mice containing a normal human *PRNP* gene develop a prion disease following the inoculation of tissue homogenates from IK patients.

GSS with the D202N Mutation and the 129V Codon on the Mutant Allele (GSS D202N, 129V)

Epidemiology and genetics. One patient has been reported in the United Kingdom.

Clinical manifestations. A description of the clinical phenotype is not currently available.

Neuropathology. The patient was neuropathologically characterized by absence of spongiform degeneration and abundant PrP-amyloid deposition in the cerebrum and cerebellum. NFTs were present in the cerebral cortex.

Characteristics of PrP^{res}. PrPres was detected in the neocortex of one patient with GSS D202N. It contained distinct bands of approximately 8 kDa, 18 to 19 kDa, and 27 to 29 kDa, a pattern very similar to that seen in GSS F198S (Piccardo et al., 1998a).

Transmissibility. No data are available.

GSS with the Q212P Mutation and the 129M Codon on the Mutant Allele (GSSQ212P, 129M)

Epidemiology and genetics. Only one family with this mutation is known at this time (Young et al., 1998). A glutamine (CAG) to proline (CCG) mutation has been found in one patient who had no family history of neurological disease. The mutation was not found in 100 control subjects (Young et al., 1998).

Clinical manifestations. Onset occurred at age 60 with gradual development of incoordination and slurring of speech. Three years following onset, the patient was found to have normal mental status, dysarthria, and ataxia. His condition began to progress more rapidly and approximately 6 years after onset, the patient entered a nursing home where he remained until his death, 8 years after onset. At the time of death, the patient was still mentally competent.

Neuropathology. Amyloid deposition was mild throughout the central nervous system. Among GSS variants, this appears to be the form with the least amount of amyloid deposits. The cerebellum was significantly less affected than in all other variants. Immunopositive PrP deposits were present in the cerebellum where they were more numerous than in the neocortex and striatum. There was axonal degeneration in the anterior and lateral corticospinal tracts throughout the spinal cord.

Characteristics and allele origin of PrP^{res}. Proteinase K digestion of brain homogenates resulted in a poorly defined smear in the 25- to 35-kDa region and two major bands of 18 to 19, and 10 kDa (Piccardo et al., 1998b).

Transmissibility. No data are available.

GSS with the Q217R Mutation and the 129V Codon on the Mutant Allele (GSSQ217R, 129V)

Epidemiology and genetics. One family of Swedish origin has been reported in the United States. A glutamine (CAG) to arginine (CGG) substitution (Hsiao et al., 1992) on a Val129 allele has been described in two patients from an American family of Swedish origin.

Clinical manifestations. The phenotype is characterized by gradual memory loss, progressive gait disturbances, parkinsonism, and dementia. The age at onset of clinical signs is 62 to 66 years. The duration of the disease is 5 to 6 years (Ghetti et al., 1994). The neurological signs may be preceded by episodes of mania or depression that respond to antidepressant medications, lithium, and neuroleptics.

Neuropathology. PrP-amyloid deposits are numerous in the cerebrum and cerebellum, while NFTs are abundant in the cerebral cortex, amygdala, substantia innominata, and thalamus.

Characteristics and allele origin of PrP^{res}. PrPres is essentially similar to that observed in F198S.

Transmissibility. Tissue homogenates from one A217G patient has been inoculated into hamsters and mice, and no transmission has occurred (Hsiao et al., 1992).

Cerebral Amyloid Angiopathy (PrP-CAA) Phenotype

PrP-CAA with the Y145Stop Mutation and the 129M Codon on the Mutant Allele (PrP-CAAY145Stop, 129M)

Epidemiology and genetics. Only one case has been reported in the literature. A tyrosine (TAT) to stop codon (TAG) substitution on a Met129 allele (Ghetti et al., 1996b) has been discovered in a Japanese patient with a clinical diagnosis of Alzheimer disease (Kitamoto et al., 1993b).

Clinical manifestations. The clinical phenotype is characterized by memory disturbance, disorientation, and a progressive, severe dementia. The EEG did not show PSDs. The age at the onset of the clinical signs was 38 years, while the duration of the disease was 21 years.

Neuropathology. Diffuse atrophy of the cerebrum and dilation of the lateral ventricles were present. Neuronal loss and gliosis were severe, whereas no spongiform changes were observed. There were PrP-amyloid deposits in the walls of small- and medium-sized parenchymal and leptomeningeal blood vessels and in the perivascular neuropil, as well as neurofibrillary lesions in the cerebral gray matter. The NFTs are composed of paired helical filaments with a periodicity of 70 to 80 nm and were decorated with monoclonal antibodies recognizing abnormally phosphorylated tau protein.

Characteristics and allele origin of PrP^{res}. PrPres cannot be detected.

Transmissibility. No data are available.

Creutzfeldt-Jakob Disease Phenotype

CJD with the E200K Mutation and the 129M Codon on the Mutant Allele (CJDE200K,129M)

Epidemiology and genetics. The epidemiology of the CJDE200K, 129M is of particular interest. The largest cluster of CJDE200K, 129M occurs among Jews of Libyan origin (Kahana et al., 1974). In this community, the incidence of CJD is 100 times higher than the worldwide incidence (50 per million as opposed to 0.5 per million), the highest incidence of CJD in the world. Although early speculation attributed the high incidence to diet or other environmental factors, a series of epidemiological studies done in Israel and other countries revealed the unusually high incidence of familial CJD in this community (Neugut et al., 1979) which asserts a genetic origin (Cathala et al., 1985; Radhakrishnan and Mousa 1988; Nisipeanu et al., 1990; Hsiao et al., 1991b; Zilber et al., 1991; Gabizon et al., 1992).

After the E200K mutation was discovered, new data were obtained regarding the penetrance and other genetic details of the disease, the biochemistry of the mutant and normal protein in these patients, and the transmission rate of prions from these patients to experimental animals. These discoveries strengthened the causative relationship between the E200K mutation and familial CJD.

Three patients homozygous for the E200K mutation (due to consanguinity) have been identified, and two more are suspected according to offspring evaluation. The clinical course in two of the homozygous patients was similar to that of heterozygotes, whereas in a third patient the course was more protracted (Chapman et al., 1992; Gabizon et al., 1993). The finding that a second mutated allele does not worsen the clinical course of the disease supports the notion that CJDE200K, 129M is a true dominant disease (Hsiao et al., 1991b).

Clinical manifestations. The CJDE200K, 129M resembles the typical form of the sporadic CJD (sCJD) and has a mean age at onset of 62 years with a range of

53 to 71, and a mean duration of less than a year with a range of 4.3 to 6 months (Kahana et al., 1991). According to another study, the mean age of onset was 55 ± 8 years and the mean duration 8 ± 18 months (Brown et al., 1991).

As in sCJD, presenting signs include cognitive impairment and psychiatric changes (80% to 83% of patients), cerebellar signs (43% to 55%), visual signs in 19%, and myoclonic jerks (12%) (Brown et al., 1991; Kahana et al., 1991). Whereas in most of the patients the course of the disease is insidious, 10% of patients present with an acute, sudden onset. During the course of the disease, all patients develop dementia as well as other cognitive and psychiatric disturbances: 73% have myoclonus, 79% cerebellar signs, and 40% seizures, while sensory and cranial nerve involvement is present in 24% (Brown et al., 1991). A puzzling clinical feature of the CJDE200K, 129M is the involvement of the peripheral nervous system, which is rare in the course of sCJD (Lope et al., 1977; Schoene et al., 1981; Guiroy et al., 1989). The peripheral neuropathy is both motor and sensory and is often accompanied by protein elevation in the cerebrospinal fluid (Sadeh et al., 1990; Neufeld et al., 1992; Antonine et al., 1996; Mastrianni, J.A., personal communication).

Electrophysiological and neuroimaging findings are also similar to those of sCJD. The typical EEG activity with PSW complexes is found in 74% to 76% of CJDE200K, 129M, as compared with approximately 56% of sCJD (Brown et al., 1986, 1991); slowing of the background is found in all patients. In some of the CJDE200K, 129M the pattern seen in routine EEG is asymmetrical and not strictly generalized (Neufeld et al., 1992). Computed tomographic (CT) scans have shown brain atrophy in one-third of the patients (Chapman et al., 1993). Brain single photon emission computed tomography (SPECT) in one patient with a normal brain CT scan showed bilateral perfusion defects (Cohen et al., 1989). All these nonspecific diagnostic findings are similar to those observed in sCJD (Galvez and Cartier, 1984).

Neuropathology. The histological changes associated with CJDE200K, 129M are very similar to those of the typical sCJD (Group 1 of Parchi et al., 1996c) and are invariably characterized by spongiosis, astrogliosis, and neuronal loss. The severity of the astrogliosis and of the neuronal loss appears to be a function of the disease duration. These lesions are severe and widely distributed in the cerebral cortex. They are also present with decreasing severity in the striatum, diencephalon, and cerebellum. Immunostaining is consistently positive throughout the brain with the punctate or "synaptic" pattern and a severity that appears to be directly related to that of the histological lesions (Young, Parchi, Gambetti, Ghetti et al., unpublished). In addition, punctate PrP immunostaining is also present in the substantia gelatinosa of the spinal cord. No PrP-positive deposits, either in the form of amyloid or nonamyloid plaques, are present. The peripheral neuropathy is both axonal and demyelinating (Chapman et al., 1993); the latter is characterized by segmental demyelination (Sadeh et al., 1990; Neufeld et al., 1992).

Characteristics and allele origin of PrP^{res}. Thc availability of paticnts homozygous for the E200K mutation has provided the opportunity to examine whether the E200K PrP^M is formed as a soluble, nonpathogenic protein and is converted into mutant PrP^{res} during the long incubation period, or whether it is initially formed as mutant PrP^{res}. The indirect evidence for conversion during the incubation period includes the similarity in age of onset of CJDE200K, 129M and sCJD. A more direct finding is the absence of PrP^{res} in the brain biopsy of a presymptomatic homozygous carrier of the E200K mutation (R. Gabizon, unpublished data).

Studies have also been conducted to determine whether both the E200K PrP^M and the PrP^C expressed by the normal allele, are converted into PrP^{res} and participate in the pathological process or if only the PrP^M is converted. PrP^M was found to be converted to PrP^{res}, whereas PrP^C did not acquire protease-resistance but became insoluble in detergents. Since insolubility is also a characteristic of PrP^{res} (Meyer et al., 1986), these findings raise the possibility that both (E200K) PrP^M and PrP^C participate in the pathogenesis of CJDE200K, 129M.

The biochemical properties of the (E200K) PrP^M have been examined in brain samples, lymphocytes, and cultured fibroblasts of CJDE200K, 129M patients, and in brain samples of transgenic mice carrying the E200K mutation. In brains of subjects with CJDE200K, 129M, the PrP^{res} has the gel migration pattern of the PrP^{res} type 1; that is, CJDE200K, 129M migrates at 21 kDa on gel and shows an underrepresentation of the unglycosylated form (Monari et al., 1994; Parchi et al., 1996a). This glycoform ratio is similar to that observed in FFI and in the new variant CJD (Monari et al., 1994; Collinge et al., 1996). It was also demonstrated that the highly glycosylated form of E200K PrP^M migrates more quickly on SDS-PAGE than does the corresponding PrP^C form (Gabizon et al., 1996), a difference that disappears following deglycosylation. This implies a difference in the glycosylation pattern between the E200K PrP^M and PrP^C, which is likely owing to the proximity of the mutation to the glycosylation site at residue N197. This conclusion is also supported by the data obtained with the E200K cell transfectants (Capellari et al., 1996).

Transmissibility. Intracerebral inoculation of CJDE200K, 129M brain homogenate has resulted in CJD transmission to apes after an incubation period of 6 years (Chapman et al., 1994). Studies with transgenic mice susceptible to human prions show transmission with specimens from familial CJDE200K, 129M patients (Telling et al., 1994). Two of three brain samples transmitted the disease following an incubation period of 170 days. The lack of transmission seen with the third sample may be attributed to codon 129 methionine/valine incompatibility (Telling et al., 1995). The exact risk of transmission of CJD from brain and other tissues, especially blood, from CJDE200K, 129M patients and healthy mutation carriers remains to be established.

CJD with the D178N Mutation and the 129V Codon on the Mutant Allele (CJDD178N, 129V)

Epidemiology. Currently, ten apparently unrelated kindreds are known (Parchi et al., 1996b), which include five American, three French, one Finnish, and one German (Medori et al., 1992; Goldfarb et al., 1992c; Parchi and Gambetti, unpublished; Goldfarb, unpublished; Kretzschmar et al., 1995).

PRNP genotype and genetic linkage. CJDD178N, 129V shares the D178N *PRNP* mutation with FFI. The genotypic difference between the two diseases resides in the codon 129 located on the mutant allele, which encodes valine in the CJDD178N, 129V and methionine in FFI (Goldfarb et al., 1992c). Genetic linkage tested in two informative kindreds yielded a lod score of 5.30, indicating that the D178N, 129V haplotype is the cause of the disease (Goldfarb et al., 1992a).

Clinical features. The mean age at onset of the disease is 39 ± 8 years (range: 26 to 47) for the 129 valine homozygous, and 49 ± 4 years ($p < 00.1$) (range: 45 to 56) for the valine/methionine heterozygous patients. The mean durations are 14 ± 4 months (range: 9 to 18) for the homozygous and 27 ± 14 months ($p < 00.5$) (range: 7 to 51) for the heterozygous patients (Kirschbaum, 1968; Goldfarb et al., 1992c; Kretzschmar et al., 1995). Clinical signs are fairly consistent and apparently independent of the zygosity at codon 129. Presentation is characterized by cognitive impairment, especially memory decrease, frequently associated with depression, irritability, and abnormal behavior. Ataxia, speech impairments with dysarthria, and aphasia, tremor, and myoclonus appear early during the course of the disease. EEG examination consistently demonstrates generalized slow wave activity without periodic complexes.

Neuropathology. There is apparently no difference between codon 129 homozygous and heterozygous subjects; however, the number of subjects examined at autopsy in which codon 129 is known is limited. The common changes of this familial CJD subtype are spongiosis associated with prominent gliosis, often in the form of gemistocytic astrocytes, and variable degrees of neuronal loss (Parchi et al., 1996b). Enlarged or ballooned neurons may be present which contain argyrophilic and Lewy body–like inclusions that immunoreact with antibodies to neurofilaments (Parchi, P. and Gambetti, P., unpublished). The topography of these lesions is consistent and fairly distinctive. The involvement of the cerebral cortex is widespread, but frontal and temporal cortices are generally more severely affected than the occipital cortex. Among the subcortical structures, the putamen and the caudate nucleus evince severe spongiosis with variable degrees of gliosis; the thalamus is minimally or moderately affected with spongiosis and gliosis, whereas the cerebellum is spared and minimal or no pathology is seen in the brain stem. The immunostaining pattern is punctate, and its intensity matches the severity of the histopathology.

Characteristics and allelic origin of the PrP^res^. All seven cases of CJDD178N, 129V examined to date have been associated with PrP^{res} that in the unglycosylated form has a gel mobility of 20 to 21 kDa, corresponding to the PrP^{res} identified as type 1 (Monari et al., 1994; Parchi et al., 1996a). The glycosylation pattern is characterized by a marked underrepresentation of the unglycosylated form, which accounts for approximately 20% of the total, while the intermediate and highly glycosylated forms are similar in amount, each accounting for approximately 40% of the total (Parchi et al., 1996a).

Determination of the allelic origin has shown that both detergent-insoluble (i.e., aggregated) PrP and PrP^{res} derive exclusively from the mutant PrP.

Animal transmissibility. CJDD178N, 129V has been transmitted to squirrel monkeys with brain tissue of seven out of ten subjects from five of six kindreds (Brown et al., 1992a., 1994b). Squirrel monkeys are homozygous for methionine at codon 129 (Schatz et al., 1995). Transmission to transgenic mice expressing the human-mouse chimera PrP (Telling et al., 1996) has failed, most likely because of the genotypic incompatibility of codon 128, corresponding to the position 129 in the human *PRNP*, in the recipient animals.

CJD with the other PRNP Mutations and the 129M Codon on the Mutant Allele (CJDV210I, 129M)

Epidemiology and genetics. The V210I mutation has been reported in ten affected subjects (Pocchiari et al., 1993; Ripoll et al., 1993; Furukawa et al., 1996; Parchi et al., 1996a; Shyu et al., 1996). Four cases, all Japanese, have been reported to have the V1801 mutation (Hitoshi et al., 1993; Kitamoto and Tateishi, 1994). The R208H mutation has been observed in only one affected subject from a kindred carrying the mutation, but negative for history of neurodegenerative disease, has been reported (Mastrianni et al., 1996). Japanese subjects have been reported to carry the M232R mutation (Hoque et al., 1996). Finally, the T183A mutation has been observed in one kindred (Nitrini et al., 1997). These five rare mutations are associated with phenotypes similar to those of CJD with or without PSW complexes at EEG and the invariable presence of spongiform degeneration.

Clinical features. The age at onset reported in nine subjects varies between 48 and 70 years of age. Although the number of subjects available is small, the polymorphism at codon 129 and a 24bp deletion polymorphism present on the nonmutant allele in one case appear to influence the age at onset (Pocchiari et al., 1993). In the four subjects 129M/M homozygous with no deletion, the disease started at a mean age of 54 years, while in the three subjects who are either 129M/V heterozygous or 129M/M homozygous but with the 24bp deletion the disease started at the mean age of 68 years. Therefore, the subjects carrying the V210I mutation that are heterozygous at codon 129 or carry the 24 bp deletion appear to have a later onset of the disease. This has also been observed in the 144 insertion, D178N and F198S mutations (Dlouhy et al., 1992; Goldfarb et al., 1992b, 1992c). The disease duration varies between 3 and 5 months (mean 4 months) (N = 4), comparable to the duration of the typical sCJD, and does not seem to be affected by the heterozygosity at the polymorphisms. However, one subject is still alive 2 years following the onset of the disease (Shyu et al., 1996). Adequate clinical data are available only in five subjects (Pocchiari et al., 1993; Ripoll et al., 1993; Furukawa et al., 1996). The presentation is reported to include memory, behavioral and gait disturbances, sudden sensory and motor hemiparesis, clumsiness, dystonic movements, and dysarthria. Subsequent common signs are myoclonus, dysarthria, mutism, and cerebellar signs. The EEG shows the typical PSW complexes in all four subjects examined (Pocchiari et al., 1993; Furukawa et al., 1996). Serial magnetic resonance imaging (MRI) in one case revealed increased signal intensity in the basal ganglia and thalamus in T2-weighted image and proton density, as well as severe brain atrophy. Diffuse white matter degeneration was present in the later stages (Shyu et al., 1996).

Neuropathology. The only two cases examined at autopsy revealed spongiosis and gliosis of cerebral cortex, and of the nuclei and the molecular layer of the cerebellum (Pocchiari et al., 1993; Ripoll et al., 1993).

Characteristics and allelic origin of the proteinase K resistant PrP (PrP^{res}). On gel electrophoresis, the unglycosylated PrP^{res} migrates at 20 to 21 kDa, corresponding to the PrP^{res} type 1 (Parchi et al., 1996a, 1996c). The ratio of the three PrP^{res} glycoforms is similar to that of the PrP^{res} associated with sporadic CJD group 1, which illustrates the relative dominance of the intermediate glycoform while the other two forms are less well represented (Parchi et al., 1996a, 1996c). Detailed studies have shown that in V210I CJD, PrP^{res} is formed by both the mutant and normal PrP as it is believed to occur in the inherited prion diseases with five or six insertion mutations (Chen et al., 1997; Silvestrini et al., 1997).

Fatal Familial Insomnia (D178N, 129M)

Epidemiology and genetics. Currently, we are aware of 24 potentially unrelated families that carry the FFI mutation (Padovani et al., 1998). To date, the FFI haplotype appears to be the third most common after the E200K-129M and P102L-129M haplotypes. In Germany, the FFI haplotype has been found to be the most common in a survey conducted between 1993 and 1997 (Kretzschmar et al., 1998).

Clinical phenotype. Several aspects of the phenotype are different in patients who are homozygous or heterozygous at codon 129 of *PRNP* (Montagna et al., 1998). The disease typically begins at a mean age of 50 years (range: 20 to 71 years) in both homozygous and heterozygous subjects. In contrast, the mean disease duration is significantly shorter in the homozygous, approximately 12 months, than in the heterozygous subjects, approximately 21 ($p < 0.004$), yet the ranges show considerable overlap. Major clinical symptoms and signs comprise sleep-wake and vigilance disturbances (insomnia, oneiric stuporous episodes with hallucinosis and episodic confusion), altered autonomic functions (systemic hypertension, irregular breathing, diaphoresis, pyrexia, impotence), and somatomotor manifestations (diplopia, dysarthria, dysphagia, ataxia/abasia, dysmetria, spontaneous and evoked myoclonus, spasticity), and tonic-clonic seizures are observed in some cases. The 129 codon polymorphism revealed that the 129 homozygous patients present prominent oneiric episodes with hallucinosis and episodic confusion as well as spontaneous and evoked myoclonus and more evident autonomic alterations (irregular breathing, hypertension) already at onset of the disease. However, these features occur later in the 129 heterozygous patients who instead presented with ataxia and dysarthria; these somatomotor disturbances worsened subsequently and remained consistently more severe in these patients. Tonic-clonic seizures were more frequent among the 129 heterozygotes. Routine EEGs are generally unhelpful.

In the 129 homozygous subjects, normal sleep is absent from the onset of the disease. Non-REM sleep is abolished and only brief REM sleep episodes occur, often in clusters, and associated with oneiric behavior.

In 129 heterozygous patients, 24-hour recordings at the onset of the disease were characterized by a relative preservation of the cyclic structure of nocturnal sleep and by the persistence of slow (< 4 Hz) EEG activity, typical of slow-wave sleep. Yet, REM sleep often showed the characteristic lack of physiological muscle atonia, and oneiric activity was present, though less prominent.

In all patients, blood pressure (BP), heart rate (HR) and norepinephrine (NE) resting plasma levels were higher than in normal controls comparable for age and sex. Moreover, circadian oscillations of BP, HR, and several hormone levels were lost or decreased.

Intellective tests and behavior remain well preserved while there are prominent vigilance alterations. Thus, FFI patients do not fulfill standard criteria for the diagnosis of dementias.

Neuroradiological and PET studies. While standard brain CT and MRI disclose nonspecific changes only late in the course, positron emission tomographic (PET) studies show marked reduction of glucose utilization in the thalamus and to a milder degree in the cingular cortex in all cases examined. The changes are more diverse and extensive.

Genotype-phenotype correlation. The heterogeneity in clinical and histopathological features that has been noted in FFI-affected subjects is consistent with the existence of two slightly different phenotypes which are determined by the genotype at codon 129 of the normal allele (Goldfarb et al., 1992c). FFI patients M/M at codon 129 manifest, on average, a more rapid course and prominent sleep and autonomic disturbances, while signs of motor and cognitive dysfunction are mild. In contrast, the subjects M/V at codon 129 have a more chronic course and manifest motor signs as a prominent clinical feature at onset, while sleep disturbances and autonomic signs are less severe.

Prion protein in FFI and CJD178. The PrP^{res} fragments associated with FFI and CJD178 differ both in the size of the core protein and in the ratio of the three PrP^{res} glycoforms (Monari et al., 1994). FFI PrP^{res} has been classified as PrP^{res} type 1. Strong evidence suggests that the size variation is the result of different degrees of NH2-terminal trimming by proteases (Bessen and Marsh, 1994; Monari et al., 1994) which, in turn, is likely to result from a different conformation of the PrP^{res}. The distinctive glycoforms ratio of FFI and CJD178 PrP^{res} is the result of the underrepresentation of the mutant unglycosylated form prior to the conversion to the PrP^{res} isoform, probably due to the pronounced instability of this D178N mutant glycoform (Petersen et al., 1996). These findings are consistent with the conclusion that the PrP^{res} associated with FFI and CJD178 have different protein conformations or distinct ligand binding interactions, or both. Therefore, the FFI and CJD178 phenotypes are likely to be determined by codon 129 on the mutant allele which, coupled with the D178N mutation, results in the expression of PrP^{res} with distinct physicochemical characteristics.

Allelic origin of PrP^{res} in inherited prion diseases. In FFI, as in CJD178,129V, both insoluble PrP and PrP^{res} derive exclusively from the PrP^{M} regardless of whether the affected subjects are homozygous or heterozygous at codon 129 (Chen et al., 1997). Whether the low amount of the PrP^{res} found in FFI is due to the monoallelic origin of PrP^{res} or of other mechanisms remains to be determined. The transmissibility to animals not carrying the FFI haplotype of diseases such as FFI in which only PrP^{M} becomes PrP^{res} is puzzling. It is unclear how PrP^{C} can be converted in the recipient animal following inoculation of FFI PrP^{res}, but this conversion does not occur spontaneously in FFI patients. The high local concentration of PrP^{res} achieved by the animal inoculation might overcome the barrier that blocks the conversion of PrP^{C} in FFI.

Transmissibility of FFI. Transgenic mice expressing a chimeric human-mouse PrP^{C} develop a prion disease 200 days after the intracerebral inoculation with a homogenate from FFI brains (Telling et al., 1996). The pathology as well as the presence of PrP^{res} are predominant in the thalamus. This transmission experiment, along with others, (Collinge et al., 1995; Tateishi et al., 1995), not only demonstrates that FFI shares transmissibility with other prion diseases but also provides important clues concerning the mechanisms of PrP^C-PrP^{res} conversion. It has been shown that the size of the PrP^{res} fragment expressed by the inoculated transgenic mice is identical in electrophoretic mobility to the PrP^{res} fragment present in the inoculum from subjects with FFI. Replication of the PrP^{res} fragment size was also observed in the transmission of sporadic and inherited CJD associated with PrP^{res} type 1. Since the recipient mice did not carry any of the donor's *PRNP* mutations, and all had the same genetic background, this remarkable finding suggests that the distinct conformations associated with PrP^{res} types 1 and 2 can be reproduced independently of the genetic information, probably on the basis of information contained in the conformation of the donor's PrP^{res}. This mechanism explains prion strain diversity by a mechanism that does not require the participation of nucleic acids.

Inherited Prion Disease with Variable Phenotype: The Insertional Mutations

Epidemiology and genetics. The wild-type *PRNP* gene has five repeating sequences between codons 51 and 91 of which the first is comprised of 27 base pairs (bp); the other four have 24 bp, and they code for a P(H/Q)GGG(G/-)WGQ nona/octapeptide. Currently, 20 families with 95 affected members have been clinically characterized, and 24 subjects were studied neuropathologically (Owen et al., 1989, 1990; Goldfarb et al., 1991; Collinge et al., 1992; Goldfarb et al., 1992b; Owen et al., 1992; Poulter et al., 1992; Duchen et al., 1993; Goldfarb et al., 1993; Isozaki et al., 1994; Mizushima et al., 1994; Cervenakova et al., 1995; Krasemann et al., 1995; Laplanche 1995; Nicholl et al., 1995; Oda et al., 1995; van Gool et al., 1995; Campbell et al., 1996; Cochran et al., 1996; Goldfarb et al., 1996; Capellari et al., 1997).

Clinical features. The disease phenotype associated with the insertion mutations is highly variable. The clinical and pathological features include the typical CJD phenotype, a phenotype more consistent with GSS, and conditions of several year duration lacking distinctive histopathology (Cochran et al., 1996; Capellari et al., 1997). This phenotypic heterogeneity appears to be related, at least in part, to the heterogeneity of the genotypes associated with these diseases, as it is substantially reduced when insertional mutations are grouped according to size. Overall, the clinical phenotypes, especially the age at onset and the duration of the disease, are related to the number of repeats. A CJD-like clinical phenotype is associated in the majority of cases with four or fewer extra repeats, whereas a GSS-like syndrome is observed in most cases carrying five or more additional repeats. The age at onset seems to correlate inversely, and the duration of disease directly, with the number of repeats. Similar data concerning disease onset and duration have been obtained with the trinucleotide repeat expansion recently characterized in Huntington's disease and in other neurological disorders (La Spada et al., 1994). However, Huntington's alleles are highly unstable when transmitted from parents to children, especially through male meioses (La Spada et al., 1994). This is not the case with the *PRNP* repeat expansion. The anticipation phenomenon that is very characteristic of the trinucleotide repeat expansion disorders is absent in the repeat-expansion families with spongiform encephalopathy. Moreover, as in other inherited prion diseases, codon 129, either on the mutant or on the normal allele, may influence the effects of *PRNP* mutation (Goldfarb et al., 1992c).

Neuropathology. The pathological features mirror quite closely the distribution of the CJD-like and GSS-like clinical phenotypes. Almost 90% (13 out of 15) of the subjects with six or fewer octapeptide inserts, which have undergone autopsy examination, demonstrate histopathological changes consistent with CJD, even if the majority of these cases have a disease duration of more than 1 year. These changes include spongiform degeneration or status spongiosus, astrogliosis, and neuronal loss in different combinations but lacking PrP amyloid plaques outside the granular cell layer of the cerebellum. In contrast, the great majority (six out of seven) of autopsied subjects with seven or more octapeptide inserts show a different histopathological phenotype. Five of these cases, in addition to one carrying a six-octapeptide insertion, have a histopathology consistent with that of GSS. In addition to various degrees of spongiosis, gliosis and neuronal loss, PrP amyloid plaques, often multicentric, are also present in the molecular layer of the cerebellum and very often in the cerebral gray matter, a distribution not seen in CJD. Rarely (1 subject from this group and 1 with six octapeptide repeat insertion out of a total of 22 autopsied cases), there are minimal or not distinctive histological changes, such as astrogliosis and neuronal loss, without spongiosis or PrP amyloid plaques.

Characteristics and allelic origin of PrP^{res}. The pattern of the gel migration and the glycoform ratios of PrP^{res} have been examined in cases with four, five, and six extra insertions (Parchi et al., 1996a). The unglycosylated PK-treated PrP^{res} from subjects with four and six extra insertions migrates on gels at 20 to 21 kDa, corresponding to the PrP^{res} type 1, whereas the corresponding PrP^{res} from the subject with the five extra insertion migrates at 18 to 19 kDa, as the PrP^{res} type 2. The ratios of the glycoforms is similar in the three insertion mutations and is characterized by the predominance of the intermediate glycoform, whereas the highly glycosylated and unglycosylated forms are less abundant and similarly represented as in the sporadic form of CJD (Parchi et al., 1996a, 1996c).

The amount and physicochemical characteristics of PrP^M and PrP^C expressed by the mutant and normal alleles have been examined in brain tissues from subjects carrying five and six extra insertions (Chen et al., 1997). PrP^M and PrP^C were differentiated because of the different size resulting in different gel mobility. The amount of PrP^M was approximately 40% lower than PrP^C with either insertion, and approximately 90% of the PrP^M is insoluble in detergents, an indication that it is aggregated. In contrast, only 50% of the total PrP^C was detergent insoluble. Since the detergent-insoluble fraction was resistant to PK digestion, it was concluded that in the insertional mutations with five and six extra repeats both PrP^M and PrP^C are converted to PrP^{res}.

Transmissibility to animals. Brain suspensions from three of four subjects with five, seven, and eight extra repeats transmitted the disease to primates after intracerebral inoculation (Brown et al., 1994b).

Cell Models of Inherited Prion Diseases

The cell models have provided considerable data and insight on the effect that *PRNP* mutations have on the metabolism and physicochemical properties of PrP^M, including its convertibility to PrP^{res}.

The characteristics of the mutant PrP have been studied in neuroblastoma cells transfected with the *PRNP* pathogenic mutations D178N, E200K, Q217R, F198S, and Y145STOP (Capellari et al., 1996; Petersen et al., 1996; Singh et al., 1997; Zanusso et al., 1997; Zaidi, S.I., Capellari, S., Gambetti, P., Petersen, R.B., unpublished). The studies of Petersen et al. (1996) and Singh et al. (1997) have demonstrated shared metabolic changes as well as changes specific to each mutation. Major common changes include the following:

1. Mutant PrP is unstable and may undergo degradation or aggregation.
2. Since PrP^M, especially the unglycosylated form, is less stable, it is transported to the plasma membrane in decreased amount. As a result, less PrP is present at the surface of mutant cells, and the ratio of the three glycoforms is altered owing to preferential underrepresentation of the unglycosylated form.
3. Overall, more PrP^M is recovered in the detergent insoluble (P2) fraction, showing that PrP^M has the tendency to aggregate.
4. A small fraction of PrP^M displays a mild resistance to digestion with PK (3.3 g/ml for 5 to 10 minutes) an order of magnitude lower than that of the PrP^{res} present in the corresponding human disease (50 g/ml for 30 minutes).
5. Most of these alterations are corrected by maintaining the mutant cells at 24°C, indicating that they are secondary to the misfolding of the mutant PrP, which is known to be partially corrected at lower temperatures (Denning et al., 1992).

In addition, transfected neuroblastoma cells display PrP^M changes more specifically associated with the individual mutations. In some mutations, abnormal PrP^M fragments are formed in significant amounts. For example, a relatively stable 32-kDa PrP^M fragment lacking the glycosylphosphatidylinositol (GPI) anchor, as well as two unstable anchor-carrying fragments, have been observed in cells transfected with the Q217R mutant construct and not with other constructs (Singh et al., 1997). Further, the Y145STOP mutation is associated with the corresponding truncated PrP^M fragment which is extremely unstable even at 15°C, suggesting a possible correlation between the extreme instability and the severity of this particular mutation (Zanusso et al., 1997). In the F198S mutation, the unglycosylated PrP^M is underrepresented not because it is degraded as in other mutations, but apparently because PrP^M is over glycosylated (Zaidi, S.I., et al., unpublished). The E200K mutation is characterized by retarded and abnormal maturation of the N-glycans, resulting in altered gel mobility of the diglycosylated or "mature" PrP^M form (Capellari et al., 1996). Finally, in transfected cells carrying the D178N, 129M, and the D178N, 129V haplotypes, corresponding to FFI and familial CJD linked to the D178N mutation, respectively, the instability of the PrP^M unglycosylated form is especially severe and results in an altered glycoform ratio at the cell surface. These changes are more conspicuous in the FFI than in the CJD178 cell model. Moreover, the unglycosylated form of PrP^M was also found to be preferentially underrepresented in the membrane fraction isolated from FFI brains but not from brains of control subjects (Petersen et al., 1996). This finding suggests that the underrepresentation of the PrP^{res} unglycosylated form in FFI and CJD178 is the result of the relative unavailability of this form for conversion to PrP^{res} resulting from the mutation, rather than the preferential conversion of the highly glycosylated forms to PrP^{res} (Petersen et al., 1996; Parchi et al., 1998a).

Comprehensively, comparable results have been obtained by Harris, Lehmann and coworkers (Lehmann and Harris 1995, 1996a, 1996b, 1997; Daude et al., 1997). These authors observed a number of characteristics shared by all these mutations, which include the following: (1) all PrP^M forms are in part hydrophobic, detergent insoluble and mildly protease resistant; and (2) PrP^M acquires these characteristics through distinct and successive steps. A finding that is at variance with that of other studies is that the GPI anchor of the mutant PrP can be cleaved by PI-PLC, but PrP^M remains attached to the cell membrane consistent with an abnormal PrP^M-membrane association (Lehmann and Harris, 1995).

Despite some differences, the cell models show that *PRNP* mutations cause profound and early changes in the metabolism and physicochemical characteristics of PrP^M. These changes appear to result from an altered conformation that PrP^M acquires co-translationally. The altered conformation confers characteristics of increased hydrophobicity, aggregability, and resistance to proteases reminiscent of those of the PrP^{res}. Whether the "PrP^{res}-like" PrP^M is already pathogenic and capable of transmitting the disease, is the direct precursor of the pathogenic PrP^{res}, or is an irrelevant early byproduct of the mutation and longer times are needed for the "real" mutant PrP^{res} to be formed, remains to be established.

PRION DISEASES ACQUIRED BY INFECTIOUS MECHANISM

This group of diseases includes kuru, the conditions acquired through medical or surgical treatments, called iatrogenic, and, potentially, the so-called new variant of CJD (vCJD), although the mode of acquisition of this variant has not yet been definitely established. According to the prion hypothesis (Prusiner, 1991), the prion diseases acquired by infection are caused by exogenous PrP^{res} that forms heterodimers with the endogenous PrP^{C}, leading to the formation of endogenous PrP^{res}. The endogenous PrP^{res} would activate an autocatalytic process, resulting in the formation of sufficient PrP^{res} to cause the disease. The mechanism by which PrP^{res} reaches the brain when the port of entry of the exogenous PrP^{res} is in the gastrointestinal (GI) tract, as in the case of kuru or, possibly, the vCJD, has not been determined. It is believed that the initial conversion occurs in the GI tract itself, where exogenous and endogenous PrP first meet. Then, in a stepwise or domino fashion, the PrP^{C}-PrP^{res} conversion continues through the lymphatic or the peripheral nervous system, or both, until it reaches the central nervous system. At this point, the high level of PrP^{C} expression and the stability of the neuronal cell population would facilitate the conversion and the accumulation of the abnormal protein.

Kuru

First described (Gajdusek and Zigas, 1957) as an epidemic among the Fore people of New Guinea, kuru (which means "shiver" or "shake") is believed to have been initiated and spread by ritualistic endocannibalism when flesh from a CJD infected individual was consumed. Kuru has virtually disappeared following the end of this ritual. The original description of kuru pathology (Klatzo et al., 1959) was based on 12 cases of ages between 5 and 50, and disease durations between 5 and 12 months. The histopathology and immunohistochemistry are similar to those of groups 3 and 4 of sporadic CJD. The "punched out" or kuru plaques appear in approximately half of the cases, and occur more frequently in the granular cell layer of the cerebellum, but are also in the cerebellar molecular layer, cerebral cortex and basal ganglia (Klatzo et al., 1959). Other types of plaques such as diffuse and "florid" are only occasionally present. The other consistent changes are neuronal loss, gliosis, and spongiosis, which, however, was only noted in subsequent examinations (Scrimgeour et al., 1983; Hainfellner et al., 1997). The lesions are present throughout the nervous system, but, as in sporadic CJD of groups 3 and 4, they are often more severe in the basal ganglia, thalamus, and brain stem. PrP immunostaining, recently performed in one case (Hainfellner et al., 1997) demonstrates a "synaptic" with strong perineuronal enhancement and a laminar distribution in the deep layers of the neocortex. In addition, it shows numerous immunoreactive PrP deposits and kuru plaques similar to groups 3 and 4 of sporadic

CJD. In agreement with these findings, PrP^{res} of type 2 has been observed in two examined cases (Parchi et al., 1997).

Iatrogenic Creutzfeldt-Jakob Disease

Since the first report of accidental human transmission of CJD following a corneal transplant in 1974 (Duffy et al., 1974), at least 176 case of iatrogenic human transmission of CJD have been recorded (Brown et al., 1994a). Differences in incubation time, phenotypic expression of the disease, and genetic susceptibility have been observed among the affected individuals. The route of infection, which is either systemic or peripheral, as in hormone administration, or by direct contact with the central nervous system, as in transplants, is believed to play a major role in determining this variability. For example, methionine homozygosity at codon 129 increases over twofold the risk for the central type of iatrogenic CJD, whereas methionine/valine heterozygosity decreases the risk almost sevenfold. In contrast, valine homozygosity increases the risk of the peripheral type by over threefold, whereas methionine homozygosity increases it only 0.2-fold (Collinge et al., 1991; Brown et al., 1994a; Deslys et al., 1994). Either PrP^{res} type 1 or type 2 are associated with iatrogenic CJD (Parchi et al., 1997). Thus, the acquired forms of human prion diseases such as iatrogenic CJD and kuru reproduce the types of PrP^{res} seen in sporadic CJD, indicating that the PrP^{res} characteristics are reproduced after human-human transmission (Parchi et al., 1997). The PrP^{res} type appears to correlate better with the codon 129 rather than with the route of transmission, and PrP^{res} type 2 codistributes with the presence of the valine codon at position 129 (Parchi et al., 1997).

The *iatrogenic CJD acquired by the central route of infection* has incubation times that vary between 16 and 28 months in subjects treated with neurosurgical procedures and intracerebral implanted electrodes, which directly expose the central nervous tissue to contamination. Incubation times are similar following corneal transplant, whereas they are much longer, 1.5 to 10 years, following dura mater implant (Brown et al., 1992b). In the cases associated with dura transplants, the most common clinical presentations appears to be a mixed cerebellar and cognitive syndrome, while an isolated cognitive impairment or cerebellar syndrome at the early stage is less common (Miyashita et al., 1991; Willison et al., 1991; Esmonde et al., 1993; Lane et al., 1994; Martinez-Loge et al., 1994; Yamada et al., 1994). The typical EEG activity is present in the majority of the cases. The course is on average 14 months with a range of 3 to 27 months. The histopathology is characterized by the presence of spongiosis, gliosis, and neuronal loss not only in the cerebral cortex and deep gray nuclei, but also in the cerebellum and brain stem. Kuru-like PrP plaques have been observed in only one case (Yamada et al., 1994).

The *iatrogenic CJD acquired by the peripheral route of infection* has a mean incubation time that varies between 5 and 17 years, according to the country in which the infection occurred, having a mean of 17, 10, and 5 years in the United

States, United Kingdom, and France, respectively, with an overall range of 4 to 30 years. In ten reported cases, the clinical presentation appears to be almost invariably a cerebellar syndrome while the typical EEG activity is lacking (Weller et al., 1986; Cochius et al., 1992; Ellis et al., 1992; Billette de Villemeur et all, 1994; Masson et al., 1994; Holmes et al., 1996). The mean duration is 15 months, with a range of 6 to 36 months. The histopathology is very similar to that of sporadic CJD of group 4, although the PrP amyloid plaques are more frequent in the iatrogenic form and are also present in subjects homozygous for methionine at codon 129.

New Variant of Creutzfeldt-Jakob Disease

In 1996, Will et al. reported ten atypical cases of CJD in the United Kingdom recorded by the British National CJD Surveillance Unit (Will et al., 1996). These cases were remarkably homogeneous and differed from the typical CJD for both clinical and histopathological features. Shortly thereafter, a similar case was observed in France (Chazot et al., 1996), and additional cases of the same variant were seen in the United Kingdom, so that the current number of confirmed vCJD cases is 15. The clinical features that distinguish these cases from the typical sporadic CJD are the early age of onset, 16 to 48 years, the relatively long duration, 7.5 to 30 months, and the presentation of behavioral changes requiring psychiatric consultation, dysesthesias and pain which are then followed by more common CJD signs such as ataxia, cognitive impairment, and myoclonus. The typical EEG activity is lacking. The pathological hallmark of the vCJD is the presence plaques encircled by a ring of spongiform vacuoles clearly visible with common histological stains. These plaques have a widespread distribution but are more prominent in the cerebral cortex, especially the occipital lobe, and the cerebellum (Ironside, 1996). These plaques, although observed in animal scrapie and named florid plaques, are only exceptionally seen in sporadic CJD, including CJD previously observed in young subjects. In addition, focal areas of spongiform changes, astrogliosis and neuronal loss are present with a widespread distribution but more prominently in the basal ganglia and thalamus (Ironside, 1996). Immunostaining for PrP demonstrates, in addition to the florid plaques, the presence of a large number of nonamyloid PrP deposits throughout the cerebral and cerebellar cortices (Ironside, 1996).

The evidence supporting the acquisition of this form from the bovine spongiform encephalopathy (BSE) is undeniable, but not definitive. The great majority of the cases of vCJD have been observed in the United Kingdom, where the human exposure to BSE is most likely to have occurred, since in the second half of 1980 there was an epidemic of this condition (Wells and Wilesmith, 1995). This would be consistent with an incubation time of 5 to 10 years, comparable to that of kuru and of the peripheral variant of iatrogenic CJD (Will et al., 1996). BSE can be transmitted to cynomologous macaque, and the histopathology of the affected animals is similar to that described in humans but distinct from that of macaques inoculated with brain tissue from subjects with sporadic CJD (Collinge et al., 1996). There is a 94.4% homology between the human and cynomologous macaque, and the macaque is homozygous methionine at codon 129 as all subjects with vCJD reported to date (Chazot et al., 1996; Collinge et al., 1996; Will et al., 1996). Moreover, the PrP^{res} associated with the vCJD has a distinctive ratio of glycoforms, similar to that observed in wild-type mice inoculated with vCJD tissue and to that of the BSE PrP^{res} itself (Lasmezas et al., 1996). Finally, BSE PrP^{res} converts to PrP^{res} human PrP^{C} in a cell-free system (Raymond et al., 1997). However, the efficiency of this conversion is not higher than that with which sheep scrapie PrP^{res} converts human PrP^{C}, although there is no evidence supporting the transmission of sheep scrapie to humans (Raymond et al., 1997). Therefore, the acquisition of vCJD from BSE, although possible, is still not proven.

SPORADIC CREUTZFELDT-JAKOB DISEASE

CJD received the current eponym from Spielmeyer. Only two of the five cases originally reported by Jakob that have been reexamined appear to fulfill the current criteria for the diagnosis of this disease (Richardson and Masters, 1995). With the observation of new cases, the phenotypic heterogeneity of the sporadic form became evident. The first variant was described by Heidenhain in 1929 and is characterized by a short disease duration and cortical blindness associated with a pathology that is more prominent in the occipital cortex (Kirschbaum, 1968; Manuelidis et al., 1987). Over the years, several other variants have been introduced, which include the myoclonic, the cerebellar or ataxic, the thalamic, and the panencephalopathic (Richardson and Masters, 1995). Moreover, cases have been reported with an atypically long disease duration, which made the diagnosis difficult.

Molecular Classification of Sporadic Creutzfeldt-Jakob Disease

Recently, the classification of the sporadic form of CJD (sCJD) has been revised on the basis of molecular features. Six distinct groups or variants have been identified on the basis of the genotype at the *PRNP* codon 129 and of the characteristics of the PrP^{res} (Parchi et al., 1996b, 1998c).

Group 1 includes both subjects who are homozygous for methionine and subjects who are methionine/valine heterozygous at codon 129 and have a PrP^{res} of type 1; that is, PrP^{res}, which has an electrophoretic mobility of approximately 20 kDa for the unglycosylated form and a distinctive ratio of the three glycoforms. Group 1, which is superlatively the most common, is associated with the typical sporadic CJD phenotype, but it also comprises the Heidenhain variant.

Group 2 includes subjects who also are homozygous for methionine at codon 129 but have PrP^{res} type 2, which is characterized by a PK-resistant unglycosylated fragment of approximately 19k Da and a glycoform ratio slightly different from that of PrP^{res} type 1. Subjects of this group show a slower progression than those of group 1, having a duration of symptoms that varies between 1 and 4 years. Patients of this group sometimes receive the diagnosis of Alzheimer's disease or of other nonacute dementing illnesses.

Groups 3 and 4 include subjects who are either heterozygous methionine/valine (group 3) or homozygous for valine at codon 129, and demonstrate PrP^{res} type 2 (group 4). They have considerable similarities between them, but are different from groups 1 and 2 as they display Kuru plaques (group 3) or plaque-like PrP deposits (both groups) as well as a pathology and a distribution of the PrP^{res} involving more the caudal regions of the brain, especially in group 4.

Current View of the Etiology of Sporadic Creutzfeldt-Jakob Disease

Group 5 includes subjects homozygous valine at codon 129 carrying PrP^{res} type 1 with a disease of relatively long duration, whereas *group 6* includes subjects with homozygous methionine at codon 129 and PrP^{res} type 2, as those of group 2 but with a clinical and pathological phenotype similar to that of FFI, suggesting that these cases are the sporadic forms of this disease or sporadic fatal insomnia (SFI) (Parchi et al., 1998b).

The finding of two types of PrP^{res} associated with distinct clinico-pathological phenotypes in subjects with the same *PRNP* haplotype has provided compelling evidence for the existence of prion strains in humans (Parchi et al. 1996c). Distinct isolates or strains of the agent causing prion disease have been originally recognized in scrapie. When passaged in PrP syngenic mice, each strain shows highly preserved characteristics such as incubation times, neuropathological profile, and distribution of PrP^{res} (Fraser and Dickinson, 1973; Bruce et al., 1991). These observations imply the existence of an informational molecule that propagates the strain-specific properties. According to the prion hypothesis, the candidate molecule is the PrP^{res} itself, and the agent strain differences are mediated by stable variations in its conformation. Although it is argued that the number of different PrP^{res} conformers required to account for the diversity of scrapie strains would be unreasonable, evidence for strain-specific differences in PrP^{res} physiochemical properties have, indeed, been recently described for scrapie (Kascsak et al., 1986), transmissible mink encephalopathy (Bessen and Marsh, 1994), and CJD (Parchi et al., 1996b, 1996c, 1997; Telling et al., 1996; Safar et al., 1998).

The striking difference in the distribution of lesions and PrP^{res} accumulation in the early stages of the disease between sporadic CJD subjects of different variants (e.g., groups 1 and 4), even in subjects with the same *PRNP* haplotype, strongly suggest that distinct neuronal population are specifically targeted at onset. This implies a role for the putative informational molecule of the agent well before the occurrence of the first PrP^{c}-PrP^{res} conversion, the event that, according to the prion hypothesis, would make PrP^{res} the carrier of strain-specific information.

The epidemiology of sporadic CJD may also provide some insights regarding the etiology of this disease. Increasing evidence indicates that codon 129 of *PRNP* modulates the susceptibility to the disease (Palmer et al., 1991; Laplanche et al., 1994; Salvatore et al., 1994; Parchi et al., 1996c; Windl et al., 1996). This theory is based on the findings that methionine and, possibly, valine homozygosity are, respectively, more prevalent among subjects with sporadic CJD than in the general population, and that methionine/valine heterozygosity is nearly three times less prevalent (Palmer et al., 1991; Laplanche et al., 1994; Salvatore et al., 1994; Windl et al., 1996). Furthermore, the frequency distribution of age at onset in sporadic CJD is symmetrical around the mean of 60 years, and does not increase exponentially after a certain age (Brown et al., 1994b). Both these characteristics of sporadic CJD are difficult to explain by a stochastic event, and anticipate clarification.

ACKNOWLEDGMENTS

This work was supported in part by PHS grants NS 29822, NS 14426, AG10133, AG08012, AG08155, AG08992, AG14359, and the Britton Fund.

REFERENCES

Adam, J., Crow, T.J., Duchen, L.W., Scaravilli, F. & Spokes, E. (1982). Familial cerebral amyloidosis and spongiform encephalopathy. J. Neurol. Neurosurg. Psychiatry 545, 37–45.

Amano, N., Yagishita, S., Yokoi, S., Itoh, Y., Kinoshita, J., Mizutani, T., Matsuishi, T. (1992). Gerstmann-Sträussler-Scheinker syndrome—a variant type: amyloid plaques and Alzheimer's neurofibrillary tangles in cerebral cortex. Acta Neuropathol. 84, 15–23.

Antonine, J.C., Laplanche, J.L., Mosnier, J.F., Beaudry, P., Chatelain, J. & Michel, D. (1996). Demyelinating peripheral neuropathy with Creutzfeldt-Jakob disease and mutation at codon 200 of the prion protein gene. Neurology 46, 1123–1127.

Baker, H.F., Duchen, L.W., Jacobs, J.M. & Ridley, R.M. (1990). Spongiform encephalopathy transmitted experimentally from Creutzfeldt-Jakob and familial Gerstmann-Sträussler-Scheinker diseases. Brain 113, 1891–1909.

Barbanti, P., Fabbrini, G., Salvatore, M., Petraroli, R., Pocchiari, M., Macchi, G. & Lenzi, G.L. (1994). No correlation between clinical heterogeneity and codon 129 polymorphism of the prion protein gene (*PRNP*) in Gerstmann-Sträussler-Scheinker syndrome (GSS) with *PRNP* codon 102 mutation. Neurobiol. Aging 15, S156.

Bessen, R.A. & Marsh, R.F. (1994). Distinct PrP properties suggest the molecular basis of strain variation in transmissible mink encephalopathy. J. Virol. 68, 7859–7868.

Billette de Villemeur, T., Gelot, A., Deslys, J.P., et al. (1994). Iatrogenic Creutzfeldt-Jakob disease in three growth hormone recipients: a neuropathological study. Neuropath. Appl. Neurobiol. 20, 111–117.

Bolton, D.C., McKinley, M.P. & Prusiner, S.B. (1982). Identification of a protein that purifies with the scrapie prion. Science 218, 1309–1311.

Borchelt, D.R., Taraboulos, A., & Prusiner, S.B. (1992) Evidence for synthesis of scrapie prion proteins in the endocytic pathway. J. Biol. Chem. 267, 16188–16199.

Borchelt, D.R., Koliatsos, V.E., Guarnieri, M., Pardo, C.A., Sisodia, S.S. & Price, D.L. (1994). Rapid anterograde axonal transport of the cellular prion glycoprotein in the peripheral and central nervous system. J. Biol. Chem. 269, 14711–14714.

Brown, P., Cathala, F., Castaigne, P. & Gajdusek, D.C. (1986). Creutzfeldt-Jakob disease: clinical analysis of a consecutive series of 230 neuropathological verified cases. Ann. Neurol. 20, 597–602.

Brown, P., Goldfarb, L.G., Gibbs, C.J., Jr. & Gajdusek, D.C. (1991). The phenotypic expression of different mutations in transmissible familial Creutzfeldt-Jakob disease. Eur. J. Epidemiol. 7, 469–476.

Brown, P., Goldfarb, L.G., Kovanen, J., Haltia, M., Cathala, F., Sulima, M., Gibbs, C.J. & Gajdusek, D.C. (1992a). Phenotypic characteristics of familial Creutzfeldt-Jakob disease associated with the codon 178 Asn PRNP mutation. Ann. Neurol. 31, 282–285.

Brown, P., Preece, M. A. & Will, R. G. (1992b). "Friendly fire" in medicine: hormones, homografts, and Creutzfeldt-Jakob disease. Lancet 340, 24–27.

Brown, P., Cervenakova, L., Goldfarb, L.G., McCombie, W.R., Rubenstein, R., Will, R.G., Pocchiari, M., Martinezlage, J.F., Scalici, C., Masullo, C., Graupera, G., Ligan, J. & Gajdusek, D.C. (1994a). Iatrogenic Creutzfeldt-Jakob disease: an example of the interplay between ancient genes and modern medicine. Neurology 44, 291–293.

Brown, P., Gibbs, C.J., Rodgers-Johnson, P., Asher, D.M., Sulima, M.P., Bacote, A., Goldfarb, L.G. & Gajdusek, D.C. (1994b). Human spongiform encephalopathy: the National Institutes of Health series of 300 cases of experimentally transmitted disease. Ann. Neurol. 35, 513–529.

Bruce, M.E., McConnel, I., Fraser, H. & Dickinson, A.G. (1991) The disease characteristics of different strains of scrapie in Sinc congenic mouse lines: implication for the nature of the agent and the host control pathogenesis. J. Gen. Virol. 72, 595–603.

Bugiani, O., Giaccone, G., Verga, L., Pollo, B., Frangione, B., Farlow, M.R., Tagliavini, F. & Ghetti, B. (1993). βPP participates in PrP-amyloid plaques of Gerstmann-Sträussler-Scheinker disease, Indiana kindred. J. Neuropathol. Exp. Neurol. 52, 64–70.

Campbell, T.A., Palmer, M.S., Will, R.G., Gibb, W.R.G., Luthert, P.J. & Collinge, J. (1996). A prion disease with a novel 96-base pair insertional mutation in the prion protein gene. Neurology 46, 761–766.

Capellari, S., Parchi, P., Urig, S.B., Richardson, S.L., Russo, C., Gambetti, P. & Petersen, R.B. (1996). E200K mutation: effect on prion protein. Soc. Neurosci Abs. 22, 711.

Capellari, S., Vital, C., Parchi, P., Petersen, R.B, Ferrer, X., Jarnier, D., Pegoraro, E., Gambetti, P. & Julien, J. (1997). Familial prion disease with a novel 144-bp insertion in the prion protein gene in a Basque family. Neurology 49, 133–141.

Cathala, F., Brown, P., LeCanuet, P. & Gajdusek, D.C. (1985). High incidence of Creutzfeldt-Jakob disease in North African immigrants to France. Neurology 35, 894–895.

Caughey, B., Race, R.E., Ernst, D., Buchmeier, M.J. & Chesebro, B. (1989). Prion protein biosynthesis in scrapie-infected and uninfected neuroblastoma cells. J. Virol. 63, 175–181.

Caughey, B.W., Dong, A., Bhat, K.S., Ernst, D., Hayes, S.F. & Caughey, W.S. (1991a). Secondary structure analysis of the scrapie associated protein PrP 27–30 in water by infrared spectroscopy. Biochemistry 30, 7672–7680.

Caughey, B., Raymond, G.J., Ernst, D. & Race, R.E. (1991b). N-terminal truncation of the scrapie-associated form of PrP by lysosomal protease(s): Implications regarding the site of conversion of PrP to the protease-resistant state. J. Virol. 65, 6597–6603.

Cervenakova, L., Goldfarb, L.G., Brown, P., Kenney, K., Cochran, E.J., Bennett, D.A., Roos, R. & Gajdusek, D.C. (1995). Three new PRNP genotypes associated with familial Creutzfeldt-Jakob disease [abstract]. Am. J. Hum. Genet. 57, A209.

Chapman, J., Brown, P., Rabey, J. M., Goldfarb, L.G., Inzelberg, R., Gibbs, C.J., Jr., Gajdusek, D.C. & Korczyn, A.D. (1992). Transmission of spongiform encephalopathies from a familial Creutzfeldt-Jakob disease patient of Jewish Libyan origin carrying the PRNP codon 200 mutation. Neurology 42, 1249–1250.

Chapman, J., Brown, P., Goldfarb, L.G., Arlazoroff, A., Gajdusek, D.C. & Korczyn, A.D. (1993). Clinical heterogeneity and unusual presentations of Creutzfeldt-Jakob disease in Jewish patients with the PRNP codon 200 mutation. J. Neurol. Neurosurg. Psychiatry 56, 1109–1112.

Chapman, J., Ben-Israel, J., Goldhammer, Y. & Korczyn, A.D. (1994). The risk of developing Creutzfeldt-Jakob disease in subjects with the *PRNP* gene codon 200 point mutation. Neurology 44, 1683–1686.

Chazot, G., Broussolle, E., Lapras, C., Blättler, T., Aguzzi, A. & Kopp, N. (1996). New variant of Creutzfeldt-Jakob disease in a 26-year-old French man. Lancet 347, 1181.

Chen, S.G., Teplow, D.B., Parchi, P., Teller, J.K., Gambetti, P., & Autilio-Gambetti, L. (1995). Truncated forms of the human prion protein in normal brain and in prion diseases. J. Biol. Chem. 270, 19173–19180.

Chen, S.G., Parchi, P., Brown, P., Zou, W., Cochran, E.J., Vnencak-Jones, C.L., Julien, J., Vital, C., Mikol, J., Lugaresi, E., Autilio-Gambetti, L. & Gambetti, P. (1997). Allelic origin of the abnormal prion proteins in familial prion diseases. Nat. Med. 3, 1009–1015.

Chesebro, B., Race, R., Wehrly, K., Nishlo, J., Bloom, M., Lechner, D., Bergstrom, S., Bobbins, K., Mayer, L., Keith, J.M., Garon, C. & Haasse, A. (1985). Identification of scrape prion protein-specific mRNA in scrapie-infected and uninfected brain. Nature 315, 331–333.

Cochius, J. I., Hyman, N. & Esiri, M.M. (1992). Creutzfeldt-Jakob disease in a recipient of human pituitary-derived gonadotrophin: a second case. J. Neurol. Neurosurg. Psychiatry 55, 1094–1095.

Cochran, E.J., Bennett, D.A., Cervenakova, L., Kenney, K., Bernard, B., Foster, N.L., Benson, D.F., Goldfarb, L.G. & Brown, P. (1996). Familial Creutzfeldt-Jakob disease with a five-repeat octapeptide insert mutation. Neurology 47, 727–733.

Cohen, D., Krausz, Y., Lossos, A., Ben-David, E. & Atlan, H. (1989). Brain SPECT imaging with Tc-99m HM-PAO in Creutzfeldt-Jakob disease. Clin. Nucl. Med. 14, 808–810.

Collinge, J., Palmer, M.S. & Dryden, A.J. (1991) Genetic predisposition to iatrogenic Creutzfeldt-Jakob diseae. Lancet 337, 1441–1442.

Collinge, J., Brown, J., Hardy, J., Mullan, M., Rossor, M. N., Baker, H., Crow, T.J., Lofthouse, R., Poulter, M., Ridley, R., Owen, F., Bennett, C., Dunn, G., Harding, A.E., Quinn, N., Doshi, B., Roberts, G.W., Honavar, M., Janota, I. & Lantos, P.L. (1992). Inherited prion disease with 144 base pair gene insertion. Brain 115, 687–710.

Collinge, J., Whittington, M.A., Sidle, K.C., Smith, C.J., Palmer, M.S., Clarke, A.R. & Jefferys, J.G.R. (1994). Prion protein is necessary for normal synaptic function. Nature 370, 295–297.

Collinge, J., Palmer, M.S., Sidle, K.C., Gowland, I., Medori, R., Ironside, J., & Lantos, P. (1995). Transmission of fatal familial insomnia to laboratory animals. Lancet 346, 569–570.

Collinge, J., Sidle, K.C.L., Meads, J., Ironside, J. & Hill, A.F. (1996). Molecular analysis of prion stain variation and the aetiology of "new variant" CJD. Nature 383, 685–690.

Creutzfeldt, H.G. (1920). Über eine eigenartige herdförmige Erkrankung des Zentralnervensystems. Z. Neurol. Psychiat. 57, 1–18.

Creutzfeldt, H.E. (1921). Über eine eigenartige herdförmige Erkrankung des Zentralnervensystems. In: Histologische und Histopathologische Arbeiten ber die Grosshirnrinde (Nissl, F. & Alzheimer, A., eds.), Vol. 6, pp. 1–48. Gustav Fischer, Jena.

Daude, N., Lehmann, S. & Harris, D.A. (1997). Identification of intermediate steps in the conversion of a mutant prion protein to a scrapie-like form in cultured cells. J. Biol. Chem. 272, 11604–11612.

DeArmond, S.J. & Prusiner, S.B. (1997). Prion diseases. In: Greenfield Neuropathology (Graham, D.I. & Lantos, P.L., ed.), pp. 235–280. Oxford University Press, London.

Denning, G.M., Anderson, M.P., Amara, J.F., Marshall, J., Smith, A.E. & Welsh, M.J. (1992). Processing of mutant cystic fibrosis transmembrane conductance regulator is temperature-sensitive. Nature 358, 761–764.

Deslys, J.P., Marce, D. & Dormont, D. (1994). Similar genetic susceptibility in iatrogenic and sporadic Creutzfeldt-Jakob disease. J. Gen. Virol. 75, 23–27.

Dlouhy, S.R., Hsiao, K., Farlow, M.R., Foroud, T., Conneally, P.M., Johnson, P., Prusiner, S.B., Hodes, M.E. & Ghetti B. (1992). Linkage of the Indiana kindred of Gerstmann-Sträussler-Scheinker disease to the prion protein gene. Nat. Genet. 1, 64–67.

Doh-ura, K., Tateishi, J., Sasaki, H., Kitamoto, T. & Sakaki, Y. (1989). Pro → Leu change at position 102 of prion protein is the most common but not the sole mutation related to Gerstmann-Sträussler syndrome. Biochem. Biophys. Res. Commun. 163, 974–979.

Duchen, L.W., Poulter, M. & Harding, A.E. (1993). Dementia associated with a 216 base pair insertion in the prion protein gene. Clinical and neuropathological features Brain 116, 555–567.

Duffy, P., Wolf, J., Collins, G., Devoe, A., Streeten, B. & Cowen, D. (1974). Possible person to person transmission of Creutzfeldt-Jakob disease. N. Engl. J. Med. 290, 692–693.

Ellis, C.J., Katifi, H. & Weller, R.O. (1992). A further British case of growth hormone induced Creutzfeldt-Jakob disease. J. Neurol. Neurosurg. Psychiatry 55, 1200–1202.

Endo, T., Groth, D., Prusiner, S.B. & Kobata, A. (1989). Diversity of oligosaccharide structures linked to asparagines of the scrapie prion protein. Biochemistry 28, 8380–8388.

Esmonde, T., Lueck, C.J., Symon, L., Duchen, L.W. & Will, R.G. (1993). Creutzfeldt-Jakob disease and lyophilised dura mater grafts: report of two cases. J. Neurol. Neurosurg. Psychiatry 56, 999–1000.

Farlow, M.R., Yee, R.D., Dlouhy, S.R., Conneally, P.M., Azzarelli, B. & Ghetti, B. (1989). Gerstmann-Sträussler-Scheinker disease. I. Extending the clinical spectrum. Neurology 39, 1446–1452.

Fraser, H. & Dickinson, A.G. (1973). Scrapie in mice. Agent-strain differences in the distribution and intensity of grey matter vacuolation. J. Comp. Path. 83, 29–40.

Furukawa, H., Kitamoto, T., Tanaka, Y., & Tateishi, J. (1995). New variant prion protein in a Japanese family with Gerstmann-Sträussler syndrome. Mol. Brain Res. 30, 385–388.

Furukawa, H., Kitamoto, T., Hashiguchi, H. & Tateishi, J. (1996). A Japanese case of Creutzfeldt-Jakob disease with a point mutation in the prion protein gene at codon 210. J. Neurol. Sci. 141, 120–122.

Gabizon, R., Kahana, E., Hsiao, K., Prusiner, S.B. & Meiner, Z. (1992). Inherited prion disease in Libyan Jews. In: Prion Diseases of Humans and Animals (Prusiner, S.B., Collinge, J., Powell, J. & Anderton, B., eds.), pp. 168–179. Ellis Horwood, London.

Gabizon, R., Rosenmann, H., Meiner, Z., Kahana, I., Kahana, E., Shugart, Y., Ott, J. & Prusiner, S.B. (1993). Mutation and polymorphism of the prion protein gene in Libyan Jews with Creutzfeldt-Jakob disease. Am. J. Hum. Genet. 53, 828–835.

Gabizon, R., Telling, G., Meiner, Z., Halimi, M., Kahana, I. & Prusiner, S.B. (1996). Insoluble wild-type and protease-resistant mutant prion protein in brains of patients with inherited prion diseases. Nat. Med. 2, 59–64.

Gajdusek, D.C. & Zigas, V. (1957) Degenerative disease of the central nervous system in New Guinea. The endemic occurrence of "Kuru" in the native population. N. Eng. J. Med. 257, 974–978.

Gajdusek, D.C., Gibbs, C.J., Jr. & Alpers, M. (1966). Experimental transmission of a Kuru-like syndrome to chimpanzees. Nature 209, 794–796.

Galvez, S. & Cartier, L. (1984). Computed tomography findings in 15 cases of Creutzfeldt-Jakob disease with histological verification. J. Neurol. Neurosurg. Psychiatry 47, 1244–1246.

Gasset, M., Baldwin, M.A., Fletterick, R.J. & Prusiner, S.B. (1993) Perturbation of the secondary structure of the scrapie prion protein under conditions associated with changes in infectivity. Proc. Natl. Acad. Sci. U. S. A. 90, 1–5.

Ghetti, B., Tagliavini, F., Masters, C.L., Beyreuther, K., Giaccone, G., Verga, L., Farlow, M.R., Conneally, P.M., Dlouhy, S.R., Azzarelli, B. & Bugiani, O. (1989). Gerstmann-Sträussler-Scheinker disease. II. Neurofibrillary tangles and plaques with PrP-amyloid coexist in an affected family. Neurology 39, 1453–1461.

Ghetti, B., Tagliavini, F., Hsiao, K., Dlouhy, S.R., Yee, R.D., Giaccone, G., Conneally, P.M., Hodes, M.E., Bugiani, O., Prusiner, S.B., Frangione, B. & Farlow, M.R. (1992). Indiana variant of Gerstmann-Sträussler-Scheinker disease. In: Prion diseases of Humans and Animals. (Prusiner, S., Collinge, J., Powell, J. & Anderton, B., eds), pp. 154–167. Ellis Horwood, New York.

Ghetti, B., Tagliavini, F., Giaccone, G., Bugiani, O., Frangione, B., Farlow, M.R. & Dlouhy, S.R. (1994). Familial Gerstmann-Sträussler-Scheinker disease with neurofibrillary tangles. Mol Neurobiol 8, 41–48.

Ghetti, B., Dlouhy, S.R., Giaccone, G., Bugiani, O., Frangione, B., Farlow, M.R. & Tagliavini, F. (1995). Gerstmann-Sträussler-Scheinker disease and the Indiana Kindred. Brain Pathol. 5, 61–75.

Ghetti, B., Piccardo, P., Frangione, B., Bugiani, O., Giaccone, G., Young, K., Prelli, F., Farlow, M.R., Dlouhy, S.R. & Tagliavini, F. (1996a). Prion protein amyloidosis. Brain Pathol. 6, 127–145.

Ghetti, B., Piccardo, P., Spillantini, M.G., Ichimiya, Y., Porro, M., Perini, F., Kitamoto, T., Tateishi, J., Seiler, C., Frangione, B., Bugiani, O., Giaccone, G., Prelli, F., Goedert, M., Dlouhy, S.R. & Tagliavini, F. (1996b). Vascular variant of prion protein cerebral amyloidosis with τ-positive neurofibrillary tangles: the phenotype of the stop codon 145 mutation in *PRNP*. Proc. Natl. Acad. Sci. U. S. A. 93, 744–748.

Ghetti, B., Young, K., Piccardo, P., Dlouhy, S.R., Pahwa, R., Lyons, K.E., Koller, W.C., Ma, M.J., DeCarli, C. & Rosenberg, R.N. (1998). Gerstmann-Sträussler-Scheinker disease (GSS) with *PRNP* A117V mutation: studies of a new family. Soc. Neurosci. Abstr. 24, 515.

Giaccone, G., Tagliavini, F., Verga, L., Frangione, B., Farlow, M.R., Bugiani, O. & Ghetti, B. (1990). Neurofibrillary tangles of the Indiana kindred of Gerstmann-Sträussler-Scheinker disease share antigenic determinants with those of Alzheimer disease. Brain Res. 530, 325–329.

Giaccone, G., Verga, L., Bugiani, O., Frangione, B., Serban, D., Prusiner, S.B., Farlow, M.R., Ghetti, B. & Tagliavini, F. (1992). Prion protein preamyloid and amyloid deposits in Gerstmann-Sträussler-Scheinker disease, Indiana kindred. Proc. Natl. Acad. Sci. U. S. A. 89, 9349–9353.

Gibbs, C.J.J., Gajdusek, D.C., Asher, D.M., Alpers, M.P., Beck, E., Daniel, P.M. & Matthews, W.B. (1968). Creutzfeldt-Jakob disease (spongiform encephalopathy): transmission to the chimpanzee. Science 161, 388–389.

Goldfarb, L.G., Brown, P., McCombie, W.R., Goldgaber, D., Swergold, G.D., Wills, P.R., Cervenakova, L., Baron, H., Gibbs, C.J. & Gajdusek, D.C. (1991). Transmissible familial Creutzfeldt-Jakob disease associated with five, seven and eight extra octapeptide coding repeats in the PRNP gene. Proc. Natl. Acad. Sci. U. S. A. 88, 10926–10930.

Goldfarb, L.G., Brown, P., Maltia, M., Cathala, F., McCombie, W.R., Kovanen, J., Cervenakova, L., Goldin, L., Nieto, A., Godec, M.S., Asher, D.M. & Gajdusek, D.C. (1992a). Creutzfeldt-Jakob disease cosegregates with the codon 178Asn PRNP mutation in families of European origin. Ann. Neurol. 31, 274–281.

Goldfarb, L.G., Brown, P., Vrbovska, A., Baron, H., McCombie, W.R., Cathala, F., Gibbs, C.J. & Gajdusek, D.C. (1992b). An insert mutation in the chromosome 20 amyloid precursor gene in a Gerstmann-Sträussler-Scheinker family. J. Neurol. Sci. 111, 189–194.

Goldfarb, L.G., Petersen, R.B., Tabaton, M., Brown, P., LeBlanc, A.C., Montagna, P., Cortelli, P., Julien, J., Vital, C., Pendlebury, W.W., Haltia, M., Wills, P.R., Hauw, J.J., McKeever, P.E., Monari, L., Schrank, B., Swergold, G.D., Autilio-Gambetti, L., Gajdusek, D.C., Lugaresi, E. & Gambetti, P. (1992c). Fatal familial insomnia and familial Creutzfeldt-Jakob disease: disease phenotype determined by a DNA polymorphism. Science 258, 806–808.

Goldfarb, L.G., Brown, P., Little, B.W., Cervenakova, L., Kenney, K., Gibbs, C.J. & Gajdusek, D.C. (1993). A new (two-repeat) octapeptide coding insert mutation in Creutzfeldt-Jakob disease. Neurology 43, 2392–2394.

Goldfarb, L.G., Cervenakova, L., Brown, P. & Gajdusek, D.C. (1996). Genotype-phenotype correlations in familial spongiform encephalopathies associated with insert mutations. In: Transmissible Subacute Spongiform Encephalopathies: Prion Diseases (Court, L. & Dodet, B., eds,), pp. 425–431. Elsevier, Paris.

Goldgaber, D., Goldfarb, L.G., Brown, P., Asher, D.M., Brown, W.T., Lin, S., Teener, J.W., Feinstone, S.M., Rubenstein, R., Kascsak, R.J., Boellaard, J.W. & Gajdusek, D.C. (1989). Mutations in familial Creutzfeldt-Jakob disease and Gerstmann-Sträussler-Scheinker's syndrome. Exp. Neurol. 106, 204–206.

Goldhammer, Y., Gabizon, R., Meiner, Z. & Sadeh, M. (1993). An Israeli family with Gerstmann-Sträussler-Scheinker disease manifesting the codon 102 mutation in the prion protein gene. Neurology 43, 2718–2719.

Griffith, J.S. (1967) Self-replication and scrapie. Nature 215, 1043–1044.

Guiroy, D.C., Shankar, S.K., Gibbs, C.J., Messenheimer, J.A., Das, S. & Gajdusek, D.C. (1989). Neuronal degeneration and neurofilament accumulation in the trigeminal ganglia in Creutzfeldt-Jakob disease. Ann. Neurol. 25, 102–106.

Hadlow, W.J. (1959) Scrapie and kuru. Lancet 2, 289–290.

Hainfellner, J.A., Brantner-Inthaler, S., Cervenakova, L., Brown, P., Kitamoto, T., Tateishi, J., Diringer, H., Liberski, P.P., Regele, H., Feucht, M., Mayr, N., Wessely, P., Summer, K., Seitelberger, F. & Budka, H. (1995). The original Gerstmann-Sträussler-Scheinker (GSS) family of Austria: divergent clinicopathological phenotypes but constant PrP genotype. Brain Pathol. 5, 201–211.

Hainfellner, J.A., Liberski, P.P., Guiroy, D.C., Cervenakova, L., Brown, P., Gajdusek, D.C. & Budka, H. (1997). Pathology and immunocytochemistry of a Kuru brain. Brain Pathol. 7, 547–553.

Hitoshi, S., Nagura, H., Yamanouchi, H. & Kitamoto, T. (1993). Double mutations at codon 180 and codon 232 of the PRNO gene in an apparently sporadic case of Creutzfeldt-Jakob disease. J. Neurol. Sci. 120, 208–212.

Holmes, S.J., Ironside, J.W. & Shalet, S.M. (1996). Neurosurgery in a patient with Creutzfeldt-Jakob disease after pituitary derived growth hormone therapy in childhood. J. Neurol. Neurosurg. Psychiatry 60, 333–335.

Hope, J., Morton, L.J.D., Farquhar, C.F., Multhaup, G., Beyreuther, K. & Kimberlin, R.H. (1986). The major polypeptide of scrapie associated fibrils (SAF) has the same size, charge distribution and N-terminal protein sequence as predicted for the normal brain protein (PrP). EMBO J. 5, 2591–2597.

Hoque, M.Z., Kitamoto, T., Furukawa, H., Muramoto, T., & Tateishi, J. (1996). Mutation in the prion protein gene at codon 232 in Japanese patients with Creutzfeldt-Jakob disease: a clinicopathological, immunohistochemical and transmission study. Acta. Neuropathol. 92, 441–446.

Hsiao, K., Baker, H.F., Crow, T.J., Poulter, M., Owen, F., Tewiliger, J., Westaway, D., Ott, J. & Prusiner, S. (1989). Linkage of a prion protein missense variant to Gertsmann-Sträussler syndrome. Nature 338, 342–345.

Hsiao, K.K., Cass, C., Schellenberg, G.D., Bird, T., Devine-Gage, E., Wisniewski, H.M. & Prusiner, S.B. (1991a). A prion protein variant in a family with the telencephalic form of Gerstmann-Sträussler-Scheinker syndrome. Neurology 41, 681–684.

Hsiao, K., Meiner, Z., Kahana, E., Cass, C., Kahana, I., Avrahami, D., Scarlato, G., Abramsky, O., Prusiner, S.B. & Gabizon, R. (1991b). Mutation of the prion protein in Libyan Jews with Creutzfeldt-Jakob disease. N. Engl. J. Med. 324, 1091–1097.

Hsiao, K., Dlouhy, S.R., Farlow, M.R., Cass, C., DaCosta, M., Conneally, P.M., Hodes, M.E., Ghetti, B. & Prusiner, S.B. (1992). Mutant prion proteins in Gerstmann-Sträussler-Scheinker disease with neurofibrillary tangles. Nat. Genet. 1, 68–71.

Ironside, J.W. (1996) Review: Creutzfeldt-Jakob disease. Brain Pathol. 6, 379–388.

Isozaki, E., Miyamoto, K., Kagamihara, Y., Hirose, K., Tanabe, H., Uchihara, T., Oda, M. & Nagashima, T. (1994). CJD presenting as frontal lobe dementia associated with a 96 base pair insertion in the prion protein gene (in Japanese). Dementia 8, 363–371.

Jakob, A. (1921a). Über eigenartige Erkrankung des Zentralnervensystems mit bemerkenswertem anatomischen Befunde (spastische Pseudosklerose-Encephalomyelopathie mit disseminierten Degenerationsherden). Dtsch. Z. Nervenheilkunde 70, 132–146.

Jakob, A. (1921b). Über eigenartige Erkrankung des Zentralnervensystems mit bemerkenswertem anatomischen Befunde (spastische Pseudosklerose-Encephalomyelopathie mit disseminierten Degenerationsherden). Z. Gesamte. Neurol. Psychiat. 64, 147–228.

Jakob, A. (1921c). Über eine der multiplen Sklerose klinisch nahestehende Erkrankung des centralnervensystems (spastische Pseudosklerose) mit bemerkenswertem anatomischem Befunde. Medizinische Klinik 13, 372–376.

Jakob, A. (1923). Spastische Pseudosklerose. In: Monographien aus dem Gesamtgebiete der Neurologie und Psychiatric (Foerster, O. & Wilmanns, K., ed.), Vol. 37, Die Extrapyramidalen Erkrankungen. Mit Besonderer Ber cksichtigung der Pathologischen Anatomie und Histologie und der Pathophysiologie der Bewegungsstörungen, section 8, pp. 215–345. Julius Springer, Berlin.

Jarrett, J.T. & Lansbury, P.T. (1993). Seeding "one-dimensional crystallization" of amyloid: a pathogenic mechanism in Alzheimer's disease and scrapie? Cell 73, 1055–1058.

Kahana, E., Milton, A., Braham, J. & Sofer, D. (1974). Creutzfeldt-Jakob disease: focus among Libyan Jews in Israel. Science 183, 90–91.

Kahana, E., Zilber, N. & Abraham, M. (1991). Do Creutzfeldt-Jakob disease patients of Jewish Libyan origin have unique clinical features? Neurology 41, 1390–1392.

Kascsak, R.J., Rubenstein, R., Merz, P.A., Carp, R.I., Robakis, N.K., Wisniewski, H.M. & Diringer, H. (1986). Immunological comparison of scrapie-associated fibrils isolated from animals infected with four different scrapie strains. J. Virol. 59, 676–683.

Kirschbaum, W.R. (1968). Jakob-Creutzfeldt Disease, pp. 1–25. Elsevier, New York.

Kitamoto, T. & Tateishi, J. (1994). Human prion diseases with variant prion protein. Philos. Trans. R. Soc. London Ser. B. 343, 391–398.

Kitamoto, T., Amano, N., Terao, Y., Nakazato, Y., Isshiki, T., Mizutani, T. & Tateishi, J. (1993a). A new inherited prion disease (PrP-P105L Mutation) showing spastic paraparesis. Ann. Neurol. 34, 808–813.

Kitamoto, T., Iizuda, R. & Tateishi, J. (1993b). An amber mutation of prion protein in Gerstmann-Sträussler-Scheinker syndrome with mutant PrP plaques. Biochem. Biophys. Res. Comm. 192, 525–531.

Kitamoto, T., Ohta, M., Dohura, K., Hitoshi, S., Terao, Y. & Tateishi, J. (1993c). Novel missense variants of prion protein in Creutzfeldt-Jakob disease or Gerstmann-Sträussler syndrome. Biochem. Biophys. Res. Comm. 191, 709–714.

Klatzo, I., Gajdusek, D. C. & Zigas, V. (1959) Pathology of kuru. Lab. Invest. 8, 799–847.

Krasemann, S., Zerr, I., Weber, T., Poser, S., Kretzschmar, H., Hunsmann, G. & Bodemer, W. (1995). Prion disease associated with a novel nine octapeptide repeat insertion in the *PRNP* gene. Mol. Brain. Res. 34, 173–176.

Kretzschmar, H.A., Stowring, L.E., Westaway, D., Stubblebine, W.H., Prusiner, S.B. & DeArmond, S.J. (1986). Molecular cloning of a human prion protein cDNA. DNA 5, 315–324.

Kretzschmar, H.A., Kufer, P., Riethmuller, G., DeArmond, S., Prusiner, S.B. & Schiffer, D. (1992). Prion protein mutation at codon-102 in an Italian family with Gerstmann-Sträussler-Scheinker Syndrome. Neurology 42, 809–810.

Kretzschmar, H.A., Neumann, M. & Stavrou, D. (1995). Codon 178 mutation of the human prion protein gene in a German family (Backer family): sequencing data from 72-year old celloidin-embedded brain tissue. Acta. Neuropathol. 89, 96–98.

Kretzschmar, H.A., Giese, A., Zerr, I., Windl, O., Schulz-Schaeffer, W.J., Skworc, K. & Poser, S. (1998). Fatal familial insomnia (FFI): the German cases. Brain Pathol. 8, 559–561.

Lane, K.L., Brown, P., Howell, D.N., Crain, B.J., Hulette, C.M., Burger, P.C. & DeArmond, S.J. (1994). Creutzfeldt-Jakob disease in a pregnant woman with an implanted dura mater graft. Neurosurgery 34, 737–740.

Laplanche, J.L., Delasnerie-Lauprêtre, N., Brandel, J.P., Chatelain, J., Beaudry, P., Alpérovitch, A., & Launay, J.-M. (1994). Molecular genetics of prion diseases in France. Neurology 44, 2347–2351.

Laplanche, J., Delasnerie-Lauprêtre, N., Brandel, J.P., Dussaucy, M. & Launay, J.M. (1995). Two novel insertions in the prion protein gene in patients with late-onset dementia. Hum. Mol. Gen. 4, 1109–1111.

Lasmezas, C.I., Deslys, J.P., Demalmay, R., Adjou, K.T., Lamoury, F., Dormont, D., Robain, O., Ironside, J. & Hauw, J.J. (1996). BSE transmission to macaques. Nature 381, 743–744.

La Spada, A.R., Paulson, H.L. & Fischbeck, K.H. (1994). Trinucleotide repeat expression in neurological disease. Ann. Neurol. 36, 814–822.

Lehmann, S. & Harris, D.A. (1995). A mutant prion protein displays an aberrant membrane association when expressed in cultured cells. J. Biol. Chem. 270, 24589–24597.

Lehmann, S. & Harris, D.A. (1996a). Two mutant prion proteins expressed in cultured cells acquire biochemical properties reminiscent of the scrapie isoform. Proc. Natl. Acad. Sci. U. S. A. 93, 5610–5614.

Lehmann, S. & Harris, D.A. (1996b). Mutant and infectious prion proteins display common biochemical properties in cultured cells. J. Biol. Chem. 271, 1633–1637.

Lehmann, S. & Harris, D.A. (1997). Blockade of glycosylation promotes acquisition of scrapie-like properties by the prion protein in cultured cells. J. Biol. Chem. 272, 21479–21487.

Lievens, P.M.-J., Quaglio, E., Piccardo, P., Tranchant, C., Warter, J.M., Bird, T., Bugiani, O., Ghetti, B. & Tagliavini, F. (1998). Distinctive PrP isoforms in Gerstmann-Sträussler-Scheinker disease with tangles. J. Neuropathol. Exp. Neurol. 57, 517.

Lope, E.S., Junquera, S.R., Martinez, A.M. & Berenguel, A.B. (1977). Acute ascending polyradiculoneuritis in a case of Creutzfeldt-Jakob disease. J. Neurol. Neurosurg. Psychiatry 40, 149–155.

Manuelidis, E.E., Fritch, W.W., Kim, J.H. & Manuelidis, L. (1987). Immortality of cell cultures derived from brains of mice and hamsters infected with Creutzfeldt-Jakob disease agent. Proc. Natl. Acad. Sci. U. S. A. 84, 871–875.

Martinez-Loge, J.F., Poza, M., Sola, J., et al. (1994). Accidental transmission of Creutzfeldt-Jakob disease by dural cadaveric grafts. J. Neurol. Neurosurg. Psychiatry 57, 1091–1094.

Masson, C., Delalande, I., Deslys, J.P., Henin, D., Falletbianco, C., Dormont, D. & Leys, D. (1994) Creutzfeldt-Jakob disease after derived human growth hormone therapy: two cases with valine 129 homozygous genotype. Neurology 44, 179.

Masters, C.L., Gajdusek, D.C. & Gibbs, C.J., Jr. (1981). Creutzfeldt-Jakob disease virus isolations from the Gerstmann-Straussler syndrome, with an analysis of the various forms of amyloid plaque deposition in the virus-induced spongiform encephalopathies. Brain 104, 559–588.

Mastrianni, J.A., Curtis, M.T., Oberholtzer, J.C., Da Costa, M.M., DeArmond, S., Prusiner, S.B. & Garbern, J. Y. (1995). Prion disease (PrP-A117V) presenting with ataxia instead of dementia. Neurology 45, 2042–2050.

Mastrianni, J.A., Iannicola, C., Myers, R.M., DeArmond, S. & Prusiner, S.B. (1996). Mutation of the prion protein gene at codon 208 in familial Creutzfeldt-Jakob disease. Neurology 47, 1305–1312.

McKinley, M.P., Bolton, D.C. & Prusiner, S.B. (1983). A protease-resistant protein is a structural component of the scrapie prion. Cell 35, 57–62.

McKinley, M.P., Taraboulos, A., Kenaga, L., Serban, D., Stieber, A., DeArmond, S.J., Prusiner, S.B. & Gonatas, N. (1991) Ultrastructural localization of scrapie prion proteins in cytoplasmic vesicles of infected cultured cells. Lab. Invest. 65, 622–630.

Medori, R., Montagna, P., Tritschler, H.J., LeBlanc, A.C., Cortelli, P., Tinuper, P., Lugaresi, E. & Gambetti, P. (1992). Fatal familial insomnia: A second kindred with mutation of prion gene at codon 178. Neurology 42, 669–670.

Merz, P.A., Somerville, R.A., Wisniewski, H.M. & Iqbal, K. (1981) Abnormal fibrils from scrapie-infected brain. Acta Neuropathol. 54, 63–74.

Meyer, R.K., McKinley, M.P., Bowman, K.A., Braunfeld, M.B., Barry, R.A. & Prusisner, S.B. (1986). Separation and properties of cellular and scrapie prion proteins. Proc. Natl. Acad. Sci. U. S. A. 83, 2310–2314.

Mirra, S.S., Young, K., Gearing, M., Jones, R., Evatt, M.L., Piccardo, P. & Ghetti, B. (1997). Coexistence of prion protein (PrP) amyloid, neurofibrillary tangles, and Lewy bodies in Gerstmann-Sträussler-Scheinker disease with prion gene (*PRNP*) mutation F198S. Brain Pathology 7, 1379.

Miyashita, K., Inuzuka, T., Kondo, H., Saito, Y., Fujita, N., Matsubara, N., Tanaka, R., Hinokuma, K., Ikuta, F. & Miyatake, T. (1991) Creutzfeldt-Jakob disease in a patient with a cadaveric dural graft. Neurology 41, 940–941.

Mizushima, S., Ishii, K. & Nishimaru, T. (1994). A case of presenile with a 168 base pair insertion in prion protein gene. (in Japanese). Dementia 8, 380–390.

Mohr, M., Tranchant, C., Heldt, N., & Warter, J. M. (1994). Alsatian variant of codon 117 form of Gerstmann-Sträussler-Scheinker syndrome: autopsic study of 3 cases. Brain Pathol. 4, 524.

Monari, L., Chen, S.C., Brown, P., Parchi, P., Petersen, R.B., Mikol, J., Gray, F., Cortelli, P., Montagna, P., Ghetti, B., Goldfarb, L.G., Gajdusek, D.C., Lugaresi, E., Gambetti, P. & Autilio-Gambetti, L. (1994). Fatal familial insomnia and familial Creutzfeldt-Jakob disease: different prion proteins determined by a DNA polymorphism. Proc. Natl. Acad. Sci. U. S. A. 91, 2839–2842.

Montagna, P., Cortelli, P., Avoni, P., Tinuper, P., Plazzi, G., Gallassi, R., Portaluppi, F., Julien, J., Vital, C., Dellale, M. B., Gambetti, P., & Lugaresi, E. (1998). Clinical features of fatal familial insomnia: phenotypic variability in relation to a polymorphism at codon 129 of the prion protein gene. Brain Pathol. 8, 515–520.

Nakazato, Y., Ohno, R., Negishi, T., Hamaguchi, K. & Arai, E. (1991). An autopsy case of Gerstmann-Sträussler-Scheinker's disease with spastic paraplegia as its principal feature. Clin. Neurol. 31, 987–992.

Neufeld, M.Y., Josiphov, J. & Korczyn, A.D. (1992). Demyelinating peripheral neuropathy in Creutzfeldt-Jakob disease. Muscle Nerve 15, 1234–1239.

Neugut, R.H., Neugut, A.I., Kahana, E., Stein, Z., & Alter, M. (1979). Creutzfeldt-Jakob disease: familial clustering among Libyan-born Israelis. Neurology 29, 225–231.

Nicholl, D., Windl, O., de Silva, R., Sawcer, S., Dempster, M, Ironside, J.W., Estibeiro, J.P., Yuill, G.M., Lathe, R., & Will, R.G. (1995). Inherited Creutzfeldt-Jakob disease in a British family associated with a novel 144 base pair insertion of the prion protein gene. J. Neurol. Neurosurg. Psychiatry 58, 65–69.

Nisipeanu, P., El Ad, B. & Korczyn, A.D. (1990). Spongiform encephalopathy in an Israeli born to immigrants from Libya [letter]. Lancet 336, 686.

Nitrini, R., Rosemberg, S., Passos-Bueno, M.R., Gambetti, P., Papadopoulos, M., Carrilho, P. M., Caramelli, P., Albrecht, S., Zatz, M. & LeBlanc, A. (1997). Familial spongiform encephalopathy with distinct clinico-pathological features associated with a novel prion gene mutation at codon 183. Ann. Neurol. 42, 138–146.

Nochlin, D., Sumi, S.M., Bird, T.D., Snow, A.D., Leventhal, C.M., Beyreuther, K. & Masters, C.L. (1989). Familial dementia with PrP-positive amyloid plaques: A variant of Gerstmann-Sträussler syndrome. Neurology 39, 910–918.

Oda, T., Kitamoto, T., Tateishi, J., Mitsuhashi, T., Iwabuchi, K., Haga, C., Oguni, E., Kato, Y., Tominaga, I., Yanai, K., Kashima, H., Kogure, T., Hori, K. & Ogino, K. (1995). Prion disease with 144 base pair insertion in a Japanese family line. Acta. Neuropathol. 90, 80–86.

Oesch, B., Westaway, D., Walchli, M., McKinley, M.P., Kent, S.B.H., Aebersold, R., Barry, R. A., Tempst, P., Teplow, D.B., Hood, L.E., Prusiner, S.B. & Weissmann, C. (1985). A cellular gene encodes scrapie PrP 27–30 protein. Cell 40, 735–746.

Owen, F., Poulter, M., Lofthouse, R., Collinge, J., Crow, T.J., Risby, D., Baker, H.F., Ridley, R.M., Hsiao, K. & Prusiner, S.B. (1989). Insertion in prion protein gene in familial Creutzfeldt-Jakob disease. Lancet 1, 51–52.

Owen, F., Poulter, M., Shah, T., Collinge, J., Lofthouse, R., Baker, H., Ridley, R., McVey, J., & Crow, T.J. (1990). An in-frame insertion in the prion protein gene in familial Creutzfeldt-Jakob disease. Mol. Brain Res. 7, 273–276.

Owen, F., Poulter, M., Collinge, J., Leach, M., Lofthouse, R., Crow, T.J. & Harding, A.E. (1992). A dementing illness associated with a novel insertion in the prion protein gene. Mol. Brain Res. 13, 155–157.

Padovani, A., D'Alessandro, M., Parchi, P., Cortelli, P., Anzola, G.P., Montagna, P., Vignolo, L.A., Petraroli, R., Pocchiari, M., Lugaresi, E. & Gambetti, P. (1998). Fatal familial insomnia in a new Italian kindred. Neurology 51, 1491–1494.

Palmer, M.S., Dryden, A.J., Hughes, J.T. & Collinge, J. (1991). Homozygous prion protein genotype predisposes to sporadic Creutzfeldt-Jakob disease. Nature 352, 340–342.

Pan, K.M., Baldwin, M., Nguyen, J., Gasset, M., Serban, A., Groth, D., Mehlhorn, I., Huang, Z.W., Fletterick, R.J., Cohen, F.E., & Prusiner, S.B. (1993) Conversion of α-helices into β-sheets features in the formation of the scrapie prion protein. Proc. Natl. Acad. Sci. U. S. A. 90, 10962–10966.

Parchi, P., Capellari, S., Chen, S.G., Ghetti, B., Mikol, J., Vital, C., Cochran, E., Trojanowski, J.Q., Dickson, D.W, Petersen, R.B. & Gambetti, P. (1996a). Similar posttranslational modifications of the prion protein in familial, sporadic and iatrogenic Creutzfeldt-Jakob disease. Soc. Neurosci. Abstr. 22, 711.

Parchi, P., Capellari, S, Sima, A.A.F., D'Amato, C., McKeever, P., Mikol, J., Brion, S., Brown, P., Chen, S.G., Petersen, R.B. & Gambetti, P. (1996b). Creutzfeldt-Jakob disease (CJD) with 178Asn mutation in the prion protein gene: neuropathological and molecular features. J Neuropathol Exp Neurol 55, 635.

Parchi, P., Castellani, R., Capellari, S., Ghetti, B., Young, K., Chen, S.G., Farlow, M., Dickson, D.W., Sima, A.A.F., Trojanowski, J.Q., Petersen, R.B. & Gambetti, P. (1996c). Molecular basis of phenotypic variability in sporadic Creutzfeldt-Jakob disease. Ann. Neurol. 39, 669–680.

Parchi, P., Capellari, S., Chen, S.G., Petersen, R.B., Gambetti, P., Kopp, N., Brown, P., Kitamoto, T., Tateishi, J., Giese, A. & Kretzschmar, H. (1997). Typing prion isoforms. Nature 386, 232–233.

Parchi, P., Chen, S.G., Brown, P., Zou, W., Capellari, S., Budka, H., Hainfellner, J., Reyes, P.F., Golden, G.T., Hauw, J.J., Gajdusek, D.C. & Gambetti. P. (1998a). Different patterns of truncated prion protein fragments correlate with distinct phenotypes in P102L Gerstmann-Sträussler-Scheinker disease. Proc. Natl. Acad. Sci. U. S. A. 95, 8322–8327.

Parchi, P., Gambetti, P., Piccardo, P.,& Ghetti, B. (1998b). Human prion diseases. In: Progress in Pathology (Kirkham, N. & Lemoine, N.R., eds.), Vol. 4, pp. 39–77. Churchill Livingstone, Edinburgh.

Parchi, P., Giese, A., Capellari, S., Brown, P., Schulz-Schaeffer, W.J., Windl, O., Budka, H., Julien, J., Kopp, N., Poser, S., Rojiani, A.M., Streichenberger, N., Vital, C., Zerr, I., Ghetti, B., Kretzschmar, H.A. & Gambetti, P. (1998c). The molecular and clinico-pathologic spectrum of phenotypes of sporadic Creutzfeldt-Jakob disease (sCJD). Neurology 50, A336.

Pattison, I.H. & Jones, K.M. (1967). The possible nature of the transmissible agent of scrapie. Vet. Rec. 80, 1–8.

Petersen, R.B., Parchi, P., Richardson, S.L., Urig, C.B. & Gambetti, P. (1996). Effect of the D178N mutation and the codon 129 polymorphism on the metabolism of the prion protein. J. Biol. Chem. 271, 12661–12668.

Piccardo, P., Ghetti, B., Dickson, D.W., Vinters, H.V., Giaccone, G., Bugiani, O., Tagliavini, F., Young, K., Dlouhy, S.R., Seiler, C., Jones, C.K., Lazzarini, A., Golbe, L.I., Zimmerman, T.R., Perlman, S.L., McLachlan, D.C., St. George-Hyslop, P.H. & Lennox, A. (1995). Gerstmann-Sträussler-Scheinker Disease (*PRNP* P102L): amyloid deposits are best recognized by antibodies directed to epitopes in PrP region 90–165. J. Neuropathol. Exp. Neurol. 54, 790–801.

Piccardo, P., Seiler, C., Dlouhy, S.R., Young, K., Farlow, M.R., Prelli, F., Frangione, B., Bugiani, O., Tagliavini, F. & Ghetti, B. (1996). Proteinase-K-resistant prion protein isoforms in Gerstmann-Sträussler-Scheinker disease (Indiana kindred). J. Neuropathol. Exp. Neurol. 55, 1157–1163.

Piccardo, P., Dlouhy, S.R., Young, K., Lievens, P., Bugiani, O., Tagliavini, F. & Ghetti, B. (1997). Prion proteins in P102L Gerstmann-Sträussler-Scheinker disease (GSS). Soc. Neurosci. Abstr. 23, 1906.

Piccardo, P., Dlouhy, S.R., Lievens, P.M.J., Young, K., Bird, T.D., Nochlin, D., Dickson, D.W., Vinters, H.V., Zimmerman, T.R., Mackenzie, I.R.A., Kish, S.J., Ang, L.-C., De Carli, C., Pocchiari, M., Brown, P., Gibbs, C.J., Gajdusek, D.C., Bugiani, O., Ironside, J., Tagliavini, F. & Ghetti, B. (1998a). Phenotypic variability of Gerstmann-Sträussler-Scheinker disease is associated with prion protein heterogeneity. J. Neuropath. Exp. Neurol. 57, 979–988.

Piccardo, P., Kish, S.J., Ang, L.C., Young, K., Bugiani, O., Tagliavini, F., Dlouhy, S.R. & Ghetti, B. (1998b). Prion protein isoforms in the new variant of Gerstmann-Sträussler-Scheinker disease Q212P. J. Neuropath. Exp. Neurol. 57, 518.

Pocchiari, M., Salvatore, M., Cutruzzola, F., Genuardi, M., Allocatelli, C. T., Masullo, C., Macchi, G., Alema, G., Galgani, S., Xi, Y.G., Petraroli, R., Silvestrini, M.C. & Brunori, M. (1993). A new point mutation of the prion protein gene in Creutzfeldt-Jakob disease. Ann. Neurol. 34, 802–807.

Poulter, M., Baker, H.F., Frith, C.D., Leach, M., Lofthouse, R., Ridley, R.M., Shah, T., Owen, F., Collinge, J., Brown, G., Hardy, J., Mullan, M.J., Harding, A.E., Bennett, C., Doshi, R. & Crow, T.H. (1992). Inherited prion disease with 144 base pair gene insertion. 1. Genealogical and molecular studies. Brain 115, 675–685.

Prusiner, S.B. (1982). Novel proteinaceous infectious particles causing scrapie. Science 216, 136–144.

Prusiner, S.B. (1991) Molecular biology of prion diseases. Science 252, 1515–1522.

Puckett, C., Concannon, P., Casey, C. & Hood, L. (1991). Genomic structure of the human prion protein gene. Am. J. Hum. Genet. 49, 320–329.

Radhakrishnan, K. & Mousa, E.M. (1988). Creutzfeldt-Jakob disease in Benghazi, Libya. Neuroepidemiology 7, 42–43.

Raymond, G.J., Hope, J., Kocisko, D.A., Priola, S.A., Raymond, L.D., Bossers, A., Ironside, J., Will, R.G., Chen, S.G., Petersen, R.B., Gambetti, P., Rubenstein R., Smits, M.A., Lansbury, P.T. & Caughey, B. (1997). Molecular assessment of the transmissibilities of BSE and scrapie to humans. Nature 388, 285–288.

Richardson, E.P., Jr. (1977). Introduction to myoclonic dementia. In: Neurological Classics in Modern Translation (Rottenberg, D.A. & Hochberg, F.H., ed.), pp. 95–96. Hafner Press, New York.

Richardson, E.P. & Masters, C.L. (1995). The nosology of Creutzfeldt-Jakob disease and conditions related to the accumulation of PrPCJD in the nervous system. Brain Pathol. 5, 33–41.

Ripoll, L., Laplanche, J.L., Salzmann, M., Jouvet, A., Planques, B., Dussaucy, M., Chatelain, J., Beaudry, P. & Launay, J.M. (1993). A new point mutation in the prion protein gene at codon 210 in Creutzfeldt-Jakob disease. Neurology 43, 1934–1938.

Rosenthal, N.P., Keesey, J., Crandall, B. & Brown, W.J. (1976). Familial neurological disease associated with spongiform encephalopathy. Arch. Neurol. 33, 252–259.

Sadeh, M., Chagnac, Y. & Goldhammer, Y. (1990). Creutzfeldt-Jakob disease associated with peripheral neuropathy. Isr. J. Med. Sci. 26, 220–222.

Safar, J., Roller, P.P., Gajdusek, D.C. & Gibbs, C.J. (1993) Conformational transitions, dissociation, and unfolding of scrapie amyloid (prion) protein. J. Biol. Chem. 268, 20276–20284.

Safar, J., Wille, H., Itri, V., Groth, D., Serban, H., Torchia, M., Choen, F.E. & Prusiner, S.B. (1998). Eight prion strains have PrPSc molecules with different conformations. Nat. Med. 4, 1157–1165.

Salvatore, M., Genuardi, M., Petraroli, R., Masullo, C., D'Alessandro, M. & Pocchiari, M. (1994). Polymorphism of the prion protein gene in Italian patients with Creutzfeldt-Jakob disease. Hum. Genet. 94, 375–379.

Schatz, H.M., Da Costa, M., Taylor, L., Cohen, F.E. & Prusiner, S.B. (1995). Prion protein gene variation among primates. J. Mol. Biol. 245, 362–374.

Schoene, W.C., Masters, C.L., Gibbs, C.J., Jr., Gajdusek, D.C., Tyler, H.R., Moore, F.D. & Dammin, G.J. (1981). Transmissible spongiform encephalopathy (Creutzfeldt-Jakob disease). Arch. Neurol. 38, 473–477.

Scrimgeour, E.M., Masters, C.L., Alpers, M.P., Kaven, J., & Gajdusek, D.C. (1983) A clinico-pathological study of case of kuru. J. Neurol. Sci. 59, 265–275.

Seitelberger, F. (1981). Straüssler's disease. Acta Neuropathol. 7, 341–343.

Shyng, S.L., Huber, M.T. & Harris, D.A. (1993) A prion protein cycles between the cell surface and an endocytic compartment in cultured neuroblastoma cells. J. Biol. Chem. 268, 15922–15928.

Shyu, W.C., Hsu, Y.D., Kao, M.C. & Tsao, W.L. (1996). Panencephalitic Creutzfeldt-Jakob disease in a Chinese family. Unusual presentation with PrP codon 210 mutation and identification by PCR-SSCP. J. Neurol. Sci. 143, 176–180.

Silvestrini, M.C., Cardone, F., Maras, B., Pucci, P., Barra, D., Brunori, M. & Pocchiari, M. (1997). Identification of the prion protein allotypes which accumulate in the brain of sporadic and familial Creutzfeldt-Jakob disease patients. Nat. Med. 3, 521–525.

Singh, N., Zanusso, G, Chen, S.G., Fujioka, H., Richardson, S., Gambetti, P. & Petersen, R.B. (1997). Prion protein aggregation reverted by low temperature in transfected cells carrying a prion protein gene mutation. J Biol Chem 272, 28461–28470.

Sparkes, R.S., Simon, M., Cohn, V.H., Fournier, R.E.K., Lem, J., Klisak, I., Heinzmann, C., Blatt, C., Lucero, M., Mohandas, T., DeArmond, S. J., Westaway, D., Prusiner, S.B., & Weiner, L.P. (1986). Assignment of the human and mouse prion protein genes to homologous chromosomes. Proc. Natl. Acad. Sci. U. S. A. 83, 7358–7362.

Speer, M.C., Goldgaber, D., Goldfarb, L.G., Roses, A.D. & Pericak-Vance, M.A. (1991). Support of linkage of Gerstmann-Sträussler-Scheinker syndrome to the prion protein gene on chromosome 20p12-pter. Genomics 9, 366–368.

Stahl, N., Borchelt, D.R., Hsiao, K. & Prusiner, S. (1987) Scrapie prion protein contains a phosphatidylinositol glycolipid. Cell 51, 229–240

Tagliavini, F., Prelli, F., Ghiso, J., Bugiani, O., Serban, D., Prusiner, S.B., Farlow, M.R., Ghetti, B. & Frangione, B. (1991). Amyloid protein of Gerstmann-Sträussler-Scheinker disease (Indiana kindred) is an 11 kd fragment of prion protein with an N-terminal glycine at codon 58. EMBO J. 10, 513–519.

Tagliavini, F., Giaccone, G., Prelli, F., Verga, L., Porro, M., Trojanowski, J., Farlow, M., Frangione, B., Ghetti, B. & Bugiani, O. (1993). A68 is a component of paired helical filaments of Gerstmann-Sträussler-Scheinker disease, Indiana kindred. Brain Res. 616, 325–328.

Tagliavini, F., Prelli, F., Porro, M., Rossi, G., Giaccone, G., Farlow, M.R., Dloughy, S.R., Ghetti, B., Bugiani, O. & Frangione, B. (1994). Amyloid fibrils in Gerstmann-Sträussler-Scheinker disease (Indiana and Swedish kindreds) express only PrP peptides encoded by the mutant allele. Cell 79, 695–703.

Tagliavini, F., Prelli, F., Porro, M., Rossi, G., Giaccone, G., Bird, T.D., Dlouhy, S.R., Young, K., Piccardo, P., Ghetti, B., Bugiani, O. & Frangione, B. (1995). Only mutant PrP participates in amyloid formation in Gerstmann-Sträussler-Scheinker disease with AlaVal substitution at codon 117. J. Neuropath. Exp. Neurol. 54, 416.

Taraboulos, A., Raeber, A.J., Borchelt, D.R., Serban, D. & Prusiner, S.B. (1992). Synthesis and trafficking of prion proteins in cultured cells. Mol. Biol. Cell. 3, 851–863.

Tateishi, J., Brown, P., Kitamoto, T., Hoque, Z.M., Roos, R., Wollman, R., Cervenakova, L., & Gajdusek, D.C. (1995). First experimental transmission of fatal familial insomnia. Nature 376, 434–435.

Tateishi, J., Kitamoto, T., Hoque, M.Z. & Furukawa, H. (1996). Experimental transmission of Creutzfeldt-Jakob disease and related diseases to rodents. Neurology 46, 532–537.

Telling, G.C., Scott, M., Hsiao, K.K., Foster, D., Yang, S.-L., Torchia, M., Sidle, K.C.L., Collinge, J., DeArmond, S.J. & Prusiner, S.B. (1994). Transmission of Creutzfeldt-Jakob disease from humans to transgenic mice expressing chimeric human-mouse prion protein. Proc. Natl. Acad. Sci. U. S. A. 9, 9936–9940.

Telling, G.C., Scott, M., Mastrianni, J., Gabizon, R., Torchia, M., Cohen, F.E., DeArmond, S.J. & Prusiner, S.B. (1995). Prion propagation in mice expressing human and chimeric PrP transgenes implicates the interaction of cellular PrP with another protein. Cell 83, 79–90.

Telling, G.C., Parchi, P., DeArmond, S.J., Cortelli, P., Montagna, P., Gabizon, R., Mastrianni, J. A., Lugaresi, E., Gambetti, P. & Prusiner, S.B. (1996). Evidence for the conformation of the pathologic isoform of the prion protein enciphering and propagating prion diversity. Science 274, 2079–2082.

Terao, Y., Hitoshi, S., Shimizu, J., Sakuta, M. & Kitamoto, T. (1992). Gerstmann-Sträussler-Scheinker disease with heterozygous codon change at prion protein codon 129. Clin. Neurol. 3, 80–883.

Tobler, I., Gaus, S.E., Deboer, T., Achermann, P., Fischer, M., Rulicke, T., Moser, M., Oesch, B., McBride, P.A. & Manson, J.C. (1996) Altered circadian activity rhythms and sleep in mice devoid of prion protein. Nature 380, 639–642.

Tranchant, C., Doh-ura, K., Steinmetz, G., Chevalier, Y., Kitamoto, T., Tateishi, J. & Warter, J. M. (1991). Mutation du codon 117 du géne du prion dans une maladie de Gerstmann-Sträussler-Scheinker. Rev. Neurol. (Paris) 147, 274–278.

Tranchant, C., Doh-ura, K., Warter, J. M., Steinmetz, G., Chevalier, Y., Hanauer, A., Kitamoto, T. & Tateishi, J. (1992). Gerstmann-Sträussler-Scheinker disease in an Alsatian family—clinical and genetic studies. J. Neurol. Neurosurg. Psychiatry 55, 185–187.

Turk, E., Teplow, D.B., Hood, L.E. & Prusiner, S.B. (1988) Purification and properties of the cellular and scrapie hamster prion proteins. Eur. J. Biochem. 176, 21–30.

van Gool, W.A., Hensels, G.W., Hoogerwaard, E.M., Wiezer, J.H.A., Wesseling, P. & Bolhuis, P.A. (1995). Hypokinesia and presenile dementia in a Dutch family with a novel insertion in the prion protein gene. Brain 118, 1565–1571.

Warter, J.M., Steinmetz, G., Heldt, N., Rumbach, L., Marescaux, C.H., Eber, A.M., Collard, M., Rohmer, F., Floquet, J., Guedenet, J.C., Gehin, P. & Weber, M. (1981). Demence pre-senile familiale syndrome de Gerstmann-Sträussler-Scheinker. Rev. Neurol. (Paris) 138, 107–121.

Weller, R.O., Steart, P.V. & Powell-Jackson, J.D. (1986). Pathology of Creutzfeldt-Jakob disease associated with pituitary-derived human growth hormone administration. Neuropath. Appl. Neurobiol. 12, 117–129.

Wells, G.A.H. & Wilesmith, J.W. (1995). The neuropathology and epidemiology of bovine spongiform encephalopathy. Brain Pathol. 5, 91–103.

Whittington, M.A., Sidle, K.C.L., Gowland, I., Meads, J., Hill, A.F., Palmer, M.S., Jefferys, J.G. R. & Collinge, J. (1995). Rescue of neurophysiological phenotype seen in PrP-null mice by transgene encoding human prion protein. Nat. Genet. 9, 197–201.

Will, R.G., Ironside, J.W., Zeidler, M., Cousens, S.N., Estibeiro, K., Alperovitch, A., Poser, S., Pocchiari, M., Hofman, A. & Smith, P.G. (1996). A new variant of Creutzfeldt-Jakob disease in the UK. Lancet 347, 921–925.

Willison, H.J., Gale, A.N. & McLaughlin, J.E. (1991) Creutzfeldt-Jakob disease following cadaveric dura mater grafts. J. Neurol. Neurosurg. Psychiatry 59, 940.

Windl, O., Dempster, M., Estibeiro, J.P., Lathe, R., Desilva, R., Esmonde, T., Will, R., Springbett, A., Campbell, T.A., Sidle, K.C.L., Palmer, M.S., & Collinge, J. (1996). Genetic basis of Creutzfeldt-Jakob disease in the United Kingdom: a systematic analysis of predisposing mutations and allelic variation in the PRNP gene. Hum. Genet. 98, 259–264.

Yamada, M., Itoh, Y., Fujigasaki, H., Naruse, S., Kaneko, K., Kitamoto, T., Tateishi, J., Otomo, E., Hayakawa, M., Tanaka, J., Matsushita, M. & Miyatake, T. (1993). A missense mutation at codon 105 with codon 129 polymorphism of the prion protein gene in a new variant of Gerstmann-Sträussler-Scheinker disease. Neurology 43, 2723–2724.

Yamada, S., Aiba, T., Endo, Y., Hara, M., Kitamoto, T., & Tateishi, J. (1994). Creutzfeldt-Jakob disease transmitted by a cadaveric dura mater graft. Neurosurgery 34, 740–744.

Yamazaki, M., Oyanagi, K., Mori, O., Kitamura, S., Ohyama, M. & Terashi, A. (1997). An autopsy case of variant Gerstmann-Sträussler-Scheinker syndrome with codon 105 mutation of the prion protein gene, showing degeneration of the pallidum, thalamus and substantia nigra, and widely distributed neurofibrillary tangles. Brain Pathol. 7, 113–115.

Young, K., Jones, C.K., Piccardo, P., Lazzarini, A., Golbe, L. I., Zimmerman, T.R., Dickson, D.W., McLachlan, D.C., St George-Hyslop, P., Lennox, A., Perlman, S., Vinters, H.V., Hodes, M.E., Dlouhy, S. & Ghetti, B. (1995). Gerstmann-Sträussler-Scheinker disease with mutation at codon 102 and methionine at codon 129 of *PRNP* in previously unreported patients. Neurology 45, 1127–1134.

Young, K., Clark, H.B., Piccardo, P., Dlouhy, S.R. & Ghetti, B. (1997). Gerstmann-Sträussler-Scheinker disease with the PRNP P102L mutation and valine at codon 129. Mol. Brain Res. 44, 147–150.

Young, K., Piccardo, P., Kish, S.J., Ang, L.C., Dlouhy, S. & Ghetti, B. (1998). Gerstmann-Sträussler-Scheinker disease (GSS) with a mutation at prion protein (PrP) residue 212. J. Neuropath. Exp. Neurol. 57, 518.

Zanusso, G., Petersen, R.B., Kanhoush, R., Gambetti, P. & Singh, N. (1997). Mutant prion protein in Gerstmann-Sträussler-Scheinker disease. Y145amber is degraded by the proteasomal pathway. Mol. Biol. Cell 8, 85.

Zilber, N., Kahana, E. & Abraham, M.P.H. (1991). The Libyan Creutzfeldt-Jakob disease focus in Israel: an epidemiologic evaluation. Neurology 41, 1385–1389.

Chapter 6

Progress in Understanding the Genetics of Epilepsy

CARL E. STAFSTROM, ASURI N. PRASAD, CHITRA PRASAD, and JOHN T. SLEVIN

Advances in Cell Aging and Gerontology
Volume 3, pages 189–241

ISBN: 0-7623-0405-7

INTRODUCTION

Epilepsy is a disorder characterized by recurrent seizures causing transient impairment of brain function, caused by a paroxysmal cortical excitation. Epilepsy can be idiopathic, in which seizures occur as unprovoked, spontaneous events, or symptomatic, in which seizures arise from a pathogenic source (e.g., trauma, neoplasia). An epileptic seizure can be provoked in anyone, and whether it occurs spontaneously or from a particular stimulus depends on the individual's seizure threshold. This threshold surely depends on several intrinsic components, some of which may be determined by genetic factors.

Consideration that epilepsy may be inherited is not new: in his monograph on the subject, Hippocrates assumed that epilepsy was inherited. However, progress in understanding the genetics of epilepsy was glacial until recent decades. With the advent of molecular genetics, there has been an exponential growth of knowledge about the genetic basis of epilepsy and epilepsy syndromes. As documented in recent reviews (Gardiner, 1990; Delgado-Escueta et al., 1994; Treiman and Treiman, 1996; Berkovic, 1997; Berkovic and Scheffer, 1997a; Dichter and Buchhalter, 1997; Prasad et al., 1999), gene defects underlying many of the inherited epilepsies have been mapped, and the pace of discovery appears to be accelerating. In parallel with advances in understanding the genetics of human epilepsies, the creation of transgenic mice that have specific genes either eliminated or overexpressed has produced several animal models of inherited epilepsy. The continued construction of transgenic animals will expand our ability to dissect mechanisms of seizure generation and propagation (Allen and Walsh, 1996; Noebels, 1996). Knowledge gained from such models may help us move from serendipity to rationality in attempts to design targeted therapies for seizure disorders. This chapter reviews the progress in mapping genes for human epilepsy, examines the relationship between gene defects and the cellular mechanisms of epilepsy, and describes some relevant animal models of genetic epilepsy.

Epileptic seizures result from excessive synchronous firing of neurons in cortical networks. A number of intricately linked mechanisms at the cellular, subcellular, and molecular levels permit recruitment and synchronization of surrounding neurons, resulting in a particular seizure phenotype. Clearly, genes and their mutations exquisitely influence these processes; however, the molecular mechanisms linking genotype with phenotype are largely unknown. Genetic factors may explain why some individuals develop epilepsy following acquired insults while others do not. Some inherited epilepsies display age-dependent expression and remission patterns, suggesting an interaction between neuronal development and seizure expression. Genes may also play a role in a person's sensitivity or resistance to antiepileptic drugs. A gene has recently been identified that may confer antiepileptic pharmacoresistance in humans (Tishler et al., 1995). MDR1, the multiple drug resistance gene, encodes a glycoprotein that functions as a membrane pump to export antiepileptic drugs out of cells. MDR1 was found to be elevated in resected brain tissue from medically intractable epilepsy patients. The importance of this finding remains to be determined. Thus, there appear to be many routes to the various epilepsy phenotypes, and many gene combinations may pave each of those routes.

Past efforts to determine the genetic basis of the epilepsies have been hampered by many problems, relating to both seizure semiology and classification. Previous studies have employed a wide variety of data collection and analysis techniques, and some have failed to adjust for patient age. Terms used to categorize epilepsies have often been imprecise. The current classification of epileptic seizures (Commission on Classification and Terminology, 1989) divides seizures into partial or generalized according to ictal onset, whereas epilepsies and epilepsy syndromes are categorized according to their region of onset (localization-related or generalized) and etiology (symptomatic or cryptogenic) (Commission on Classification, 1989). The latter scheme differentiates between epilepsy syndromes according to the mode of onset, etiology, clinical behavior, electroencephalogram (EEG) patterns, natural history, response to treatment, and eventual outcome. However, the current classifications may prove incomplete as the complex genetic basis of many epilepsies becomes unraveled. For example, seizures with generalized onset were believed to be inherited, whereas those with partial onset were considered to be largely symptomatic. It is now recognized that several partial epilepsies have a genetic component as well (Ryan, 1995). Along with the growing realization that many epilepsies are inherited is the parallel recognition that EEG traits also have a genetic basis (Pedley, 1991; Doose, 1997). Significant proportions of asymptomatic relatives of persons with epilepsy have EEG abnormalities. Indeed, as the primary method for assessing the brain's electrical activity, the EEG has been used as an assay of cortical hyperexcitability among families of probands (e.g., spike wave discharges, photosensitivity, and focal sharp waves). The development of epileptic seizures in a genetically predisposed individual is likely the result of a complex interaction between genetic and exogenous factors.

Two major approaches to understanding genetic epilepsies have evolved: (1) the experimental dissection of the genetic bases of hyperexcitability, and (2) attempts to map mutations underlying human epilepsies. It is likely that separate pathophysiological mechanisms, and therefore different genes, underlie the determination of seizure threshold, synchronization, and spread beyond a focus, generalization, and termination. Alone or in combination, such factors as voltage-gated membrane channels, excitatory and inhibitory synaptic transmission in specific cortical networks, neuromodulation, and neuronal network connectivity can predispose to hyperexcitability. Furthermore, understanding the genetic mechanisms controlling neuronal migration, synaptic connectivity, nuclear and cytoplasmic signaling cascades, effector molecules, and proteins mediating vesicle mobilization and docking will further refine our understanding of the molecular basis of hyperexcitability. Much of the progress in these areas has come from the study of animal models, including spontaneous seizure phenotypes in rodents and transgenic mice carrying a null mutation or overexpressing a gene involved in the regulation of excitability (Noebels, 1996).

PATTERNS OF INHERITANCE IN HUMAN EPILEPSY

Patterns of human inheritance may involve traditional single gene (mendelian) traits (autosomal or X linked, dominant or recessive), or more complex patterns such as polygenic or multifactorial inheritance. A genetic disorder may also arise because of an aberration of part or all of a chromosome, mutation of the mitochondrial genome or an abnormally expanded (repeated) DNA sequence. Each of these genetic patterns accounts for certain human epilepsies.

Single gene defects have now been identified for several human epilepsies (see Table 1). Complicating the identification of single gene defects in epilepsy is the variability in phenotypic seizure expression between affected individuals, even within the same family. Genotypic heterogeneity has been convincingly demonstrated even in several well-recognized clinical syndromes, such as juvenile myoclonic epilepsy (JME) (Whitehouse et al., 1993b; Delgado-Escueta et al., 1994) and benign familial neonatal convulsions (BFNC) (Lewis et al., 1993). Mutations at different loci can produce a similar phenotype, while allelic variation at the same locus can produce different phenotypes. Other forms of familial epilepsy are likely to be polygenic, that is, resulting from the additive effects of multiple gene mutations. Genetic and environmental factors can interact such that a gene may lower the susceptibility to epilepsy, whereas an environmental factor is required for seizure expression, as in photosensitive epilepsy (Figure 1) (Doose and Gerken, 1973; Naquet et al., 1995).

Chromosomal anomalies, including trisomies, deletions, duplications, inversions, ring chromosomes, and translocations, usually result in multiple somatic abnormalities and may comprise a recognizable electroclinical seizure phenotype (e.g., Angelman's syndrome). Small supernumerary marker chromosomes are frequently encountered in clinical cytogenetics, and are designated "extra structurally abnormal chromosomes" (ESACs) (Hook and Cross, 1987). Although previously believed to be benign, it is now thought that ESACs may have phenotypic consequences. An example of an ESAC anomaly associated with profound seizures is inversion of chromosome 15 (inv dup 15) (Battaglia et al., 1997).

Nonmendelian genetic patterns such as mitochondrial inheritance, triplet nucleotide repeats, and minisatellite expansions are also emerging as mechanisms by which epilepsy can be inherited. Mitochondrial DNA and its mutations are transmitted to offspring only by the mother (the ovum contributes all the mitochondria to the zygote), and both male and female offspring can be affected. The pattern of segregation and threshold effect determines the severity and variability of the phenotype in an affected individual. Mutant mitochondria may segregate in varying proportions in different tissues and organs during replication, a phenomenon known as heteroplasmy. Differences in the ratio of mutant to normal mitochondria in a

Table 1. Inherited Human Epilepsies: Seizure Types, Clinical Characteristics, Gene Loci, Putative Molecular Mechanisms

Epilepsy Type	*Type of Seizure*	*Clinical and EEG Characteristics*	*Chromosomal Locus and Mode of Inheritance*	*Putative Mechanism*	*References*
Idiopathic Epilepsy with Generalized Seizures as a Major Component					
Juvenile myoclonic epilepsy (JME)	Myoclonic, GTCS, absence	Age of onset 8–25 yr, diffuse polyspike-multispike waves (3.5–6 Hz)	Autosomal dominant variable penetrance 6p21.2-p11 15q14 MIM 254770	Unknown ? $\alpha 7$ subunit of nicotinic Ach receptor	1 2, 3 4
Benign familial neonatal convulsions (EBN1)	Mixed-type with tonic posturing, ocular symptoms, apnea	Day 3 in onset, seizures remit by 6 weeks in 68%; normal development	Autosomal dominant 20q13.3 MIM 121200	KCQN2 K^+ channel subtype	5
Benign familial neonatal convulsions (EBN2)	Mixed-type	Seizures remit by 6-24 mo	Autosomal dominant 8q24 MIM 121201	KCQN3 K^+ channel subtype	6 7
Northern epilepsy sndrome	GTCS	Age of onset 5–10 yr, progressive mental retardation, seizures stop by age 35 yrs	Autosomal recessive 8pter-p22 MIM 600143	Unknown (?) Cathepsin B localizes to 8p22	8
Idiopathic generalized epilepsy (IGE)	GTCS	Normal intellectual development, generalized 3 Hz spike and wave complexes	Autosomal MIM 600669 8q24 ? locus at 15q11.2–q12	Unknown syntenic with mouse Chr. 15 at the stg locus Evidence of linkage to GABA(A) $\alpha 5$, $\beta 3$ and $\gamma 3$ subunit gene cluster on Chr. 15	9 10
Generalized epilepsy with febrile seizures (FS) plus	FS, FS+ absences, FS+ myoclonic seizures, FS + atonic, FS + myoclonic astatic epilepsy	Median age of onset 1 yr Febrile seizures persist beyond 6, but cease by 11, yr of age	?	Unknown	11

(*continued*)

Table 1. Continued

Epilepsy Type	*Type of Seizure*	*Clinical and EEG Characteristics*	*Chromosomal Locus and Mode of Inheritance*	*Putative Mechanism*	*References*
Idiopathic Epilepsy with Partial Seizures as a Major Component					
Partial epilepsy	SP, CPS, and SGTC	Age of onset 8–10 yr, nonspecific auditory disturbances, normal intelligence	Autosomal dominant 10q22–q24 MIM 600512	Unknown	12
Autosomal dominant frontal lobe epilepsy (ADNFLE)	SPS	Age of onset 8 yr, nocturnal seizures, occur in clusters with aura; frontal lobe-origin hyperkinetic, tonic movements mistaken for sleep disturbances	Autosomal dominant 20q13.2–13.3 MIM 600513	(Ser → Phe substitution at codon 248); mutation in α4 subunit of the nicotinic Ach receptor	13 14 15
Benign childhood epilepsy with centrotemporal spikes (BCECTS)	Partial	Age of onset 5–10 yr, unilateral motor or sensory seizures relating to sleep; focal spikes in Rolandic regions, with horizontal dipole	Complex inheritance No linkage	Unknown	16
Benign infantile familial convulsions (BIFC)	Partial	Age of onset 4–8 mo, parieto-occipital foci Associated paroxysmal choreoathetosis in French kindred	Autosomal dominant 9q11-13 in Italian families, 16q in French kindred	Unknown	17 18 19
Chromosomal aberrations with cerebral dysgenesis and prominent seizures					
Tuberous sclerosis (TSC1)	Infantile spasms	Seizures, mental retardation, hydrocephaly secondary to cerebral glial nodules, hamartomas	Autosomal dominant 9q34 MIM 191100	Hamartin ? functions as a tumor suppressor	20 21

Tuberous sclerosis (TSC2)	Infantile spasms	Mental and motor retardation; intractable seizures; facial angiofibroma, skin changes; renal, cardiac, and neural tumors	Autosomal dominant 16p13.3 MIM191092	Native tuberin bears structural homology to Rap1GAP, functions as a tumor suppressor gene	22 23, 24
X-linked lissencephaly	Infantile spasms, mixed seizures	Lissencephaly in hemizygous males, subcortical band heterotopia in heterozygous females; mental retardation	X-linked dominant Xq22.3–q23 MIM 300067	Neuronal migration defects	25 26
Periventricular heterotopia	Heterogeneous types of epilepsy GTCS, TLE	Variable age, no cognitive or neurological markers, heterotopias on imaging	X-linked dominant Xq28 MIM 300049	Candidate genes: GABA receptor α3 subunit and L1CAM	27
Lissencephaly I	Infantile spasms, mixed seizures	Characteristic facies, motor and mental retardation, neuronal migrational defect, deficient opercularisation	Autosomal recessive 17p13.3 MIM 247200	LIS1 gene codes for PAF acetylhydrolase, a neuroregulatory molecule	28
Trisomy 21 (Down syndrome)	Infantile spasms, mixed seizures	Classic phenotypic features, mental retardation, hypotonia	Trisomy 21 MIM 190685 21q22.3	Contiguous gene syndrome, several gene candidates overlapping DS critical region	29 30
Wolff-Hirschorn (4p-) syndrome	Myoclonic, unilateral or generalized Atypical absence	Multiple anomalies, characteristic electroclinical features (see text)	4p-deletion *4p16.3 MIM 194190	? $GABA_A$ receptor dysfunction	31
Angelman's syndrome	GTCS NCSE	Characteristic electroclinical features (see text)	Imprinting, uniparental disomy 15q11–q12 MIM 105830	$GABA_A$ receptor β3 subunit, ?$GABA_A$ receptor α5 dysfunction UBE3A gene deletion	32 33 34 35, 36

(continued)

Table 1. Continued

Epilepsy Type	*Type of Seizure*	*Clinical and EEG Characteristics*	*Chromosomal Locus and Mode of Inheritance*	*Putative Mechanism*	*References*
Fragile X	GTCS SPS	Centrotemporal spikes resemble the Rolandic trait	Triplet expansion Xq28 MIM 309550	FMR-1	37 38
Trisomy 12p	GTCS Myoclonic	Characteristic facies, mental retardation, hypotonia, cerebral malformations; 3 Hz spike-wave complexes	Trisomy due to a breakpoint distal to 12p12	?Glyceraldehyde-3-PO_4 Dehydrogenase also maps to 12p	39
Inverted duplication 15	Atypical absence, tonic, atonic, tonic-clonic	Characteristic facies, autistic, moderate/severe mental retardation, hyperactive; diffuse and multifocal EEG abnormalities, slow spike/polyspike waves	Inv dup(15) (pter-q12~13:q12~13-pter) MIM ?	? tetrasomy of genes may alter GABA receptor activity	40 41
Ring chromosome 20 epilepsy syndrome	Prolonged nonconvulsive status	Characteristic electroclinical features described	Locus of fusion p13q13, p13q13.3, p13q13.33	ADNFLE, BFNC and CHNRA4 map to neighboring regions	42
Progressive Myoclonic Epilepsies (EPM)					
Unverricht-Lundborg type (EPM1)	GTCS, myoclonic	Seizures begin early, with mental deterioration, ataxia; photosensitive myoclonus; spike wave discharges precede seizures	Autosomal recessive 21q22.3 MIM 254800	Cystatin B (family of protease inhibitors) (Arg 68ter and G-C,-1) Unstable minisatellite expansion	43 44
Dentatorubropallidoluysian atrophy (DRPLA)	Myoclonic	Myoclonic epilepsy, dementia, ataxia, choreoathetosis	Autosomal dominant 12p13.31 MIM 125370	Triplet repeat expansion (CAG)n ?Aberrant glyceraldehyde-3-PO_4 dehydrogenase	45

Myoclonic epilepsy of Lafora (MELF)	GTCS, myoclonic	Age of onset age 15 yr, rapid mental deterioration, psychosis, blindness; PAS + bodies in myoepithelial cells of eccrine glands	Autosomal recessive 6q23.q25 MIM 254780	Unknown	46
Myoclonic epilepsy with ragged red fibers (MERRF)	Myoclonic	Myopathy with ragged red fibers, sensorineural deafness	Mitochondrial tRNA mutation (8344 A-G, 8356 T-C) MIM 545000	Respiratory chain enzyme deficiencies	47 48 49
Gaucher's disease type III	Myoclonic, GTCS	Childhood, early adulthood seizures, ataxia, supranuclear gaze palsies	Autosomal recessive 1q21 MIM 231000	Glucocerebrosidase deficiency	50
Sialidoses I	Myoclonic	Age of onset 8-15 yr, vertex positive spikes, stimulus-sensitive myoclonic seizures, ataxia, visual failure with cherry red spots	Autosomal recessive 6p21.3 MIM 256550	Alpha-neuraminidase deficiency	51 52
Neuronal ceroid lipofuscinoses (CLN2)	Myoclonic, GTCS, atonic, atypical absence	Age of onset 2.5-4 yr, photoparoxysmal response at < 3 Hz stimulation	Autosomal recessive 11p15 15q21-23 (Finnish variant—CLN5) MIM 204500	Gene encodes for a pepstatin-insensitive lysosomal peptidase	53
Neuronal ceroid lipofuscinoses (CLN3)	Atypical absence, myoclonic, GTCS	Age of onset 4-10 yr, visual loss, ataxia, dementia	Autosomal recessive 16p12.1 MIM 204200	Gene encodes a lysozomal N-glycosylated 438 amino acid protein of unknown function	54, 55
Metabolic Disorders with Prominent Seizures					
Mitochondrial encephalomyopathy, lactic acidosis, stroke (MELAS)	Mixed seizures, GTCS	Headaches, strokes, dementia , seizures, sensorineural hearing loss and blindness	Mitochondrial tRNA mutation (3243 A-G) (3271-3273 T-A) MIM 540000	Respiratory chain enzyme deficiencies, defective energy metabolism	56 57

(continued)

Table 1. Continued

Epilepsy Type	*Type of Seizure*	*Clinical and EEG Characteristics*	*Chromosomal Locus and Mode of Inheritance*	*Putative Mechanism*	*References*
Pyridoxine dependency	Neonatal seizures Status epilepticus	Intrauterine or neonatal-onset refractory seizures, responsive to high dose pyridoxine	Autosomal recessive 2q31 MIM 266100	Decreased GABA formation due to impaired glutamic acid decarboxylase	58 59 60
Biotinidase deficiency	Neonatal seizures	Seizures, sensorineural hearing loss, ataxia, hypotonia cutaneous	Autosomal recessive 3p25 MIM 253260	Multiple carboxylase deficiency, GABA-transporter at 3p25	61 62
Nonketotic hyperglycinemia (NKH 1-4)	Myoclonic seizures	Neonatal onset, hypotonia, hyporeflexia and apnea; burst supression	Autosomal recessive 9p22 MIM 238300	Glycine cleavage system abnormality, ? overactive NMDA receptor	63 64
Other Disorders					
Rasmussen's encephalitis	Epilepsia partialis continua	Progressive seizures and cerebral hemiatrophy, hemiplegia and dementia	GluR3 gene maps to Xq25-q26 MIM 305915	GluR3 antibodies cause immune-mediated inflammation	65 66
Intractable, drug-resistant epilepsy		Increased resistance to anticonvulsants	MDR1 gene maps to 7q MIM 171050	Increased expression of p-glycoprotein in patients with intractable seizures	67

Notes: Chr., chromosome; CPS, ; FS, febrile seizures; GTCS, generalized tonic-clonic seizures; NCSE, ; SGTC, ; SP, ; SPS, ; TLE, .

1 (Greenberg et al., 1988), 2 (Serratosa et al., 1996), 3 (Weissbecker et al., 1991), 4 (Elmslie et al., 1997), 5 (Singh et al., 1998), 6 (Charlier et al., 1998), 7 (Lewis et al., 1996), 8 (Tahvanainen et al., 1994), 9 (Zara et al., 1995), 10 (Sander et al., 1997b), 11 (Scheffer and Berkovic, 1997), 12 (Ottman et al., 1995), 13 (Scheffer et al., 1994), 14 (Phillips et al., 1995), 15 (Phillips and Mulley, 1997), 16 (Heijbel et al., 1975), 17 (Vigevano et al., 1992), 18 (Guipponiet al., 1997), 19 (Szepetowski et al., 1997), 20 (van Slegtenhorst et al., 1997), 21 (Green et al., 1994), 22 (Sampson and Harris, 1994), 23 (Wienecke et al., 1995), 24 (Wienecke et al., 1997), 25 (Srivistava et al., 1996), 26 (Ross M.E., et al., 1997b), 27 (Eksioglu et al., 1996), 28 (Dobyns et al., 1993), 29 (Korenberg, 1993), 30 (Korenberg, 1995), 31 (Sgro et al., 1995), 32 (Wagstaff et al., 1991), 33 (Sugimoto et al., 1992), 34 (Matsumoto et al., 1992), 35 (Sutcliffe et al., 1997), 36 (Matsuura et al., 1997), 37 (Musumeci et al., 1988), 38 (Musumeci et al., 1991), 39 (Guerrini et al., 1990), 40 (Bundey et al., 1994), 41 (Battaglia et al., 1997), 42 (Inoue et al., 1997), 43 (Pennacchio and Myers, 1996), 44 (Virtaneva et al., 1997), 45 (Koide et al., 1994), 46 (Serratosa et al., 1995), 47 (Shoffner et al., 1990), 48 (Silvestri et al., 1992), 49 Human mitochondrial genome database: http://www.gen.emory.edu/mitomap.html (1995), 50 (Tsuji et al., 1987), 51 (Oohira et al., 1985), 52 (Bonten et al., 1996), 53 (Sharp et al., 1997), 54 (Consortium, 1995), 55 (Jarvela et al., 1998), 56 (Goto et al., 1990), 57 (Shoffner et al., 1995), 58 (Bu et al., 1992), 59 (Krishnamoorthy, 1983), 60 (Goutieres and Aicardi, 1985), 61 (Pomponio et al., 1995), 62 (Norrgard et al., 1997), 63 (Isobe et al., 1994), 64 (Shoffner et al., 1990), 65 (McNamara et al., 1996), 66 (Andrews and McNamara, 1996), 67 (Tishler et al., 1995). All MIM numbers were obtained through an Internet search: World Wide Web URL:

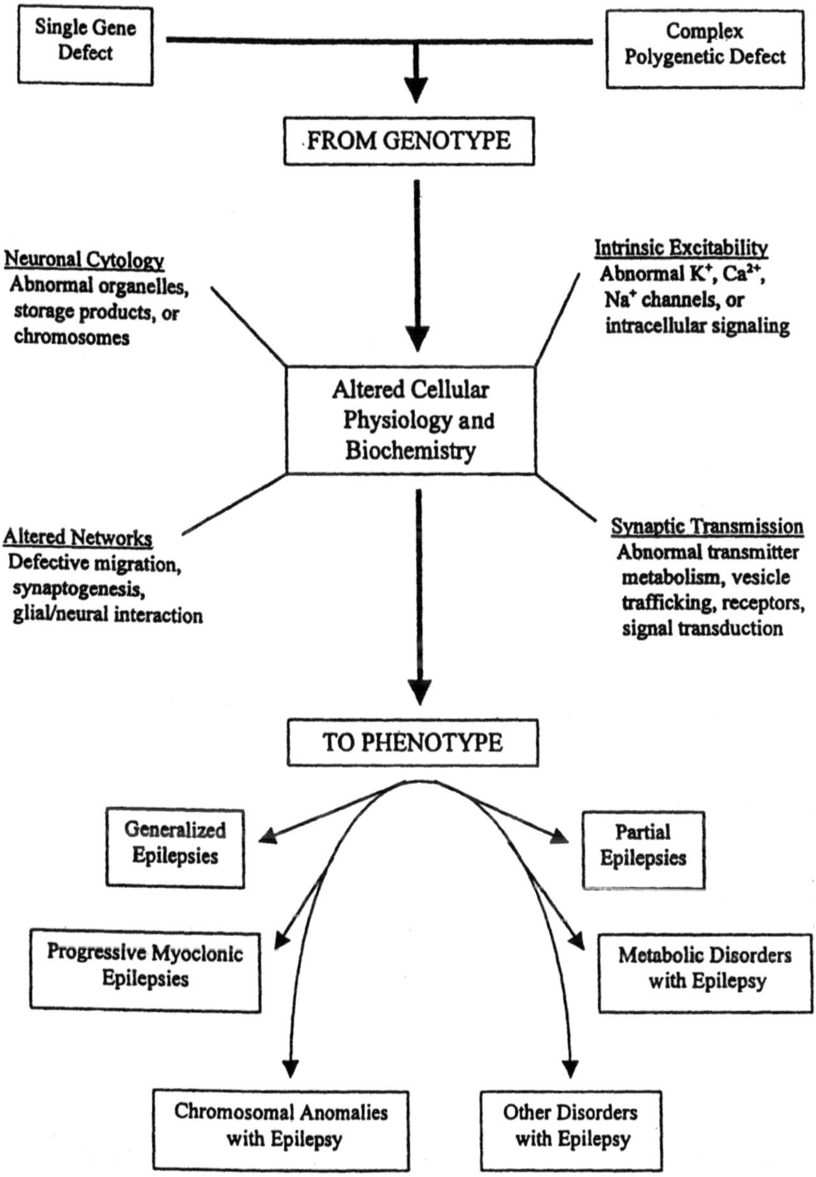

Figure 1. The many roads from gene defects to epilepsy phenotypes.

given tissue are partly responsible for the wide variation in clinical phenotype in mitochondrial disorders. Clinical signs and symptoms occur when a certain "threshold" amount of mitochondrial DNA becomes mutated. Examples of mitochondrial disorders with prominent seizures include myoclonic epilepsy with ragged red fibers (MERRF), mitochondrial encephalomyopathy lactic acidosis and stroke (MELAS), neuropathy ataxia and retinitis pigmentosa (NARP) and Leigh's syndrome (see later discussion).

The recently defined triplet expansion syndromes involve amplification of the number of repeats in a DNA sequence, rather than a change in the nucleotide sequence per se. This phenomenon underlies genetic anticipation (the worsening of phenotype over successive generations). Examples of triplet expansion syndromes with prominent seizures are the fragile X syndrome and dentatorubropallidoluysian atrophy (DRPLA) (Bates and Lehrach, 1994; O'Donnell and Zoghbi, 1995). Another new mechanism for mutations may be inheritance of stable expansions of DNA called "minisatellites" sequences (tandem repeats); these do not contribute to the function of a gene but are repeated within the gene. Expansion of minisatellites has been implicated in progressive myoclonic epilepsy of the Unvericht-Lundborg type (EPM1) (Virtaneva et al., 1997), although this finding is controversial (Lalioti et al., 1997).

MAPPING HUMAN EPILEPSY GENES

As previously indicated, the second approach to elucidating the genetic basis of epilepsy involves mapping the mutations underlying human seizure disorders. Steps in the process of finding an epilepsy gene are outlined in Figure 2. The first step is to define the seizure phenotype as specifically as possible; the second is to determine whether the epilepsy has a heritable basis, including its mode of inheri-

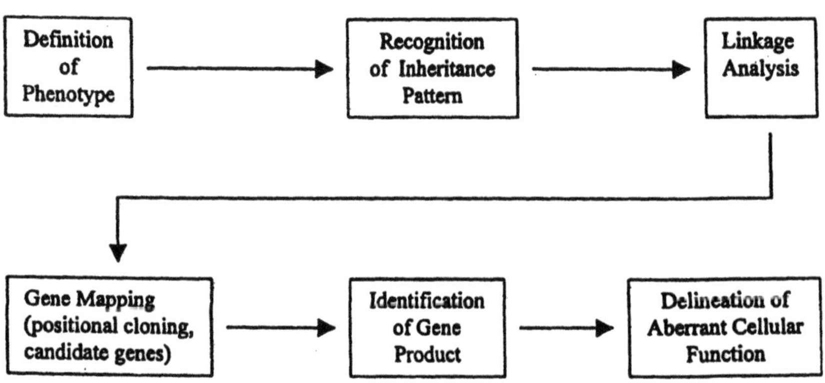

Figure 2. Finding genes in inherited disorders and understanding disease mechanisms.

tance and degree of penetrance. Next, the responsible genetic locus is identified; linkage analysis has traditionally been used for this purpose. Tightly linked genes on the same chromosome tend to be transmitted as a unit, and are less likely to be separated during meiotic recombination. Linkage analysis evaluates the "tightness of linkage"; that is, the extent to which the gene of interest is inherited along with known DNA markers in that chromosomal region. The strength of linkage, which is a measure of the relative distance separating two loci, is indicated by the "lod" score (logarithm of the odds); a value of greater than 3.0 (> 1000:1 odds) suggests significant linkage, while a value less than –2 indicates lack of linkage. Linkage analysis is a laborious process and ideally requires a large family with multiple affected individuals over several generations. To avoid errors introduced by sampling methods, cases must be ascertained carefully, including accurate classification of seizure phenotype, differentiation of genetic from nongenetic factors in seizure etiology, and appreciation of complex genetic patterns including nonmendelian inheritance (Lander and Schork, 1994). Furthermore, incomplete and age-dependent penetrance results in intra- and interfamilial variability that can confound even the most careful observations. It is often difficult to identify families with sufficiently large numbers of affected individuals to carry out a linkage analysis. Despite these practical hurdles, linkage analysis and gene mapping have been successful in a growing list of human epilepsies.

Once linkage is established, the next step is to identify and map the precise gene involved. This can be approached in either of three ways: positional cloning, the candidate gene approach, and the positional candidate approach. In positional cloning, the goal is to determine the gene or genes responsible for a naturally occurring epilepsy, making no a priori assumptions about the function of the gene. Positional cloning is best suited for mendelian disorders. Linkage is sought to known chromosomal markers and the flanking regions around the marker are mapped for genes and their mutations. Using the positional cloning approach, several epilepsies have been mapped, including the progressive myoclonic epilepsy of Unverricht-Lundborg, juvenile neuronal ceroid lipofuscinosis, autosomal dominant nocturnal frontal lobe epilepsy, Miller-Dieker syndrome, and tuberous sclerosis. Each of these is discussed later in this chapter. Certain disorders with complex inheritance have been linked using positional cloning as well, such as JME and childhood absence epilepsy. In these disorders, interactions between several genes and environmental factors make linkage analysis difficult. Positional cloning has also been used in several mouse models of tonic clonic and generalized spike wave epilepsies to identify the genes responsible for epilepsy predisposition (Noebels, 1996).

In the candidate gene approach, a gene with a known or suspected role in cellular excitability is manipulated (overexpressed or knocked out) by creating a transgenic or knockout animal model. A broad range of such genes can be envisioned, such as those that code for ion channel proteins, neurotransmitter receptors and transporters, membrane pumps, and so on. Some of the genes found to be responsible for epilepsy in animal models have been surprises, coding for proteins previously not suspected to play a major role in governing excitability, such as synapsins, neurotransmitter vesicle proteins, and serotonin receptors (Noebels, 1996). The candidate gene approach does not require any linkage analysis, nor rely upon map position. With the availability of rapid sequencing technology and bioinformatics, it is possible to considerably speed up the process by combining linkage analysis with a survey of potential candidate genes in the linked region. This is the so-called positional candidate approach (Collins, 1991, 1995). Such an approach was utilized in the identification of gene defects underlying two forms of benign neonatal convulsions (Charlier et al, 1998; Singh et al, 1998).

Finally, the protein encoded by the affected gene must be identified and its biological role established. This process will enable development of prenatal tests for the genetic defect, permit genetic counseling when appropriate and possibly lead to therapies such as gene replacement or targeted pharmacological intervention (McNamara, 1994).

Although a detailed discussion of the technical aspects and difficulties in mapping genes is beyond the scope of this chapter, some inherent problems must be mentioned. Cases must be ascertained carefully, including accurate classification of seizure phenotype, differentiation of genetic from nongenetic factors in seizure etiology, and appreciation of the complex genetic patterns including nonmendelian inheritance (Lander and Schork, 1994). Furthermore, incomplete and age-dependent penetrance results in intrafamilial and interfamilial variability that can confound even the most careful observations. It is often difficult to identify families with sufficiently large numbers of affected individuals to carry out a linkage analysis. These confounding variables can cause false positive or negative results. Despite these practical hurdles, linkage analysis and gene mapping have been successful in several human epilepsies (see Table 1).

GENE DEFECTS IN HUMAN EPILEPSY

Several genes implicated in human epilepsy have been identified. Table 1 summarizes the gene loci, mode of inheritance, and key features of some of these disorders. The genetic epilepsies are discussed in detail under the following categories: (1) idiopathic generalized epilepsies (IGEs), (2) partial epilepsies, (3) progressive myoclonic epilepsies, (4) chromosomal anomalies, and (5) metabolic disorders.

Idiopathic Generalized Epilepsies

A genetic basis is widely accepted for IGEs, which include JME, juvenile absence epilepsy (JAE), childhood absence epilepsy (CAE), and epilepsy with generalized tonic-clonic seizures (GTCS). The clinical and EEG characteristics of IGEs overlap considerably. Each syndrome has an age-dependent onset and normal EEG background. Affected individuals usually have normal intelligence and a normal neurological examination. The concordance rate for IGE among monozygotic twins is over 75% (Berkovic et al., 1998). Efforts to map the gene loci have been complicated by intra- and interfamilial variability in seizure phenotype, and many individuals express EEG abnormalities without ever developing seizures. Furthermore, age-dependent onset and disappearance of clinical and EEG features complicate case ascertainment.

Juvenile Myoclonic Epilepsy

JME is an IGE with age-dependent onset, usually in the early teens. Patients often present with myoclonic jerks especially in the early morning shortly after awakening. Generalized tonic-clonic seizures may also occur. The EEG shows generalized 4- to 6-Hz ("fast") polyspike wave discharges. Approximately 12% of first-degree relatives have EEG abnormalities such as generalized spike wave complexes (Delgado-Escueta et al., 1989).

JME has a complex inheritance, possibly involving interaction among several genes (Greenberg et al., 1988; Delgado-Escueta et al., 1994). Initial linkage studies of families with JME suggested localization at or near the HLA-linked susceptibility gene on chromosome 6p, designated "EJM1." Several studies (Greenberg et al., 1988; Delgado-Escueta et al., 1989; Durner et al., 1991) have reported a linkage to the 6p locus whereas others (Whitehouse et al., 1993a, 1993b; Delgado-Escueta et al., 1994; Elmslie et al., 1996) have failed to demonstrate such a linkage. It has not yet been possible to clarify the role of the 6p region in seizure susceptibility, or to correlate what phenotypic trait this region confers. Other loci for JME have now been identified. Linkage has been demonstrated to the $\alpha7$ subunit of the nicotinic cholinergic receptor (nAChR) located on chromosome 15q14 in the majority of the 34 families studied (Elmslie et al., 1997). Using nonparametric methods, Zara et al. (1995) have shown linkage of JME to a locus at 8q24, but not to 6p.

The existence of genetic and clinical heterogeneity in JME is firmly established. The accumulated evidence supports JME and the subclinical EEG abnormality (fast spike wave trait) as inherited in an autosomal dominant fashion with variable penetrance. More families must be studied to delineate the specific genetic mechanisms of JME.

Absence Epilepsies

CAE occurs in children 5 to 10 years of age, manifesting as multiple daily brief absence seizures with characteristic generalized 3-Hz spike wave discharges. The seizures usually respond to ethosuximide or valproate and often resolve during the teenage years. JAE initially appears in preadolescents or adolescents; clinical absences are less frequent but generalized tonic-clonic seizures are more common than in CAE. Generalized 3- or 4- to 5-Hz spike wave discharges accompany the clinical seizures in JAE. There is considerable clinical overlap between CAE and JAE and also among other IGEs such as JME.

The absence epilepsies have a clearly genetic basis, probably inherited as autosomal dominant with age-dependent, high penetrance (Delgado-Escueta et al., 1990; Buoni et al., 1998). There is no linkage of JAE or CAE to EJM1 on chromosome 6p (Serratosa et al., 1996; Sander et al., 1997b). Recently, using a candidate gene approach, Sander and colleagues showed a significant allelic association of JAE with a noncoding sequence of the kainate-selective GluR5 (GRIK1) receptor on chromosome 21q22.1 (Sander et al., 1997a). The study supports a JAE-related susceptibility gene at the GRIK1 locus. Given the importance of excessive central nervous system (CNS) excitability in the pathogenesis of epilepsy, an altered glutamate receptor gene could contribute to the generalized cortical hyperexcitability in this IGE.

In the past few years, significant insights into the molecular and genetic mechanisms underlying absence seizures have been gained. This has come about through electrophysiological and molecular biological investigation of pathophysiology of absence and the study of genetic mouse models. The anatomical substrate of absence synchronous burst firing includes neocortical pyramidal cells, thalamic relay neurons, and gamma amino butyric acid (GABA)-ergic neurons of the nucleus reticularis thalami (Steriade et al., 1993). These latter neurons synchronize the burst firing between the reciprocally connected neocortical and thalamic relay neurons. This appears to involve a complex interplay among pre- and postsynaptic $GABA_B$ receptors and $GABA_A$ receptors, that themselves may show developmental differences depending on their cortical or thalamic location (Hosford et al., 1997).

Investigations of the genetic absence epilepsy rat from Strasbourg (GAERS) indicate that the increased basal GABA in ventraolateral thalamus of this model is intrinsic rather than a consequence of seizures (Richards et al., 1995). $GABA_A$ and $GABA_B$ receptor affinity and density are no different between GAERS and control rats (Knight and Bowery, 1992), suggesting a role for modified transduction, perhaps at the presynaptic $GABA_B$ receptor. Further understanding of the pathological alteration within this network that allows for the burst-firing of absence seizures has been gained through scrutiny of single-locus mice mutations, including the tottering, slow-wave and lethargic models discussed in a later section.

Benign Familial Neonatal Convulsions

A genetic basis is now recognized for some epilepsies that begin early in life. BFNC is a rare form of idiopathic generalized epilepsy that occurs in full-term newborns in the first few days of life, hence their designation as "third day fits" (Cunniff et al., 1988; Ronen et al., 1993). Focal or generalized seizures occur, sometimes accompanied by apnea, but infants are normal between seizures. The interictal EEG is normal, but the ictal pattern shows generalized abnormalities, with generalized voltage attenuation followed by slow waves, spikes, or burst suppression (Andrews and Stafstrom, 1993; Hirsch et al., 1993; Ronen et al., 1993). Seizures usually subside over weeks to months, but up to 15% of patients subsequently develop epilepsy. Intellectual and motor development are usually normal (Plouin, 1998). This disorder is usually inherited in an autosomal dominant manner, although an autosomal recessive form has also been reported (Schiffmann et al., 1991). Therefore, the condition shows phenotypic and genotypic heterogeneity.

Three loci are now recognized, one on chromosome 20q13.3 (EBN1) (Leppert et al., 1989), another on chromosome 8q24 (EBN2) (Lewis et al., 1993), and a third locus (Lewis et al., 1996) that is not yet identified but does not map to the other two known loci. Previously, the gene coding for the α4 subunit of the neuronal nicotinic acetylcholine receptor (CHRNA4) was mapped in the candidate region for BFNC (Beck et al., 1994). However, recent work has revealed that the responsible mutation affects the gene for a subfamily of voltage-gated potassium channels (KCNQ) that are responsible for repolarization of the action potential. The gene for EBN1, called KCNQ2 (Singh et al., 1998), codes for one potassium channel component, while the gene for EBN2, termed KCNQ3 (Charlier et al., 1998), codes for another portion of the potassium channel. When expressed in *Xenopus* oocytes, KCNQ2 facilitated development of potassium currents, whereas expression of the mutant channel showed no such currents (Biervert et al., 1998). By impairing the restoration of ionic balance after neuronal firing, potassium channel dysfunction could lead to repetitive neuronal firing and a state of heightened neuronal excitability. It remains to be explained why a fixed genetic deficit in potassium channel structure would not lead to continuous, lifelong seizures, rather than intermittent seizures which tend to remit over time. This demonstration of altered channel structure is a pivotal advance, since it is the first genetic defect that specifically targets an excitability mechanism.

BFNC should be distinguished from a sporadic, nongenetic form of neonatal convulsions termed "fifth day fits" (Dehan et al., 1977; Navelet et al., 1977). In this disorder, seizures occur in healthy term infants, often on the fourth or fifth days of life. The seizures last for about 24 hours, are refractory to drug therapy, and no etiology is established. The outcome is variable (Pryor et al., 1981; North et al., 1989). A third form of benign familial convulsions early in life, termed benign familial infantile convulsions, manifests as seizures with partial origin and are described later in this chapter (Vigevano et al., 1992).

Northern Epilepsy Syndrome

A newly described autosomal recessive childhood epilepsy with mental deterioration has been described in Northern Finland (Hirvasniemi et al., 1994). Onset is between 5 and 10 years of age, with generalized tonic-clonic and complex partial seizures increasing in frequency up until puberty, after which seizures appear to partially remit. The EEG shows progressive slowing of the background with relatively sparse epileptiform activity (Lang et al., 1997). Progressive cognitive decline begins 2 to 5 years after the onset of seizures, leading to mental retardation by middle age. By linkage analysis, the gene locus has been assigned to the telomeric region of chromosome 8p. The gene for the protease cathepsin B is located in the same region (Tahvanainen et al., 1994). It is a mystery why this particular gene defect leads to the epileptic phenotype. Cathepsins are widely distributed throughout the nervous system, with no known special relationship to the control of excitability. It is possible that the defective protease may lead to accumulation of an abnormal or toxic product within neurons, eventually leading to their degeneration, with epilepsy as a secondary phenomenon.

Febrile Seizures

Febrile seizures, classified as "situation-related," affect 2% to 5% of children between the ages of 6 months and 5 years (Commission on Classification, 1989). Febrile seizures are associated with a low risk of subsequent epilepsy, although some children with febrile seizures, especially complex ones, may develop temporal lobe epilepsy later in life (Nelson and Ellenberg, 1978; VanLandingham et al., 1998).

A family history of febrile seizures is often identified, suggesting a common environmental or genetic basis. Twin studies (Tsuboi and Endo, 1991) indicate a genetic basis and a complex polygenic inheritance pattern is currently favored. In a large segregation analysis of 467 families, a polygenic pattern of inheritance was strongly suggested in families of probands with a single febrile convulsion. However in families of probands with multiple febrile convulsions, evidence supports a single major locus model with nearly dominant seizure susceptibility (Rich et al., 1987). Linkage analysis suggests that a locus on chromosome 8q may be involved, but this conclusion is tentative because the study assumed very low penetrance (60%).

Recently, in a large multiplex family of more than 2,000 persons, Scheffer and Berkovic identified a second clinical pattern of febrile seizures. In this family, a variety of dominantly inherited epilepsy phenotypes occurs, designated by the authors as "generalized epilepsy with febrile seizures plus" (Scheffer and Berkovic, 1997). The most common phenotype, termed "febrile seizures plus" (FS+), consists of multiple febrile seizures in infancy and persistence of afebrile seizures above 6 years of age. Febrile and afebrile seizures cease by adolescence. Other phenotypes identified in this family included FS plus absences, FS plus myoclonic seizures, and FS plus atonic seizures. The authors concluded that febrile seizures display genetic and phenotypic heterogeneity, and suggested that, at least in this family, a major, autosomal dominant epilepsy susceptibility gene was highly penetrant and could be modified by other genes, giving rise to the varied phenotypes. In one family with FS+, linkage was established to 19q13.1, with a mutation in a voltage-gated sodium channel subunit (Wallace et al., 1998).

There is no genetic animal model of febrile seizures, but experimental models may allow investigation of the various mechanisms by which fever can induce hyperexcitability (Baram et al., 1997). Nearly all processes governing neuronal excitability are affected by temperature and are under genetic control, so fever may cause a seizure by a wide variety of mechanisms (Buchhalter, 1993). For example, among the potential candidate genes on chromosome 8q are those for corticotrophin-releasing hormone (CRH) and calbindin. CRH causes seizures in infant rats only within a specific time window (Baram and Schultz, 1991). Calbindin is a calcium-binding protein that is important in buffering neurons from the effects of excess intracellular calcium, and has been shown to protect neurons against excitotoxicity (Mattson et al., 1991). Calbindin levels are modulated by seizure activity (Lee et al., 1997), and studies of brain tissue from human epilepsy patients indicate that levels of calbindin are decreased, suggesting that calbindin could play a role in the pathogenesis of seizure-induced neuronal degeneration (Magloczky et al., 1997).

Partial Epilepsies

Most partial or focal epilepsies are assumed to not be inherited. However, relatives of probands with partial epilepsy have an increased risk of epilepsy compared to the general population, suggesting a genetic influence in at least some partial epilepsies. A defective gene may cause localized cortical hyperexcitability if it is expressed only in certain vulnerable neuronal regions, or if it has widespread expression but its effects are modified by local factors (Ryan, 1995).

Benign Epilepsy of Childhood with Central-Temporal Spikes

There is a strong genetic influence in this common syndrome, also known as benign rolandic epilepsy (Bray and Wiser, 1964; Heijbel et al., 1975; Degen and Degen, 1990). Clinically, children have normal development and neurological examinations, with an EEG that shows a normal background and uni- or bilateral independent central-temporal spikes. Seizures consist of some combination of facial twitching, drooling, and speech arrest, sometimes with ipsilateral upper extremity motor or sensory involvement. Seizures most often occur at night, and nocturnal seizures are more likely to secondarily generalize. The seizures characteristically remit in the teenage years (Lerman and Kivity, 1975).

Benign epilepsy of childhood with central-temporal spikes (BECTS) and its EEG trait are considered to be inherited as autosomal dominant with age-dependent penetrance (Heijbel et al., 1975). There is an increased incidence of both focal and generalized EEG abnormalities among first-degree relatives and, interestingly, a far greater frequency of generalized abnormalities. These observations suggest that there is an inherited susceptibility to both focal and generalized cortical hyperexcitability that only occasionally manifests as clinical seizures. Doose (1997) considers typical BECTS to be a special case in a broad spectrum of clinical manifestations due to this genetic excitability defect. Important candidate genes that have been excluded are those for EJM1 on chromosome 6p (Whitehouse et al., 1993a) and the fragile X locus (Rees et al., 1993; Whitehouse et al., 1993a).

Autosomal dominant nocturnal frontal lobe epilepsy. Autosomal dominant nocturnal frontal lobe epilepsy (ADNFLE) was first described in an Australian kindred of 27 affected individuals spanning six generations (Scheffer et al., 1994; Phillips et al., 1995). Patients have normal neurological examinations and intelligence. The seizures begin in mid-childhood and occur in clusters during drowsiness or sleep. Symptoms vary in severity among family members. Affected individuals have frequent nocturnal motor seizures (tonic or hyperkinetic), often preceded by an aura; cases are often misdiagnosed as nightmares or sleep disorders. The interictal EEG is normal; ictally, seizures originate from the frontal lobes (Oldani et al., 1996, 1998).

ADNFLE is transmitted in an autosomal dominant fashion with incomplete penetrance. There is considerable intrafamilial variability in clinical expression. Using exclusion mapping, that is, by systematic elimination of candidate genes mapping to the same region, the disorder was assigned a chromosomal locus of 20q13.2 (Steinlein et al., 1995). A missense mutation (serine replacing phenylalanine at codon 248 in the α4 subunit of the nicotinic acetylcholine receptor [CHRNA4]) mapped to this region in all 21 affected members of one ADNFLE family (Steinlein et al., 1995). However, there is allelic heterogeneity in this condition, as other families have shown different mutations in the CHRNA4 gene (Steinlein et al., 1997). Therefore, this disorder shows genetic but not clinical heterogeneity. The ADNFLE locus on chromosome 20q13.2 overlaps that of the EBN1 locus, and initially both disorders were thought to involve cholinergic receptor dysfunction; it is now known that EBN1 codes for a potassium channel (see earlier discussion). Nevertheless, the relationship between cholinergic receptor function and seizure susceptibility is an intriguing one. Cholinergic agents are involved in seizure modulation (Wasterlain and Fairchild, 1985; Shytle et al., 1995). In reconstitution experiments in *Xenopus* oocytes, nicotinic cholinergic receptors with mutant α4 subunits had faster desensitization kinetics than wild-type receptors, more quickly becoming unresponsive to acetylcholine stimulation (Weiland et al., 1996). Reconstituted mutant receptors also show a significantly reduced calcium permeability (Kuryatov et al., 1997). It is conceivable that an alteration of cholinergic receptor function modulates sleep and arousal at the thalamocortical level, somehow causing a lowered threshold for spontaneous seizures (Crespel et al., 1998). The ADNFLE locus is not linked to the homologous El mouse locus (Lopes-Cendes et al., 1995).

Familial temporal lobe epilepsy. This recently described syndrome consists of familial occurrence of mild typical temporal lobe seizures with age-dependent onset. Some seizures originate from the lateral temporal lobe, others from the mesial temporal lobe. The hippocampal pathology seen in typical temporal lobe epilepsy is not seen in the familial variety (Berkovic and Scheffer, 1997b; Cendes et al., 1998). Ottman and colleagues described a single family containing 11 affected individuals who presented with partial seizures and nonspecific auditory disturbances beginning in adolescence (Ottman et al., 1995). Using linkage analysis, the epilepsy susceptibility gene mapped to chromosome 10q22-24. Several candidate genes of potential interest in this region may affect the susceptibility to epilepsy in this family, including those for $\beta 1$- and $\alpha 2a$-adrenergic receptors, glutamate metabolism, and a subunit of calcium-calmodulin protein kinase (Ottman et al., 1995). In other families with this syndrome, the 10q locus has been excluded. Berkovic et al. (1996) described 38 patients from 13 families who presented with mild simple or complex partial seizures, and suggested an age-dependent autosomal dominant inheritance. Cendes and colleagues (1998) studied 36 persons from 11 families, again with simple or complex partial seizures, but found that not all patients had benign seizures. Therefore, there may be genetic and clinical heterogeneity in this syndrome, which needs to be elucidated further. The condition may be analogous to the El mouse model (Rise et al., 1991; Mutoh et al., 1993).

Recently, other autosomal dominant partial epilepsies have been identified in clinical studies of large families: autosomal dominant partial epilepsy with variable foci (Berkovic and Scheffer, 1997b), autosomal dominant rolandic epilepsy with speech dyspraxia (Scheffer et al., 1995), and benign infantile familial convulsions (BIFC) (Vigevano et al., 1992). In none of these inherited syndromes has a mutation been identified.

Benign Familial Infantile Convulsions

In BFIC, partial seizures with secondary generalization begin between 4 and 8 months of age, occur in clusters, are characterized by head and eye deviation, and are followed by generalization with tonic-clonic activity. Ictal recordings confirm a focal origin in the parietal occipital region. The seizures are usually easily controlled with medication, and seizures later in life are rare (Caraballo et al., 1997). The disorder is transmitted as an autosomal dominant and is not linked to the BFNC gene on chromosome 20q13.2 (Malafosse et al., 1994). Linkage to chromosome 19q11-13 was demonstrated in five Italian families (Guipponi et al., 1997) and to chromosome 16 in four French families (Szepetowski et al., 1997), confirming genetic heterogeneity.

Progressive Myoclonic Epilepsies

The progressive myoclonic epilepsies constitute a group of seizure disorders with phenotypic features of myoclonic seizures, generalized seizures, ataxia, and cognitive defects occurring in variable combinations that progress over time. These conditions, which are often difficult to distinguish clinically, include progressive myoclonic epilepsy of Unverricht-Lundborg (EPM1); myoclonic epilepsy with Lafora bodies (MELF); mitochondrial disorders, especially MERRF; sialidosis; neuronal ceroid lipofuscinosis (CLN); and DRPLA.

Unverricht-Lundborg Disease

Progressive myoclonus epilepsy of the Unverricht-Lundborg type (EPM1), also known as Baltic myoclonic epilepsy, is an autosomal recessive condition with onset around age 10 years. It is characterized by progressive disability from stimulus sensitive myoclonus and generalized tonic-clonic seizures, with mild mental deterioration and ataxia developing late in the course (Norio and Koskiniemi, 1979). EPM1 has been reported almost exclusively in a genetically homogeneous population, permitting studies using linkage disequilibrium to target the gene defect to a small segment of chromosome 21q22.3 (Pennacchio et al., 1996). The gene (CST6) codes for cystatin B, a cysteine protease inhibitor. This mutation results from an unstable 15- to 18-mer minisatellite (Virtaneva et al., 1997) or a dodecamer repeat expansion (Lalioti et al., 1997) in the putative promoter region of the CST6 gene. Cystatin B mRNA is markedly reduced in EPM1 patients. As a ubiquitous protease inhibitor, cystatin B has no direct relevance to any known or suspected epilepsy mechanism, which has led to speculation that its primary role may be in neuronal degeneration, with epilepsy being a secondary effect (Allen and Walsh, 1996). A promising mouse model has been created, with the gene for cystatin B disrupted; this model may shed light on the mechanisms by which this loss of function mutation causes epilepsy (Pennacchio et al., 1997).

Lafora Body Disease

Progressive MELF closely resembles EPM1 clinically, but linkage to the EPM1 locus has been ruled out. This condition presents in late childhood or adolescence and leads to a fatal outcome within a decade of initial symptoms. Seizures are initially generalized tonic-clonic, absence, and atonic. Later, myoclonic jerks (often asymmetrical) appear and become progressively severe (Berkovic et al., 1993). Periodic acid-Schiff (PAS) positive cytoplasmic inclusion bodies are found in brain, skeletal muscle, liver, and skin and are considered diagnostic, although their pathological significance is unknown. In nine families with Lafora body disease confirmed by biopsy, a 17 cM region on chromosome 6q23-25 has been defined that contains the disease gene (Serratosa et al., 1995).

Neuronal Ceroid Lipofuscinosis and Sialidosis

Sialidosis and neuronal ceroid lipofuscinosis are neurometabolic storage disorders in which seizures occur as a secondary phenomenon. Characteristic inclusion bodies or storage material accumulates within neurons, resulting in their degeneration. Sialidosis, also known as cherry red spot myoclonus, results from a mutation in the gene encoding glycoprotein specific alpha neuraminidase. In human lysosomes, neuraminidase (sialidase) is conjugated with β-galactosidase and protective protein/cathepsin A. The neuraminidase cleaves terminal sialic acid residues from such sialoglycoconjugates. Deficiency of the gene, which is located on chromosome 6p21.3 (Pryor et al., 1981; Oohira et al., 1985; Pshezhetsky et al., 1997), causes a clinical syndrome of myoclonic epilepsy and mental retardation. Why myoclonic seizures occur in this disorder is not readily apparent, but it is intriguing that the gene for another cathepsin (B) is defective in the northern epilepsy syndrome.

The CLN are a group of progressive neurodegenerative disorders in children and young adults, which share clinical and morphological features. The phenotype includes seizures, motor disturbances, visual impairment, dementia, and familial occurrence in an autosomal recessive fashion. The ultrastructural appearance of the intracellular lipopigment accumulations is characteristic in each subtype.

Several NCL subgroups have been identified based on age at presentation (Boustany and Kolodny, 1989; Boustany, 1992; Consortium, 1995; Goebel, 1996) and each subtype is a distinct genetic entity. The gene for the infantile form (INCL), CLN1, is localized to chromosome 1p32 and encodes palmitoyl-protein thioesterase, a lysosomal enzyme that is involved in lipid modification of proteins by removing fatty acyl groups from cysteine residues (Vesa et al., 1995). The gene for the classical late infantile form (LINCL), CLN2, maps to chromosome 11p15 (Sharp et al., 1997). Mutations in the gene encoding a pepstatin-insensitive lysosomal peptidase have been identified in LINCL patients and its enzymatic activity is deficient in CLN2 autopsy specimens (Sleat et al., 1997). CLN3, the juvenile form of NCL (also known as Batten disease), is the commonest NCL subtype; the gene defect localizes to 16p12.1, which codes for a protein whose function is not yet known, but probably represents a lysosomal protein (Consortium, 1995). The CLN3 protein appears to be an N-glycosylated single-chain polypeptide of 438 amino acids (Jarvela et al., 1998). Twenty-three different mutations affecting this gene have now been identified (Munroe et al., 1997), providing clear evidence of genetic heterogeneity. The Finnish variant (a subtype of CLN2, termed CLN5) maps to 13q21-23 (Savukoski et al., 1998). The adult variant, Kufs disease (CLN 4), begins in the third or fourth decade with progressive dementia, seizures, myoclonus, and ataxia; blindness and retinal degeneration are not features of this form. The mechanism by which the gene defects in NCL lead to epilepsy are unclear, but the mutations in each subtype appear to represent a defect in either a lysosomal enzyme or lysosomal membrane protein.

Dentatorubralpallidoluysian Atrophy (DRPLA)

DRPLA is a recently identified disorder caused by expansion of an unstable trinucleotide (CAG, more than 49 repeats) within a gene of unknown function on chromosome 12p (Koide et al., 1994). DRPLA exhibits genetic anticipation, variable penetrance and heterogeneity in severity and phenotype. Clinically, DRPLA presents with variable combinations of myoclonus, epilepsy, cerebellar ataxia, choreoathetosis, and dementia. The gene codes for polyglutamine tracts and the defective protein has been termed atrophin. Atrophin is one of the proteins that interacts with glyceraldehyde-3-phosphate dehydrogenase (GAPDH), a key enzyme in glycolysis, possibly inhibiting its function (Burke et al., 1996; Ross, C.A., et al., 1997a). Whether this represents an important mechanism leading to seizures or neuronal degeneration remains to be proven.

Myoclonic Epilepsy with Ragged Red Fibers

Some forms of myoclonic epilepsy are the result of deficiencies in mitochondrial energy production. MERRF is the prototype of such a disorder. In MERRF, myoclonic seizures are prominent, in combination with a myopathy and sensorineural deafness. There is considerable variability in the clinical phenotype.

An adenine to guanine transition mutation at nucleotide pair 8344 in human mitochondrial DNA (mtDNA) has been identified in the majority of patients with MERRF (Shoffner et al., 1990). The mutation creates a specific restriction site on the tRNA (Lys) gene, producing defects in complex I and IV enzymes of the oxidative-phosphorylation (OXPHOS) system (Wallace et al., 1994). The proportion of normal and mutant mitochondrial (mt) DNAs in affected tissues at a particular age appears to determine the severity of symptoms (Shoffner et al., 1991). Myoclonic seizures occur when the percentage of mutated mtDNA reaches a critical threshold. Although the 8344 mutation is the most common, it is not observed in 10% to 20% of cases, suggesting genetic heterogeneity. Another mutation involves a thymine to cystine transition at nucleotide position 8356 (Silvestri et al., 1992). These findings suggest that tRNA(Lys) alterations play a specific role in the pathogenesis of the MERRF syndrome and support the notion that different mutations can lead to the same phenotype. Deficient energy production or utilization could lead to seizures in a variety of ways; because maintenance of ionic gradients across neuronal membranes is energy dependent, an imbalance of could result in hyperexcitability.

The progressive myoclonic epilepsies provide a paradigm for a common phenotypic expression for multiple defects at a molecular level resulting in myoclonic seizures. Defects in neuronal energy production (MERRF and DRPLA), neuronal degeneration resulting from storage (sialidosis, MELF, and CLN3) and cellular mechanisms involving protease inhibitors (cystatin B in EPM1) may all result in a similar seizure phenotype. There are no animal models of progressive myoclonic epilepsies.

Chromosomal Anomalies and Cerebral Dysgenesis Syndromes with Prominent Seizures

Disorders of neuronal migration are associated with about 25% of recurrent childhood epilepsies, and up to 40% of brains of patients with primary generalized epilepsy show developmental malformations at postmortem examination, ranging from subtle to gross lesions (Meencke and Janz, 1985). Furthermore, studies of surgically resected specimens from patients with intractable seizures show a 30% to 50% incidence of developmental malformations, especially in children (Farrell, 1993). High-resolution magnetic resonance imaging is leading to greater recognition of such developmental malformations in persons with epilepsy. Cerebral dysgenesis is common in many syndromes with chromosomal aberrations such as Down syndrome and tuberous sclerosis. Understanding the molecular genetic basis for cerebral dysgenesis will provide an important avenue for increasing the understanding of human epilepsy.

Certain chromosomal aberrations have been associated with characteristic electroclinical syndromes of seizures (e.g., infantile spasms in Down syndrome, myoclonic status in Angelman's syndrome). The search to map constructs matching the genotype to phenotype may yield additional information on critical regions on chromosomes where mutations predispose to the development of epilepsy. Genes involved in signal transduction have been reported in two inherited disorders associated with cerebral malformations, tuberous sclerosis and lissencephaly of the Miller-Dieker type.

Tuberous Sclerosis Complex

Tuberous sclerosis (TSC) is an autosomal dominant disorder associated with seizures and mental retardation. Hamartomas develop in various regions of the body, including the brain. Two different genetic loci have been implicated in TSC, on chromosomes 9q and 16p (Kandt et al., 1992; Sampson and Harris, 1994). Van Slegtenhorst and colleagues (1997) have identified the TSC1 gene on chromosome 9q34. The TSC1 gene encodes hamartin, a widely expressed protein of 130 kD, and is postulated to function as a tumor suppressor gene (van Slegtenhorst et al., 1997). The tuberous sclerosis-2 gene (TSC2, 16p13.3), encodes tuberin, a protein that bears limited homology to the catalytic domain of a GTPase activating protein (Rap1GAP), which is an important cellular signaling molecule. Loss of normal tuberin function leads to constitutive activation of Rap1 in patients with TSC that could lead to aberrant neuronal growth and cellular pathology (Wienecke et al., 1995), and thereby play a role in seizure expression. Intense staining of tuberin, using immunohistochemical techniques, was observed in the small blood vessels of many organs, including the kidney, skin, adrenal gland, and brain. High levels of tuberin are expressed in cortical neurons and cerebellar Purkinje cells. These findings imply that loss of function mutations in TSC2 might lead to the development of highly vascularized tumors, subcortical tubers, and focal atrophy of the cerebellar cortex, all common features of TSC (Wienecke et al., 1997).

Lissencephaly

Lissencephalies (meaning "smooth brain") are a form of neuronal migration disorder, an important cause of refractory epilepsy and mental retardation. One such migration disorder, the Miller-Dieker lissencephaly syndrome (MDS), results in an abnormal four-layered cortex. Clinical features include microcephaly, facial dysmorphisms (narrow forehead, long philtrum, upturned nares, retrognathus), and seizures. Deletions of a critical region of the gene LIS1, on chromosome 17p13.3, appear to be responsible for most cases (Dobyns et al., 1993). Lissencephaly without facial dysmorphisms has also been observed and is referred to as the isolated lissencephaly sequence (ILS). Fluorescent in situ hybridization in several patients with MDS and ILS shows that the mutations for both disorders lie within the LIS1 critical region (Chong et al., 1997). ILS and MDS appear to be a contiguous gene syndrome with other genes distal to LIS1 playing a role in the development of facial dysmorphic features in the MDS phenotype. LIS1 shows homology to genes involved in signal transduction and cerebral development. LIS1 encodes a subunit of brain platelet-activating factor (PAF) acetylhydrolase, that inactivates PAF, a neuroregulatory molecule (Hattori et al., 1994) that plays an important role in neuronal excitability. In a mouse model, LIS1 gene transcripts are strongly expressed in the cortical plate and provide a good model for the study of signal transduction pathways in normal and abnormal cerebral development (Reiner et al., 1995). PAF released from synaptic terminals can alter growth cone morphology and destabilize synaptic connections. PAF modulates excitatory synaptic transmission, neuronal plasticity and memory (Bazan et al., 1997), possibly leading to hyperexcitable neuronal circuits.

In addition to the recessively inherited lissencephalies, X-linked forms have been described that occur in hemizygous males (carrying the abnormal X chromosome). Heterozygous females manifest with a different, distinct neuronal migration defect termed subcortical band heterotopia (SBH). Boys with X-linked lissencephaly phenotypically resemble patients with classic lissencephaly, whereas females with SBH have milder mental retardation, fewer behavioral problems, and less severe epilepsy. SBH and LIS can be inherited alone or together in a single pedigree. A new genetic locus, XLIS, has been mapped by linkage analysis to Xq21-q24 and the critical region has been narrowed to Xq22.3-q23 by high-resolution chromosome analysis. SBH and X-linked lissencephaly are possibly caused by mutation of a single XLIS gene with complete absence of the gene product in hemizygous males producing LIS, while the milder SBH phenotype in heterozygous females results from random X-inactivation (lyonization) (Ross, M.E., et al., 1997). The gene, recently identified as doublecortin, encodes a protein widely expressed in the developing brain (des Portes et al., 1998), mutations of which cause the clinical syndrome. The protein encoded by doublecortin likely plays a critical role in signal transduction (Gleeson et al., 1998), probably involving the nonreceptor tyrosine kinase cascade.

Isolated periventricular heterotopia (PH) is another recently recognized clinical entity with X-linked dominant inheritance (Kamuro and Tenokuchi, 1993). PH malformations consist of well-differentiated cortical neurons and glia forming nodules in the subependymal zone along the ventricular surface. The condition is often confused with tuberous sclerosis because the lesions do not enhance with gadolinium on a magnetic resonance image (MRI). However, the mental retardation, extracranial hamartomas, and depigmented patches present in TSC are not seen in PH. Individuals with PH are at high risk for epilepsy, although they may have no other neurological signs or external stigmata. Partial, secondarily generalized, and mixed seizures are common, although some affected adults are seizure free (Dubeau et al., 1995). Seizures beginning in infancy have also been reported (Jardine et al., 1996). A high frequency of spontaneous abortions has been reported in the families studied, compatible with X-linked dominant inheritance with lethality in males (Huttenlocher et al., 1994). In one family, PH was mapped to Xq28 by linkage analysis; this region comprises several candidate genes of interest (Eksioglu et al., 1996), including the neural L1 cell adhesion molecule (L1CAM) gene and the $\alpha 3$ subunit of the GABA receptor (GABARA3). The PH gene represents an important epilepsy susceptibility locus, and possibly plays a key role in normal cortical development. A sporadic form of nodular heterotopia has also been described, distinguishable from the X-linked form by occurrence in males, later onset of seizures, and fewer nodules (Raymond et al., 1994).

Chromosomal Anomalies

Distinctive electroclinical features have been reported for several chromosomal anomalies, including fragile X syndrome, Angelman's syndrome, inversion duplication of chromosome 15, Wolf-Hirschhorn (4p-) syndrome, and trisomy 12p. There is also an increased seizure frequency in trisomy 21 (Down syndrome, DS).

Fragile X. Fragile X syndrome is the most common genetic cause of mental retardation. Affected individuals have moderate mental retardation, language impairment, deficient social and behavioral skills, and hyperactivity. Dysmorphic features, sometimes not noticeable until after puberty, include macrocephaly, a long narrow face with large ears and a prominent forehead, and macroorchidism. Seizures of a variety of types, especially generalized tonic-clonic, occur in 20% to 40% of affected individuals (Wisniewski et al., 1991). Seizure severity is usually mild and seizure frequency and severity often decline with age. Children with fragile X have a characteristic EEG (Musumeci et al., 1988), with centrotemporal spikes similar to those seen in BECTS (Musumeci et al., 1988, 1991; Wisniewski et al., 1991). However, the FraX locus has been excluded as a candidate gene for BECTS (Rees et al., 1993).

Fragile X syndrome is associated with genetic anticipation, the worsening of phenotype with successive generations. It is caused by a shutdown of transcription of the FMR1 gene, located in the Xq27.3 region. Shutdown is associated with expansion of a CGG triplet nucleotide within the 5′ untranslated region of the gene and abnormal methylation of a CpG island 250 base pairs proximal to this repeat (Sutcliffe et al., 1992). The normal chromosome contains no more than 52 copies whereas the full mutation has more than 200 base pair repeats. However, it is the degree of methylation, not the quantity of repeats, that correlates with the severity of mental retardation (Steyaert et al., 1996; Mornet et al., 1998) and inversely with quantity of gene product (Sutcliffe et al., 1992). A similar correlation with seizure frequency and severity has not been reported. This loss-of-function mutation may have clinically important consequences for cerebral development in affected children (Kluger et al., 1996). The PH syndrome described earlier also maps to the Xq28 region, whereas X-linked lissencephaly maps to the nearby Xq22-24 region.

Angelman's syndrome. Angelman's syndrome (AS) is a condition characterized by mental retardation, seizures, and dysmorphic features (acquired microcephaly, large protruding tongue and jaw); additional clinical features include a pleasant disposition, lack of speech, ataxic gait, and peculiar jerky limb movements (hence the nickname "Happy Puppet syndrome") (Angelman, 1965; Williams et al., 1995). No consistent brain structural abnormality is described, although mild cortical atrophy and delayed myelination are common.

More than 90% of children with AS develop seizures, usually by the third year of life. The seizure characteristics change with age, often beginning with infantile spasms in infancy, and later evolving to a Lennox-Gastaut picture with atypical absences, myoclonic, astatic, and atonic seizures (Viani et al., 1995; Minassian et al., 1998). Seizures in AS often remit after puberty, providing additional support for their age-dependent nature. Affected children have characteristic EEG patterns consisting of interictal 2 to 3 Hz high voltage bi-anterior bursts slow waves, often with notched components, especially in sleep. Large amplitude 3 to 6 Hz bi-posterior slow spike wave complexes unrelated to drowsiness are also frequently seen (Matsumoto et al., 1992; Sugimoto et al., 1992); the EEG pattern changes with age (Boyd et al., 1988).

Genomic imprinting, whereby the phenotype is determined by the parent from whom the mutation is inherited, produces AS. AS results from a variety of genetic mechanisms, the most common of which is a maternally inherited deletion of chromosome 15q11-13, accounting for about 70% of cases. If the same chromosomal region is deleted in the paternal germ line, an entirely different clinical disorder, Prader-Willi syndrome (PWS), results. Rarely, AS can occur because of uniparental disomy of chromosome 15, whereby both copies of chromosome 15 are inherited from the father (UPD). In this case, AS arises because neither of the mother's 15q11-13 is expressed, a situation functionally identical to the more common circumstance in which that region is deleted. Other rare causes of AS are imprinting mutations in which the methylation pattern of 15q11-13 is abnormal, and mutations of the gene for ubiquitin protein ligase (UBE3A), which lies near GABRB3 (Kishino et al., 1997).

The AS critical region, 15q11-13, comprises a number of genes which may explain the high predilection for epilepsy in affected children. Genes that map to this region and encode GABA subunits β3 and α5 (GABARA3, GABARA5) were initially suggested as candidates for the disorder (Wagstaff et al., 1991), but the phenotype can also be produced by a mutation not including the region. Wagstaff and colleagues showed that the gene for E6-AP ubiquitin-protein ligase (UBE3A) maps to the critical region for AS (Sutcliffe et al., 1997), a finding that was confirmed independently (Kishino et al., 1997; Matsuura et al., 1997). Mutation of UBE3A results in dysfunction of the ubiquitin-dependent proteolytic pathway. UBE3A mutation, UPD, or an imprinting methylation abnormality are sufficient to produce the AS phenotype, but seizures and EEG abnormalities seem to be more severe if the 15q11-13 region is deleted (Minassian et al., 1998). These observations suggest that one or more genes in the deleted region, including those for GABA receptors, may modify the AS phenotype caused by the lack of maternal UBE3A function. Dysfunction of GABA receptors could alter inhibitory neurotransmission sufficiently to increase cortical excitability and enhance seizure propensity. This appealing hypothesis remains to be proven directly.

A proposed animal model of AS has been produced by knocking out the $GABA_A$ β3 receptor subunit. Such mice develop widespread neurological dysfunction and seizures (Homanics et al., 1997), and hippocampal neurons isolated from these mutant mice have decreased responses to GABA evoked currents (Krasowski et al., 1998). Thus the Angelman critical region on chromosome 15 may bear a gene whose dysfunction contributes to the intractable seizure phenotype seen in AS. The interesting age-dependence of cortical hyperexcitability in AS may be explained by altered ontogentic expression of GABA receptor subunits (Ma et al., 1993). Another mouse model of AS, involving UPD, may be more relevant since the critical genes are not expressed (Cattanach et al., 1997).

Inversion duplication of chromosome 15 (inv dup 15) syndrome. Supernumerary marker chromosomes are small, extra pieces of a chromosomal material attached anywhere along a chromosome. Such "extra structurally abnormal chromosomes" (ESACs) can cause a variety of clinical syndromes, including some with epilepsy as a predominant feature. For example, inversion duplication of chromosome 15 (inv dup 15) results in tetrasomy 15p and partial tetrasomy 15q. The syndrome is characterized by mild phenotypic abnormalities (downslanting palpebral fissures, epicanthus, low-set ears), severe pervasive developmental delay, aggressiveness, hyperactivity, and intractable seizures (Leana-Cox et al., 1994; Battaglia et al., 1997). The seizures can be of several types, including generalized tonic-clonic, tonic, atonic atypical absence, and infantile spasms (Bingham et al., 1996; Battaglia et al., 1997). The predilection for seizures in the inv dup 15 syndrome may be related to altered GABA receptor function (Battaglia et al., 1997). By FISH analysis, it was shown that inv dup 15 contains the Prader-Willi/Angelman syndrome (PWS/AS) critical region, so among the duplicated genes are GABRA5 and GABRB3 (Leana-Cox et al., 1994). As previously discussed, alteration of GABA receptor subunits can lead to decreased cortical inhibition that may allow expression of an epileptic phenotype.

Wolf-Hirschhorn syndrome. Wolf-Hirschhorn syndrom (WHS [4p-]) is an autosomal disorder that arises from partial deletion of the short arm of chromosome 4p (Guthrie et al., 1971). Affected children have severe mental retardation, cardiac defects, and multiple craniofacial anomalies, including microcephaly, hypertelorism, and fish-like mouths. Seizures are reported in 50% to 100% of cases, with onset usually in the first year. Partial motor, generalized tonic-clonic, and myoclonic seizures have been most commonly reported. There is a stereotyped electrographical and clinical picture, characterized by generalized or unilateral myoclonic seizures accompanied by a pattern of centroparietal or parietotemporal sharp waves; high-voltage atypical spike-wave complexes, often elicited by eye closure; bursts of diffuse spikes and waves; and frequent jerks (Sgro et al., 1995). In the few available autopsy cases, cortical developmental abnormalities such as neuronal heterotopias and gyral abnormalities have been described (Gottfried et al., 1981; Guthrie et al., 1971).

The WHS critical region has been confined to a locus of 750 kbase pairs within 4p16.3 (Altherr et al., 1997), a particularly gene-rich area. Several potentially expressed sequences have been found throughout the critical region, but none have been studied in detail except for FGFR3, which encodes fibroblast growth factor receptor 3. FGFR3 is expressed predominantly in the germinal epithelium of the neural tube and may, in tandem with FGFR1, contribute to stem cell differentiation and migration. However, at birth and in the adult brain, FGFR3 is expressed diffusely and localized in cells with morphological characteristics of glia (a pattern distinctly different from the discrete neuronal expression of FGFR1) (Peters et al., 1993). A role for basic fibroblast growth factor (FGF) in modulating neuronal excitability and vulnerability to seizure-induced degeneration is suggested by data showing that basic FGF can suppress glutamate-induced elevations of intracellular calcium levels and prevent excitotoxicity in hippocampal neurons (Mattson et al., 1989; 1993; Cheng et al., 1995).

Some of the seizure types observed in patients with the 4p- syndrome are consistent with those arising from heterotopias. However, the presence of generalized spike wave discharges in virtually all affected individuals is more consonant with models of epileptogenesis resulting from reduced GABAergic inhibition. Genes encoding the α and $\beta 1$ $GABA_A$ subunit receptors are located nearby on 4p13-p12. Reduction of their gene product or unmasking of deleterious alleles on the normal chromosome could contribute to the high epileptogenicity of this syndrome.

Trisomy 12p. Trisomy 12p is associated with severe mental retardation, craniofacial anomalies, and seizures that are usually of the generalized tonic-clonic or myoclonic types. In three cases of trisomy 12p, the EEG consisted of generalized 3-Hz spike wave discharges (Guerrini et al., 1990). All of these infants died because of associated malformations, and all were confirmed to have cortical malformations. Another disorder with myoclonic seizures, DRPLA, localizes to the same region of chromosome 12p. Thus, alteration of one or more genes localized to 12p may lead to increased susceptibility to generalized spike wave abnormalities and myoclonic seizures. Three voltage-gated potassium channels are located in the 12p13 region, possibly explaining the predisposition to hyperexcitability (Guerrini et al., 1998).

Down syndrome. DS (trisomy 21) is the most common chromosomal cause of mental retardation. Seizures occur in 5% to 10% of persons with DS, higher than the expected frequency in the general population (Tatsuno et al., 1984; Pueschel et al., 1991; Stafstrom et al., 1991). The high frequency of seizures in DS is likely due to a combination of inherent structural anomalies (brain dysgenesis) and common medical complications of the syndrome (cardiac defects with hypoxic ischemia, immune dysfunction with frequent infections, etc.) (Stafstrom et al., 1991). There is an age-related incidence of epilepsy in DS, peaking in the first year and again in the fourth and fifth decades (Veall, 1974). The early peak is contributed to by perinatal medical complications and by the increased incidence of infantile spasms in DS (Stafstrom and Konkol, 1994). The increased seizure occurrence later in life is coincident with the development of the neuropathological Alzheimer-like changes seen in essentially all DS individuals by middle age (Wisniewski et al., 1985; Lai and Williams, 1989; Evenhuis, 1990). About 10% of Alzheimer's disease patients develop seizures, whereas as many as 75% of older DS individuals develop seizures, suggesting that there is some factor inherent to the DS brain that favors hyperexcitability later in life. One possible mechanism underlying increased hyperexcitability in Alzheimer's disease and DS is that there is a decrease in levels of the secreted form of β-amyloid precursor protein which has been shown to suppress neuronal excitability and decrease neuronal vulnerability to excitotoxicity (Mattson et al., 1993; Furukawa et al., 1996).

Several possible factors contribute to the pathophysiology of the DS brain to enhance seizure susceptibility. Cortical dysgenesis is present, with fewer GABAergic interneurons in critical neocortical layers (Ross et al., 1984; Wisniewski et al., 1984). Dendritic spines have altered morphology and lower density, in a pattern that favors hyperexcitability (Becker et al., 1986; Stafstrom, 1993). Intrinsic abnormalities of the neuronal membrane may also contribute to hyperexcitability in DS. Studies using fetal DS dorsal root ganglia (DRG) have demonstrated decreased action potential threshold (Scott et al., 1981), faster action potential rise time due to altered sodium channel activation kinetics (Caviedes et al., 1990), and a variety of neurotransmitter abnormalities (reviewed in (Stafstrom, 1993)).

Additional support for hyperexcitability in DS brains comes from the mouse trisomy 16 model; the distal part of mouse chromosome 16 contains genes homologous to those found on human chromosome 21 (Holtzman et al., 1992; Lacey-Casem and Oster-Granite, 1994). Some electrophysiological features seen in human DS DRG have been confirmed in the trisomy 16 mouse, including rapid action potential rise (Ault et al., 1989). However, more recent studies on hippocampal neurons from trisomy 16 mice have shown that there is reduced current through voltage-gated sodium channels due to reduced expression of these channels (Stoll and Galdzicki, 1996). Thus, the explanation of hyperexcitability is unclear, and may reflect altered signal transduction distal to membrane ionic channels.

Down syndrome is now recognized as a contiguous gene disorder (Korenberg et al., 1990). Among putative genes residing on chromosome 21 that are of potential relevance for epilepsy are those for the ionotropic (kainate type) glutamate receptor (GRIK1), copper-zinc superoxide dismutase (SOD1), and the interferon a and b receptor (INFAR). An increased dosage of the gene SOD1 could lead to membrane oxidative damage, and trisomy 16 mice overproduce superoxide anions (Colton et al., 1990). It is intriguing that the genes for both Unverricht-Lundborg disease (EPM1) (Lehesjoki et al., 1991) and for the glutamate subunit GluR5 (Eubanks et al., 1993) localize to the distal long arm of chromosome 21, within the critical DS region. Thus, genetic factors may play an important role in the expression of myoclonic seizures in DS. The role of specific genes in the development of seizures in DS remains to be clarified.

Metabolic Disorders with Prominent Seizures

Metabolic disorders typically produce diffuse cerebral disturbances and multi-system dysfunction. Many metabolic disorders are accompanied by recurrent seizures, which may occur through a variety of mechanisms, including alteration in cerebral energy metabolism, neurotransmitter production or turnover, and dysfunction of intracellular organelles. Some metabolic defects present with seizures in early childhood.

Pyridoxine Dependency

Pyridoxine dependency is a very rare autosomal recessive disorder associated with severe intractable seizures that usually begin in the first few hours or days of life, but have developed as late as age 14 months (Krishnamoorthy, 1983; Goutieres and Aicardi, 1985). The seizures are refractory to all antiepileptic drugs, but are exquisitely sensitive to the intravenous infusion of pyridoxine (vitamin B_6). In such cases, the clinical seizures and EEG abnormalities improve within minutes to hours. To maintain seizure control, pyridoxine must be provided throughout the patient's life.

The lowered seizure threshold in B_6-dependent patients is presumed due to reduced GABA, resulting from diminished activity of glutamic acid decarboxylase (GAD) (Scriver and Whelan, 1969), which requires pyridoxal-5′-phosphate (PLP) as a co-factor to convert glutamic acid to GABA. In a single autopsied case, GABA levels were reduced and glutamic acid levels increased in frontal and occipital poles compared to controls; PLP levels were reduced in frontal pole (Lott et al., 1978). A mouse model with the gene for tissue-nonspecific alkaline phosphatase (TNAP) knocked out has decreased intracellular PLP levels and reduced brain GABA levels (Waymire et al., 1995). This model produces a lethal seizure phenotype that can be rescued with pyridoxal treatment. Whether the biochemical defect associated with pyridoxine dependency is directly related to abnormalities of GAD or its access to sufficient intracellular PLP may be determined when more is learned about the GAD gene, known to map to chromosome 2q31 (Bu et al., 1992).

Biotinidase Deficiency

Biotinidase deficiency (BTD) is an autosomal recessive disorder with a combined incidence for the profound and partial deficiency of about 1 in 60,000. Detection is critical since its manifestations can be avoided by administration of exogenous biotin. The deficient enzyme normally recycles biotin to its free form, which is essential for fatty acid, carbohydrate, and amino acid metabolism (Sutherland et al., 1991; Wolf and Heard, 1991). In the absence of sufficient biotin, carboxylases remain in their inactive form, producing a multiple carboxylase deficiency. The phenotype includes seizures, hypotonia, sensorineural hearing loss, ataxia, diarrhea, and cutaneous manifestations (rash, alopecia) in early infancy. A screen for biotinidase deficiency is included in the routine newborn screen in some states.

The seizures of BTD are usually generalized tonic-clonic, or more rarely, infantile spasms or myoclonic seizures (Salbert et al., 1993). The seizures are resistant to standard antiepileptic drugs but most respond well to administration of biotin. BTD is caused by a mutation on chromosome 3p25 (Pomponio et al., 1995); half of symptomatic children have a seven base pair deletion and a three base pair insertion in one of the alleles of the biotinidase gene, producing a truncated peptide. Other mutations have also been identified (Norrgard et al., 1997; Pomponio et al., 1997). A GABA transporter gene mapping to 3p25-p24 may be a candidate gene for the seizure phenotype (Huang et al., 1995).

Nonketotic Hyperglycinemia

Nonketotic hyperglycinemia (NKH) is a primary disorder of glycine metabolism in which accumulation of glycine in the cerebrospinal fluid is particularly prominent. This disorder manifests in the newborn period as a myoclonic epileptic encephalopathy, with extreme lethargy, poor feeding, hypotonia, and apnea, often leading to intractable seizures and coma. The EEG shows a burst suppression pattern (Dalla Bernardina et al., 1983). However, NKH is phenotypically heterogenous and a small number of patients have atypical, milder presentations. The metabolic lesion of NKH is in the glycine cleavage system (GCS), a complex mitochondrial enzyme system with four enzyme components: P-, T-, H-, and L-protein. The vast majority of patients with NKH are deficient of P-protein, the gene for which maps to chromosome 9p23-24 (Isobe et al., 1994); this enzyme catalyses the decarboxylation of glycine. However, several mutations have been identified for the T- and one for the H-protein gene. Treatment has traditionally been directed toward reducing cerebrospinal fluid glycine levels with sodium benzoate, which is presumed to conjugate with glycine to form hippuric acid that is renally excreted (MacDermot et al., 1981).

Immunochemical and in situ hybridization analyses have demonstrated strong GCS expression in rat hippocampus, olfactory bulb, and cerebellum (Kure et al., 1997), a distribution similar to the N-methyl-D-aspartic acid (NMDA) receptor that has a binding site for glycine. Based on the concept that the neurological disturbance in NKH may be caused by overstimulation of the NMDA receptor allosterically activated (Wroblewski et al., 1989) by high concentrations of glycine, treatment with NMDA antagonists has been attempted. There is some evidence that early treatment with NMDA modulators, such as dextromethorphan (Kure et al.,

1997), may reduce late neurological complications and is possibly as important as treatment with sodium benzoate (Boneh et al., 1996).

Mitochondrial Encephalopathy, Lactic Acidosis, and Stroke-like Episodes

As previously introduced, mitochondrial cytopathies represent a unique nonmendelian mechanism of inheritance. These disorders occur as a result of mutations in mitochondrial DNA, resulting in impaired energy metabolism. Neurological manifestations, particularly seizures, are prominent features of mitochondrial syndromes. MELAS is a disorder of mitochondrial function resulting from tRNA mutations (*MELAS 3243G) (Goto et al., 1990; Shoffner et al., 1995). The core clinical features of MELAS are seizures and recurrent metabolic strokes, with less common features including encephalomyopathy with deafness, ataxia and dementia, progressive external ophthalmoplegia, limb weakness, developmental delay, and a variety of abnormalities of other organ systems. The proportion of mutant mitochondria is higher in those with strokes and seizures (Morgan-Hughes et al., 1995). Although seizures initially occur as a consequence of acidotic episodes, the brain damage induced by recurrent seizures creates an epileptic substrate that independently supports seizures.

ANIMAL MODELS OF GENETIC EPILEPSIES

Animal models of the genetic epilepsies are found in numerous species, including mice, rats, gerbils, baboons, chickens, and dogs (Buchhalter, 1993). Such models are of great interest since they permit consideration of the genetic and biochemical factors that predispose to seizures, as described earlier for several types of inherited epilepsy. Many have been well characterized in regard to seizure morphology, EEG features, neuropathology, and response to antiepileptic drugs. Whether or not these fortuitous mutant models correspond to naturally occurring genes for human epilepsies, they do define critical sites of vulnerability in the array of mechanisms that maintain the balance between excitation and inhibition.

Spontaneous Single Gene Mutations

Of the various mutant animal epilepsies, the largest and best-characterized group comprises those caused by single gene mutations in the mouse, most of which are inherited as autosomal recessive traits. From these murine models, it has become apparent that single gene mutations can cause an entire epilepsy phenotype or multiple phenotypes within individuals, different defects can produce similar phenotypes, and interactions among epilepsy genes may either mask an epilepsy phenotype or produce intermediate epilepsy syndromes. Multiple examples of these phenomena have been provided in the previous sections. Some of the better-studied mutant models are discussed next.

Tottering Mouse

The tottering mutant mouse expresses a phenotype that includes absence seizures, accompanied by bilateral synchronous 6- to 7-Hz spike-wave activity, and intermittent myoclonic seizures (Noebels and Sidman, 1979). The tottering gene has been localized to mouse chromosome 8 on the background of C57bl6 mice; two other alleles of this gene exist, leaner and rolling (Noebels, 1995). These mice have significant hypertrophy of noradrenergic fibers originating from locus ceruleus and treatment of neonatal animals with the catecholaminergic neurotoxin, 6-OHDA, blocks the hypertrophy, EEG changes and absence seizures (Noebels, 1995).

However, more recently it has also been demonstrated that the gene mutated in these mice encodes the α1A subunit of voltage-dependent P/Q calcium channels (Fletcher et al., 1996). Studies indicate tottering mouse has a missense mutation that leads to a nonconservative Pro-to-Leu amino acid substitution close to the pore-forming loop of the second of four homologous transmembrane repeat regions of the α1A subunit (Ophoff et al., 1998). Equal amounts of α1A transcripts are present in normal and tottering mice, suggesting a gain-of-function mechanism affecting P/Q channel pore activity (Doyle et al., 1997). Curiously, four different missense mutations have been identified in the homologous human gene of families bearing a syndrome characterized by episodic ataxia and hemiplegic migraine (Ophoff et al., 1996). One of these human mutations is located within the pore-forming hairpin structure very close to the mouse tottering mutation. Phenotypic expression does not include seizures, however, EEGs of patients have not been described. At present it is unknown if there are allelic variations of this channelopathy in humans that express an epileptic phenotype.

Lethargic Mouse

An autosomal recessive mutation on chromosome 2 of the lethargic mouse (*lh/lh*) produces an absence epilepsy with a 5- to 6-Hz spike-wave EEG pattern. Although the specific genetic defect in this animal is unknown, an enhanced expression of neocortical and thalamic $GABA_B$ receptors has been reported. $GABA_B$-mediated inhibition of GABA release appears enhanced in neocortex but blunted in thalamus (Lin et al., 1995). Additionally, inhibitory postsynaptic potentials of ventrobasilar thalamic neurons induced by $GABA_B$-mediated K^+ conductances are similar in lethargic and wild-type mice (Caddick and Hosford, 1996). These observations suggest a more complex pattern of changes in the mutant than that expected for a generalized enhanced expression of receptor. This model may prove to be superior to the "gold standard" high-dose pentylenetetrazol model in predicting efficacy of putative antiepileptic drugs against human absence seizures (Hosford and Wang, 1997).

Slow-Wave Epilepsy Mouse

This spontaneous mouse mutant has a form of absence epilepsy associated with a 3- to 4.5 Hz spike-wave discharge, similar to the pattern seen in human absence epilepsy, in addition to ataxia and tonic-clonic seizures. Mutants show selective neuronal death in the cerebellum and brain stem but otherwise are healthy. Using positional cloning strategies, a mutation in the Nhe1 gene located on the distal portion of chromosome 4, was demonstrated responsible for the absence seizures (Cox et al., 1997). Nhe1 is a ubiquitous and extensively studied protein that apparently plays a homeostatic role in nearly all cells by mediating the electroneutral 1:1 exchange of Na^+ and H^+ across the plasma membrane, thereby helping maintain intracellular pH and cell volume. The role of this mutation within the context of imbalanced inhibition/excitation or synchronized thalmocortical burst firing, the putative mechanism behind absence epilepsy, remains unknown. It is noteworthy that hyperventilation, which raises serum pH, has been used by clinicians for decades as a provocative test to induce absence seizures.

Stargazer Mouse

The stargazer mouse mutation causes prolonged absence seizures associated with spike-wave discharges, an ataxic gait, and vestibular problems, including a distinctive head-tossing motion. A genetic map of the stargazer region has been constructed on mouse chromosome 15; the DNA interval including the stargazer locus has been narrowed to 150 kb (Letts et al., 1997). Pharmacological investigation (Aizawa et al., 1997) indicated that $GABA_B$ receptors play a significant role in the pathogenesis of the absence seizures in this model, although suppression of spike-wave discharges by the NMDA antagonist, MK-801, suggest aberrant excitatory mechanisms may be involved as well. Electrophysiological studies (Di Pasquale et al., 1997) have demonstrated an autonomous increase in cortical network excitability, suggesting that the genetic defect lowers the threshold for aberrant thalamocortical spike wave oscillations in vivo, and thereby causes the behavioral absence, consonant with current concepts of human absence epilepsy.

Transgenic Animals in the Study of Epilepsy

Of reciprocal benefit to both molecular geneticists and epileptologists has been the surprising surfeit of seizure phenotypes among transgenic mice, some examples of which are presented in Table 2. For the former, this visible marker of cerebral dysfunction provides a ready guide to help direct studies to understand the biochemical expression of these genetic alterations. To the latter, this represents a unique opportunity to rapidly increase the substrate upon which to direct focused studies to delineate mechanisms of epileptogenesis. The challenge is to know which genes to target for mutations. As a primary consideration, one would select genes whose products are involved with the balance of neural excitation and inhibition. The choice becomes more challenging with the necessary modification that the balance defect must lead to intermittent cortical dysrhythmia. Despite these constraints, a plethora of epilepsy models can be expected to surface from geneticists' laboratories. Transgenic expression of human mutations in mouse models will

Table 2. Selected Animal Models of Epilepsy

Animal Model	*Phenotype*	*Resulting Effect*	*Reference*
Altered Intrinsic Neuronal Excitability			
Mouse homolog of Shaker gene	Lethal epilepsy	Elimination of a delayed rectifier K^+ channel α subunit → increased transmitter release	1
Weaver mice	Ataxia and spontaneous seizures	Missense mutation for a protein coding for a G protein coupled inward rectifier channel (GIRK2)	2
Altered Synaptic Transmission			
Rats given EAAC1 anti-sense oligonucleotides (GLT-1)	Facial twitching, freezing behavior, clonic seizures, paresis	Neuronal glutamate transporter (EAAC1)	3, 4
Synapsin I and II knock-out mice	Seizures precipitated by sensory stimuli	Synaptic vesicle proteins altered (↓ number of releasable vesicles) → less robust NT release; ↓ hippocampal PTP not LTP	5
Transgenic mice CaMK II α subunit null mutation	Limbic seizures	Hippocampal hyperexcitability, altered evoked neurotransmitter release; ↓ LTP not PTP	6
Tissue nonspecific alkaline phosphatase (TNAP) mutant mice	Lethal seizure phenotype	Defect in metabolism of pyridoxal-5-phosphate, reduced brain GABA levels	7
Postsynaptic Effects			
Transgenic mice: Q/R editing deficient (GluR-B allele)	Lethal limbic seizures and neuronal degeneration	Altered Ca permeability of AMPA receptors, ↓ GluRB → abnormal immediate early gene response	8
Transgenic mice lacking $5HT_{2c}$ receptors	Overweight with spontaneous lethal seizures (fifth week of life)	Lowered seizure threshold and progression of seizure activity	9
Transgenic mice lacking Inositol1,4,5-triphosphate (IP_3R1)	Ataxia and tonic clonic seizures	?	10
Transgenic jerky mice	Partial seizures with secondary generalization	Altered nuclear regulatory DNA binding protein	11
Altered Neuronal Network			
Growth cone associated protein (GAP 43) over expression	Spontaneous seizures at age 4 weeks	Hippocampal mossy fiber axonal sprouting, interneuron loss	12
Myelin proteolipid protein (PLP) over expression	Demylinating syndrome with tremor and convulsions	Altered oligodendroglial function	13
Tissue plasminogen activator (tPA) deficient mice	Resistance to neuronal degeneration and drug-induced seizures	Involved in neuronal plasticity	15

(continued)

Table 2. Continued

Animal Model	*Phenotype*	*Resulting Effect*	*Reference*
Altered Seizure Threshold			
Brain derived neurotrophic factor (BDNF) deficient mice	Suppressed kindling development	Establishes synaptic efficacy and hyperexcitability	14

Notes: 1 (Smart et al., 1998), 2 (Patil et al., 1995), 3 (Rothstein et al., 1996), 4 (Tanaka et al., 1997), 5 (Rosahl et al., 1995), 6 (Butler et al., 1995), 7 (Waymire et al., 1995), 8 (Brusa et al., 1995), 9 (Tecott et al., 1995), 10 (Matsumoto et al., 1996), 11 (Toth et al., 1995), 12 (Aigner et al., 1995), 13 (Kagawa et al., 1994), 14 (Kokaia et al., 1995), 15 (Tsirka et al., 1995).

permit the critical epileptogenic steps to be elucidated and ordered, whereas the models themselves will serve as test systems for pharmacologic interventions. The following sections and Table 2 describe a few of the many new models, more of which are being described with increasing frequency.

Intrinsic Excitability

The balance between excitation and inhibition exists along a hierarchy, the first at the level of intrinsic neuronal excitability. Critical elements among determinants of neuronal excitability are the genes encoding ion channels. The potassium channel mutation of BFNC was described in an earlier section (Biervert et al., 1998). Mice lacking the voltage-gated potassium channel α m,asubunit, K(V)1.1, display frequent spontaneous limbic and tonic-clonic seizures throughout adult life (Smart et al., 1998) and Table 2. Other support for some genetic epilepsies as ion channel disorders include calcium channel defects in certain mouse models, described earlier, and evidence of altered brain sodium channel transcript levels in human epileptic foci (Lombardo et al., 1996). All of these are potential genes to target for further study, including strategies for their deletion, overexpression, and site-directed mutation.

Synaptic Transmission

The second level in the excitation/inhibition balance hierarchy involves synaptic information transfer. This includes potential aberrations of vesicle proteins (Rosahl et al., 1995); proteins for neurotransmitter synthesis and transport (see Table 2 and examples in prior sections); second messengers, including kinases (Butler et al., 1995) and protein phosphatases, acting at pre- and postsynaptic (Matsumoto et al., 1996) sites; and synaptic receptors. Examples of transgenic animals with seizure-associated receptor abnormalities include those with alteration of the GluRB (AMPA-type) receptor subunit (Brusa et al., 1995) and targeted deletion of the 5-HT_{2c} gene that encodes a G-protein coupled serotonin receptor

(Tecott et al., 1995). As described previously, a nicotinic acetylcholine receptor subunit defect is associated with ADNFLE, an idiopathic partial epilepsy.

Altered Networks

The highest hierarchical level involves the development and maintenance of neural networks, including the balance of excitability. Postnatal overexpression of growth cone-associated protein, GAP-43, induces mossy fiber axon terminal sprouting after 4 weeks of age with subsequent onset of spontaneous seizures (Aigner et al., 1995). Convulsions occur in a glial mutant in which wild-type copies of the proteolipid protein gene are overexpressed, leading to a dysmyelination syndrome and oligodendroglial cell death (Kagawa et al., 1994).

SUMMARY

The progress made in the last decade in targeting human epilepsy genes has been phenomenal. Although these advances shed some light on our understanding of human epilepsies, they probably raise more questions than answers at present. Many gene defects and mechanisms can result in a single phenotype, while many differing phenotypes may result from a single gene defect. Some gene defects may alter specific neurotransmitter function and postsynaptic function; however, many appear to produce alterations at multiple points: the presynaptic level, early development of synaptic connections, intracellular signaling mechanisms with resulting cerebral developmental malformations, failure of cytoprotective mechanisms, and altered cellular metabolism with resultant degeneration of neurons. The chromosomal aberrations highlight the association of seizure phenotypes with certain chromosomes and regions within the human genome. Developing constructs for region-specific gene expression and linking genotype to the phenotype are the challenges for molecular biologists. Such advances will provide a neurobiological basis for understanding the mechanisms underlying development of human epilepsies. The challenge to clinicians and molecular neurobiologists is to unravel this tangled web and translate these advances into practical applications with implications for early diagnosis, genetic counseling and therapy for patients with seizures.

ACKNOWLEDGMENT

The authors thank Drs. Rosemary Boustany and Joseph Wagstaff for helpful comments on selected portions of the manuscript. Dr. Jeffrey Noebels critiqued the entire manuscript.

REFERENCES

Aigner, L., Arber, S., Kapfhammer, J.P., Laux, T., Schneider, C., Botteri, F., Brenner, H.R. & Caroni, P. (1995). Overexpression of the neural growth-associated protein GAP-43 induces nerve sprouting in the adult nervous system of transgenic mice. Cell 83, 269–278.

Aizawa, M., Ito, Y. & Fukuda, H. (1997). Pharmacological profiles of generalized absence seizures in lethargic, stargazer and gamma-hydroxybutyrate-treated model mice. Neurosci. Res. 29, 17–25.

Allen, K.M. & Walsh, C. (1996). Shaking down new epilepsy genes. Nat. Med. 2, 516–518.

Altherr, M.R., Wright, T.J., Denison, K., Perez-Castro, A.V. & Johnson, V.P. (1997). Delimiting the Wolf-Hirschhorn syndrome critical region to 750 kilobase pairs. Am. J. Med. Genet. 71, 47–53.

Andrews, P.I. & McNamara, J.O. (1996). Rasmussen's encephalitis: an autoimmune disorder? Curr. Opin. Neurobiol. 6, 673–678.

Andrews, P.I. & Stafstrom, C.E. (1993). Ictal EEG findings in an infant with benign familial neonatal convulsions. J. Epilepsy 6, 174–179.

Angelman, H. (1965). 'Puppet' children. Dev. Med. Child Neurol. 7, 681–688.

Ault, B., Caviedes, P. & Rapoport, S.I. (1989). Neurophysiological abnormalities in cultured dorsal root ganglion neurons from the trisomy-16 mouse fetus, a model for Down syndrome. Brain Res. 485, 165–170.

Baram, T.Z. & Schultz, L. (1991). Corticotropin-releasing hormone is a rapid and potent convulsant in the infant rat. Brain Res Dev Brain Res 61, 97–101.

Baram, T.Z., Gerth, A. & Schultz, L. (1997). Febrile seizures: an appropriate-aged model suitable for long-term studies. Brain Res. Dev. Brain Res. 98, 265–270.

Bates, G. & Lehrach, H. (1994). Trinucleotide repeat expansions and human genetic disease. Bioessays 16, 277–284.

Battaglia, A., Gurrieri, F., Bertini, E., Bellacosa, A., Pomponi, M.G., Paravatou-Petsotas, M., Mazza, S. & Neri, G. (1997). The inv dup(15) syndrome: a clinically recognizable syndrome with altered behavior, mental retardation, and epilepsy. Neurology 48, 1081–1086.

Bazan, N.G., Packard, M.G., Teather, L. & Allan, G. (1997). Bioactive lipids in excitatory neurotransmission and neuronal plasticity. Neurochem. Intl. 30, 225–231.

Beck, C., Moulard, B., Steinlein, O., Guipponi, M., Vallee, L., Montpied, P., Baldy-Moulnier, M. & Malafosse, A. (1994). A nonsense mutation in the alpha4 subunit of the nicotinic acetylcholine receptor (CHRNA4) cosegregates with 20q-linked benign neonatal familial convulsions (EBNI) [published erratum appears in Neurobiol Dis 1995 Jun;2(3):189]. Neurobiol. Dis. 1, 95–99.

Becker, L.E., Armstrong, D.L. & Chan, F. (1986). Dendritic atrophy in children with Down's syndrome. Ann. Neurol. 20, 520–526.

Berkovic, S.F. (1997). Epilepsy genes and the genetics of epilepsy syndromes: the promise of new therapies based on genetic knowledge. Epilepsia 38, S32–36.

Berkovic, S.F. & Scheffer, I.E. (1997a). Epilepsies with single gene inheritance. Brain Dev. 19, 13–18.

Berkovic, S.F. & Scheffer, I.E. (1997b). Genetics of human partial epilepsy. Curr. Opin. Neurol. 10, 110–114.

Berkovic, S.F., Cochius, J., Andermann, E. & Andermann, F. (1993). Progressive myoclonus epilepsies: clinical and genetic aspects. Epilepsia 34, S19–30.

Berkovic, S.F., McIntosh, A., Howell, R.A., Mitchell, A., Sheffield, L.J. & Hopper, J.L. (1996). Familial temporal lobe epilepsy: a common disorder identified in twins. Ann. Neurol. 40, 227–235.

Berkovic, S.F., Howell, R.A., Hay, D.A. & Hopper, J.L. (1998). Epilepsies in twins: genetics of the major epilepsy syndromes. Ann. Neurol. 43, 435–445.

Biervert, C., Schroeder, B. C., Kubisch, C., Berkovic, S.F., Propping, P., Jentsch, T.J. & Steinlein, O.K. (1998). A potassium channel mutation in neonatal human epilepsy. Science 279, 403–406.

Bingham, P.M., Spinner, N.B., Sovinsky, L., Zackai, E.H. & Chance, P.F. (1996). Infantile spasms associated with proximal duplication of chromosome 15q. Pediatr. Neurol. 15, 163–165.

Boneh, A., Degani, Y. & Harari, M. (1996). Prognostic clues and outcome of early treatment of nonketotic hyperglycinemia. Pediatr. Neurol. 15, 137–141.

Bonten, E., van der Spoel, A., Fornerod, M., Grosveld, G. & d'Azzo, A. (1996). Characterization of human lysosomal neuraminidase defines the molecular basis of the metabolic storage disorder sialidosis. Genes Dev. 10, 3156–3169.

Boustany, R.M. (1992). Neurology of the neuronal ceroid-lipofuscinoses: late infantile and juvenile types. Am. J. Med. Genet. 42, 533–535.

Boustany, R.M. & Kolodny, E.H. (1989). Neurological progress. The neuronal ceroid lipofuscinoses: a review. Rev. Neurol. 145, 105–110.

Boyd, S.G., Harden, A. & Patton, M.A. (1988). The EEG in early diagnosis of the Angelman (happy puppet) syndrome. Eur. J. Pediatr. 147, 508–513.

Bray, P.F. & Wiser, W.C. (1964). Evidence for a genetic etiology of temporal-central abnormalities in focal epilepsy. N. Eng. J. Med. 271, 926–933.

Brusa, R., Zimmermann, F., Koh, D.S., Feldmeyer, D., Gass, P., Seeburg, P.H. & Sprengel, R. (1995). Early-onset epilepsy and postnatal lethality associated with an editing-deficient GluR-B allele in mice. Science 270, 1677–1680.

Bu, D.F., Erlander, M.G., Hitz, B.C., Tillakaratne, N.J., Kaufman, D.L., Wagner-McPherson, C.B., Evans, G.A. & Tobin, A.J. (1992). Two human glutamate decarboxylases, 65-kDa GAD and 67-kDa GAD, are each encoded by a single gene. Proc. Natl. Acad. Sci. U. S. A. 89, 2115–2119.

Buchhalter, J.R. (1993). Animal models of inherited epilepsy. Epilepsia 34, S31–41.

Bundey, S., Hardy, C., Vickers, S., Kilpatrick, M.W. & Corbett, J.A. (1994). Duplication of the 15q11–13 region in a patient with autism, epilepsy and ataxia. Dev. Med. Child Neurol. 36, 736–742.

Buoni, S., Grosso, S., Di Cosmo, G., Di Bartolo, R.M., Di Marco, V. & Fois, A. (1998). Segregation analysis in typical absence epilepsy. J. Child. Neurol. 13, 89–93.

Burke, J.R., Enghild, J.J., Martin, M.E., Jou, Y.S., Myers, R.M., Roses, A.D., Vance, J.M. & Strittmatter, W.J. (1996). Huntingtin and DRPLA proteins selectively interact with the enzyme GAPDH [see comments]. Nat. Med. 2, 347–350.

Butler, L.S., Silva, A.J., Abeliovich, A., Watanabe, Y., Tonegawa, S. & McNamara, J.O. (1995). Limbic epilepsy in transgenic mice carrying a Ca2+/calmodulin-dependent kinase II alpha-subunit mutation. Proc. Natl. Acad. Sci. U. S. A. 92, 6852–6855.

Caddick, S.J. & Hosford, D.A. (1996). GABAB-activated gK+ in thalamic neurons in the lethargic (lh/lh) mouse model of generalized absence seizures. Neurosci.Lett. 205, 29–32.

Caraballo, R., Cersosimo, R., Galicchio, S. & Fejerman, N. (1997). [Benign infantile familial convulsions]. Rev. Neurol. 25, 682–684.

Cattanach, B.M., Barr, J.A., Beechey, C.V., Martin, J., Noebels, J. & Jones, J. (1997). A candidate model for Angelman syndrome in the mouse [published erratum appears in Mamm. Genome 1997;8(11):877]. Mamm. Genome 8, 472–478.

Caviedes, P., Ault, B. & Rapoport, S.I. (1990). The role of altered sodium currents in action potential abnormalities of cultured dorsal root ganglion neurons from trisomy 21 (Down syndrome) human fetuses. Brain Res. 510, 229–236.

Cendes, F., Lopes-Cendes, I., Andermann, E. & Andermann, F. (1998). Familial temporal lobe epilepsy: a clinically heterogeneous syndrome. Neurology 50, 554–557.

Charlier, C., Singh, N.A., Ryan, S.G., Lewis, T.B., Reus, B.E., Leach, R.J. & Leppert, M. (1998). A pore mutation in a novel KQT-like potassium channel gene in an idiopathic epilepsy family [see comments]. Nat. Genet. 18, 53–55.

Cheng, B., Furukawa, K., O'Keefe, J.A., Goodman, Y., Kihiko, M., Fabian, T. & Mattson, M.P. (1995). Basic fibroblast growth factor selectively increases AMPA-receptor subunit GluR1 protein level and differentially modulates Ca^{2+} responses to AMPA and NMDA in hippocampal neurons. J. Neurochem. 65, 2525–2536.

Chong, S.S., Pack, S.D., Roschke, A.V., Tanigami, A., Carrozzo, R., Smith, A.C., Dobyns, W.B. & Ledbetter, D.H. (1997). A revision of the lissencephaly and Miller-Dieker syndrome critical regions in chromosome 17p13.3. Hum. Mol. Genet. 6, 147–155.

Collins, F.S. (1991). Of needles and haystacks: finding human disease genes by positional cloning. Clin. Res. 39, 615–623.

Collins, F.S. (1995). Positional cloning moves from perditional to traditional. Nat. Genet. 9, 347–350.

Colton, C.A., Yao, J.B., Gilbert, D. & Oster-Granite, M.L. (1990). Enhanced production of superoxide anion by microglia from trisomy 16 mice. Brain. Res. 519, 236–242.

Commission on Classification and Terminology, I.L.A.E. (1989). Proposal for revised classification of epilepsies and epileptic syndromes. Epilepsia 30, 389–399.

Consortium, T. (1995). Isolation of a novel gene underlying Batten disease, CLN3. The International Batten Disease Consortium. Cell 82, 949–957.

Cox, G.A., Lutz, C.M., Yang, C.L., Biemesderfer, D., Bronson, R.T., Fu, A., Aronson, P.S., Noebels, J.L. & Frankel, W.N. (1997). Sodium/hydrogen exchanger gene defect in slow-wave epilepsy mutant mice [published erratum appears in Cell 1997 Dec 12;91(6):861]. Cell 91, 139–148.

Crespel, A., Baldy-Moulinier, M. & Coubes, P. (1998). The relationship between sleep and epilepsy in frontal and temporal lobe epilepsies: practical and physiopathologic considerations. Epilepsia 39, 150–157.

Cunniff, C., Wiedlin, N. & Jones, K.L. (1988). Autosomal dominant benign neonatal seizures. Am. J. Med. Genet. 30, 963–966.

Dalla Bernardina, B., Dulac, O., Fejerman, N., Dravet, C., Capovilla, G., Bondavalli, S., Colamaria, V. & Roger, J. (1983). Early myoclonic epileptic encephalopathy (E.M.E.E.). Eur. J. Pediatr. 140, 248–252.

Degen, R. & Degen, H.E. (1990). Some genetic aspects of rolandic epilepsy: waking and sleep EEGs in siblings. Epilepsia 31, 795–801.

Dehan, M., Quillerou, D., Navelet, Y., D'Allest, A.M., Vial, M., Retbi, J.M., Lelong-Tissier, M.C. & Gabilan, J.C. (1977). [Convulsions in the fifth day of life: a new syndrome?]. Arch. Fr. Pediatr. 34, 730–742.

Delgado-Escueta, A.V., Greenberg, D.A., Treiman, L., Liu, A., Sparkes, R.S., Barbetti, A., Park, M.S. & Terasaki, P.I. (1989). Mapping the gene for juvenile myoclonic epilepsy. Epilepsia 30, S8–18; discussion S24–17.

Delgado-Escueta, A.V., Greenberg, D., Weissbecker, K., Liu, A., Treiman, L., Sparkes, R., Park, M.S., Barbetti, A. & Terasaki, P.I. (1990). Gene mapping in the idiopathic generalized epilepsies: juvenile myoclonic epilepsy, childhood absence epilepsy, epilepsy with grand mal seizures, and early childhood myoclonic epilepsy. Epilepsia 31, S19–29.

Delgado-Escueta, A.V., Serratosa, J.M., Liu, A., Weissbecker, K., Medina, M.T., Gee, M., Treiman, L.J. & Sparkes, R.S. (1994). Progress in mapping human epilepsy genes. Epilepsia 35, S29–40.

des Portes, V., Pinard, J.M., Billuart, P., Vinet, M.C., Koulakoff, A., Carrie, A., Gelot, A., Dupuis, E., Motte, J., Berwald-Netter, Y., Catala, M., Kahn, A., Beldjord, C. & Chelly, J. (1998). A novel CNS gene required for neuronal migration and involved in X-linked subcortical laminar heterotopia and lissencephaly syndrome. Cell 92, 51–61.

Di Pasquale, E., Keegan, K.D. & Noebels, J.L. (1997). Increased excitability and inward rectification in layer V cortical pyramidal neurons in the epileptic mutant mouse Stargazer. J. Neurophysiol. 77, 621–631.

Dichter, M.A. & Buchhalter, J.R. (1997). The genetic epilepsies. In: The Molecular and Genetic Basis of Neurological Disease. (Dichter, M.A. & Buchhalter, J.R., eds.), pp. 757–783. Butterworth-Heinemann, Boston.

Dobyns, W.B., Reiner, O., Carrozzo, R. & Ledbetter, D.H. (1993). Lissencephaly. A human brain malformation associated with deletion of the LIS1 gene located at chromosome 17p13. J.A.M.A. 270, 2838–2842.

Doose, H. (1997). Genetic EEG traits in the pathogenesis of the epilepsies. J Epilepsy 10, 97–110.

Doose, H. & Gerken, H. (1973). On the genetics of EEG-anomalies in childhood. IV. Photoconvulsive reaction. Neuropadiatrie 4, 162–171.

Doyle, J., Ren, X., Lennon, G. & Stubbs, L. (1997). Mutations in the Cacnl1a4 calcium channel gene are associated with seizures, cerebellar degeneration, and ataxia in tottering and leaner mutant mice. Mamm. Genome. 8, 113–120.

Dubeau, F., Tampieri, D., Lee, N., Andermann, E., Carpenter, S., Leblanc, R., Olivier, A., Radtke, R., Villemure, J.G. & Andermann, F. (1995). Periventricular and subcortical nodular heterotopia. A study of 33 patients. Brain 118, 1273–1287.

Durner, M., Sander, T., Greenberg, D.A., Johnson, K., Beck-Mannagetta, G. & Janz, D. (1991). Localization of idiopathic generalized epilepsy on chromosome 6p in families of juvenile myoclonic epilepsy patients. Neurology 41, 1651–1655.

Eksioglu, Y.Z., Scheffer, I.E., Cardenas, P., Knoll, J., DiMario, F., Ramsby, G., Berg, M., Kamuro, K., Berkovic, S.F., Duyk, G.M., Parisi, J., Huttenlocher, P.R. & Walsh, C.A. (1996). Periventricular heterotopia: an X-linked dominant epilepsy locus causing aberrant cerebral cortical development. Neuron 16, 77–87.

Elmslie, F.V., Williamson, M.P., Rees, M., Kerr, M., Kjeldsen, M.J., Pang, K.A., Sundqvist, A., Friis, M.L., Richens, A., Chadwick, D., Whitehouse, W.P. & Gardiner, R.M. (1996). Linkage analysis of juvenile myoclonic epilepsy and microsatellite loci spanning 61 cM of human chromosome 6p in 19 nuclear pedigrees provides no evidence for a susceptibility locus in this region. Am. J. Hum. Genet. 59, 653–663.

Elmslie, F.V., Rees, M., Williamson, M.P., Kerr, M., Kjeldsen, M.J., Pang, K.A., Sundqvist, A., Friis, M.L., Chadwick, D., Richens, A., Covanis, A., Santos, M., Arzimanoglou, A., Panayiotopoulos, C.P., Curtis, D., Whitehouse, W.P. & Gardiner, R.M. (1997). Genetic mapping of a major susceptibility locus for juvenile myoclonic epilepsy on chromosome 15q. Hum. Mol. Genet. 6, 1329–1334.

Eubanks, J.H., Puranam, R.S., Kleckner, N.W., Bettler, B., Heinemann, S.F. & McNamara, J.O. (1993). The gene encoding the glutamate receptor subunit GluR5 is located on human chromosome 21q21.1–22.1 in the vicinity of the gene for familial amyotrophic lateral sclerosis. Proc. Natl. Acad. Sci. U. S. A. 90, 178–182.

Evenhuis, H.M. (1990). The natural history of dementia in Down's syndrome. Arch. Neurol. 47, 263–267.

Farrell, K. (1993). Classifying epileptic syndromes: problems and a neurobiologic solution. Neurology 43, S8–11.

Fletcher, C.F., Lutz, C.M., O'Sullivan, T.N., Shaughnessy, J.D., Jr., Hawkes, R., Frankel, W.N., Copeland, N.G. & Jenkins, N.A. (1996). Absence epilepsy in tottering mutant mice is associated with calcium channel defects. Cell 87, 607–617.

Furukawa, K., Barger, S.W., Blalock, E. & Mattson, M.P. (1996) Activation of K^+ channels and suppression of neuronal activity by secreted β-amyloid precursor protein. Nature 379, 74–78.

Gardiner, R.M. (1990). Genes and epilepsy. J. Med. Genet. 27, 537–544.

Gleeson, J.G., Allen, K.M., Fox, J.W., Lamperti, E.D., Berkovic, S., Scheffer, I., Cooper, E.C., Dobyns, W.B., Minnerath, S.R., Ross, M.E. & Walsh, C.A. (1998). Doublecortin, a brain-specific gene mutated in human X-linked lissencephaly and double cortex syndrome, encodes a putative signaling protein. Cell 92, 63–72.

Goebel, H.H. (1996). The neuronal ceroid-lipofuscinoses. Semin. Pediatr. Neurol. 3, 270–278.

Goto, Y., Nonaka, I. & Horai, S. (1990). A mutation in the tRNA(Leu)(UUR) gene associated with the MELAS subgroup of mitochondrial encephalomyopathies [see comments]. Nature 348, 651–653.

Gottfried, M., Lavine, L. & Roessmann, U. (1981). Neuropathological findings in Wolf-Hirschhorn (4p-) syndrome. Acta Neuropathol. 55, 163–165.

Goutieres, F. & Aicardi, J. (1985). Atypical presentations of pyridoxine-dependent seizures: a treatable cause of intractable epilepsy in infants. Ann. Neurol. 17, 117–120.

Green, A.J., Johnson, P.H. & Yates, J.R. (1994). The tuberous sclerosis gene on chromosome 9q34 acts as a growth suppressor. Hum. Mol. Genet. 3, 1833–1834.

Greenberg, D.A., Delgado-Escueta, A.V., Widelitz, H., Sparkes, R.S., Treiman, L., Maldonado, H.M., Park, M.S. & Terasaki, P.I. (1988). Juvenile myoclonic epilepsy (JME) may be linked to the BF and HLA loci on human chromosome 6. Am. J. Med. Genet. 31, 185–192.

Guerrini, R., Bureau, M., Mattei, M.G., Battaglia, A., Galland, M.C. & Roger, J. (1990). Trisomy 12p syndrome: a chromosomal disorder associated with generalized 3-Hz spike and wave discharges. Epilepsia 31, 557–566.

Guerrini, R., Gobbi, G., Genton, P., Bonanni, P., & Carrazo, R. (1998). Chromosomal abnormalities. In: Epilepsy: A Comprehensive Textbook. (Engel, Jr., J. & Pedley, T.A., Eds.), pp. 2533–2546, Lippincott-Raven, Philadelphia.

Guipponi, M., Rivier, F., Vigevano, F., Beck, C., Crespel, A., Echenne, B., Lucchini, P., Sebastianelli, R., Baldy-Moulinier, M. & Malafosse, A. (1997). Linkage mapping of benign familial infantile convulsions (BFIC) to chromosome 19q. Hum. Mol. Genet. 6, 473–477.

Guthrie, R.D., Aase, J.M., Asper, A.C. & Smith, D.W. (1971). The 4p-syndrome. A clinically recognizable chromosomal deletion syndrome. Am. J. Dis. Child 122, 421–425.

Hattori, M., Adachi, H., Tsujimoto, M., Arai, H. & Inoue, K. (1994). Miller-Dieker lissencephaly gene encodes a subunit of brain platelet-activating factor acetylhydrolase [corrected] [published erratum appears in Nature 1994 Aug 4;370(6488):391]. Nature 370, 216–218.

Heijbel, J., Blom, S. & Rasmuson, M. (1975). Benign epilepsy of childhood with centrotemporal EEG foci: a genetic study. Epilepsia 16, 285–293.

Hirsch, E., Velez, A., Sellal, F., Maton, B., Grinspan, A., Malafosse, A. & Marescaux, C. (1993). Electroclinical signs of benign neonatal familial convulsions. Ann. Neurol. 34, 835–841.

Hirvasniemi, A., Lang, H., Lehesjoki, A.E. & Leisti, J. (1994). Northern epilepsy syndrome: an inherited childhood onset epilepsy with associated mental deterioration. J. Med. Genet. 31, 177–182.

Holtzman, D.M., Bayney, R.M., Li, Y.W., Khosrovi, H., Berger, C.N., Epstein, C.J. & Mobley, W.C. (1992). Dysregulation of gene expression in mouse trisomy 16, an animal model of Down syndrome. EMBO J. 11, 619–627.

Homanics, G.E., DeLorey, T.M., Firestone, L.L., Quinlan, J.J., Handforth, A., Harrison, N.L., Krasowski, M.D., Rick, C.E., Korpi, E.R., Makela, R., Brilliant, M.H., Hagiwara, N., Ferguson, C., Snyder, K. & Olsen, R.W. (1997). Mice devoid of gamma-aminobutyrate type A receptor beta3 subunit have epilepsy, cleft palate, and hypersensitive behavior. Proc. Natl. Acad. Sci. U. S. A. 94, 4143–4148.

Hook, E.B. & Cross, P.K. (1987). Extra structurally abnormal chromosomes (ESAC) detected at amniocentesis: frequency in approximately 75,000 prenatal cytogenetic diagnoses and associations with maternal and paternal age [published erratum appears in Am J Hum Genet 1988 Jan;42(1):210]. Am. J. Hum. Genet. 40, 83–101.

Hosford, D.A. & Wang, Y. (1997). Utility of the lethargic (lh/lh) mouse model of absence seizures in predicting the effects of lamotrigine, vigabatrin, tiagabine, gabapentin, and topiramate against human absence seizures. Epilepsia 38, 408–414.

Hosford, D.A., Caddick, S.J. & Lin, F.H. (1997). Generalized epilepsies: emerging insights into cellular and genetic mechanisms. Curr. Opin. Neurol. 10, 115–120.

Huang, F., Shi, L.J., Heng, H.H., Fei, J. & Guo, L.H. (1995). Assignment of the human GABA transporter gene (GABATHG) locus to chromosome 3p24–p25. Genomics 29, 302–304.

Huttenlocher, P.R., Taravath, S. & Mojtahedi, S. (1994). Periventricular heterotopia and epilepsy. Neurology 44, 51–55.

Inoue, Y., Fujiwara, T., Matsuda, K., Kubota, H., Tanaka, M., Yagi, K., Yamamori, K. & Takahashi, Y. (1997). Ring chromosome 20 and nonconvulsive status epilepticus. A new epileptic syndrome. *Brain* 120, 939–953.

Isobe, M., Koyata, H., Sakakibara, T., Momoi-Isobe, K. & Hiraga, K. (1994). Assignment of the true and processed genes for human glycine decarboxylase to 9p23-24 and 4q12. Biochem. Biophys. Res. Commun. 203, 1483–1487.

Jardine, P.E., Clarke, M.A. & Super, M. (1996). Familial bilateral periventricular nodular heterotopia mimics tuberous sclerosis. Arch. Dis. Child 74, 244–246.

Jarvela, I., Sainio, M., Rantamaki, T., Olkkonen, V.M., Carpen, O., Peltonen, L. & Jalanko, A. (1998). Biosynthesis and intracellular targeting of the CLN3 protein defective in Batten disease. Hum. Mol. Genet. 7, 85–90.

Kagawa, T., Ikenaka, K., Inoue, Y., Kuriyama, S., Tsujii, T., Nakao, J., Nakajima, K., Aruga, J., Okano, H. & Mikoshiba, K. (1994). Glial cell degeneration and hypomyelination caused by overexpression of myelin proteolipid protein gene. Neuron 13, 427–442.

Kamuro, K. & Tenokuchi, Y. (1993). Familial periventricular nodular heterotopia. Brain Dev. 15, 237–241.

Kandt, R.S., Haines, J.L., Smith, M., Northrup, H., Gardner, R.J., Short, M.P., Dumars, K., Roach, E.S., Steingold, S., Wall, S., et al. (1992). Linkage of an important gene locus for tuberous sclerosis to a chromosome 16 marker for polycystic kidney disease. Nat. Genet. 2, 37–41.

Kishino, T., Lalande, M. & Wagstaff, J. (1997). UBE3A/E6-AP mutations cause Angelman syndrome [published erratum appears in Nat Genet 1997 Apr;15(4):411]. Nat. Genet. 15, 70–73.

Kluger, G., Bohm, I., Laub, M.C. & Waldenmaier, C. (1996). Epilepsy and fragile X gene mutations. Pediatr. Neurol. 15, 358–360.

Knight, A.R. & Bowery, N.G. (1992). GABA receptors in rats with spontaneous generalized nonconvulsive epilepsy. J. Neural Transm. Suppl. 35, 189–196.

Koide, R., Ikeuchi, T., Onodera, O., Tanaka, H., Igarashi, S., Endo, K., Takahashi, H., Kondo, R., Ishikawa, A., Hayashi, T., et al. (1994). Unstable expansion of CAG repeat in hereditary dentatorubral-pallidoluysian atrophy (DRPLA). Nat. Genet. 6, 9–13.

Kokaia, M., Ernfors, P., Kokaia, Z., Elmer, E., Jaenisch, R. & Lindvall, O. (1995). Suppressed epileptogenesis in BDNF mutant mice. Exp. Neurol. 133, 215–224.

Korenberg, J.R. (1993). Toward a molecular understanding of Down syndrome. Prog. Clin. Biol. Res. 384, 87–115.

Korenberg, J.R. (1995). Mental modelling [news; comment]. Nat. Genet. 11, 109–111.

Korenberg, J.R., Kawashima, H., Pulst, S.M., Ikeuchi, T., Ogasawara, N., Yamamoto, K., Schonberg, S.A., West, R., Allen, L., Magenis, E., et al. (1990). Molecular definition of a region of chromosome 21 that causes features of the Down syndrome phenotype. Am. J. Hum. Genet. 47, 236–246.

Krasowski, M.D., Rick, C.E., Harrison, N.L., Firestone, L.L. & Homanics, G.E. (1998). A deficit of functional GABA(A) receptors in neurons of beta 3 subunit knockout mice. Neurosci. Lett. 240, 81–84.

Krishnamoorthy, K.S. (1983). Pyridoxine-dependency seizure: report of a rare presentation. Ann. Neurol. 13, 103–104.

Kure, S., Tada, K. & Narisawa, K. (1997). Nonketotic hyperglycinemia: biochemical, molecular, and neurological aspects. Jpn. J. Hum. Genet. 42, 13–22.

Kuryatov, A., Gerzanich, V., Nelson, M., Olale, F. & Lindstrom, J. (1997). Mutation causing autosomal dominant nocturnal frontal lobe epilepsy alters Ca2+ permeability, conductance, and gating of human alpha4beta2 nicotinic acetylcholine receptors. J. Neurosci. 17, 9035–9047.

Lacey-Casem, M.L. & Oster-Granite, M.L. (1994). The neuropathology of the trisomy 16 mouse. Crit. Rev. Neurobiol. 8, 293–322.

Lai, F. & Williams, R.S. (1989). A prospective study of Alzheimer disease in Down syndrome. Arch. Neurol. 46, 849–853.

Lalioti, M.D., Scott, H.S. & Antonarakis, S.E. (1997). What is expanded in progressive myoclonus epilepsy? [letter]. Nat. Genet. 17, 17.

Lander, E.S. & Schork, N.J. (1994). Genetic dissection of complex traits [published erratum appears in Science 1994 Oct 21;266(5184):353]. Science 265, 2037–2048.

Lang, A.H., Hirvasniemi, A. & Siivola, J. (1997). Neurophysiological findings in the northern epilepsy syndrome. Acta Neurol. Scand. 95, 1–8.

Leana-Cox, J., Jenkins, L., Palmer, C.G., Plattner, R., Sheppard, L., Flejter, W.L., Zackowski, J., Tsien, F. & Schwartz, S. (1994). Molecular cytogenetic analysis of inv dup(15) chromosomes, using probes specific for the Prader-Willi/Angelman syndrome region: clinical implications. Am. J. Hum. Genet. 54, 748–756.

Lee, S., Williams, J., Lothman, E.W., Szele, F.G., Chesselet, M.F., Sapolsky, R.M., Mattson, M.P. & Christakos, S. (1997) Early induction of mRNA for calbindin-D28k and BDNF but not NT-3 in rat hippocampus after kainic acid treatment. Mol. Brain Res. 47, 183–194.

Lehesjoki, A.E., Koskiniemi, M., Sistonen, P., Miao, J., Hastbacka, J., Norio, R. & de la Chapelle, A. (1991). Localization of a gene for progressive myoclonus epilepsy to chromosome 21q22. Proc. Natl. Acad. Sci. U. S. A. 88, 3696–3699.

Leppert, M., Anderson, V.E., Quattlebaum, T., Stauffer, D., O'Connell, P., Nakamura, Y., Lalouel, J.M. & White, R. (1989). Benign familial neonatal convulsions linked to genetic markers on chromosome 20. Nature 337, 647–648.

Lerman, P. & Kivity, S. (1975). Benign focal epilepsy of childhood. A follow-up study of 100 recovered patients. Arch. Neurol. 32, 261–264.

Letts, V.A., Valenzuela, A., Kirley, J.P., Sweet, H.O., Davisson, M.T. & Frankel, W.N. (1997). Genetic and physical maps of the stargazer locus on mouse chromosome 15. Genomics 43, 62–68.

Lewis, T.B., Leach, R.J., Ward, K., O'Connell, P. & Ryan, S.G. (1993). Genetic heterogeneity in benign familial neonatal convulsions: identification of a new locus on chromosome 8q. Am. J. Hum. Genet. 53, 670–675.

Lewis, T.B., Shevell, M.I., Andermann, E., Ryan, S.G. & Leach, R.J. (1996). Evidence of a third locus for benign familial convulsions. J. Child Neurol. 11, 211–214.

Lin, F.H., Wang, Y., Lin, S., Cao, Z. & Hosford, D.A. (1995). GABAB receptor-mediated effects in synaptosomes of lethargic (lh/lh) mice. J. Neurochem. 65, 2087–2095.

Lombardo, A.J., Kuzniecky, R., Powers, R.E. & Brown, G.B. (1996). Altered brain sodium channel transcript levels in human epilepsy. Brain Res. Mol. Brain Res. 35, 84–90.

Lopes-Cendes, I., Phillips, H.A., Scheffer, I.E., Mulley, J.C., Desbiens, R., Andermann, E., Cendes, F., Verret, S., Andermann, F., Berkovic, S. F., et al. (1995). Genetic linkage studies in familial frontal epilepsy: exclusion of the human chromosome regions homologous to the El-1 mouse locus. Epilepsy Res. 22, 227–233.

Lott, I.T., Coulombe, T., Di Paolo, R.V., Richardson, E.P., Jr. & Levy, H.L. (1978). Vitamin B6-dependent seizures: pathology and chemical findings in brain. Neurology 28, 47–54.

Ma, W., Saunders, P.A., Somogyi, R., Poulter, M.O. & Barker, J.L. (1993). Ontogeny of GABAA receptor subunit mRNAs in rat spinal cord and dorsal root ganglia. J. Comp. Neurol. 338, 337–359.

MacDermot, K.D., Nelson, W., Soutter, V., Towne, D. & Schulman, J.D. (1981). Glycine and benzoate conjugation and glycine acyltransferase activity in the developing and adult rat: possible relationships to nonketotic hyperglycinemia. Dev. Pharmacol. Ther. 3, 150–159.

Magloczky, Z., Halasz, P., Vajda, J., Czirjak, S. & Freund, T.F. (1997). Loss of Calbindin-D28K immunoreactivity from dentate granule cells in human temporal lobe epilepsy. Neuroscience 76, 377–385.

Malafosse, A., Beck, C., Bellet, H., Di Capua, M., Dulac, O., Echenne, B., Fusco, L., Lucchini, P., Ricci, S., Sebastianelli, R., et al. (1994). Benign infantile familial convulsions are not an allelic form of the benign familial neonatal convulsions gene. Ann. Neurol. 35, 479–482.

Matsumoto, A., Kumagai, T., Miura, K., Miyazaki, S., Hayakawa, C. & Yamanaka, T. (1992). Epilepsy in Angelman syndrome associated with chromosome 15q deletion. Epilepsia 33, 1083–1090.

Matsumoto, M., Nakagawa, T., Inoue, T., Nagata, E., Tanaka, K., Takano, H., Minowa, O., Kuno, J., Sakakibara, S., Yamada, M., Yoneshima, H., Miyawaki, A., Fukuuchi, Y., Furuichi, T., Okano, H., Mikoshiba, K. & Noda, T. (1996). Ataxia and epileptic seizures in mice lacking type 1 inositol 1,4,5-trisphosphate receptor. Nature 379, 168–171.

Matsuura, T., Sutcliffe, J.S., Fang, P., Galjaard, R.J., Jiang, Y.H., Benton, C.S., Rommens, J.M. & Beaudet, A.L. (1997). De novo truncating mutations in E6-AP ubiquitin-protein ligase gene (UBE3A) in Angelman syndrome. Nat. Genet. 15, 74–77.

Mattson, M.P., Murrain, M., Guthric, P.B. & Kater, S.B. (1989) Fibroblast growth factor and glutamate: opposing actions in the generation and degeneration of hippocampal neuroarchitecture. J. Neurosci. 9, 3728–3740.

Mattson, M.P., Rychlik, B., Chu, C. & Christakos, S. (1991) Evidence for calcium-reducing and excitoprotective roles for the calcium binding protein (calbindin-D28k) in cultured hippocampal neurons. Neuron 6, 41–51.

Mattson, M.P., Cheng, B., Culwell, A., Esch, F., Lieberburg, I. & Rydel, R.E. (1993) Evidence for excitoprotective and intraneuronal calcium-regulating roles for secreted forms of β-amyloid precursor protein. Neuron 10, 243–254.

McNamara, J.O. (1994). Identification of genetic defect of an epilepsy: strategies for therapeutic advances. Epilepsia 35, S51–57.

McNamara, J.O., Patel, M., He, X.P., Janumpalli, S. & Whitney, K.D. (1996). Glutamate receptor autoimmunity in Rasmussen's encephalitis. Cold Spring Harb. Symp. Quant. Biol 61, 327–332.

Meencke, H.J. & Janz, D. (1985). The significance of microdysgenesia in primary generalized epilepsy: an answer to the considerations of Lyon and Gastaut. Epilepsia 26, 368–371.

Minassian, B.A., DeLorey, T.M., Olsen, R.W., Philippart, M., Bronstein, Y., Zhang, Q., Guerrini, R., Van Ness, P., Livet, M.O. & Delgado-Escueta, A.V. (1998). Angelman syndrome: correlations between epilepsy phenotypes and genotypes. Ann. Neurol. 43, 485–493.

Morgan-Hughes, J.A., Sweeney, M.G., Cooper, J.M., Hammans, S.R., Brockington, M., Schapira, A.H., Harding, A.E. & Clark, J.B. (1995). Mitochondrial DNA (mtDNA) diseases: correlation of genotype to phenotype. Biochim Biophys Acta 1271, 135–140.

Mornet, E., Chateau, C., Simon-Bouy, B. & Serre, J.L. (1998). The intermediate alleles of the fragile X CGG repeat in patients with mental retardation [In Process Citation]. Clin. Genet. 53, 200–201.

Munroe, P.B., Mitchison, H.M., O'Rawe, A.M., Anderson, J.W., Boustany, R.M., Lerner, T.J., Taschner, P.E., de Vos, N., Breuning, M.H., Gardiner, R.M. & Mole, S.E. (1997). Spectrum of mutations in the Batten disease gene, CLN3. Am. J. Hum. Genet. 61, 310–316.

Musumeci, S.A., Colognola, R.M., Ferri, R., Gigli, G.L., Petrella, M.A., Sanfilippo, S., Bergonzi, P. & Tassinari, C.A. (1988). Fragile-X syndrome: a particular epileptogenic EEG pattern. Epilepsia 29, 41–47.

Musumeci, S.A., Ferri, R., Elia, M., Colognola, R.M., Bergonzi, P. & Tassinari, C.A. (1991). Epilepsy and fragile X syndrome: a follow-up study. Am. J. Med. Genet. 38, 511–513.

Mutoh, K., Ito, M., Tsuda, H., Shiraishi, H., Oguro, K., Shirasaka, Y., Okuno, T. & Mikawa, H. (1993). Depth EEG in mutant epileptic El mice: demonstration of secondary generalization of the seizure from the hippocampus. Electroencephalogr. Clin. Neurophysiol. 86, 205–212.

Naquet, R., Silva-Barrat, C. & Menini, C. (1995). Reflex epilepsy in the Papio-papio baboon, particularly photosensitive epilepsy. Ital. J. Neurol. Sci. 16, 119–125.

Navelet, Y., D'Allest, A.M., Dehan, M. & Lelong-Tissier, M.C. (1977). [Are convulsions on the fifth day of life a distinct clinical and electrophysiological entity? (author's transl)]. Rev. Electroencephalogr. Neurophysiol. Clin. 7, 366–370.

Nelson, K.B. & Ellenberg, J.H. (1978). Prognosis in children with febrile seizures. Pediatrics 61, 720–727.

Noebels, J.L. (1995). Single locus mutations in mice expressing generalized spike-wave absence epilepsies. Ital. J. Neurol. Sci. 16, 107–111.

Noebels, J.L. (1996). Targeting epilepsy genes. Neuron 16, 241–244.

Noebels, J.L. & Sidman, R.L. (1979). Inherited epilepsy: spike-wave and focal motor seizures in the mutant mouse tottering. Science 204, 1334–1336.

Norio, R. & Koskiniemi, M. (1979). Progressive myoclonus epilepsy: genetic and nosological aspects with special reference to 107 Finnish patients. Clin. Genet. 15, 382–398.

Norrgard, K.J., Pomponio, R.J., Swango, K.L., Hymes, J., Reynolds, T.R., Buck, G.A. & Wolf, B. (1997). Mutation (Q456H) is the most common cause of profound biotinidase deficiency in children ascertained by newborn screening in the United States. Biochem. Mol. Med. 61, 22–27.

North, K.N., Storey, G.N. & Henderson-Smart, D.J. (1989). Fifth day fits in the newborn. Aust. Paediatr. J. 25, 284–287.

O'Donnell, D.M. & Zoghbi, H.Y. (1995). Trinucleotide repeat disorders in pediatrics. Curr. Opin. Pediatr. 7, 715–725.

Oldani, A., Zucconi, M., Ferini–Strambi, L., Bizzozero, D. & Smirne, S. (1996). Autosomal dominant nocturnal frontal lobe epilepsy: electroclinical picture [see comments]. Epilepsia 37, 964–976.

Oldani, A., Zucconi, M., Asselta, R., Modugno, M., Bonati, M.T., Dalpra, L., Malcovati, M., Tenchini, M.L., Smirne, S. & Ferini-Strambi, L. (1998). Autosomal dominant nocturnal frontal lobe epilepsy. A video-polysomnographic and genetic appraisal of 40 patients and delineation of the epileptic syndrome. Brain 121, 205–223.

Oohira, T., Nagata, N., Akaboshi, I., Matsuda, I. & Naito, S. (1985). The infantile form of sialidosis type II associated with congenital adrenal hyperplasia: possible linkage between HLA and the neuraminidase deficiency gene. Hum. Genet. 70, 341–343.

Ophoff, R.A., Terwindt, G.M., Vergouwe, M.N., van Eijk, R., Oefner, P.J., Hoffman, S.M., Lamerdin, J.E., Mohrenweiser, H.W., Bulman, D.E., Ferrari, M., Haan, J., Lindhout, D., van Ommen, G.J., Hofker, M.H., Ferrari, M.D. & Frants, R.R. (1996). Familial hemiplegic migraine and episodic ataxia type-2 are caused by mutations in the Ca2+ channel gene CACNL1A4. Cell 87, 543–552.

Ophoff, R.A., Terwindt, G.M., Frants, R.R. & Ferrari, M.D. (1998). P/Q-type Ca2+ channel defects in migraine, ataxia and epilepsy. Trends Pharmacol. Sci .19, 121–127.

Ottman, R., Risch, N., Hauser, W.A., Pedley, T.A., Lee, J.H., Barker-Cummings, C., Lustenberger, A., Nagle, K.J., Lee, K.S., Scheuer, M.L., et al. (1995). Localization of a gene for partial epilepsy to chromosome 10q [see comments]. Nat. Genet.10, 56–60.

Patil, N., Cox, D.R., Bhat, D., Faham, M., Myers, R.M. & Peterson, A.S. (1995). A potassium channel mutation in weaver mice implicates membrane excitability in granule cell differentiation [see comments]. Nat. Genet. 11, 126–129.

Pedley, T.A. (1991). The use and role of EEG in the genetic analysis of epilepsy. Epilepsy Res. Suppl. 4, 31–44.

Pennacchio, L.A. & Myers, R.M. (1996). Isolation and characterization of the mouse cystatin B gene [letter]. Genome Res. 6, 1103–1109.

Pennacchio, L.A., Lehesjoki, A.E., Stone, N.E., Willour, V.L., Virtaneva, K., Miao, J., D'Amato, E., Ramirez, L., Faham, M., Koskiniemi, M., Warrington, J.A., Norio, R., de la Chapelle, A., Cox, D.R. & Myers, R.M. (1996). Mutations in the gene encoding cystatin B in progressive myoclonus epilepsy (EPM1) [see comments]. Science 271, 1731–1734.

Pennacchio, L.A., Higgins, K.M., Scott, M.P. & Myers, R.M. (1997). Toward an animal model for progressive myoclonus epilepsy (EPM1): targeted disruption of the mouse cystatin B gene. Epilepsia 38, 122.

Peters, K., Ornitz, D., Werner, S. & Williams, L. (1993). Unique expression pattern of the FGF receptor 3 gene during mouse organogenesis. Dev. Biol. 155, 423–430.

Phillips, H.A. & Mulley, J.C. (1997). SSCP variants within the alpha 4 subunit of the neuronal nicotinic acetylcholine receptor gene. Clin. Genet. 51, 135–136.

Phillips, H.A., Scheffer, I.E., Berkovic, S.F., Hollway, G.E., Sutherland, G.R. & Mulley, J.C. (1995). Localization of a gene for autosomal dominant nocturnal frontal lobe epilepsy to chromosome 20q 13.2 [letter] [see comments]. Nat. Genet. 10, 117–118.

Plouin, P. (1998). Benign familial neonatal convulsions and benign idiopathic neonatal convulsions. In: Epilepsy: A Comprehensive Textbook (Engel, Jr., J. & Pedley, T.A., Eds.), pp. 2247–2255, Lippincott-Raven, Philadelphia.

Pomponio, R.J., Reynolds, T.R., Cole, H., Buck, G.A. & Wolf, B. (1995). Mutational hotspot in the human biotinidase gene causes profound biotinidase deficiency [see comments]. Nat. Genet. 11, 96–98.

Pomponio, R.J., Norrgard, K.J., Hymes, J., Reynolds, T.R., Buck, G.A., Baumgartner, R., Suormala, T. & Wolf, B. (1997). Arg538 to Cys mutation in a CpG dinucleotide of the human biotinidase gene is the second most common cause of profound biotinidase deficiency in symptomatic children. Hum. Genet. 99, 506–512.

Prasad, A.N., Prasad, C., & Stafstrom, C.E. (1999). Recent advances in the genetics of epilepsy: Insights from human and animal studies. Epilepsia 40, 1329–1352.

Pryor, D.S., Don, N. & Macourt, D.C. (1981). Fifth day fits: a syndrome of neonatal convulsions. Arch. Dis. Child 56, 753–758.

Pshezhetsky, A.V., Richard, C., Michaud, L., Igdoura, S., Wang, S., Elsliger, M.A., Qu, J., Leclerc, D., Gravel, R., Dallaire, L. & Potier, M. (1997). Cloning, expression and chromosomal mapping of human lysosomal sialidase and characterization of mutations in sialidosis. Nat. Genet. 15, 316–320.

Pueschel, S.M., Louis, S. & McKnight, P. (1991). Seizure disorders in Down syndrome. Arch. Neurol. 48, 318–320.

Raymond, A.A., Fish, D.R., Stevens, J.M., Sisodiya, S.M., Alsanjari, N. & Shorvon, S.D. (1994). Subependymal heterotopia: a distinct neuronal migration disorder associated with epilepsy. J. Neurol. Neurosurg. Psychiatry 57, 1195–1202.

Rees, M., Diebold, U., Parker, K., Doose, H., Gardiner, R.M. & Whitehouse, W.P. (1993). Benign childhood epilepsy with centrotemporal spikes and the focal sharp wave trait is not linked to the fragile X region. Neuropediatrics 24, 211–213.

Reiner, O., Albrecht, U., Gordon, M., Chianese, K.A., Wong, C., Gal-Gerber, O., Sapir, T., Siracusa, L.D., Buchberg, A.M., Caskey, C.T., et al. (1995). Lissencephaly gene (LIS1) expression in the CNS suggests a role in neuronal migration. J .Neurosci. 15, 3730–3738.

Rich, S.S., Annegers, J.F., Hauser, W.A. & Anderson, V.E. (1987). Complex segregation analysis of febrile convulsions. Am. J. Hum. Genet. 41, 249–257.

Richards, D.A., Lemos, T., Whitton, P.S. & Bowery, N.G. (1995). Extracellular GABA in the ventrolateral thalamus of rats exhibiting spontaneous absence epilepsy: a microdialysis study. J. Neurochem. 65, 1674–1680.

Rise, M.L., Frankel, W.N., Coffin, J.M. & Seyfried, T.N. (1991). Genes for epilepsy mapped in the mouse. Science 253, 669–673.

Ronen, G.M., Rosales, T.O., Connolly, M., Anderson, V.E. & Leppert, M. (1993). Seizure characteristics in chromosome 20 benign familial neonatal convulsions. Neurology 43, 1355–1360.

Rosahl, T.W., Spillane, D., Missler, M., Herz, J., Selig, D.K., Wolff, J.R., Hammer, R.E., Malenka, R.C. & Sudhof, T.C. (1995). Essential functions of synapsins I and II in synaptic vesicle regulation [see comments]. Nature 375, 488–493.

Ross, C.A., Becher, M.W., Colomer, V., Engelender, S., Wood, J.D. & Sharp, A.H. (1997). Huntington's disease and dentatorubral-pallidoluysian atrophy: proteins, pathogenesis and pathology. Brain Pathol. 7, 1003–1016.

Ross, M.E., Allen, K.M., Srivastava, A.K., Featherstone, T., Gleeson, J.G., Hirsch, B., Harding, B.N., Andermann, E., Abdullah, R., Berg, M., Czapansky-Bielman, D., Flanders, D.J., Guerrini, R., Motte, J., Mira, A.P., Scheffer, I., Berkovic, S., Scaravilli, F., King, R.A., Ledbetter, D.H., Schlessinger, D., Dobyns, W.B. & Walsh, C.A. (1997). Linkage and physical mapping of X-linked lissencephaly/SBH (XLIS): a gene causing neuronal migration defects in human brain. Hum. Mol. Genet 6, 555–562.

Ross, M.H., Galaburda, A.M. & Kemper, T.L. (1984). Down's syndrome: is there a decreased population of neurons? Neurology 34, 909–916.

Rothstein, J.D., Dykes–Hoberg, M., Pardo, C.A., Bristol, L.A., Jin, L., Kuncl, R.W., Kanai, Y., Hediger, M.A., Wang, Y., Schielke, J.P. & Welty, D.F. (1996). Knockout of glutamate transporters reveals a major role for astroglial transport in excitotoxicity and clearance of glutamate. Neuron 16, 675–686.

Ryan, S.G. (1995). Partial epilepsy: chinks in the armour [news; comment]. Nat. Genet. 10, 4–6.

Salbert, B.A., Pellock, J.M. & Wolf, B. (1993). Characterization of seizures associated with biotinidase deficiency. Neurology 43, 1351–1355.

Sampson, J.R. & Harris, P.C. (1994). The molecular genetics of tuberous sclerosis. Hum. Mol. Genet. 3, 1477–1480.

Sander, T., Hildmann, T., Kretz, R., Furst, R., Sailer, U., Bauer, G., Schmitz, B., Beck-Mannagetta, G., Wienker, T.F. & Janz, D. (1997a). Allelic association of juvenile absence epilepsy with a GluR5 kainate receptor gene (GRIK1) polymorphism. Am. J. Med. Genet. 74, 416–421.

Sander, T., Kretz, R., Williamson, M.P., Elmslie, F.V., Rees, M., Hildmann, T., Bianchi, A., Bauer, G., Sailer, U., Scaramelli, A., Schmitz, B., Gardiner, R.M., Janz, D. & Beck-Mannagetta, G. (1997b). Linkage analysis between idiopathic generalized epilepsies and the GABA(A) receptor alpha5, beta3 and gamma3 subunit gene cluster on chromosome 15. Acta Neurol. Scand. 96, 1–7.

Savukoski, M., Klockars, T., Holmberg, V., Santavuori, P., Lander, E.S. & Peltonen, L. (1998). CLN5, a novel gene encoding a putative transmembrane protein mutated in Finnish variant late infantile neuronal ceroid lipofuscinosis [In Process Citation]. Nat. Genet. 19, 286–288.

Scheffer, I.E. & Berkovic, S.F. (1997). Generalized epilepsy with febrile seizures plus. A genetic disorder with heterogeneous clinical phenotypes. Brain 120, 479–490.

Scheffer, I.E., Bhatia, K.P., Lopes-Cendes, I., Fish, D.R., Marsden, C.D., Andermann, F., Andermann, E., Desbiens, R., Cendes, F., Manson, J.I., et al. (1994). Autosomal dominant frontal epilepsy misdiagnosed as sleep disorder [see comments]. Lancet 343, 515–517.

Scheffer, I.E., Jones, L., Pozzebon, M., Howell, R.A., Saling, M.M. & Berkovic, S.F. (1995). Autosomal dominant rolandic epilepsy and speech dyspraxia: a new syndrome with anticipation. Ann Neurol. 38, 633–642.

Schiffmann, R., Shapira, Y. & Ryan, S.G. (1991). An autosomal recessive form of benign familial neonatal seizures. Clin. Genet. 40, 467–470.

Scott, B.S., Petit, T.L., Becker, L.E. & Edwards, B.A. (1981). Abnormal electric membrane properties of Down's syndrome DRG neurons in cell culture. Brain Res 254, 257–270.

Scriver, C.R. & Whelan, D.T. (1969). Glutamic acid decarboxylase (GAD) in mammalian tissue outside the central nervous system, and its possible relevance to hereditary vitamin B6 dependency with seizures. Ann. N. Y. Acad. Sci. 166, 83–96.

Serratosa, J.M., Delgado-Escueta, A.V., Posada, I., Shih, S., Drury, I., Berciano, J., Zabala, J.A., Antunez, M.C. & Sparkes, R.S. (1995). The gene for progressive myoclonus epilepsy of the Lafora type maps to chromosome 6q. Hum. Mol. Genet. 4, 1657–1663.

Serratosa, J.M., Delgado-Escueta, A.V., Medina, M.T., Zhang, Q., Iranmanesh, R. & Sparkes, R.S. (1996). Clinical and genetic analysis of a large pedigree with juvenile myoclonic epilepsy. Ann. Neurol. 39, 187–195.

Sgro, V., Riva, E., Canevini, M.P., Colamaria, V., Rottoli, A., Minotti, L., Canger, R. & Dalla Bernardina, B. (1995). 4p(-) syndrome: a chromosomal disorder associated with a particular EEG pattern. Epilepsia 36, 1206–1214.

Sharp, J.D., Wheeler, R.B., Lake, B.D., Savukoski, M., Jarvela, I.E., Peltonen, L., Gardiner, R. M. & Williams, R.E. (1997). Loci for classical and a variant late infantile neuronal ceroid lipofuscinosis map to chromosomes 11p15 and 15q21–23. Hum. Mol. Genet. 6, 591–595.

Shoffner, J.M., Lott, M.T., Lezza, A.M., Seibel, P., Ballinger, S.W. & Wallace, D.C. (1990). Myoclonic epilepsy and ragged-red fiber disease (MERRF) is associated with a mitochondrial DNA tRNA(Lys) mutation. Cell 61, 931–937.

Shoffner, J.M., Lott, M.T. & Wallace, D.C. (1991). MERRF: a model disease for understanding the principles of mitochondrial genetics. Rev. Neurol. 147, 431–435

Shoffner, J.M., Bialer, M.G., Pavlakis, S.G., Lott, M., Kaufman, A., Dixon, J., Teichberg, S. & Wallace, D.C. (1995). Mitochondrial encephalomyopathy associated with a single nucleotide pair deletion in the mitochondrial tRNALeu(UUR) gene. Neurology 45, 286–292.

Shytle, R.D., Borlongan, C.V. & Sanberg, P.R. (1995). Nicotine blocks kainic acid-induced wet dog shakes in rats. Neuropsychopharmacology 13, 261–264.

Silvestri, G., Moraes, C.T., Shanske, S., Oh, S.J. & DiMauro, S. (1992). A new mtDNA mutation in the tRNA(Lys) gene associated with myoclonic epilepsy and ragged-red fibers (MERRF). Am. J. Hum. Genet. 51, 1213–1217.

Singh, N.A., Charlier, C., Stauffer, D., DuPont, B.R., Leach, R.J., Melis, R., Ronen, G.M., Bjerre, I., Quattlebaum, T., Murphy, J.V., McHarg, M.L., Gagnon, D., Rosales, T.O., Peiffer, A., Anderson, V.E. & Leppert, M. (1998). A novel potassium channel gene, KCNQ2, is mutated in an inherited epilepsy of newborns [see comments]. Nat. Genet. 18, 25–29.

Sleat, D.E., Donnelly, R.J., Lackland, H., Liu, C.G., Sohar, I., Pullarkat, R.K. & Lobel, P. (1997). Association of mutations in a lysosomal protein with classical late-infantile neuronal ceroid lipofuscinosis. Science 277, 1802–1805.

Smart, S.L., Lopantsev, V., Zhang, C.L., Robbins, C.A., Wang, H., Chiu, S.Y., Schwartzkroin, P.A., Messing, A. & Tempel, B.L. (1998). Deletion of the K(V)1.1 potassium channel causes epilepsy in mice. Neuron 20, 809–819.

Srivistava, A.K., Ross, M.E., Allen, K.M., Featherstone, T., Christopher, J., Gleeson, G.J., Young, J.H., Amderman, E., Schlessinger, D., Walsh, C.A. & Dobyns, W.B. (1996). X-linked lissencephaly and SBH (XLIS): mapping of a novel neuronal migration gene. Am. J. Hum. Genet. 59, A55.

Stafstrom, C.E. (1993). Epilepsy in Down syndrome: clinical aspects and possible mechanisms. Am. J. Ment. Retard. 98, 12–26.

Stafstrom, C.E. & Konkol, R.J. (1994). Infantile spasms in children with Down syndrome. Dev. Med. Child. Neurol. 36, 576–585.

Stafstrom, C.E., Patxot, O.F., Gilmore, H.E. & Wisniewski, K.E. (1991). Seizures in children with Down syndrome: etiology, characteristics and outcome. Dev. Med. Child. Neurol. 33, 191–200.

Steinlein, O.K., Mulley, J.C., Propping, P., Wallace, R.H., Phillips, H.A., Sutherland, G.R., Scheffer, I.E. & Berkovic, S.F. (1995). A missense mutation in the neuronal nicotinic acetylcholine receptor alpha 4 subunit is associated with autosomal dominant nocturnal frontal lobe epilepsy. Nat. Genet. 11, 201–203.

Steinlein, O.K., Magnusson, A., Stoodt, J., Bertrand, S., Weiland, S., Berkovic, S.F., Nakken, K.O., Propping, P. & Bertrand, D. (1997). An insertion mutation of the CHRNA4 gene in a family with autosomal dominant nocturnal frontal lobe epilepsy. Hum. Mol. Genet. 6, 943–947.

Steriade, M., McCormick, D.A. & Sejnowski, T.J. (1993). Thalamocortical oscillations in the sleeping and aroused brain. Science 262, 679–685.

Steyaert, J., Borghgraef, M., Legius, E. & Fryns, J.P. (1996). Molecular-intelligence correlations in young fragile X males with a mild CGG repeat expansion in the FMR1 gene. Am. J. Med. Genet. 64, 274–277.

Stoll, J. & Galdzicki, Z. (1996). Reduced expression of voltage-gated sodium channels in neurons cultured from trisomy 16 mouse hippocampus. Int. J. Dev. Neurosci. 14, 749–760.

Sugimoto, T., Yasuhara, A., Ohta, T., Nishida, N., Saitoh, S., Hamabe, J. & Niikawa, N. (1992). Angelman syndrome in three siblings: characteristic epileptic seizures and EEG abnormalities. Epilepsia 33, 1078–1082.

Sutcliffe, J.S., Nelson, D.L., Zhang, F., Pieretti, M., Caskey, C.T., Saxe, D. & Warren, S.T. (1992). DNA methylation represses FMR-1 transcription in fragile X syndrome. Hum. Mol. Genet. 1, 397–400.

Sutcliffe, J.S., Jiang, Y.H., Galijaard, R.J., Matsuura, T., Fang, P., Kubota, T., Christian, S.L., Bressler, J., Cattanach, B., Ledbetter, D.H. & Beaudet, A.L. (1997). The E6-Ap ubiquitin-protein ligase (UBE3A) gene is localized within a narrowed Angelman syndrome critical region [letter]. Genome Res. 7, 368–377.

Sutherland, S.J., Olsen, R.D., Michels, V., Schmidt, M.A. & O'Brien, J.F. (1991). Screening for biotinidase deficiency in children with unexplained neurologic or developmental abnormalities. Clin. Pediatr. (Phila) 30, 81–84.

Szepetowski, P., Rochette, J., Berquin, P., Piussan, C., Lathrop, G.M. & Monaco, A.P. (1997). Familial infantile convulsions and paroxysmal choreoathetosis: a new neurological syndrome linked to the pericentromeric region of human chromosome 16. Am. J. Hum. Genet. 61, 889–898.

Tahvanainen, E., Ranta, S., Hirvasniemi, A., Karila, E., Leisti, J., Sistonen, P., Weissenbach, J., Lehesjoki, A.E. & de la Chapelle, A. (1994). The gene for a recessively inherited human childhood progressive epilepsy with mental retardation maps to the distal short arm of chromosome 8. Proc. Natl. Acad. Sci. U. S. A. 91, 7267–7270.

Tanaka, K., Watase, K., Manabe, T., Yamada, K., Watanabe, M., Takahashi, K., Iwama, H., Nishikawa, T., Ichihara, N., Kikuchi, T., Okuyama, S., Kawashima, N., Hori, S., Takimoto, M. & Wada, K. (1997). Epilepsy and exacerbation of brain injury in mice lacking the glutamate transporter GLT-1. Science 276, 1699–1702.

Tatsuno, M., Hayashi, M., Iwamoto, H., Suzuki, Y. & Kuroki, Y. (1984). Epilepsy in childhood Down syndrome. Brain Dev. 6, 37–44.

Tecott, L.H., Sun, L.M., Akana, S.F., Strack, A.M., Lowenstein, D.H., Dallman, M.F. & Julius, D. (1995). Eating disorder and epilepsy in mice lacking 5-HT2c serotonin receptors [see comments]. Nature 374, 542–546.

Tishler, D.M., Weinberg, K.I., Hinton, D.R., Barbaro, N., Annett, G.M. & Raffel, C. (1995). MDR1 gene expression in brain of patients with medically intractable epilepsy. Epilepsia 36, 1–6.

Toth, M., Grimsby, J., Buzsaki, G. & Donovan, G.P. (1995). Epileptic seizures caused by inactivation of a novel gene, jerky, related to centromere binding protein-B in transgenic mice [published erratum appears in Nat Genet 1996 Jan;12(1):110]. Nat. Genet. 11, 71–75.

Treiman, L.J. & Treiman, D.M. (1996). Genetic aspects of epilepsy. In: The Treatment of Epilepsy: Principles and Practice. (Treiman, L.J. & Treiman, D.M., eds.), pp. 151–164. Williams & Wilkins, Baltimore.

Tsirka, S.E., Gualandris, A., Amaral, D.G. & Strickland, S. (1995). Excitotoxin-induced neuronal degeneration and seizure are mediated by tissue plasminogen activator. Nature 377, 340–344.

Tsuboi, T. & Endo, S. (1991). Genetic studies of febrile convulsions: Analysis of twin and family data. Epilepsy Res. 4 (Suppl.), 119–128.

Tsuji, S., Choudary, P.V., Martin, B.M., Stubblefield, B.K., Mayor, J.A., Barranger, J.A. & Ginns, E.I. (1987). A mutation in the human glucocerebrosidase gene in neuronopathic Gaucher's disease. N. Engl. J. Med. 316, 570–575.

van Slegtenhorst, M., de Hoogt, R., Hermans, C., Nellist, M., Janssen, B., Verhoef, S., Lindhout, D., van den Ouweland, A., Halley, D., Young, J., Burley, M., Jeremiah, S., Woodward, K., Nahmias, J., Fox, M., Ekong, R., Osborne, J., Wolfe, J., Povey, S., Snell, R.G., Cheadle, J.P., Jones, A.C., Tachataki, M., Ravine, D., Kwiatkowski, D.J., et al. (1997). Identification of the tuberous sclerosis gene TSC1 on chromosome 9q34. Science 277, 805–808.

VanLandingham, K.E., Heinz, E.R., Cavazos, J.E. & Lewis, D.V. (1998). Magnetic resonance imaging evidence of hippocampal injury after prolonged focal febrile convulsions [see comments]. Ann. Neurol. 43, 413–426.

Veall, R.M. (1974). The prevalance of epilepsy among mongols related to age. J. Ment. Defic. Res. 18, 99–106.

Vesa, J., Hellsten, E., Verkruyse, L.A., Camp, L.A., Rapola, J., Santavuori, P., Hofmann, S.L. & Peltonen, L. (1995). Mutations in the palmitoyl protein thioesterase gene causing infantile neuronal ceroid lipofuscinosis. Nature 376, 584–587.

Viani, F., Romeo, A., Viri, M., Mastrangelo, M., Lalatta, F., Selicorni, A., Gobbi, G., Lanzi, G., Bettio, D., Briscioli, V., et al. (1995). Seizure and EEG patterns in Angelman's syndrome. J. Child Neurol. 10, 467–471.

Vigevano, F., Fusco, L., Di Capua, M., Ricci, S., Sebastianelli, R. & Lucchini, P. (1992). Benign infantile familial convulsions [see comments]. Eur. J. Pediatr. 151, 608–612.

Virtaneva, K., D'Amato, E., Miao, J., Koskiniemi, M., Norio, R., Avanzini, G., Franceschetti, S., Michelucci, R., Tassinari, C.A., Omer, S., Pennacchio, L.A., Myers, R.M., Dieguez-Lucena, J.L., Krahe, R., de la Chapelle, A. & Lehesjoki, A.E. (1997). Unstable minisatellite expansion causing recessively inherited myoclonus epilepsy, EPM1. Nat. Genet. 15, 393–396.

Wagstaff, J., Knoll, J.H., Fleming, J., Kirkness, E.F., Martin-Gallardo, A., Greenberg, F., Graham, J.M., Jr., Menninger, J., Ward, D., Venter, J.C., et al. (1991). Localization of the gene encoding the GABAA receptor beta 3 subunit to the Angelman/Prader-Willi region of human chromosome 15. Am. J. Hum. Genet. 49, 330–337.

Wallace, D.C., Lott, M.T., Shoffner, J.M. & Ballinger, S. (1994). Mitochondrial DNA mutations in epilepsy and neurological disease. Epilepsia 35, S43–50.

Wallace, R.H., Wang, D.W., Singh, R., Scheffer, I.E., George, A.L., Phillips, H.A., Saar, K., Reis, A., Johnson, E.W., Sutherland, G.R., Berkovic, S.F., & Mulley, J.C. (1998). Febrile seizures and generalized epilepsy associated with a mutation in the Na^{+}-channel β1 subunit gene SCN1B. Nat. Genet. 19, 366–370.

Wasterlain, C.G. & Fairchild, M.D. (1985). Transfer between chemical and electrical kindling in the septal-hippocampal system. Brain Res. 331, 261–266.

Waymire, K.G., Mahuren, J.D., Jaje, J.M., Guilarte, T.R., Coburn, S.P. & MacGregor, G.R. (1995). Mice lacking tissue non-specific alkaline phosphatase die from seizures due to defective metabolism of vitamin B-6. Nat. Genet. 11, 45–51.

Weiland, S., Witzemann, V., Villarroel, A., Propping, P. & Steinlein, O. (1996). An amino acid exchange in the second transmembrane segment of a neuronal nicotinic receptor causes partial epilepsy by altering its desensitization kinetics. FEBS Lett. 398, 91–96.

Weissbecker, K.A., Durner, M., Janz, D., Scaramelli, A., Sparkes, R.S. & Spence, M.A. (1991). Confirmation of linkage between juvenile myoclonic epilepsy locus and the HLA region of chromosome 6. Am. J. Med. Genet. 38, 32–36.

Whitehouse, W., Diebold, U., Rees, M., Parker, K., Doose, H. & Gardiner, R.M. (1993a). Exclusion of linkage of genetic focal sharp waves to the HLA region on chromosome 6p in families with benign partial epilepsy with centrotemporal sharp waves. Neuropediatrics 24, 208–210.

Whitehouse, W.P., Rees, M., Curtis, D., Sundqvist, A., Parker, K., Chung, E., Baralle, D. & Gardiner, R.M. (1993b). Linkage analysis of idiopathic generalized epilepsy (IGE) and marker loci on chromosome 6p in families of patients with juvenile myoclonic epilepsy: no evidence for an epilepsy locus in the HLA region. Am. J. Hum. Genet. 53, 652–662.

Wienecke, R., Konig, A. & DeClue, J.E. (1995). Identification of tuberin, the tuberous sclerosis-2 product. Tuberin possesses specific Rap1GAP activity. J. Biol. Chem. 270, 16409–16414.

Wienecke, R., Maize, J.C., Jr., Reed, J.A., de Gunzburg, J., Yeung, R.S. & DeClue, J.E. (1997). Expression of the TSC2 product tuberin and its target Rap1 in normal human tissues. Am. J. Pathol. 150, 43–50.

Williams, C.A., Zori, R.T., Hendrickson, J., Stalker, H., Marum, T., Whidden, E. & Driscoll, D.J. (1995). Angelman syndrome. Curr. Probl. Pediatr. 25, 216–231.

Wisniewski, K.E., Laure-Kamionowska, M. & Wisniewski, H.M. (1984). Evidence of arrest of neurogenesis and synaptogenesis in brains of patients with Down's syndrome [letter]. N. Engl. J. Med. 311, 1187–1188.

Wisniewski, K.E., Dalton, A.J., McLachlan, C., Wen, G.Y. & Wisniewski, H.M. (1985). Alzheimer's disease in Down's syndrome: clinicopathologic studies. Neurology 35, 957–961.

Wisniewski, K.E., Segan, S.M., Miezejeski, C.M., Sersen, E.A. & Rudelli, R.D. (1991). The Fra(X) syndrome: neurological, electrophysiological, and neuropathological abnormalities. Am. J. Med. Genet. 38, 476–480.

Wolf, B. & Heard, G.S. (1991). Biotinidase deficiency. Adv. Pediatr. 38, 1–21.

Wroblewski, J.T., Fadda, E., Mazzetta, J., Lazarewicz, J.W. & Costa, E. (1989). Glycine and D-serine act as positive modulators of signal transduction at N-methyl-D-aspartate sensitive glutamate receptors in cultured cerebellar granule cells. Neuropharmacology 28, 447–452.

Zara, F., Bianchi, A., Avanzini, G., Di Donato, S., Castellotti, B., Patel, P.I. & Pandolfo, M. (1995). Mapping of genes predisposing to idiopathic generalized epilepsy. Hum. Mol. Genet. 4, 1201–1207.

Chapter 7

Cerebrovascular Disease

LAROY PENIX and DOUGLAS LANSKA

Advances in Cell Aging and Gerontology
Volume 3, pages 243–286

ISBN: 0-7623-0405-7

INTRODUCTION

In a sense, all diseases result from the interaction of genetic predisposition with environmental factors (McKusick, 1980; Miura, 1997). Some diseases, such as homocystinuria or sickle cell anemia, are typically considered to be purely genetic (i.e., the result of a mutation of a single gene), yet the expression and severity of the disease process in individuals is also influenced by environmental factors. Other diseases have traditionally been thought to result solely from environmental causes, but genetic influences probably play a role in these as well. Indeed, even susceptibility to infectious diseases (Abel and Dessein, 1997) and to drug and alcohol abuse are determined in part by genetic factors (Tarter, 1995; Ferguson and Goldberg, 1997).

Genetic diseases—that is diseases in which genetic factors play a major role—can be categorized into three broad groups, as follows:

1. *Mendelian disorders* are caused primarily by mutation at a single genetic locus and have typical patterns of inheritance.
2. *Chromosomal disorders* are caused by a missegregation of chromosomes during gametogenesis or meiosis (McKusick, 1980) and are generally not inherited.
3. *Polygenic disorders* are caused by a combination of factors at several genetic loci. Heritability is the proportion of phenotypic expression that is caused by genetic factors.

Epidemiological studies, and developments in the cellular and molecular mechanisms of stroke pathophysiology, have begun to unravel the role of genetic influences in cerebrovascular disease. Most strokes probably result from the interaction

Table 1. Clinical Stroke Subtypes

Ischemic Stroke	
	Large artery atherothrombotic[a]
	Small penetrating artery
	Cardioembolic
	Stroke of other determined etiology
	Stroke of undetermined etiology
Hemorrhagic Stroke	
	Intracerebral hemorrhage
	Subarachnoid hemorrhage

Note: [a]Involving Middle, Anterior or Posterior Cerebral artery, Vertebrobasilar or major branches.

of environmental influences with several polygenic phenotypes, but a minority, albeit important, subgroup of strokes have a definitive genetic etiology caused by a single gene with a mendelian pattern of inheritance.

Background

Stroke is a major cause of morbidity and mortality in Western society. It is the third leading cause of death and a leading cause of disability in the United States (Caplan, 1993). There are between 500 and 550,000 new strokes per year in the United States with approximately 150,000 new stroke-related deaths per year (National Stroke Association, 1995) and an estimated 3 million stroke survivors. Although, most stroke victims recover, the majority are faced with a broad spectrum of disabilities, including inability to ambulate and cognitive impairment.

Clinical Stroke Subtypes

Table 1 shows a simple scheme for the clinical classification of stroke subtype. There are two major categories of stroke: ischemic and hemorrhagic. Ischemic stroke occurs when there is obstruction of a cerebral artery that results in death of all cell types fed by that artery, whereas hemorrhagic stroke occurs when there is rupture of a blood vessel with resulting direct mechanical injury to surrounding brain tissue or other secondary damage. Ischemic strokes account for approximately 80% to 90% of stroke subtypes in the United States, while hemorrhagic strokes (subarachnoid and intracerebral hemorrhage) account for the remainder (Sacco, 1994).

STROKE PATHOPHYSIOLOGY

Ischemic Neuronal Injury

Stroke is fundamentally a secondary consequence of what are primarily vascular disease processes affecting cerebral arteries. All cerebral infarctions, for example, are caused by total or near total interruption of blood flow, leading to ischemia and hypoxia in the part of the brain supplied by the responsible vessel. Although there are many causes of ischemic brain injury, each cause ultimately results in the same pathophysiological response. Reduction of local cerebral blood flow below a critical threshold results in characteristic metabolic derangements that eventually lead to the demise of neurons and other supporting cells in the brain.

Ischemia in experimental animals produces local brain hypoxia, which results in energy failure and a lack of metabolic substrates for normal cellular function. These conditions cause massive increases in extracellular glutamate concentrations (Benveniste et al., 1984), resulting from synaptic release of glutamate from neuronal terminals (Rothman, 1984). An increased concentration of glutamate in the synaptic cleft produces a marked increase in intracellular calcium concentration (Choi, 1995), followed by proteolysis, cellular membrane disruption, and neuronal necrosis if the ischemic insult lasts long enough (Pulsinelli et al., 1982). This sequence of alterations of cellular physiology takes place regardless of the cause of ischemic neuronal injury.

Atherosclerosis and Arterial Thrombosis

Atherosclerotic changes in arterial walls cause the vascular endothelium to shift from its normally antithrombotic state to a prothrombotic state, which in turn leads to thrombus formation (Scharf and Harker, 1987). Less commonly, other pathological processes—such as mechanical disruption of the arterial wall, and some metabolic and inflammatory diseases (Ross, 1993; Pearson, 1994; Rabbani and Loscalzo, 1994)—can also injure the arterial endothelium and lead to thrombus formation.

Intra-arterial thrombosis is a pathological aberration of normal hemostasis (Rabbani and Loscalzo, 1994). Hemostasis involves the interaction of the coagulation cascade with the aggregation of activated platelets. When endothelial cells are injured, platelets adhere to the vessel wall. Upon activation, these adherent platelets trigger platelet-platelet aggregation by way of fibrinogen binding to platelet surface proteins. A thrombus made up solely of platelet-platelet aggregation is, however, unstable. Stability is provided to this confluence of platelets by the enzymatic formation of a fibrin network. Fibrin is formed as the final step in the blood coagulation cascade as the result of proteolytic cleavage of fibrinogen by the protease thrombin (Rabbani and Loscalzo, 1994). Thus, through each step in the processes leading to thrombus formation, one encounters possible influences of genetic diseases.

Lipohyalinosis

Vascular hyalinosis refers to a nonspecific, eosinophilic, homogeneous refractile material in the intima or other layers of small arteries (Manning et al., 1974). The term lipohyalinosis refers to the presence of lipid-staining hyalin-like material in the walls of small arteries of usually less than 200-μm diameter, associated with thinning of connective tissue in the vessel wall (Fisher, 1982). Russel proposed that lypohyalinosis was caused by chronic hypertension (Russell, 1975). In autopsy studies of patients with lacunar infarctions, Fisher (1979) demonstrated lipohyalinosis of the deep penetrating cerebral arteries supplying the region of infarction. For decades, this was considered to be the most important factor in the development of subcortical strokes. More recently, however, Lammie and colleagues (1997) demonstrated that lipohyalinosis occurs much less frequently than earlier suspected; in a study of 70 consecutive brain autopsies in patients with evidence of small vessel disease, the most frequent finding was concentric hyaline wall thickening, whereas lipohyalinosis and fibrinoid necrosis were seen only rarely.

Microaneuryms of Small Cerebral Arteries and Arterioles

Charcot and Bouchard (1869) first described the association of microaneurysms of the small penetrating arteries of the brain with intracerebral hemorrhage. The frequency of small penetrating arterial microaneurysms increases with hypertension and aging (Weir, 1987). Cole and Yates (1967) found microaneurysms in 46 of 100 hypertensive patients at autopsy compared to only 7 of 100 age- and sex-matched normotensive patients. Of the 20 patients who died of massive cerebral hemorrhage, 18 showed these microaneurysms. Microaneurysms were present in 71% of hypertensive patients aged 65 to 69, but in only 10% (2 out of 21) of hypertensive patients aged less than 50. Fisher (1971) found an association of small miliary aneurysms in penetrating arteries with lipohyalinosis, which he proposed might be a cause of hypertensive intracerebral hemorrhages. Mauro and colleagues (1980) suggested that the angiopathy of small penetrating cerebral vessels resulting from microaneurysms and lipohyalinosis has genetics, hemodynamics, or both as the primary operant factors.

STROKE ETIOLOGY

Most strokes are caused by occlusion of a cerebral artery either from a thrombus that develops at the site of occlusion, or from a thromboembolus originating from a proximal atherothrombotic arterial site (Caplan, 1993). Atherosclerosis is thought to result from exposure to one or more recognized risk factors, including hypertension, diabetes mellitus, cigarette smoking, hyperlipidemia, and hyperhomocystinemia (the latter being just recently recognized as an independent vascular risk factor). Each of these classic stroke risk factors is, in turn, caused by a combination of environmental and genetic (polygenic) factors. Less frequent (but still relatively common) stroke subtypes are nonatheromatous arterial occlusion, cardiogenic thromboembolism, and ischemic injury secondary to brain hemorrhage. There are also a great number of rare metabolic, inflammatory, connective tissue, hematological, and coagulation disorders that can impair arterial perfusion and lead to ischemic brain injury.

Etiology of Ischemic Strokes

The etiology of a particular stroke is usually determined by a combination of clinical history and examination, brain imaging, and laboratory test results. The process begins with neuroanatomical and cerebrovascular localization of the infarction based on the clinical history and neurological examination, complemented with either brain computed tomography (CT) or magnetic resonance imaging (MRI). The localization indicates which cerebral artery (or arteries) are responsible for the brain infarction, and allows classification of the stroke subtype. Then, an etiologic diagnosis (or differential diagnosis) is made using the results of judiciously chosen laboratory tests in conjunction with the clinical knowledge base of the examining physician. Table 2 shows a comprehensive, though by no means complete, list of etiologies of ischemic stroke.

Cardioembolic stroke occurs when a piece of thrombotic material breaks off from the heart and travels in the arterial tree until it lodges in a cerebral artery, occluding it, and resulting in infarction of the brain fed by that artery. These strokes are most frequently caused by thrombus formation in the left atrium (usually as a result of chronic atrial fibrillation) or on the left ventricular wall after a myocardial infarction. Less frequently, they are caused by mitral or aortic valvular disease or from a shunt from the venous side of the circulation to the arterial side through a patent foramen ovale or an atrial septal defect (Caplan, 1993). There are many different etiologies of cardiomyopathy, valvular disease, and septal defects that can lead to cardiogenic emboli, some of which have genetic causes. The genetic factors that lead to coronary artery disease and cardiomyopathies have been reviewed recently (Coonar and McKenna, 1997; Peyser, 1997) and are beyond the scope of this chapter. Although many of the risk factors for ischemic stroke and ischemic heart disease are similar, accumulating evidence indicates that some of the pathophysiological mechanisms may be different (Fisher et al., 1993).

Much of the discussion in this chapter focuses on the major etiological categories of ischemic stroke. These include atherothrombotic, nonatheromatous arterial thrombosis, arteriopathies, and metabolic disorders which together account for approximately 60% of all stroke types.

Etiology of Hemorrhagic Strokes

Hemorrhagic stroke can be subdivided into intracerebral hemorrhage, subarachnoid hemorrhage, subdural hematoma, and epidural hematoma. The latter two categories are seen mostly in association with head trauma or coagulopathy and are not always associated with brain injury. Thus, they are not covered in this chapter.

Several types of intracerebral and subarachnoid hemorrhage may be caused by alteration of the structural integrity of walls of large cerebral arteries and small penetrating intraparenchymal arteries. Table 3 outlines a general etiologic scheme for hemorrhagic stroke, some of which are hereditary.

Table 2. Etiology of Ischemic Strokes

Type of Stroke	*Category*
I. Arterial Occlusive Disorders	
A. Extracranial internal carotid artery atherothrombotic	E
B. Intracranial in situ atherothrombotic	E
C. Nonatheromatous arterial thrombosis	
1. Disorders of hemostasis (coagulopathies)	
a. Familial	
• Protein S deficiency	M
• Protein C deficiency	M
• Antithrombin III deficiency	M
• Factor V Leiden mutation	M
• Dysfibrinogenemias	M
b. Acquired	
• Antiphospholipid syndrome	S
• Malignancy	S
• Myeloproliferative disorders	S
• Nephrotic syndrome	S
• Liver disease	S
2. Hemoglobinopathies	M
3. Lipohylinosis	E
4. Inflammatory	E
a. Autoimmune vasculitis	S
b. Central nervous system infection	E
D. Arteriopathies	
1. CADASIL	M
2. Cerebral amyloid angiopathy	M and S
3. Connective tissue disorders	M, F
4. Neurocutaneous syndromes	M, F
E. Metabolic disorders	
1. MELAS syndrome	M
2. Fabry's disease	M
3. Leigh syndrome	M
II. Cardioembolic Disorders	**S, E, and M**

Notes: E, strong environmental influence; M, identified mendelial genetic defect; F, familial disorder without identified genetic defect; S, sporadic.

CADASIL, cerebral autosomal dominant arteriopathy with subcortical infarcts and leukoencephalopathy; MELAS, mitochondrial myopathy, encephalopathy, lactic acidosis, and stroke-like episodes.

Table 3. Etiology of Hemorrhagic Stroke

Type of Stroke	*Category*
I. Intracerebral Hemorrhage	
A. Hypertensive	E
B. Amyloid angiopathy	M and S
C. Arteriovenous malformations	S
II. Subarachnoid Hemorrhage	
A. Saccular aneurysm	S
B. Familial intracranial aneurysms	F
C. Connective tissue disorders	M
D. Cerebral aneurysms associated with polycystic kidney disease	F

Note: E, strong environmental influence; M, identified Mendelian genetic defect; F, familial disorder without identified genetic defect; S, sporadic.

EPIDEMIOLOGY OF STROKE

Twin Studies

The Twin Registry of the National Academy of Sciences-National Research Council (NAS-NRC), which included 15,948 pairs of male twins born between 1917 and 1927, was used to study the genetic and environmental contributions to the development of stroke (Brass et al., 1992). A questionnaire covering vascular risk factors, cardiac events, and stroke was mailed to registrants in 1985. Responses were received from 9,475 individuals, including 2,722 complete twin pairs (1,382 monozygotic twin pairs, 1,221 dizygotic twin pairs, and 119 twin pairs of unknown zygosity). The results of the responses from the twin pairs are shown in Table 4. Although the cumulative stroke incidence was identical (3.1%) for both monozygotic twins (131 of 4,220) and dizygotic twins (143 of 4,585), the proband concordance was 17.7% for monozygotic twins and 3.6% for dizygotic twins. The higher concordance in monozygotic than dizygotic twins suggests that there is a genetic contribution to the development of stroke. However, the limited number of

Table 4. Response to Stroke Question in NAS-NRC Twin Registry Study

Response	*Monozygotic*	*Dizygotic*	χ^2	*Relative Risk*
Both answered no	1,271	1,128		
Only one answered yes	65	53		
Both answered yes	7	1	4.94 (df = 1)	4.3 ($p < 0.05$)

strokes in twin pairs precludes a reliable estimate of the magnitude of this contribution (Brass et al., 1992).

Geographical Differences in Stroke Incidence

There are marked regional differences in stroke incidence, hospitalization rates, and mortality in the United States, with states of the southeastern United States having much higher rates than those in the north or west (Lanska and Kryscio, 1994). These differences could not be explained by differences in death certification practices, differences in the accuracy of the diagnosis of stroke, or variations in the standards of medical care (Lanska and Peterson, 1996). Neither could migration effects explain the differences seen (Lanska and Peterson, 1995). Many authors have attributed the regional differences in stroke incidence to socioeconomic factors (Ahmed et al., 1989; Kenton, 1991; Navarro, 1991). A more recent analysis of more than 400,000 participants in the National Longitudinal Mortality Study estimated that less than 16% of the excess stroke mortality seen in the southeastern United States was attributable to socioeconomic status (Howard et al., 1997). Howard and colleagues have also suggested that the excess differences seen in this region are likely caused by unmeasured lifestyle influences, social resources, and genetic factors (Howard et al., 1995).

The World Health Organization Monitoring Trends and Determinants in Cardiovascular Disease Project (WHO MONICA) has determined age-standardized stroke incidence rates in 16 European countries, China, and Japan (Thorvaldsen et al., 1995). The incidence rate was generally higher in eastern European than in western countries with Finland having the highest rate. China also had a very high incidence, particularly among women. These worldwide variations of stroke incidence are not necessarily the result of genetic differences, however, and may simply reflect variation in diagnostic and reporting practices, or in exposure to environmental factors. Further systematic analysis of these geographical clusters of increased stroke incidence and mortality with an emphasis on population genetics is sorely needed.

Ethnic Differences in Stroke Incidence

Several studies in the United States have now demonstrated racial differences in stroke incidence, with African-American men consistently having the highest rate (Schoenberg et al., 1986; Kittner et al., 1993). Further support for genetic differences of stroke susceptibility between the races is the relative difference in distribution of stroke subtype among different ethnic groups. Caplan and colleagues (1986) found that blacks tend to have intracranial large and medium artery stenoses as a cause of ischemic cerebral infarction more often than do whites. Sacco and colleagues (1995) found that whereas the proportion of extracranial atherosclerotic stroke was similar between blacks, whites, and Hispanics in northern Manhattan, there was an increased incidence of intracranial atherosclerotic stroke in blacks and

Hispanics compared to whites. Inzitari and colleagues (1990) used multivariate analysis to compare prevalence of risk factors with ischemic stroke subtype in whites, blacks, and Orientals. They found that Orientals had the lowest prevalence of vascular risk factors with a lower frequency of hypertension than either blacks or whites. They also found that race was an independent and strong predictor of the location of cerebrovascular lesions. Leung and colleagues (1993), in a study of the pattern of vascular distribution of cerebral atherosclerosis, found that the extent of intracranial atherosclerosis was greater whereas the frequency of extracranial stenosis was less in Hong Kong Chinese than in Caucasians. Taken together, these studies suggest a strong influence of ethnic origin on the incidence, severity and type of cerebrovascular disease. It is yet to be determined whether these ethnic differences are due to genetic or environmental factors.

Kuller (1991) has noted that obese African-Americans have very high risks of hypertension, diabetes, stroke, and heart failure. In contrast, Native Americans and Hispanics, particularly in the Southwest, who are equally or more obese, do not seem to develop hypertension or stroke as frequently but do acquire diabetes as often, if not more often than African-Americans. Genetic factors may be the most likely explanation of these differences (Kuller, 1991). Although cultural influences such as diet and acceptance of medical intervention as well as socioeconomic factors such as access to preventative health care undoubtedly play some role, there appears to be a significant contribution of genetic factors. Thus, further studies focused on population genetics and identification of specific molecular markers of these differences are needed.

HERITABILITY OF ISCHEMIC STROKE RISK FACTORS

Hypertension

Early epidemiologic studies recognized the overwhelming importance of environmental factors contributing to the development of essential hypertension (Williams et al., 1980). Over the past 20 years, however, our understanding of the genetic contributions to the development of hypertension has advanced considerably. A review of family and twin studies by Corvol and colleagues (1997) found that about 30% of hypertension can be attributed to genetic (polygenic) factors and about 50% to environmental influence.

There are several identified mendelian inherited hypertension syndromes, all of which are rare (Caulfield et al., 1997). Glucocorticoid-suppressible hyperaldosteronism is an autosomal dominantly inherited form of severe, early-onset hypertension associated with high levels of aldosterone, 18-hydroxycortisol, and 18-oxocortisol, as well as suppressed plasma renin activity (Sutherland et al., 1966). An increased incidence of cerebral hemorrhage has been reported in some families (Caulfield et al., 1997). Lifton and colleagues (1992) have identified a chimeric 11-β-hydroxylase/aldosterone synthase gene that causes this condition. Liddle's syndrome is another rare familial form of severe hypertension, with hypokalemia, suppressed plasma renin activity, and low aldosterone concentrations when challenged with a low-sodium diet. This disorder is caused by a mutation in the β subunit of the epithelial sodium channel (Shimkets et al., 1994). Another severe inherited form of hypertension, hypertension and brachydactyly syndrome, is transmitted as an autosomal dominant trait, and has been mapped to chromosome 12 in a Turkish kindred (Schuster et al., 1996).

Hypertension has also been associated with the following genetic defects: a common DNA sequence variant within the angiotensinogen gene linked to the determination of circulating angiotensinogen levels (Kunz et al., 1997), and an angiotensin II type 1 receptor $A^{1166}C$ polymorphism (Davis and Roberts, 1997). Interestingly, despite much continued speculation to the contrary, to date there have been no genetic determinants of hypertension that show a specific susceptibility in the black population (Cooper and Rotimi, 1997). This would imply that the increased frequency of stroke in blacks is not caused by a genetic predisposition to the development of hypertension necessarily but to other factors such as an increased environmental risk factor burden or to an as yet unidentified genetic predisposition.

Diabetes Mellitus

Data from the Framingham Study (Wolf et al., 1991), the National Health and Nutrition Examination Survey (Kittner et al., 1990), and the Honolulu Heart Study (Abbott et al., 1987) each showed roughly a twofold increase in relative risk for stroke in patients with diabetes mellitus. Diabetes is a heterogeneous group of disorders with glucose intolerance as a common feature and, like anemia, is not a diagnostic term but a description of symptoms and laboratory abnormalities that can have many distinct etiologies (Scheuner et al., 1997). Familial aggregation and twin studies have shown that there is both strong genetic and environmental influence in diabetes. There are now at least 75 genetic syndromes associated with diabetes mellitus or glucose intolerance (Scheuner et al., 1997). These include syndromes associated with pancreatic degeneration, hereditary endocrine disorders, inborn errors of metabolism, syndromes with nonketotic insulin resistance, acanthosis nigricans, hereditary neuromuscular syndromes, progeroid syndromes, mitochondrial disorders, obesity syndromes, cytogenetic disorders, and miscellaneous syndromes. There are 74 known single gene mutations that cause these conditions (McKusick, 1994; Scheuner et al., 1997).

Type I or insulin-dependent diabetes mellitus (IDDM) was first found to have a genetic locus contained within the HLA region that conferred disease susceptibility (Field and Tobias, 1997). Since then seven new loci associated with susceptibility have been identified. The occurrence of more than one of these genes seems to increase the likelihood of an individual developing diabetes.

Type II or noninsulin-dependent diabetes mellitus (NIDDM) is a multifactorial hyperglycemia with a strong environmental influence combined with a variable genetic susceptibility. An autosomal dominant form of maturity-onset NIDDM has been associated with at least four genes (Velho and Froguel, 1997).

Smoking

Of the so-called classic stroke risk factors, smoking is the one most thought to be a pure environmental risk. However, recent studies suggest that nicotine dependency does have a strong genetic influence with a heritability index of 53% (Pomerleau, 1995). This influence is thought to be conveyed through a higher initial sensitivity to nicotine, a more rapid development of tolerance, and more extensive self-administration in genetically susceptible individuals.

Hyperhomocystinemia

Homocystinuria is a genetic disorder of metabolism first described about 35 years ago (Carson et al., 1963). Phenotypically, patients have a marfanoid body habitus (i.e., tall stature, arachnodactyly, pectus excavatum), mental retardation, skeletal deformities, lens ectopia, and premature atherosclerosis (Mudd and Levy, 1983). Patients usually succumb to myocardial infarction, ischemic stroke, or pulmonary embolism.

Homocysteine—a 4-carbon thiol-containing amino acid—binds covalently with other homocysteine or cysteine molecules via sulfide bonds to produce the homocysteinyl moieties homocystine and cysteine-homocysteine, respectively. By convention, the combined serum concentrations of homocysteine and the two homocysteinyl moieties are referred to as the homocyst(e)ine level.

Patients with homocystinuria have exceedingly high concentrations of homocystine in their urine and homocyst(e)ine in the blood. Elevated blood concentrations of homocystine have been shown to injure vascular endothelium in vivo (Harker et al., 1974) and in vitro in cell culture systems (Wall et al., 1980; Weimann et al., 1980). This injury of endothelial cells leads to monocyte adherence and the subsequent development of atherosclerosis, and shifts the normally antithrombotic state of the endothelial lining to a more prothrombotic state.

There are three main pathways of homocysteine metabolism: transsulfuration of homocysteine to cysteine by cystathionine β-synthase (a vitamin B_6–dependent enzyme), remethylation of homocysteine to methionine by methionine synthase (a vitamin B_{12}–dependent enzyme), and transmethylation of *S*-adenosyl-methionine by multiple methyltransferases (folate dependent enzymes) (Ueland and Refsum, 1989). Cystathionine β-synthase (CBS) deficiency is an autosomal dominant defect and is the most common genetic cause of homocystinuria (Mudd and Levy, 1983). It has also been shown that a heterozygous single gene defect of CBS can lead to elevated serum levels of homocyst(e)ine (Clarke et al., 1991). Homocysteinemia (caused either by a heterozygous genetic defect or by a vitamin deficiency) can be seen in up to 30% of patients with premature arterial occlusive disease (Boers et al., 1985).

More recently, it has been shown that hyperhomocysteinemia is an independent risk factor for ischemic stroke (Coull et al., 1990; Clarke et al., 1991; Brattström et al., 1992; Perry et al., 1995). Coull and colleagues (1990) found that about 30% of 99 patients with acute ischemic stroke, transient ischemic attack, or stroke risk factors alone had a serum homocyst(e)ine level two standard deviations above the mean value for control patients. Clarke and colleagues (1991) found that 42% of 38 patients with cerebrovascular disease had hyperhomocysteinemia (Clarke et al., 1991). Selhub and colleagues (1995) showed that there was an association of high homocyst(e)ine levels with carotid artery stenosis in elderly patients. Boers and colleagues (1985) found that 30% of patients with premature atherosclerosis (angiographically confirmed arteriosclerotic lesions at age under 50) with no medical stroke risk factors (absence of diabetes, hyperlipidemia, and hypertension) were heterozygous for homocysteinemia due to CBS deficiency. The frequency of the homozygous CBS deficiency has been reported to be 1/60,000 to 1/200,000 (Mudd and Levy, 1983; Ueland and Refsum, 1989). Therefore, it has been estimated that the heterozygous carrier rate should occur in approximately 1/70 to 1/225 individuals (Boers et al., 1985; Tsai et al., 1996).

Hyperlipidemia

Hyperlipidemia is associated with an increased incidence of coronary artery disease (CAD) (Brewer et al., 1996) and it is well established now that use of lipid lowering drugs can decrease CAD incidence and mortality (Holme, 1995; Brewer et al., 1996; Watts and Burke, 1996). An association between hyperlipidemia and cerebrovascular disease has been less clear, however, partly owing to a paucity of studies using stroke occurrence as a primary outcome variable (Gorelick et al., 1997; Stoy, 1997). Two recent meta-analysis studies (Hebert et al., 1997; Bucher et al., 1998), and a review of epidemiological and pathophysiological studies (Gorelick et al., 1997), indicate that there is a strong positive association between hyperlipidemia and ischemic atherothrombotic stroke incidence. Contrary to these findings, however, Dyker and colleagues (1977) found an association of poor stroke outcome with lower serum cholesterol concentration in 977 patients with acute stroke. This finding was independent of stroke type, vascular territory, and patient age. There are also two cohort studies that have shown an increased risk of hemorrhagic stroke in the subgroups of patients with the lowest cholesterol concentration (Frank et al., 1992; Neaton et al., 1992). Several studies have shown a positive association of hyperlipidemia with carotid plaque progression (Tonstad et al., 1996; Hodis et al., 1997; Wilson et al., 1997), and preliminary data indicate that treatment of hypercholesterolemia with lipid-lowering agents can reverse the intimal-medial thickening of carotid arteries (Salonen et al., 1995; Wendelhag et al., 1995).

Table 5. Genetically Transmitted Lipid Disorders and Their Association with Atherosclerosis

Disorder	*Inheritance*	*Genetic Defect*	*Frequency*	*Association with Atherogenesis*
Hyperlipoproteinemias				
Familial hypercholesterolemia	AD	LDL receptor	1/500	Definitely
Familial defective Apolipoprotein B-100	AD	Apo B	1/500	Definitely
Familial combined hyperlipidemia	AD	Unknown	3-5/1,000	
Familial dysbetalipoproteinemia	AD or AR	Apo E	1/5,000	
Familial hyperchylomicronemia				
Apolipoprotein C-II deficiency	AR	Apo C-II	Rare	None established
Lipoprotein lipase deficiency	AR	Lipoprotein lipase	Rare	None established
Hepatic lipase deficiency	AR	Hepatic lipase	Rare	
Elevated lipoprotein (A)	AD	—	20%	
β-Sitosterolemia	AR	Unknown	Rare	
Hypolipoproteinemias				
Tangier disease	AD	Unknown	Rare	None established
Apolipoprotein A-I deficiency	AR	Apo A-I	Rare	Definitely
Familial hypoalphalipoproteinemia	AR	Unknown	Rare	

Notes: AD, autosomal dominant; AR, autosomal recessive.

Adapted from: Brewer, H.B., Jr., Santamarina-Fojo, S.M. & Hoeg, J.M. (1996). Genetic dyslipoproteinemias. In: Atherosclerosis and Coronary Artery Disease Fuster, V., Ross, R. & Topol, E.J., eds.), p. XX. Lippincott-Raven, Philadelphia.

Several specific genetically transmitted lipid disorders have an increased frequency of accelerated atherogenesis, whereas others have not been definitely associated with an increased frequency and others still appear to confer a protection against atherogenesis (Table 5) (Durrington, 1995b). The most important genetically determined hyperlipidemia that we have been able to identify is familial hypercholesterolemia (Durrington, 1995a). This is an autosomal dominant disorder with increased serum cholesterol, which is due primarily to an increase in low-density lipoprotein (LDL). Clinical evidence of ischemic heart disease (IHD) can occur as early as the mid-twenties in heterozygotes but may develop in childhood in homozygotes with 100% of men and 74% of women manifesting IHD by the seventh decade (Durrington, 1995a).

Apolipoprotein E (Apo E) has been implicated as a major cause of the genetic basis of Alzheimer's disease and cardiovascular disease (Contois et al., 1996; Roses, 1997). There have been recent conflicting reports about an association of increased ischemic stroke risk in subjects with the Apo E epsilon-4 allele. Higuchi and colleagues (1996) found that the occurrence of the Apo E-4 gene conferred no increased incidence of Binswanger's disease (subcortical ischemic leukoencephalopathy) or non-Binswanger's vascular dementia. Kessler and colleagues (1997) recently found an association of the Apo E-4 allele with stroke caused by large vessel extracranial cerebrovascular disease (> 70% stenosis in the symptomatic extracranial artery) but not with lacunar stroke or cardioembolic stroke or stroke of indeterminate etiology (Kessler et al., 1997).

The Apo E-2 genotype may be associated with an increased risk of cerebral amyloid angiopathy (CAA) (Nicoll et al., 1997). Nicoll and colleagues determined the Apo E genotype in 36 patients (17 with Alzheimer's disease [AD]) who presented with intracerebral hemorrhage (ICH) associated with histologically confirmed CAA and in 104 subjects without hemorrhage (61 with AD). The frequency of Apo E-4 was 0.29 in those with ICH and AD versus 0.08 in those with ICH but without AD, compared to 0.35 in the controls with AD alone and 0.13 with no AD and no ICH. The frequency of Apo E-2 was 0.21 in those with ICH and AD versus 0.35 in those with ICH but without AD, compared to 0.05 in the controls with AD alone and 0.14 with no AD and no ICH. Their conclusions were that the Apo E-2 gene may be a risk factor for cerebral hemorrhage resulting from CAA and that Apo E-4 was a risk factor for concomitant AD but not a risk factor for CAA-related hemorrhage.

HEREDITARY ISCHEMIC LEUKOENCEPHALOPATHIES

Hereditary Multi-Infarct Dementia

Sourander and Walinder (1977) described a family with five individuals from three successive generations who developed progressive dementia secondary to multiple small subcortical infarcts despite an absence of the classic stroke risk factors. Three of the affected family members were female. The mean age of onset was about 34 years (range: 29 to 38) and the average duration of illness was about 9.7 years (range: 0.5 to 5). Gross neuropathological findings were similar in three of the cases with multiple small areas of necrosis and postnecrotic cystic lesions in the basal ganglia, thalamus, periventricular white matter, and pons. There was a variable degree of cortical atrophy, but only one patient had a cortical infarction, which was 1.5 cm in diameter and located in the cerebellum. Microscopic examination showed widespread occlusive vascular changes, most prominently in small muscular arteries and arterioles of the leptomeninges, basal ganglia, thalamus, mesencephalon, pons, and cerebellum. The most common alteration in these vessels consisted of concentric thickening of the vascular wall with narrowing of the arterial lumen caused by a fibrous proliferation in the subendothelial region or hyaline degeneration of the intima. Fibrinoid necrosis of the intima was seen in two of the cases. Occasional perivascular inflammatory infiltration and intramural edema was seen. No recent thrombus formation was identified, but there were occasional changes consistent with organized thrombosis and sometimes with signs of recanalization. Histochemical analysis demonstrated the most consistent finding as a strong periodic acid-Schiff (PAS) reaction for carbohydrates within the intimal thickening and occasionally the media. Lipid extraction had no effect but acetylation prevented the PAS-positive reaction, leading the investigators to suggest that the PAS positivity may be owing to accumulation of acid mucopolysaccharides. Subependymal deposits of cholesterol, cholesteryl esters, and triglycerides were seen infrequently and only in a few small arteries of the basal ganglia. No evidence of amyloid deposition was seen. Similar microvascular changes were seen in the spleen in two of the autopsied cases.

Sonninen and Savontaus (1987) described a similar disease process in 16 individuals of a family in four successive generations with an autosomal dominant pattern of transmission. Nine of the affected family members were female. The mean age of onset was 46 years (range: 28 to 68) and the average duration 10.6 years (range: 1 to 29). Three affected members had diabetes mellitus, two had hypertension, and two hyperlipidemia (one had all three risk factors). Two affected members had migraine headaches. Brain CT imaging was performed on four affected members and one nonaffected person. Three were reported to have identical findings with marked bilateral hypodensities in the periventricular white matter and basal ganglia with completely normal gray matter. The nonaffected individual had a normal CT. Postmortem examination was performed on one affected individual only. Microscopic findings were not available at the time of the published report. Gross examination of the brain showed multiple small necrotic and postnecrotic cystic lesions with softening in the basal ganglia, thalamus, and the periventricular white matter.

CADASIL: Cerebral Autosomal Dominant Arteriopathy with Subcortical Infarcts and Leukoencephalopathy

In 1991, Tournier-Lasserve and colleagues reported a syndrome of subcortical strokes and leukoencephalopathy transmitted in an autosomal dominant pattern in 9 members of a French family. All nine had recurrent strokes, and some had episodic migraine-like headaches, pseudobulbar palsy, or dementia. Vascular risk factors, including hypertension, were rarely present. The pathological findings in one of the clinically affected family members included multiple small deep white matter infarcts with a widespread vasculopathy of the penetrating small arteries of the deep white matter (Baudrimont et al., 1993). The descriptive name "cerebral autosomal dominant arteriopathy with subcortical infarcts and leukoencephalopathy" (CADASIL) was coined for this disease entity in 1993 (Tournier-Lasserue et al., 1993).

The ischemic events reported in CADASIL include transient ischemic attacks, and minor or major strokes starting between the third and sixth decade with a mean age of onset in the mid-40s (Bousser and Tournier-Lasserve, 1994). The clinical presentation includes stroke, dementia, migraine with aura, mood disturbance (depression or manic episodes), and seizures. Several patients have progressed to develop pseudobulbar palsy and subcortical dementia.

Two principle abnormalities are seen on MRI (Bousser and Tournier-Lasserve, 1994). Firstly, there are well-demarcated, small areas of abnormal signal (decreased on T1-weighted and increased on T2-weighted images) that have the appearance of infarctions. Secondly, there are areas of increased signal intensity on T2-weighted images in the white matter of the cerebral hemispheres. The lesions were relatively symmetrical, appear to spare the U fibers and concentrate in the antero-lateral regions and in the external capsule (Bousser and Tournier-Lasserve, 1994). These changes are seen in all patients with clinical symptoms and are also seen in a large number of asymptomatic patients. Cortical and cerebellar lesions are unusual. Cerebral angiography shows no consistent abnormal findings.

Ruchoux and Maurage (1997) reviewed the gross and microscopic pathological findings of cases of subcortical small vessel cerebral infarction from 19 published reports since 1977. Ten of these reports are of patients who have been diagnosed with CADASIL, two are of patients with sporadic multiple subcortical infarcts, and six others are of patients with an autosomal dominant familial pattern of subcortical ischemic stroke, which may represent CADASIL. The latter reports included the hereditary multi-infarct dementia cases reported by Sourander and Walinder (1977) and Sonninen and Savontaus (1987). The features common to all these reported cases are thickening of deep white matter and leptomeningeal arterial walls, and a smudgy and granular appearance of the media with a loss of smooth muscle cell nuclei. Partial or complete occlusion of arterial lumens was rare, and when present was caused by concentric intimal proliferation. PAS-positive material was strongly present in two and mild in three of the five CADASIL studies in which it was performed. Amyloid staining by either the Congo red method or with specific immunohistochemistry showed no positive staining in any cases.

Electron microscopic studies demonstrate granular osmophilic material (GOM) within the media of arteries in the deep white matter and the leptomeninges. This finding has now been reported by many investigators using both brain autopsy material (Baudrimont et al., 1993; Gray et al., 1994; Gutierrez-Molina et al., 1994; Ruchoux et al., 1995; Malandrini et al., 1996) or biopsy specimens (Lammie et al., 1995). More recently GOM deposits have also been described in vessel walls of muscle biopsy (Ruchoux et al., 1995), nerve biopsy (Schroder et al., 1995), and skin biopsy (Ebke et al., 1997) specimens from patients with CADASIL. Ebke and colleagues (1997) performed skin biopsy on eight subjects from a German kindred with CADASIL and compared the findings to skin biopsies from five patients with sporadic leukoencephalopathies. The skin biopsies showed GOM in all eight members of the kindred, including a 22-year-old woman who had a normal MRI and was asymptomatic except for migraine-like headaches. Genetic linkage analysis demonstrated that this subject carried the disease haplotype. No GOM was found in patients with sporadic leukoencephalopathy. Ebke and colleagues concluded that skin biopsy is helpful in the diagnosis of CADASIL, and they also suggested that skin biopsy may be used to define the carrier status of individuals in CADASIL families. The latter suggestion, however, presents ethical problems akin to those associated with presymptomatic testing for Huntington's disease.

Genetic linkage analysis in two unrelated families in France with CADASIL first localized the defect to chromosome 19q12 (Tournier-Lasserve et al., 1993). The linkage analysis was performed using cerebral MRI data in patients above age 35 to determine phenotype penetrance. Abnormalities of genes coding for proteins that could possibly contribute to arteriopathic changes such as elastin, fibrillins, collagen and the amyloid precursor protein (APP) were first excluded. A sequential study of the entire human genome located the defective gene to chromosome 19 within a 14 cM interval between the D19S221 and D19S215 loci (Tournier-Lasserve et al., 1993), with no evidence of genetic heterogeneity. More recently, the human Notch-3 gene was mapped to the CADASIL critical region and mutations of this gene were identified in patients with CADASIL (Joutel et al., 1996). Further investigation was performed on 50 unrelated subjects with CADASIL using conformational polymorphism, heteroduplex, and sequence analysis, compared to 100 healthy control subjects (Joutel et al., 1997). The results showed missense mutations in the epidermal growth factor–like (EGF-like) repeats in the extracellular domain of the Notch-3 gene in 45 of the CADASIL patients. No mutations were seen in the 100 normal control subjects. There was clustering of the mutations within the two exons encoding the first five EGF-like repeats in 32 of the patients. These mutations lead to the loss or gain of a cysteine residue in the EGF protein, which could then lead to abnormal disulfide bridging within the EGF molecule.

We are currently following several members of a family with CADASIL in the Department of Neurology and the Memory Disorders Clinic of the Sanders-Brown

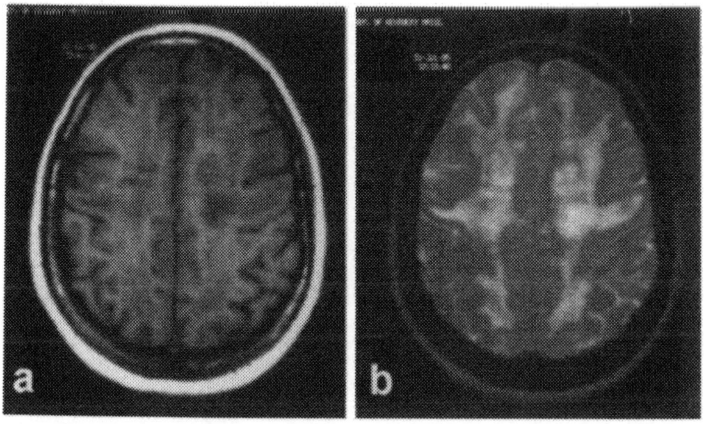

Figure 1. Magnetic resonance image of proband with CADASIL. Confluent diffuse hypointensities on T1-weighted (**a**) and hyperintensities on T2-weighted (**b**) images are seen in the subcortical white matter.

Center on Aging of the University of Kentucky. The propositus is a 59-year-old woman with a history of multiple lacunar infarcts and progressive dementia. She was first seen in the university hospital in March 1990, when she presented with acute onset of dysarthria and difficulty swallowing. She had a previous episode of sudden onset of ataxia, which resolved after several days. She had no history of hypertension, diabetes mellitus, ischemic heart disease, or hyperlipidemia. Her

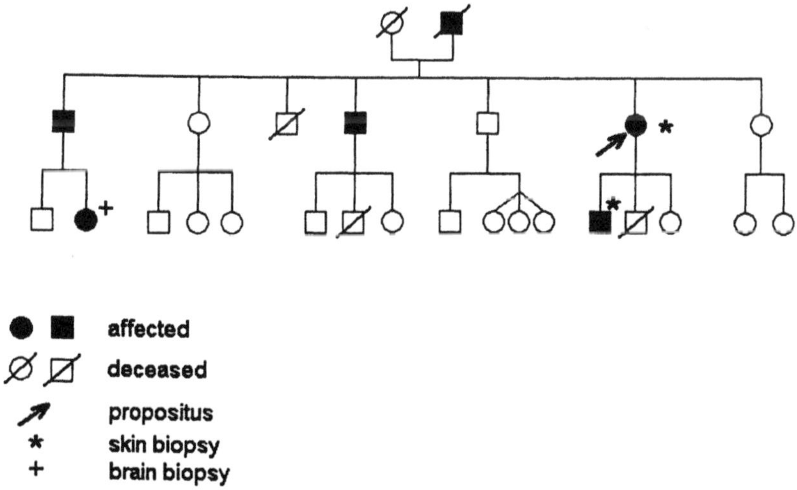

Figure 2. Pedigree of a kindred with CADASIL. Subjects who were fully examined clinically, and by skin biopsy and magnetic resonance imaging are indicated by asterisk.

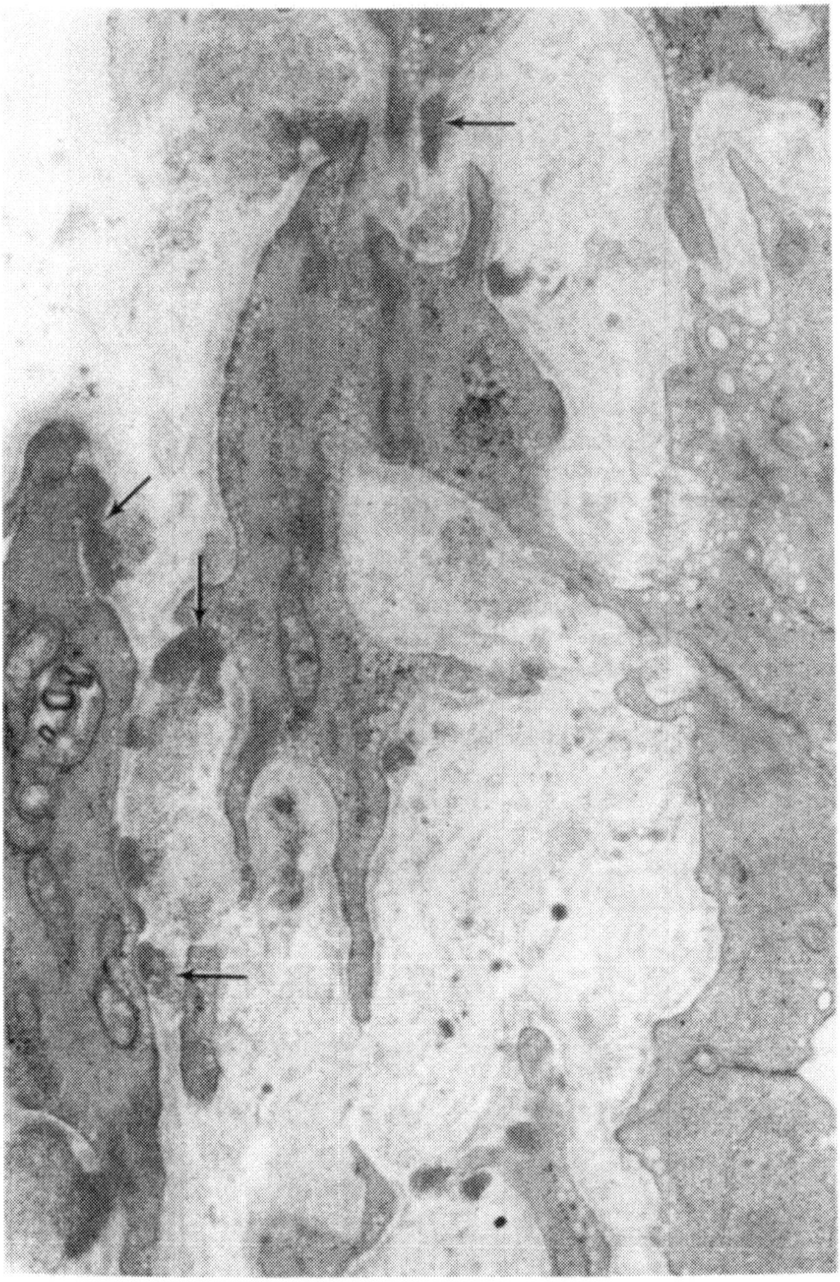

Figure 3. Electron micrograph of skin biopsy showing multiple granular osmophilic deposits (*arrows*) adjacent to vascular smooth muscle cells. Granular osmophilic deposits are closely related to the basal lamina. Note numerous caveolae. (×24,000).

family history was strongly positive for stroke. A brother had been diagnosed with "Binswanger's disease" (multifocal subcortical ischemic leukoencephalopathy) with onset in his late 40s, which progressed and required nursing home placement in his late 50s. An axial brain MRI of the propositus is illustrated in Figure 1. The pedigree of three generations is illustrated in Figure 2. The patient and her son underwent skin biopsy at the University of Kentucky. A niece underwent brain biopsy in a teaching hospital in Ohio. The skin biopsies were both positive for GOM, as illustrated in Figure 3.

Jen and colleagues (1997) have recently described the clinical manifestations, neuropathological, and neuroimaging findings of a stroke syndrome affecting 11 members of three generations of a Chinese-American family. The disorder has been called hereditary endotheliopathy with retinopathy, nephropathy, and stroke (HERNS), and is characterized by focal neurologic deficits secondary to ischemic cerebral infarction occurring in the third or fourth decade. Clinical signs and symptoms included migrainous headaches, psychiatric disturbances, hemiparesis, dysarthria, and apraxia. MRI and CT imaging showed contrast-enhancing subcortical lesions with surrounding edema consistently in affected individuals. Microscopic analysis of small vessels in the brain, kidney, stomach, appendix, omentum, and skin demonstrated a distinctive pattern of multilaminated vascular basement membranes. Retinopathy was characterized by perifoveal microangiopathic telangiectases. Linkage to the CADASIL locus on chromosome 19 was ruled out. The pattern of transmission was consistent with an autosomal dominant syndrome.

CEREBRAL AMYLOID ANGIOPATHY

CAA is a clinicopathological entity characterized by an acellular thickening of the small- and medium-sized arteries of the brain associated with mental deterioration and an increased incidence of intracerebral hemorrhage (Vinters, 1987). It has been recognized for the past 90 years (Okazaki et al., 1979) and is not necessarily associated with systemic amyloidosis (Jellinger, 1977). CAA was first referred to as congophilic angiopathy by Pantelakis because of the Congo red stain, which shows a characteristic yellow-green birefringence under polarized light (Pantelakis, 1954). Affected vessels show a characteristic "double-barrel" lumen with amyloid deposition in either the inner or outer media of the "double-barrel" (Vinters, 1987). These vessels may also demonstrate fibrinoid necrosis, and microaneurysm formation may occur (Vinters, 1987). Primary CAA has been described in two distinct familial cerebral hemorrhage syndromes, Alzheimer's disease, Down syndrome, sporadic CAA, and in normal aging.

In the mid-1970s several authors noted the association of CAA with intracerebral hemorrhage. Torack reported three patients who developed intracerebral hemorrhage after surgical procedures who had evidence of amyloid deposition in intracerebral vessels (Torack, 1975). Two of these patients had undergone shunt placement for treatment of dementia. Jellinger described the neuropathological findings in 15 cases of CAA from two large autopsy series (Jellinger, 1977). One series consisted of 400 consecutive nontraumatic cerebral hemorrhage cases of which eight (2%) were cases of "atypical" hematomas associated with congophilic angiopathy. In another series of 1,010 subjects with dementia over the age of 55, seven cases of intracerebral hemorrhage were identified. Ojemann and Heros (1983) estimated that CAA accounts for 5% to 10% of primary nontraumatic brain hemorrhages.

In all 15 subjects with CAA from both of Jellinger's series (Jellinger, 1977), there were amyloid deposits in the pial-cortical arteries, arterioles, and intracerebral capillaries identified by positive Congo red staining. However, the pathological features differed somewhat between subjects who were normotensive and those with coexistent hypertension. In the seven normotensive subjects, only the amyloid change was seen in the small arteries and arterioles, whereas in the eight subjects with hypertension, hyalin angiopathy was also present. In these latter cases, there was frequent duplication of the wall of small pial arteries and perforating cortical vessels often accompanied by onion skin intimal proliferation or thickening of the vessel walls with deposition of hyalin or fibrinoid material with subsequent fibrosis. Only one of the series showed extracerebral amyloid.

CAA is seen frequently in patients with Alzheimer's disease. Mandybur (1975) found evidence of CAA in 13 of 15 patients with pathological evidence of Alzheimer's disease (senile plaques and neurofibrillary tangles) and clinical dementia. He found that amyloid deposition in small cerebral arteries is associated with the presence of senile plaques of the type seen in Alzheimer's disease but not with the presence of neurofibrillary tangles (Mandybur, 1975). In the latter series of Jellinger described earlier, of 92 cases of Alzheimer's disease identified neuropathologically, two had coincident intracerebral hemorrhage and 72% showed amyloid angiopathy (Jellinger, 1977). Glenner and colleagues (1981) found evidence of CAA in 89% of patients with pathologically proven Alzheimer's disease. They proposed that CAA was a major feature in the pathogenesis of Alzheimer's disease (Glenner et al., 1981). Okazaki and colleagues (1979) examined the brains of 23 patients with primary CAA. Ten of the patients were demented. In all 23 cases, there were varying amounts of perivascular or independent senile plaques in the cerebral cortex. However, neurofibrillary tangles were absent or limited to the hippocampal region in all but two cases.

Icelandic Amyloid Angiopathy

The clinical description of familial intracerebral hemorrhage in several Icelandic families was first reported by Árnason in 1935 (Árnason, 1935). He studied 1,003 individuals from 17 families and designated 10 of the families as "diseased" and 7 as "healthy" (Gudmundsson et al., 1972). In the diseased families, cerebral hemorrhage was the cause of death in 79 cases, whereas only 5 cases of death related to cerebral hemorrhage were found in the healthy families. In the diseased families, 59% of those with cerebral hemorrhage died before age 40 and only 5.4% died at age 70 years or greater. In contrast, four of five of those dying of cerebral hemorrhage in the healthy families died after age 70 (Gudmundsson et al., 1972). Árnason proposed that a single dominant gene could account for this disease.

Gudmundsson and colleagues (1972), with a detailed pedigree and description of the neuropathological findings, reported further examination of one of these families. The pedigree included 159 individuals in five generations with 18 cases of definite cerebral hemorrhage identified and 4 deceased individuals with probable cerebral hemorrhage determined by interviewing family members. Of those with definite cerebral hemorrhage, seven were women. None of the patients showed signs of hypertension. The patients survived from several hours up to 13 years after the first hemorrhage. Many of the patients had recurring hemorrhages with progressive loss of mental functions.

Autopsy findings were reported on five of these patients. Congo red staining identified green birefringent material within the walls of cerebral arteries that also stained a homogeneous red color with hematoxylin, thus indicating the presence of amyloid deposition. Senile plaques were not seen. Gudmundsson and colleagues (1972) proposed that intracerebral hemorrhage in this family was a hereditary illness with a dominant pattern of transmission that exhibited complete penetrance in all except for two cases. In addition they proposed that the amyloid deposit occurred first and that this then led to a weakening of the arterial wall with subsequent rupture.

More recently, it has been determined that the Icelandic amyloid angiopathy does have an autosomal dominant pattern of inheritance (Jensson et al., 1987). Patients with this disease typically present with the onset of intracerebral hemorrhage or thrombotic stroke. The age of onset is usually less than 40 years. Amino acid sequence analysis of amyloid fibrils isolated from leptomeningeal blood vessels obtained at autopsy from three patients with HCHWA-I (hereditary cerebral hemorrhage with amyloidosis, Icelandic type) showed homology to human gamma trace protein (Cohen et al., 1983). Gamma trace was later identified as a cysteine proteinase inhibitor and was given the name cystatin C (Barrett et al., 1984). Further neuropathological evaluation of brains from these patients has identified cystatin C immunoreactivity which co-localizes with hyalinization on light microscopy characterized by congophilic staining and green birefringence under polarized light (Jensson et al., 1987). It is now thought that the amyloid deposited in cerebral vessels in HCHWA-I is a variant of cystatin C with an amino acid substitution of glutamine for leucine at position 58 and beginning at the 11th amino terminal residue (Ghiso et al., 1986). Levy and colleagues (1989) isolated the gene encoding cystatin C from genomic DNA libraries made from normal tissue and from the brain of an Icelandic woman with HCHWA who died at 38 years of age. The gene from the Icelandic woman was identical except for a point mutation (substitution of a CAG instead of CTG in the normal gene) in the second exon. Figure 4 shows double-labled immunohistochemical staining of archival autopsy brain tissue from a patient with HCHWA-I or what Wang and colleagues (Wang et al., 1997) call hereditary cystatin C amyloid angiopathy (HCCAA). Cystatin C amyloid deposition is especially prominent in the vascular adventitia of leptomeningeal arteries (Figure 4, black staining), whereas cystatin C was found mainly in the media of affected vessel walls in the brain parenchyma (not shown).

Hereditary Cerebral Hemorrhage with Amyloidosis, Dutch Type

Wattendorff and colleagues (1982) described 11 patients in two generations of a single Dutch family with intracerebral hemorrhage. Microscopic examination was performed on six autopsy cases and one biopsy specimen from members with proven cerebral hemorrhage. There was hyalin thickening of the walls of small meningeal and cortical arteries in each. In five of these samples, amyloid was present in these hyalinized vessels. The mode of inheritance was thought to be

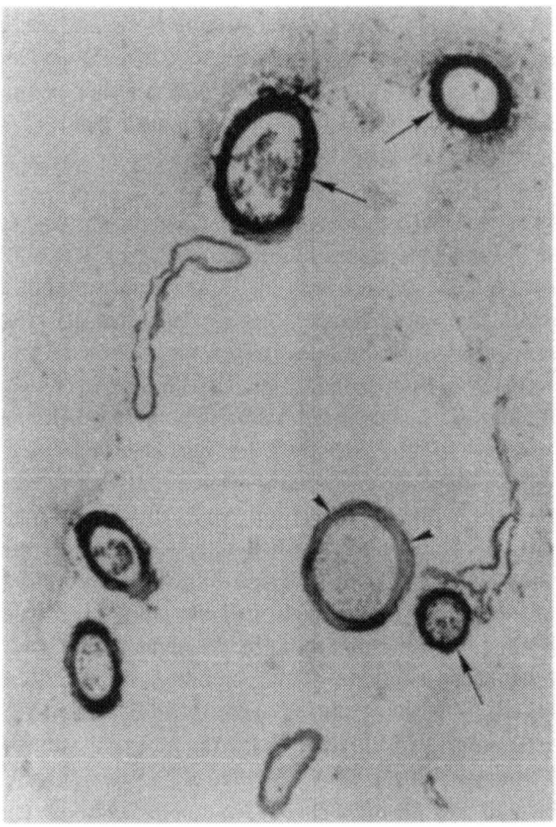

Figure 4. Double-label immunohistochemical stain of cerebral leptomeningeal arteries from a patient with HCHWA-I. Cystatin C amyloid positive staining is black (arrows) and smooth muscle actin positive staining is gray (arrow heads) (×120).

autosomal dominant and the authors speculated that this represented an arteriopathy similar to that described by Gudmunsson and colleagues (1972). Levy and colleagues (1990) identified a point mutation at position 1852 of the amyloid precursor protein gene in four patients with hereditary cerebral hemorrhage with amyloidosis, Dutch type (HCHWA-D). They later confirmed that this single point mutation was present in 22 HCHWA-D patients from three pedigrees whereas it was absent in nonaffected family members and randomly selected normal Dutch individuals (Bakker et al., 1991). The point mutation in the amyloid precursor protein gene leads to a single amino acid substitution of glutamine for glutamic acid at position 22 of the β-amyloid (Aβ) protein (Bakker et al., 1991).

The mutant form of Aβ protein that causes HCHWA-D has been shown to have an accelerated fibril formation in vitro compared to the wildtype Aβ (Wisniewski et al., 1991). The $A\beta_{1-42}$ protein has been shown to be toxic to human cultured smooth muscle cells and endothelial cells as well as to neurons (Mattson, 1997). Blanc and colleagues have shown that subtoxic levels of $A\beta_{25-35}$ causes increased permeability across endothelial cell monolayer cultures and impaired glucose transport, whereas toxic levels cause apoptotic cell death (Blanc et al., 1997). Kalaria (1977) demonstrated a marked loss of smooth muscle cells in larger vessels and absence or attenuation of capillary endothelium in regions of neocortex with abundant Aβ deposits in subjects with and without Alzheimer's disease. Freshly solubilized $A\beta_{1-42}$ causes degeneration of human leptomeningeal smooth muscle cells in culture (Davis-Salinas et al., 1995). Interestingly, the application of exogenous $A\beta_{1-42}$ led to a marked increase in cellular amyloid β precursor protein (AβPP) as well. Thomas and colleagues found that Aβ interferes with endothelium-dependent smooth muscle cell relaxation in rat aortic ring preparations in vitro and also caused degeneration of endothelial cells demonstrated by electron microscopy (Thomas et al., 1996). The shorter $A\beta_{1-40}$ protein, however, does not produce these responses (Davis-Salinas and Van Nostrand, 1995). Davis and colleagues have shown that the HCHWA-D mutant variety of the 1-40 Aβ protein induces marked toxicity in human leptomeningeal smooth muscle cells in vitro and leads to cellular degeneration and marked increase in intracellular levels of Aβ peptide (Davis and Van Nostrand, 1996).

HCHWA-D and familial Alzheimer's disease have some similarities. They are both autosomal dominant disorders, both are associated with progressive cognitive decline, and both are associated with cerebral deposition of Aβ protein. However, there are distinct differences in the two diseases. The cognitive changes in HCHWA-D are associated with the strokes and the dementia is a subcortical type unlike the cortical dementia seen in Alzheimer's disease. The amyloid deposition in HCHWA-D is primarily in the cerebral vessels and in early parenchymal plagues, whereas the deposition in Alzheimer's disease is seen primarily in senile plaques and neurofibrillary tangles (Wisniewski and Frangione, 1992). Though the early plaques are a frequent feature in HCHWA-D, neurofibrillary tangles are seen rarely and then are limited to the hippocampus (Okazaki et al., 1979).

HERITABLE HEMORRHAGIC STROKE SYNDROMES

Familial Cerebral Aneurysms

There are two lines of evidence that suggest the possibility of a genetic basis for some cerebral aneurysms (Schievink, 1997b). These are the familial aggregation of cerebral aneurysms and the association of cerebral aneurysms with hereditary connective tissue disorders. Schievink reported the pathological findings of 28 patients with aneurysmal subarachnoid hemorrhage (SAH) undergoing autopsy over a 3-year period (Schievink et al., 1997a). Three of these had at least one first-degree relative with intracranial aneurysms documented by angiography, surgery, or postmortem examination. Two of the three patients with a family history of intracranial aneurysms showed degeneration of elastic fibers and increased ground substance in the media of systemic arteries. None of these changes were seen in the 25 patients without first-degree relatives with intracranial aneurysms. The authors state that these findings suggest an underlying arteriopathy involving the tunica media in familial cerebral aneurysms but not in sporadic aneurysmal disease.

Vascular Malformations

Pettigrew and colleagues (1989) reported a kindred of 75 individuals with hereditary cavernous hemangiomas. The most frequent presenting symptoms were seizures and headache. Only four patients required surgical resection of the cavernous hemangioma and hematoma and only two patients died of hemorrhagic complications of the hemangiomas.

HERITABLE HEMATOLOGICAL DISORDERS ASSOCIATED WITH STROKE

Hematological disorders can be the primary cause of stroke in about 1% of all ischemic strokes and in up to 4% of stroke in young patients (Hart and Kanter, 1990). Stroke caused by genetic hematological etiologies can be the result of hemoglobinopathies, hereditary deficiencies of coagulation inhibitors, abnormalities in clotting factors, or hereditary abnormalities of fibrinolysis. Table 6 lists some of the more commonly seen hematological disorders, their pattern of inheritance, and their relative association with cerebrovascular disease.

Hemoglobinopathies

Sickle cell anemia is an autosomal recessive hemoglobinopathy caused by a single point substitution of thymine for adenine in the DNA codon, resulting in a substitution of valine for glutamic acid in the sixth amino acid position of β globin molecules (Beutler, 1995). Deoxygenation of blood causes molecular aggregation of sickle hemoglobin (HbS), which forms a gel and thereby distorts red blood cells

Table 6. Hereditary Hematological Disorders Associated with Stroke

Disorder	*Inheritance*	*Stroke Type*	*Association with Stroke*
Hemoglobinopathies			
Sickle cell anemia	AR	Ischemic, hemorrhagic	~15% prevalence
Sickle cell trait	AR	Ischemic	≅ to general black population
Coagulopathies			
Protein S deficiency	AD	Ischemic, CVT	Very rare
Protein C deficiency	AD	Ischemic, CVT	Very rare
Antithrombin III deficiency	AD	Ischemic, CVT	Rare
Factor V Leiden	AD	CVT	Rare to nonexistent
Clotting Factor Abnormalities			
Factor VII deficiency	XLR	Hemorrhagic	Uncommon
Factor VIII deficiency	AR	Hemorrhagic	Common
Factor IX deficiency	XLR	Hemorrhagic	Common
Factor X deficiency	AR	Hemorrhagic	Occasional
Factor XI deficiency	AR	Hemorrhagic	Rare
Factor XIII deficiency	AR	Hemorrhagic	Common
Abnormalities in Fibrinolysis			
Dysfibrinogenemia	AD, AR	Ischemic	Rare

Note: AD, autosomal dominant; AR, autosomal recessive; CVT, cerebral vein thrombosis; XLR, X-linked recessive.

giving them a sickle appearance. Approximately 8% of African-Americans are heterozygous for HbS and about 1 in 650 will have sickle cell anemia (Beutler, 1995). It has been estimated that up to 10% of patients with homozygous sickle cell disease will develop strokes (Powars et al., 1978; Pavlakis et al., 1989). Approximately 75% of these are the result of occlusive disease and 25% are caused by intracranial hemorrhage (Adams et al., 1998). Many patients with sickle cell anemia have clinically silent strokes (Pavlakis et al., 1988). A few isolated case reports of ischemic stroke in patients with heterozygous HbS have been published (Greenberg and Massey, 1985; Reyes, 1989), but it is suspected that the incidence of stroke in these patients is no greater than that seen in the African-American population in general (Hart and Kanter, 1990).

Hereditary Deficiencies of Coagulation Inhibitors

Deficiencies of antithrombin III (AT-III), protein C, protein S, and the factor V Leiden mutation, and dysfibrinogenemia are inherited disorders that predispose to thrombosis. AT-III is a protease inhibitor that inhibits thrombin and factor Xa (Hirsh et al., 1994). Protein C is a vitamin K–dependent protein that is activated by thrombin and thereby inactivates activated factor V and activated factor VII on platelet and endothelial surfaces (Hirsh et al., 1994). It also stimulates fibrinolysis. Protein S is a vitamin K–dependent protein that is a required cofactor for activation of protein C. Thus, all three of these proteins are major physiological inhibitors of coagulation and deficiencies in either results in a tendency for thrombosis. By far, the most common clinical presentations of these so-called thrombophilias are venous thrombosis and pulmonary embolism. There have been a few scattered case reports of ischemic stroke thought to be caused by AT-III (Martinelli et al., 1993; Okajima et al., 1993), protein C (Grewal and Goldberg, 1990; Kato et al., 1995), or protein S (Girolami et al., 1989; Davous et al., 1990) deficiency. Martinez and colleagues (1993) measured AT-III, protein C, and protein S in 60 patients with acute ischemic stroke under the age of 45 years at the time of hospital admission and 3 months later. They found protein C deficiency in three, protein S deficiency in two, and AT-III deficiency in five. Of these, one protein S–deficient and two AT-III–deficient patients were thought to have an acquired defect whereas all others were thought to have a heterozygous genetic defect. There have been no case-control studies or large consecutive clinical series to specifically determine the incidence of ischemic stroke caused by AT-III, protein C, or protein S deficiency. Therefore, our knowledge of such a causal relationship is limited to these anecdotal case reports and an etiological role has not been firmly established (Hart and Kanter, 1990; Tatlisumak and Fisher, 1996). Numerous reports now implicate AT-III (Tuite et al., 1993) and protein C (Wintzen et al., 1985; Deschiens et al., 1996) deficiencies as independent risk factors for cerebral vein thrombosis.

A single point mutation of factor V with arginine 506 replaced by glutamine was called factor V Leiden (Bertina et al., 1994; Dahlback, 1995). The presence of this point mutation alters the factor V molecule at the location where activated protein C would ordinarily cleave and inactivate factor V (Longstreth et al., 1998). This leads to activated protein C resistance and has been associated with an increased incidence in venous thrombosis and pulmonary embolism (Samama et al., 1996). The factor V Leiden gene mutation is present in approximately 6% of the U.S. population (Ridker et al., 1995) and therefore is the most frequent cause of thrombophilia (Samama et al., 1996). Simioni and colleagues (1995) published a report of three cases of acute ischemic stroke in patients with activated protein C resistance due to the factor V Leiden mutation. Bontempo and colleagues (1997) reported the results of 1,376 consecutive factor V Leiden mutation assays performed in a single coagulation laboratory. They found 270 subjects who were positive for the mutation, with 12 being homozygous and 258 heterozygous. They were able to obtain clinical historical information either from the ordering physician or the patient on 166 subjects with the identified mutation. Of these 114 (69%) had thrombotic events including 75 (66%) with deep venous thrombosis, 31 (27%) with pulmonary embolus, and 11 (10%) with stroke or TIA (Bontempo et al., 1997). Nowak-Göttl and colleagues (1996) assayed blood from 19 children with venous thrombosis, 18 with arterial thrombosis (16 with ischemic stroke), and 196 age-matched healthy controls. They found 38% of the children with arterial thrombosis had the factor V Leiden gene mutation compared to 52% of those with venous thrombosis and 5.1% of controls. Since these earlier reports there have been several large series that show no statistical association between factor V Leiden mutation and ischemic stroke (Ridker et al., 1995; van der Bom et al., 1996; Haan et al., 1997; Iniesta et al., 1997; Sanchez et al., 1997; Cushman et al., 1998; Longstreth et al., 1998). Combining the results from these seven studies, these investigators found the factor V Leiden gene in 37 of 783 (4.7%) subjects with either an ischemic stroke or transient ischemic attack, compared to 110 of 1,964 (5.6%) controls. However, there does appear to be an association of the factor V Leiden mutation with cerebral venous thrombosis (Thomas et al., 1994; Zuber et al., 1996).

Abnormalities in Clotting Factors

Deficiencies of factors VII, VIII, IX, X, XI, and XIII have all been reported to be associated with hemorrhagic stroke (Natowicz and Kelley, 1987) (Table 6). Of these, factor VIII deficiency is the most common, affecting about 1 in 10,000 male children (Natowicz and Kelley, 1987). About 7.5% of patients with hemophilia (factor VIII deficiency) can be expected to suffer an intracranial hemorrhage (de Tezanos et al., 1992).

Abnormalities of Fibrinolysis

Dysfibrinogenemia refers to the presence of abnormal fibrinogen molecules in plasma (McDonagh et al., 1994). There have been about 300 varieties of congenital dysfibrinogenemia discovered thus far, with identification of 83 structural defects (Martinez, 1997). Most patients with dysfibrinogenemia are asymptomatic, although both bleeding and thrombosis have been described in several varieties. In a recent review of the 250 reported cases of familial dysfibrinogenemia, Haverkate and Samama (1995) determined that 55% are asymptomatic, 25% have bleeding tendency, and 20% have thrombotic complications. Again, deep venous thrombosis and pulmonary embolism are the most common thrombotic disorders seen with dysfibrinogenemia; however, fibrinogens Haifa and Copenhagen have been associated with stroke caused by arterial thrombosis. Cases of intracerebral hemorrhage (al-Fawaz and Gader, 1992) and ischemic stroke from carotid artery thrombosis (Quattrone et al., 1983) have been reported to occur in patients with congenital dysfibrinogenemia.

OTHER HERITABLE DISORDERS ASSOCIATED WITH STROKE

Heritable Connective Tissue Disorders

The heritable connective tissue disorders pseudoxanthoma elasticum, Ehlers-Danlos syndrome type IV, Marfan's syndrome, and osteogenesis imperfecta have all been shown to be associated with an increased risk of stroke. Polycystic kidney disease has a strong association with intracranial aneurysm formation. It has been estimated that up to 5% of intracranial arterial aneurysms occur in association with one of the heritable connective tissue diseases (Schievink, 1997a).

Pseudoxanthoma elasticum (PXE) is a hereditary abnormality of connective and elastic tissue with both autosomal dominant and recessive patterns of inheritance and an estimated prevalence of 1 in 160,000 (Lebwohl et al., 1982). Initially the skin becomes thickened with yellowish papules and plaques, but with time becomes redundant and loose. Arteries become tortuous with premature calcification, intimal thickening, and saccular and fusiform aneurysms (Caplan, 1993). There is frequent association with cerebral artery aneurysm formation (Munyer and Margulis, 1981) and premature arterial occlusive disease of the large cerebral vessels (Goto, 1975). There have also been case reports of carotid-cavernous sinus fistula formation (Rios-Montenegro et al., 1972) and arteriovenous malformations (Chalk et al., 1989).

Ehlers-Danlos syndrome (EDS) includes a heterogeneous group of genetic disorders with different but overlapping clinical presentations that all share a common pathophysiological mechanism of defects in collagen (Caplan, 1993). EDS type IV is an autosomal dominant disorder caused by mutations of the COL3A1 gene which encodes for type III procollagen. Thus, patients with this disorder will have a reduction of type III collagen, which has been shown to cause marked fragility of cerebral arteries (Pope et al., 1991). EDS type IV is associated with carotid-cavernous fistulae, intracranial aneurysms, and cervical artery dissections (Schievink et al., 1990, 1994; Mattar et al., 1994). North and colleagues (1995) reviewed clinical records of 202 patients with EDS type IV. They identified 19 in this group who had cerebrovascular complications, which included subarachnoid hemorrhage from ruptured cerebral aneurysms, spontaneous carotid-cavernous sinus fistulae, and dissection of cervical arteries. The mean age of presentation with these cerebrovascular events was 28.3 years (range: 17–48).

Marfan's syndrome is a pleiotropic autosomal dominant disorder of connective tissue. With an estimated occurrence of 4 to 6 per 100,000 individuals it is the most common of the heritable connective tissue disorders (Caplan, 1993). Clinically, these patients have long thin limbs, arachnodactyly, pectus excavatum chest deformities, joint laxity, and subluxation of the ocular lens. Marfan's syndrome is caused by a variety of mutations of the fibrillin-1 gene on chromosome 15q21.1 (Dietz and Pyeritz, 1995). Fibrillin is the major constituent of extracellular microfibrils, which are primary components of collagen found throughout the body (Hayward and Brock, 1997). Frequent vascular complications of Marfan's syndrome are mitral valve prolapse, and aortic root dilatation with aortic aneurysms and dissection. Cerebral (both intracranial and extracranial) artery aneurysm formation (Hainsworth and Mendelow, 1991; Ohyama et al., 1992) and arterial dissections (Schievink et al., 1994) have also been reported in Marfan's syndrome. Schievink and colleagues (1997b) reported the autopsy findings of seven patients with Marfan's syndrome. Two of the seven had intracranial aneurysms. The first of these, a 32-year-old man had a solitary unruptured saccular supraclinoid carotid aneurysm. The other, a 20-year-old man, had four aneurysms, two ruptured and two unruptured. On microscopic examination, both patients had fragmentation of the internal elastic lamina and the second also demonstrated intimal proliferation and medial degeneration. The authors proposed that the findings confirmed an association of cerebral arterial aneurysms with Marfan's syndrome. Van den Berg and colleagues reviewed the records of 135 patients diagnosed with Marfan's syndrome followed in a Marfan's clinic and 826 patients with a diagnosis of cerebral aneurysm or subarachnoid hemorrhage who were admitted to the neurology or neurosurgery services of the same hospital over a 2-year period (van den Berg et al., 1996). None of the patients visiting the Marfan's clinic had a symptomatic cerebral aneurysm and none of them were admitted to the hospital with subarachnoid hemorrhage.

Osteogenesis imperfecta (OI) is a group of inherited disorders involving the connective tissue that causes a generalized decrease in bone mass (osteopenia) (Prockop et al., 1998). There are four known types. Type I OI is the mildest form, has an autosomal dominant pattern of inheritance, and has an incidence of about 1 in 30,000. Types II, III, and IV are much more severe, have an overall incidence of about 1 in 20,000, and are sporadic or rarely autosomal recessive, autosomal recessive or dominant, or autosomal dominant, respectively (Prockop et al., 1998). Cerebrovascular complications are much less likely to occur in OI than in PXE or EDS type IV (Schievink et al., 1994). However, there are case reports of multiple cervical artery dissections (Mayer et al., 1996), spontaneous carotid-cavernous sinus fistulae (de Campos et al., 1982), and subarachnoid hemorrhage as a result of ruptured cerebral aneurysm (Narvaez et al., 1996).

Autosomal dominant polycystic kidney disease (ADPKD) accounts for approximately 10% of all cases of end-stage renal disease (ESRD) (Asplin and Coe, 1998). The penetrance is nearly 100% in carriers surviving into the eighth decade and the prevalence is estimated to be 1 in 300 to 1 in 1,000 in the United States. ADPKD is a genetically heterogeneous disorder with three different forms identified (Asplin and Coe, 1998). ADPKD-1 is the most common form accounting for about 90% of cases. The gene for ADPKD-1 has been localized to the short arm of chromosome 16 (van Dijk et al., 1995) and encodes for an as yet unknown 4,300-amino acid protein (Asplin and Coe, 1998). In a linkage analysis study by Torra and colleagues (1996), examination of 336 subjects from 49 families with ADPKD showed that ADPKD-1 had an average age of diagnosis of 27.4 years, age of onset of ESRD of 53.4 years, and age of diagnosis of hypertension of 34.8 years. ADPKD-2, which localizes to the short arm of chromosome 4 (van Dijk et al., 1995), appears to be a milder form of disease with an average age of diagnosis of 41.4 years, age of onset of ESRD of 72.7 years, and age of diagnosis of hypertension of 49.7 years (Torra et al., 1996). There is a strong association with intracranial aneurysms and subarachnoid hemorrhage in both types (van Dijk et al., 1995). The estimated prevalence of asymptomatic aneurysms in ADPKD is about 15% with an annual rupture rate of 1.6% (Butler et al., 1996). Case reports of cranial artery dissections (Larranaga et al., 1995) and arteriovenous malformations (Tsuruda et al., 1996) have been reported in association with ADPKD.

Neurocutaneous Syndromes

Neurofibromatosis I (NF1) is an autosomal dominant disorder and is the most common of the neurocutaneous syndromes occurring in about 1 in 3,000 persons (Love, 1994). There are several mechanisms by which NF1 can cause stroke. Arteriovenous fistulae have been described in patients with NF1 (Cluzel et al., 1994). Neurofibromatosis has also been reported to be associated with intracranial arterial occlusive disease and with intracranial aneurysm formation (Schievink et al., 1994). Tuberous sclerosis is an autosomal dominant phakomatosis that has been reported to cause cerebrovascular disease of several types. Rare cases of ischemic stroke have been reported in patients with tuberous sclerosis secondary to emboli from cardiac tumors (Kandt et al., 1985; Konkol et al., 1986) and the rare occurrence of cerebral aneurysms have been reported in patients with tuberous sclerosis (Blumenkopf and Huggins, 1985; Brill et al., 1985). Osler-Rendu-Weber syndrome (hereditary hemorrhagic telangiectasia) is an autosomal dominant disorder that is associated with small arteriovenous vascular anomalies that affect the skin, mucous membranes, gastrointestinal tract, genitourinary tract, and occasionally the brain or spinal cord (Adams and Victor, 1993).

Hereditary Disorders of Metabolism Associated with Stroke

Mitochondrial myopathy, encephalopathy, lactic acidosis, and stroke-like episodes (MELAS) syndrome, Fabry's disease, and Leigh's disease have each been associated with increased frequency of stroke. MELAS is one of the mitochondrial encephalomyopathies caused by point mutations in the mitochondrial DNA (Pavlakis et al., 1984). The most common of these mutations is an $A \rightarrow G$ substitution at position 3243 of the mitochondrial tRNALeu (UUR) gene (Flierl et al., 1997). Seizures, headaches, and intellectual deterioration characterize the clinical presentation of MELAS, with a frequent occurrence of ischemic cortical infarctions, which primarily affect the parieto-occipital or cerebellar region (Caplan, 1993).

Fabry's disease is an X-linked disorder of glycosphingolipid metabolism that causes strokes, renal failure, painful dysesthesias, and cutaneous angiokeratomas (Caplan, 1993). The genetic defect causes a deficiency in α-galactosidase A, which leads to the deposition of glycophingolipids in blood vessel walls (Mendez et al., 1997). Most cases are homozygous men but heterozygous women are occasionally affected (Caplan, 1993). Utsumi and colleagues (1997) reviewed the records of 60 patients with Fabry's disease (45 homozygous men and 15 heterozygous women) and found 7 who had arterial thrombosis, 6 of whom had ischemic strokes.

Leigh's disease (subacute necrotizing encephalomyelopathy) is a heterogeneous group of disorders caused by point mutations of the mitochondrial DNA. Deficiencies in the respiratory chain complexes have been found in patients with Leigh's syndrome, with involvement of complexes IV and I being the most common etiologies (Morris et al., 1996). Several forms have been described, including both autosomal recessive or X-linked recessive varieties (Love, 1994). The X-linked type has been found to be caused by a T → G mutation of the 8993 gene encoding ATPase 6 (Rahman et al., 1996) and by a valine tRNA mutation at base pair 1644 in an adult-onset variety (Chalmers et al., 1997). There is evidence also of a less severely effected form of Leigh's disease with a T → C mutation at the 8993 gene location (Santorelli et al., 1996). Several subjects with de novo mutations have been identified by screening other family members for this defect (Tulinius et al., 1995; Seller et al., 1997). Heteroplasmic deletions have also been described as a cause of Leigh's disease (Rahman et al., 1996). Clinically, these patients have early-onset stroke-like episodes in addition to muscle weakness, ophthalmoplegia with ragged red fibers, retinitis pigmentosa, and progressive myoclonic seizures (Zeviani et al., 1996).

CONCLUDING REMARKS

There are myriad genetic causes for ischemic and hemorrhagic stroke. Of these, three can be considered primary stroke syndromes, producing brain injury without significant involvement of other organ systems. These are CADASIL and the two hereditary cerebral amyloid angiopathies—the Dutch and Icelandic types. These diseases serve as prototypical examples of genetic influence on stroke with identified genetic loci and specific pathological findings. The numerous other secondary causes of stroke are caused by other medical conditions. Of these, the most important are hypertension, hyperlipidemias and homocystinemia, a variety of hematological diseases, the heritable connective tissue disorders, and several genetic metabolic disorders. The common pathophysiological mechanisms for the hereditary stroke syndromes are worsening of atherosclerotic processes, mechanical disruption of the structural integrity of cerebral blood vessels, or interference with normal thrombotic and hemostatic processes of the coagulation cascade.

SUMMARY

Stroke is the result of irreversible neuronal injury resulting from interruption of blood flow to a part of the brain or spinal cord. Modifiable risk factors (e.g., hypertension, diabetes, and smoking) and nonmodifiable risk factors (e.g., age, race, and sex) for stroke have been identified. Accumulating evidence indicates that these traditional stroke risk factors have some genetic determinants. In addition, many genetic diseases may result in stroke in a subgroup of patients (e.g., hemoglobinopathies, coagulopathies, connective tissue disorders, and metabolic disorders). There are also several genetic diseases in which stroke is a key component of the clinical presentation, such as CADASIL, and two distinct but similar forms of cerebral amyloid angiopathy. Identification of specific mechanisms involved in genetically determined stroke syndromes may help elucidate the pathogenesis of more common multifactorially determined stroke syndromes, and may also suggest new strategies for stroke prevention and treatment.

ACKNOWLEDGMENTS

The authors thank the following individuals for their contribution: William R. Markesbery, M.D., provided photomicrographs of skin biopsy materials processed by his laboratory; Zhen Z. Wang and Harry V. Vinters provided photomicrographs of immunostained cerebral blood vessels from patients with HCHWA-I; Lawrence Goldstick, M.D., and Michelle Gaier Rush, M.D., provided clinical information on some members of the CADASIL kindred; and Sherry Williams provided editorial assistance.

REFERENCES

Abbott, R.D., Donahue, R.P., MacMahon, S.W., Reed, D.M. & Yano, K. (1987). Diabetes and the risk of stroke. The Honolulu Heart Program J.A.M.A. 257, 949–952.

Abel, L. & Dessein, A.J. (1997). The impact of host genetics on susceptibility to human infectious diseases. Curr. Opin. Immunol. 9, 509–516.

Adams, R.D. & Victor, M. (1993). Developmental diseases of the nervous system. In: Principles of Neurology, pp. 1010–1055. McGraw-Hill, New York.

Adams, R.J., McKie, V.C., Brambilla, D., Carl, E., Gallagher, D., Nichols, F.T., Roach, S., Abboud, M., Berman, B., Driscoll, C., Files, B., Hsu, L., Hurlet, A., Miller, S., Olivieri, N., Pegelow, C., Scher, C., Vichinsky, E., Wang, W., Woods, G., Kutlar, A., Wright, E., Hagner, S., Tighe, F. & Waclawiw, M.A.. (1998). Stroke prevention trial in sickle cell anemia. Control. Clin. Trials. 19, 110–129.

Ahmed, O., Neser, W. & Sharma, R. (1989). Sociodemographic indicators of stroke mortality J. Natl. Med. Assoc. 81, 653–658.

Al-Fawaz, I.M. & Gader, AM. (1992). Severe congenital dysfibrinogenemia (fibrinogen-Riyadh): a family study. Acta Haematol. 88, 194–197.

Árnason, A. (1935). Apoplexie und ihre Vererbung. Acta Psychiat. Neurol. Suppl. VII.

Asplin, J.R. & Coe, F.L. (1998). Hereditary tubular disorders. In: Harrison's Principles of Internal Medicine (Fauci, A.S., Braunwald, E., Isselbacher, K.J., Wilson, J.D., Martin, J.B., Kasper, D.L., Hauser, S.L., & Longo, D.L., eds.), pp. 1562–1569. McGraw Hill, New York.

Bakker, E., Van Broeckhoven, C., Haan, J., Voorhoeve, E., van Hul, W., Levy, E., Lieberburg, I., Carman, M.D., van Ommen, G.J. & Frangione, B. (1991). DNA diagnosis for hereditary cerebral hemorrhage with amyloidosis (Dutch type). Am. J. Hum. Genet. 49, 518–521.

Barrett, A.J., Davies, M.E. & Grubb, A. (1984). The place of human γ-trace (cystatin C) amongst the cysteine proteinase inhibitors. Biochem. Biophys. Res. Commun. 120, 631–636.

Baudrimont, M., Dubas, F., Joutel, A., Tournier-Lasserve, E. & Bousser, M.G. (1993). Autosomal dominant leukoencephalopathy and subcortical ischemic stroke: a clinicopathological study. Stroke 24, 122–125.

Benveniste, H., Drejer, J., Schousboe, A. & Diemer, N.H. (1984). Elevation of the extracellular concentrations of glutamate and aspartate in rat hippocampus during transient cerebral ischemia monitored by intracerebral microdialysis. J. Neurochem. 43, 1369–1374.

Bertina, R.M., Koeleman, B.P., Koster, T., Rosendaal, F.R., Dirven, R.J., de Ronde, H., van der Velden, P.A. & Reitsma, P.H. (1994). Mutation in blood coagulation factor V associated with resistance to activated protein C. Nature 369, 64–67.

Beutler, E. (1995). The sickle cell diseases and related disorders. In: Williams Hematology (Beutler, E., Lichtman, M.A., Coller, B.S., & Kipps, T.J., eds.), pp. 616–650. McGraw-Hill, New York.

Blanc, E.M., Toborek, M., Mark, R.J., Hennig, B. & Mattson, M.P. (1997). Amyloid β-peptide induces cell monolayer albumin permeability, impairs glucose transport, and induces apoptosis in vascular endothelial cells. J. Neurochem. 68, 1870–1881.

Blumenkopf, B. & Huggins, M.J. (1985). Tuberous sclerosis and multiple intracranial aneurysms: case report. Neurosurgery 17, 797–800.

Boers, G.H., Smals, A.G., Trijbels, F.J., Fowler, B., Bakkeren, J.A., Schoonderwaldt, H.C., Kleijer, W.J. & Kloppenborg, P.W. (1985). Heterozygosity for homocystinuria in premature peripheral and cerebral occlusive arterial disease. N. Engl. J. Med. 313, 709–715.

Bontempo, F.A., Hassett, A.C., Faruki, H., Steed, D.L., Webster, M.W. & Makaroun, M.S. (1997). The factor V Leiden mutation: spectrum of thrombotic events and laboratory evaluation. J. Vasc. Surg. 25, 271–275.

Bousser, M.G. & Tournier-Lasserve, E. (1994). Summary of the proceedings of the First International Workshop on CADASIL. Paris, May 19–21, 1993. Stroke 25, 704–707.

Brass, L.M., Isaacsohn, J.L., Merikangas, K.R. & Robinette, C.D. (1992). A study of twins and stroke. Stroke 23, 221–223.

Brattström, L., Lindgren, A., Israelsson, B., Malinow, M.R., Norrving, B., Upson, B. & Hamfelt, A. (1992). Hyperhomocysteinaemia in stroke: prevalence, cause, and relationships to type of stroke and stroke risk factors. Eur. J. Clin. Invest. 22, 214–221.

Brewer, H.B., Jr., Santamarina-Fojo, S.M. & Hoeg, J.M. (1996). Genetic dyslipoproteinemias. In: Atherosclerosis and Coronary Artery Disease (Fuster, V., Ross, R. & Topol, E.J., eds.), pp. 69–88. Lippincott-Raven, Philadelphia.

Brill, C.B., Peyster, R.G., Hoover, E.D. & Keller, M.S. (1985). Giant intracranial aneurysm in a child with tuberous sclerosis: CT demonstration. J. Comput. Assist. Tomogr. 9, 377–380.

Bucher, H.C., Griffith, L.E. & Guyatt, G.H. (1998). Effect of HMGcoA reductase inhibitors on stroke. A meta-analysis of randomized, controlled trials. Ann. Intern. Med. 128, 89–95.

Butler, W.E., Barker, F.G. & Crowell, R.M. (1996). Patients with polycystic kidney disease would benefit from routine magnetic resonance angiographic screening for intracerebral aneurysms: a decision analysis. Neurosurgery 38, 506–515.

Caplan, L.R. (1993). Stroke: A Clinical Approach, 2nd ed. Butterworth-Heinemann, Stoneham, MA.

Caplan, L.R., Gorelick, P.B. & Hier, D.B. (1986). Race, sex and occlusive cerebrovascular disease: a review. Stroke 17, 648–655.

Carson, N.A., Cusworth, D.C., Dent, C.E., Field, C.M.B., Neill, D.W. & Westall, R.G. (1963). Homocysteinuria: a new inborn error of metabolism associated with mental deficiency. Arch. Dis. Child. 38, 425–436.

Caulfield, M., Bouloux, P.M. & Munroe, P. (1997). Progress in determining the genes for hypertension, insulin resistance, and dyslipidemia. Ann. N. Y. Acad. Sci. 827, 110–117.

Chalk, J.B., Patterson, M.C. & Pender, M.P. (1989). An intracranial arteriovenous malformation and palatal myoclonus related to pseudoxanthoma elasticum. Aust. N. Z. J. Med. 19, 141–143.

Chalmers, R.M., Lamont, P.J., Nelson, I., Ellison, D.W., Thomas, N.H., Harding, A.E. & Hammans, S.R. (1997). A mitochondrial DNA tRNA(Val) point mutation associated with adult onset Leigh syndrome. Neurology 49, 589–592.

Charcot, J.-M. & Bouchard, C. (1868). Nouvelles recherches sur la pathogenie de l'hémorrhagie cérébrale. Arch. Physiol. 1, 110–127, 643–665, 725–734.

Choi, D.W. (1995). Calcium: still center-stage in hypoxic-ischemic neuronal death. Trends. Neurosci. 18, 58–60.

Clarke, R., Daly, L., Robinson, K., Naughten, E., Cahalane, S., Fowler, B. & Graham, I. (1991). Hyperhomocysteinemia: an independent risk factor for vascular disease. N. Engl. J. Med. 324, 1149–1155.

Cluzel, P., Pierot, L., Leung, A., Gaston, A., Kieffer, E. & Chiras, J. (1994). Vertebral arteriovenous fistulae in neurofibromatosis: report of two cases and review of the literature. Neuroradiology 36, 321–325.

Cohen, D.H., Feiner, H., Jensson, O. & Frangione, B. (1983). Amyloid fibril in hereditary cerebral hemmorrhage with amyloidosis (HCHWA) is related to the gastroentero-pancreatic neuroendocrine protein, gamma trace. J. Exp. Med. 158, 623–628.

Cole, F.M. & Yates, P. (1967). Intracerebral microaneurysms and small cerebrovascular lesions. Brain 90, 759–767.

Contois, J.H., Anamani, D.E. & Tsongalis, G.J. (1996). The underlying molecular mechanism of apolipoprotein E polymorphism: relationships to lipid disorders, cardiovascular disease, and Alzheimer's disease. Clin. Lab. Med. 16, 105–123.

Coonar, A.S. & McKenna, W.J. (1997). Molecular genetics of familial cardiomyopathies. Adv. Genet. 35, 285–324.

Cooper, R. & Rotimi, C. (1997). Hypertension in blacks. Am. J. Hypertens. 10, 804–812.

Corvol, P., Soubrier, F. & Jeunemaitre, X. (1997). Molecular genetics of the renin-angiotensin-aldosterone system in human hypertension. Pathol. Biol. 45, 229–239.

Coull, B.M., Malinow, M.R., Beamer, N., Sexton, G., Nordt, F. & de Garmo, P. (1990). Elevated plasma homocyst(e)ine concentration as a possible independent risk factor for stroke. Stroke 21, 572–576.

Cushman, M., Rosendaal, F.R., Psaty, B.M., Cook, E.F., Valliere, J., Kuller, L.H. & Tracy, R.P. (1998). Factor V Leiden is not a risk factor for arterial vascular disease in the elderly: results from the Cardiovascular Health Study. Thromb. Haemost. 79, 912–915.

Dahlback, B. (1995). Resistance to activate protein C, the Arg506 to Gln mutation in the factor V gene, and venous thrombosis. Functional tests and DNA-based assays, pros and cons. Thromb. Haemost. 73, 739–742.

Davis, G.K. & Roberts, D.H. (1997). Molecular genetics of the renin-angiotensin system: implications for angiotensin II receptor blockade. Pharmacol. Ther. 75, 43–50.

Davis, J. & Van Nostrand, W.E. (1996). Enhanced pathologic properties of Dutch-type mutant amyloid beta-protein. Proc. Natl. Acad. Sci. U. S. A. 93, 2996–3000.

Davis-Salinas, J. & Van Nostrand, W.E. (1995). Amyloid beta-protein aggregation nullifies its pathologic properties in cultured cerebrovascular smooth muscle cells. J. Biol. Chem. 270, 20887–20890.

Davis-Salinas, J., Saporito-Irwin, S.M., Cotman, C.W. & Van Nostrand, W.E. (1995). Amyloid beta-protein induces its own production in cultured degenerating cerebrovascular smooth muscle cells. J. Neurochem. 65, 931–934.

Davous, P., Horellou, M.H., Conard, J. & Samama, M. (1990). Cerebral infarction and familial protein S deficiency. Stroke 21, 1760–1761.

de Campos, J.M., Ferro, M.O., Burzaco, J.A. & Boixados, J.R. (1982). Spontaneous carotid-cavernous fistula in osteogenesis imperfecta. J. Neurosurg. 56, 590–593.

de Tezanos, P., Fernandez, J. & Perez, B.P. (1992). Update of 156 episodes of central nervous system bleeding in hemophiliacs. Haemostasis 22, 259–267.

Deschiens, M.A., Conard, J., Horellou, M.H., Ameri, A., Preter, M., Chedru, F., Samama, M.M. & Bousser, M.G. (1996). Coagulation studies, factor V Leiden, and anticardiolipin antibodies in 40 cases of cerebral venous thrombosis. Stroke 27, 1724–1730.

Dietz, H.C. & Pyeritz, R.E. (1995). Mutations in the human gene for fibrillin-1 (FBN1) in the Marfan syndrome and related disorders. Hum. Mol. Genet. 4, 1799–1809.

Durrington, P.N. (1995a). Familial hypercholesterolaemia. In: Hyperlipidaemia Diagnosis and Management, pp. 108–139. Butterworth-Heinemann, Oxford.

Durrington, P.N. (1995b). Genetics of lipoprotein disorders and coronary atheroma. In: Hyperlipidaemia Diagnosis and Management, pp. 361–384. Butterworth-Heinemann, Oxford.

Dyker, A.G., Weir, C.J. & Lees, K.R. (1997). Influence of cholesterol on survival after stroke: retrospective study. B. M. J. 314, 1584–1588.

Ebke, M., Dichgans, M., Bergmann, M., Voelter, H.U., Rieger, P., Gasser, T. & Schwendemann, G. (1997). CADASIL: skin biopsy allows diagnosis in early stages. Acta Neurol. Scand. 95, 351–357.

Ferguson, R.A. & Goldberg, D.M. (1997). Genetic markers of alcohol abuse. Clin. Chim. Acta 257, 199–250.

Field, L.L. & Tobias, R. (1997). Unravelling a complex trait: the genetics of insulin-dependent diabetes mellitus. Clin. Invest. Med. 20, 41–49.

Fisher, C.M. (1971). Cerebral miliary aneurysms in hypertension. Am. J. Pathol. 66, 313–330.

Fisher, C.M. (1979). Capsular infarcts: the underlying vascular lesions. Arch. Neurol. 36, 65–73.

Fisher, C.M. (1982). Lacunar strokes and infarcts: a review. Neurology 32, 871–876.

Fisher, M., Martin, A., Cosgrove, M. & Norris, J.W.. (1993). The NASCET-ACAS plaque project. Stroke 24 (suppl. 1), I-24-I-25.
Flierl, A., Reichmann, H. & Seibel, P. (1997). Pathophysiology of the MELAS 3243 transition mutation. J. Biol. Chem. 272, 27189–27196.
Frank, J.W., Reed, D.M., Grove, J.S. & Benfante, R. (1992). Will lowering population levels of serum cholesterol affect total mortality? Expectations from the Honolulu Heart Program. J. Clin. Epidemiol. 45, 333–346.
Ghiso, J., Jensson, O. & Frangione, B. (1986). Amyloid fibrils in hereditary cerebral hemorrhage with amyloidosis of Icelandic type is a variant of gamma-trace basic protein (cystatin C). Proc. Natl. Acad. Sci. U. S. A. 83, 2974–2978.
Girolami, A., Simioni, P., Lazzaro, A.R. & Cordiano, I. (1989). Severe arterial cerebral thrombosis in a patient with protein S deficiency (moderately reduced total and markedly reduced free protein S): a family study. Thromb. Haemost. 61, 144–147.
Glenner, G.G., Henry, J.H. & Fujihara, S. (1981). Congophilic angiopathy in the pathogenesis of Alzheimer's degeneration. Ann. Pathol. 1, 120–129.
Gorelick, P.B., Schneck, M., Berglund, L.F., Feinberg, W. & Goldstone, J. (1997). Status of lipids as a risk factor for stroke. Neuroepidemiology. 16, 107–115.
Goto, K. (1975). Involvement of central nervous system in pseudoxanthoma elasticum Folia. Psychiatr. Neurol. Jpn. 29, 263–277.
Gray, F., Robert, F., Labrecque, R., Chretien, F., Baudrimont, M., Fallet-Bianco, C., Mikol, J. & Vinters, H.V. (1994). Autosomal dominant arteriopathic leuko-encephalopathy and Alzheimer's disease. Neuropathol. Appl. Neurobiol. 20, 22–30.
Greenberg, J. & Massey, E.W. (1985). Cerebral infarction in sickle cell trait. Ann. Neurol. 18, 354–355.
Grewal, R.P. & Goldberg, M.A. (1990). Stroke in protein C deficiency. Am. J. Med. 89, 538–539.
Gudmundsson, G., Hallgrímsson, J., Jónasson, T.A. & Bjarnason, O. (1972). Hereditary cerebral hæmorrhage with amyloidosis. Brain 95, 387–404.
Gutierrez-Molina, M., Caminero, R.A., Martinez, G.C., Arpa, G.J., Morales, B.C. & Amer, G. (1994). Small arterial granular degeneration in familial Binswanger's syndrome. Acta Neuropathol. (Berl.) 87, 98–105.
Haan, J., Kappelle, L.J., de Ronde, H., Ferrari, M.D. & Bertina, R.M. (1997). The factor V Leiden mutation (R506Q) is not a major risk factor for migrainous cerebral infarction. Cephalalgia 17, 605–607.
Hainsworth, P.J. & Mendelow, A.D. (1991). Giant intracranial aneurysm associated with Marfan's syndrome: a case report. J. Neurol. Neurosurg. Psychiatry 54, 471–472.
Harker, L.A., Slichter, S.J., Scott, C.R. & Ross, R. (1974). Homocysteinemia. Vascular injury and arterial thrombosis. N. Engl. J. Med. 291, 537–543.
Hart, R.G. & Kanter, M.C. (1990). Hematologic disorders and ischemic stroke: a selected review. Stroke 21, 1111–1121.
Haverkate, F. & Samama, M. (1995). Familial dysfibrinogenemia and thrombophilia. Report on a study of the SSC Subcommittee on Fibrinogen. Thromb. Haemost. 73, 151–161.
Hayward, C. & Brock, D.J. (1997). Fibrillin-1 mutations in Marfan syndrome and other type-1 fibrillinopathies. Hum. Mutat. 10, 415–423.
Hebert, P.R., Gaziano, J.M., Chan, K.S. & Hennekens, C.H. (1997). Cholesterol lowering with statin drugs, risk of stroke, and total mortality. An overview of randomized trials. J. A. M. A. 278, 313–321.
Higuchi, S., Arai, H., Nakagawa, T., Muramatsu, T, Sasaki, H. & Trojanowski, J.Q. (1996). The apolipoprotein E gene in Binswanger's disease and vascular dementia. Clin. Genet. 50, 459–461.
Hirsh, J., Salzman, E.W., Marder, V.J., & Colman, R.W. (1994). Overview of the thrombotic process and its therapy. In: Hemostasis and Thrombosis: Basic Principles and Clinical Practice (Colman, R.W., Hirsh, J., Marder, V.J., & Salzman, J.B., eds.), pp. 1151–1163. J.B. Lippincott, Philadelphia.

Hodis, H.N., Mack, W.J., Dunn, M., Liu, C., Selzer, R.H. & Krauss, R.M. (1997). Intermediate-density lipoproteins and progression of carotid arterial wall intima-media thickness. Circulation 95, 2022–2026.

Holme, I. (1995). Effects of lipid-lowering therapy on total and coronary mortality. Curr. Opin. Lipidol. 6, 374–378.

Howard, G., Russell, G.B., Anderson, R., Evans, G.W., Morgan, T., Howard, V.J. & Burke, G.L. (1995). Role of social class in excess black stroke mortality. Stroke 26, 1759–1763.

Howard, G., Anderson, R., Johnson, N.J., Sorlie, P., Russell, G. & Howard, V.J. (1997). Evaluation of social status as a contributing factor to the stroke belt region of the United States. Stroke 28, 936–940.

Iniesta, J.A., Corral, J., Fernandez-Pardo, J., Gonzalez-Conejero, R. & Vicente, V. (1997). Factor-V (Arg506 → Gln) mutation in ischemic cerebrovascular disease. Haemostasis 27, 105–111.

Inzitari, D., Hachinski, V.C., Taylor, D.W. & Barnett, H.J.M. (1990). Racial differences in the anterior circulation in cerebrovascular disease: How much can be explained by risk factors? Arch. Neurol. 47, 1080–1084.

Jellinger, K. (1977). Cerebrovascular amyloidosis with cerebral hemorrhage. J. Neurol. 214, 195–206.

Jen, J., Cohen, A.H., Yue, Q., Stout, J.T., Vinters, H.V., Nelson, S. & Baloh, R.W. (1997). Hereditary endotheliopathy with retinopathy, nephropathy, and stroke (HERNS). Neurology 49, 1322–1330.

Jensson, O., Gudmundsson, G., Arnason, A., Blöndal, H., Petursdottir, I., Thorsteinsson, L., Grubb, A., Löfberg, H., Cohen, D. & Frangione, B. (1987). Hereditary cystatin C (γ-trace) amyloid angiopathy of the CNS causing cerebral hemorrhage. Acta Neurol. Scand. 76, 102–114.

Joutel, A., Corpechot, C., Ducros, A., Vahedi, K., Chabriat, H., Mouton, P., Alamowitch, S., Domenga, V., Cecillion, M., Marechal, E., Maciazek, J., Vayssiere, C., Cruaud, C., Cabanis, E.A., Ruchoux, M.M., Weissenbach, J., Bach, J.F., Bousser, M.G. & Tournier-Lasserve, E. (1996). Notch3 mutations in CADASIL, a hereditary adult-onset condition causing stroke and dementia. Nature 383, 707–710.

Joutel, A., Vahedi, K., Corpechot, C., Troesch, A., Chabriat, H., Vayssiere, C., Cruaud, C., Maciazek, J., Weissenbach, J., Bousser, M.G., Bach, J.F. & Tournier-Lasserve, E. (1997). Strong clustering and stereotyped nature of Notch3 mutations in CADASIL patients. Lancet 350, 1511–1515.

Kalaria, R.N. (1997). Cerebrovascular degeneration is related to amyloid-beta protein deposition in Alzheimer's disease. Ann. N. Y. Acad. Sci. 826, 263–271.

Kandt, R.S., Gebarski, S.S. & Goetting, M.G. (1985). Tuberous sclerosis with cardiogenic cerebral embolism: magnetic resonance imaging. Neurology 35, 1223–1225.

Kato, H., Shirahama, M., Ohmori, K. & Sunaga, T. (1995). Cerebral infarction in a young adult associated with protein C deficiency. A case report. Angiology. 46, 169–173.

Kenton, E.J. (1991). Access to neurological care for minorities. Arch. Neurol. 48, 480–483.

Kessler, C., Spitzer, C., Stauske, D., Mende, S., Stadlmuller, J., Walther, R. & Rettig, R. (1997). The apolipoprotein E and beta-fibrinogen G/A-455 gene polymorphisms are associated with ischemic stroke involving large-vessel disease. Arterioscler. Thromb. Vasc. Biol. 17, 2880–2884.

Kittner, S.J., White, L.R., Losonczy, K.G., Wolf, P.A. & Hebel, J.R. (1990). Black-white differences in stroke incidence in a national sample. The contribution of hypertension and diabetes mellitus. J. A. M. A. 264, 1267–1270.

Kittner, S.J., McCarter, R.J., Sherwin, R., Sloan, M.A., Stern, B.J., Johnson, C.J., Buchholz, D., Seipp, M.J. & Price, T.R. (1993). Black-white differences in stroke risk among young adults. Stroke 24 (suppl. I), I-13-I-15.

Konkol, R.J., Walsh, E.P., Power, T. & Bresnan, M.J. (1986). Cerebral embolism resulting from an intracardiac tumor in tuberous sclerosis. Pediatr. Neurol. 2, 108–110.

Kuller, L.H. (1991). Cardiovascular diseases and stroke in African-Americans and other racial minorities in the United States. A statement for health professionals. Introduction. Circulation 83, 1463–1465.

Kunz, R., Kreutz, R., Beige, J., Distler, A. & Sharma, A.M. (1997). Association between the angiotensinogen 235T-variant and essential hypertension in whites: a systematic review and methodological appraisal. Hypertension 30, 1331–1337.

Lammie, G.A., Rakshi, J., Rossor, M.N., Harding, A.E. & Scaravilli, F. (1995). Cerebral autosomal dominant arteriopathy with subcortical infarcts and leukoencephalopathy (CADASIL)—confirmation by cerebral biopsy in 2 cases. Clin. Neuropathol. 14, 201–206.

Lammie, G.A., Brannan, F., Slattery, J. & Warlow, C. (1997). Nonhypertensive cerebral small-vessel disease. An autopsy study. Stroke 28, 2222–2229.

Lanska, D.J. & Kryscio, R. (1994). Geographic distribution of hospitalization rates, case fatality, and mortality from stroke in the United States. Neurology 44, 1541–1550.

Lanska, D.J. & Peterson, P.M. (1995). Effects of interstate migration on the geographic distribution of stroke mortality in the United States. Stroke 26, 554–561.

Lanska, D.J. & Peterson, P.M. (1996). Comparison of additive and multiplicative models of regional variation in the decline of stroke mortality in the United States. Stroke 27, 1055–1059.

Larranaga, J., Rutecki, G.W. & Whittier, F.C. (1995). Spontaneous vertebral artery dissection as a complication of autosomal dominant polycystic kidney disease. Am. J. Kidney Dis. 25, 70–74.

Lebwohl, M.G., Distefano, D., Prioleau, P.G., Uram, M., Yannuzzi, L.A. & Fleischmajer, R. (1982). Pseudoxanthoma elasticum and mitral-valve prolapse. N. Engl. J. Med. 307, 228–231.

Leung, S.Y., Ng, T.H., Yuen, S.T., Lauder, I.J. & Ho, F.C. (1993). Pattern of cerebral atherosclerosis in Hong Kong Chinese. Severity in intracranial and extracranial vessels. Stroke 24, 779–786.

Levy, E., Lopez-Otin, C., Ghiso, J., Geltner, D. & Frangione, B. (1989). Stroke in Icelandic patients with hereditary amyloid angiopathy is related to a mutation in the cystatin C gene, an inhibitor of cysteine proteases. J. Exp. Med. 169, 1771–1778.

Levy, E., Carman, M.D., Fernandez-Madrid, I.J., Power, M.D., Lieberburg, I., Van Duinen, S.G., Bots, G.T., Luyendijk, W. & Frangione, B. (1990). Mutation of the Alzheimer's disease amyloid gene in hereditary cerebral hemorrhage, Dutch type. Science 248, 1124–1126.

Lifton, R.P., Dluhy, R.G., Powers, M., Rich, G.M., Cook, S., Ulick, S. & Lalouel, J.M. (1992). A chimaeric 11-beta-hydroxylase/aldosterone synthase gene causes glucocorticoid-remediable aldosteronism and human hypertension. Nature 355, 262–265.

Longstreth, W.T. Jr., Rosendaal, F.R., Siscovick, D.S., Vos, H.L., Schwartz, S.M., Psaty, B.M., Raghunathan, T.E., Koepsell, T.D. & Reitsma, P.H. (1998). Risk of stroke in young women and two prothrombotic mutations: factor V Leiden and prothrombin gene variant (G20210A). Stroke 29, 577–580.

Love, B.B. (1994). Rare genetic disorders predisposing to stroke. In: Stroke in Children & Young Adults (Biller, J., Mathews, K.D. & Love, B.B., eds.), pp. 147–164. Butterworth-Heinemann, Newton, MA.

Malandrini, A., Carrera, P., Palmeri, S., Cavallaro, T., Fabrizi, G.M., Villanova, M., Fattapposta, M., Vismara, L., Brancolini, V., Tanganelli, P., Cali, A., Morocutti, C., Zeviani, M., Ferrari, M. & Guazzi, G.C. (1996). Clinicopathological and genetic studies of two further Italian families with cerebral autosomal dominant arteriopathy. Acta Neuropathol. (Berl.) 92, 115–122.

Mandybur, T.I. (1975). The incidence of cerebral amyloid angiopathy in Alzheimer's disease. Neurology 25, 120–126.

Manning, P.J., Lee, S.S., & Yamanaka, W.K. (1974). Coronary lipohyalinosis. Dietary induction in guinea pigs. Atherosclerosis 20, 437–445.

Martinelli, I., Moia, M., Arbini, A., Tripodi, A. & Mannucci, P.M. (1993). Ischemic stroke in congenital (type II C) defective antithrombin III. Int. J. Clin. Lab. Res. 23, 212–214.

Martinez, H.R., Rangel-Guerra, R.A. & Marfil, L.J. (1993). Ischemic stroke due to deficiency of coagulation inhibitors. Report of 10 young adults. Stroke 24, 19–25.

Martinez, J. (1997). Congenital dysfibrinogenemia. Curr. Opin. Hematol. 4, 357–365.

Mattar, S.G., Kumar, A.G. & Lumsden, A.B. (1994). Vascular complications in Ehlers-Danlos syndrome. Am. Surg. 60, 827–831.

Mattson, M.P. (1997). Cellular actions of beta-amyloid precursor protein and its soluble and fibrillogenic derivatives. Physiol. Rev. 77, 1081–1132.

Mauro, A.J., Johnson, E.S., Chikos, P.M. & Alvord, E.C., Jr. (1980). Lipohyalinosis and miliary microaneurysms causing cerebral hemorrhage in a patient with moyamoya. A clinicopathological study. Stroke 11, 405–412.

Mayer, S.A., Rubin, B.S., Starman, B.J. & Byers, P.H. (1996). Spontaneous multivessel cervical artery dissection in a patient with a substitution of alanine for glycine (G13A) in the alpha 1 (I) chain of type I collagen. Neurology 47, 552–556.

McDonagh, J., Carrell, N. & Lee, M.H. (1994). Dysfibrinogenemia and other disorders of fibrinogen structure and function. In: Hemostasis and Thrombosis: Basic Principles and Clinical Practice (Colman, R.W., Hirsh, J., Marder, V.J. & Salzman, E.W., eds.), pp. 314–334. J.B. Lippincott, Philadelphia.

McKusick, V.A. (1980). Medical genetics: general considerations. In: The Principle and Practice of Medicine (Harvey, A.M., Johns, R.J., McKusick, V.A., Owens, A.H. & Ross, R.S., eds.), pp. 427–428. Appleton-Century-Crofts, New York.

McKusick, V.A. (1994). Mendelian Inheritance in Man, 11th ed. Johns Hopkins University Press, Baltimore.

Mendez, M.F., Stanley, T.M., Medel, N.M., Li, Z. & Tedesco, D.T. (1997). The vascular dementia of Fabry's disease. Dement. Geriatr. Cogn. Disord. 8, 252–257.

Miura, D.S. (1997). Genetic factors in human hypertension. Curr. Opin. Nephrol. Hypertens. 6, 215–218.

Morris, A.A., Leonard, J.V., Brown, G.K., Bidouki, S.K., Bindoff, L.A., Woodward, C.E., Harding, A.E., Lake, B.D., Harding, B.N., Farrell, M.A., Bell, J.E., Mirakhur, M. & Turnbull, D.M. (1996). Deficiency of respiratory chain complex I is a common cause of Leigh disease. Ann. Neurol. 40, 25–30.

Mudd, S.H. & Levy, H.L. (1983). Disorders of transsulfuration. In: The Metabolic Basis of Inherited Disease (Stanbury, J.B., Wingarden, J.B., Fredrickson, D.S., Goldstein, J.L. & Brown, M.S., eds.), pp. 522–559. McGraw-Hill, New York.

Munyer, T.P. & Margulis, A.R. (1981). Pseudoxanthoma elasticum with internal carotid artery aneurysm. Am. J. Roentgenol. 136, 1023–1024.

Narvaez, J., Narvaez, J.A., Majos, C., Clavaguera, M.T. & Alegre-Sancho, J.J. (1996). Subarachnoid haemorrhage secondary to ruptured cerebral aneurysm in a patient with osteogenesis imperfecta. Br. J. Rheumatol. 35, 1332–1333.

National Stroke Association. (1995). The Brain at Risk—Understanding and Preventing Stroke. 1995. (Pamphlet).

Natowicz, M. & Kelley, R.I. (1987). Mendelian etiologies of stroke. Ann. Neurol. 22, 175–192.

Navarro, V. (1991). Race or class or race and class: growing mortality differentials in the United States. Int. J. Health Serv. 21, 229–235.

Neaton, J.D., Blackburn, H., Jacobs, D., Kuller, L., Lee, D.J., Sherwin, R., Shih, J., Stamler, J. & Wentworth, D. (1992). Serum cholesterol level and mortality findings for men screened in the Multiple Risk Factor Intervention Trial. Multiple Risk Factor Intervention Trial Research Group. Arch. Intern. Med. 152, 1490–1500.

Nicoll, J.A., Burnett, C., Love, S., Graham, D.I., Dewar, D., Ironside, J.W., Stewart, J. & Vinters, H.V. (1997). High frequency of apolipoprotein E epsilon 2 allele in hemorrhage due to cerebral amyloid angiopathy. Ann. Neurol. 41, 716–721.

North, K.N., Whiteman, D.A., Pepin, M.G. & Byers, P.H. (1995). Cerebrovascular complications in Ehlers-Danlos syndrome type IV. Ann. Neurol. 38, 960–964.

Nowak-Göttl, U., Koch, H.G., Aschka, I., Kohlhase, B., Vielhaber, H., Kurlemann, G., Oleszcuk-Raschke, K., Kehl, H.G., Jurgens, H. & Schneppenheim, R. (1996). Resistance to activated protein C (APCR) in children with venous or arterial thromboembolism. Br. J. Haematol. 92, 992–998.

Ohyama, T., Ohara, S., & Momma, F. (1992). Aneurysm of the cervical internal carotid artery associated with Marfan's syndrome—case report. Neurol. Med. Chir.(Tokyo) 32, 965–968.

Ojemann, R.G. & Heros, R.C. (1983). Spontaneous brain hemorrhage. Stroke 14, 468–475.

Okajima, K., Abe, H., Maeda, S., Motomura, M., Tsujihata, M., Nagataki, S., Okabe, H. & Takatsuki, K. (1993). Antithrombin III Nagasaki (Ser116-Pro): a heterozygous variant with defective heparin binding associated with thrombosis. Blood 81, 1300–1305.

Okazaki, H., Reagan, T.J. & Campbell, R.J. (1979). Clinicopathologic studies of primary cerebral amyloid angiopathy. Mayo Clin. Proc. 54, 22–31.

Pantelakis, S. (1954). Un type particulier d'angiopathie sénile du système nerveux central: L'angiopathie congophile. Topographie et fréquence. Monatsschr. Psychiatr. Neurol. 128, 219–256.

Pavlakis, S.G., Phillips, P.C., DiMauro, S., De Vivo, D.C. & Rowland, L.P. (1984). Mitochondrial myopathy, encephalopathy, lactic acidosis, and strokelike episodes: a distinctive clinical syndrome. Ann. Neurol. 16, 481–488.

Pavlakis, S.G., Bello, J., Prohovnik, I., Sutton, M., Ince, C., Mohr, J.P., Piomelli, S., Hilal, S. & De Vivo, D.C. (1988). Brain infarction in sickle cell anemia: magnetic resonance imaging correlates. Ann. Neurol. 23, 125–130.

Pavlakis, S.G., Prohovnik, I., Piomelli, S. & DeVivo, D.C. (1989). Neurologic complications of sickle cell disease. Adv. Pediatr. 36, 247–276.

Pearson, J.D. (1994). Vessel wall interactions regulating thrombosis. Br. Med. Bull. 50, 776–788.

Perry, I.J., Refsum, H., Morris, R.W., Ebrahim, S.B., Ueland, P.M. & Shaper, A.G. (1995). Prospective study of serum total homocysteine concentration and risk of stroke in middle-aged British men. Lancet 346, 1395–1398.

Pettigrew, L., Immken, L. & Hejtmancik, J. (1989). Natural history of familial cavernous hemangioma [abstract]. Stroke 20, 139A.

Peyser, P.A. (1997). Genetic epidemiology of coronary artery disease. Epidemiol. Rev. 19, 80–90.

Pomerleau, O.F. (1995). Individual differences in sensitivity to nicotine: implications for genetic research on nicotine dependence. Behav. Genet. 25, 161–177.

Pope, F.M., Kendall, B.E., Slapak, G.I., Kapoor, R., McDonald, W.I., Compston, D.A., Mitchell, R., Hope, D.T., Millar-Craig, M.W. & Dean, J.C. (1991). Type III collagen mutations cause fragile cerebral arteries. Br. J. Neurosurg. 5, 551–574.

Powars, D., Wilson, B., Imbus, C., Pegelow, C. & Allen, J. (1978). The natural history of stroke in sickle cell disease. Am. J. Med. 65, 461–471.

Prockop, D.J., Kuivaniemi, H. & Tromp, G. (1998). Inherited disorders of connective tissue. In: Harrison's Principles of Internal Medicine (Fauci, A.S., Braunwald, E., Isselbacher, K.J., Wilson, J.D., Martin, J.B., Kasper, D.L., Hauser, S.L. & Longo, D.L., eds.), pp. 2183–2194. McGraw Hill, Inc., New York.

Pulsinelli, W.A., Brierly, J.B. & Plum, F. (1982). Temporal profile of neuronal damage in a model of transient forebrain ischemia. Ann. Neurol. 11, 491–498.

Quattrone, A., Colucci, M., Donati, M.B., Meduri, M., Mussoni, L., Roncaglioni, M.C., Semeraro, N., Carlomagno, S. & Bonavita, V. (1983). Stroke in two young siblings with congenital dysfibrinogenemia. Ital. J. Neurol. Sci. 4, 229–232.

Rabbani, L.E. & Loscalzo, J. (1994). The relationship between thrombosis and atherosclerosis. In: Thrombosis and Hemorrhage (Loscalzo, J. & Schafer, A.I., eds.), pp. 769–789. Blackwell Scientific Publications, Inc., Cambridge, MA.

Rahman, S., Blok, R.B., Dahl, H.H., Danks, D.M., Kirby, D.M., Chow, C.W., Christodoulou, J. & Thorburn, D.R. (1996). Leigh syndrome: clinical features and biochemical and DNA abnormalities. Ann. Neurol. 39, 343–351.

Reyes, M.G. (1989). Subcortical cerebral infarctions in sickle cell trait. J. Neurol. Neurosurg. Psychiatry 52, 516–518.

Ridker, P.M., Hennekens, C.H., Lindpaintner, K., Stampfer, M.J., Eisenberg, P.R. & Miletich, J.P. (1995). Mutation in the gene coding for coagulation factor V and the risk of myocardial infarction, stroke, and venous thrombosis in apparently healthy men. N. Engl. J. Med. 332, 912–917.

Rios-Montenegro, E.N., Behrens, M.M. & Hoyt, W.F. (1972). Pseudoxanthoma elasticum. Association with bilateral carotid rete mirabile and unilateral carotid-cavernous sinus fistula. Arch. Neurol. 26, 151–155.

Roses, A.D. (1997). Apolipoprotein E, a gene with complex biological interactions in the aging brain. Neurobiol. Dis. 4, 170–185.

Ross, R. (1993). The pathogenesis of atherosclerosis: a perspective for the 1990s. Nature 362, 801–809.

Rothman, S. (1984). Synaptic release of excitatory amino acid neurotransmitter mediates anoxic neuronal death. J. Neurosci. 4, 1884–1891.

Ruchoux, M.M. & Maurage, C.A. (1997). CADASIL: Cerebral autosomal dominant arteriopathy with subcortical infarcts and leukoencephalopathy. J. Neuropathol. Exp. Neurol. 56, 947–964.
Ruchoux, M.M., Guerouaou, D., Vandenhaute, B., Pruvo, J.P., Vermersch, P. & Leys, D. (1995). Systemic vascular smooth muscle cell impairment in cerebral autosomal dominant arteriopathy with subcortical infarcts and leukoencephalopathy. Acta Neuropathol. (Berl.) 89, 500–512.
Russell, R.W. (1975). How does blood-pressure cause stroke? Lancet 2, 1283–1285.
Sacco, R.L. (1994). Classification of stroke. In: Clinical Atlas of Cerebrovascular Disorders (Fisher, M., Ed.), pp. 2.1–2.25. Mosby-Year Book Europe, London.
Sacco, R.L., Kargman, D.E., Gu, Q. & Zamanillo, M.C. (1995). Race-ethnicity and determinants of intracranial atherosclerotic cerebral infarction: the Northern Manhattan Stroke Study. Stroke 26, 14–20.
Salonen, R., Nyyssonen, K., Porkkala, E., Rummukainen, J., Belder, R., Park, J.S. & Salonen, J.T. (1995). Kuopio Atherosclerosis Prevention Study (KAPS). A population- based primary preventive trial of the effect of LDL lowering on atherosclerotic progression in carotid and femoral arteries. Circulation 92, 1758–1764.
Samama, M.M., Simon, D., Horellou, M.H., Trossaert, M., Elalamy, I. & Conard, J. (1996). Diagnosis and clinical characteristics of inherited activated protein C resistance. Haemostasis 26 (suppl 4), 315–330.
Sanchez, J., Roman, J., de la Torre, M.J., Velasco, F. & Torres, A. (1997). Low prevalence of the factor V Leiden among patients with ischemic stroke. Haemostasis 27, 9–15.
Santorelli, F.M., Mak, S.C., Vazquez-Memije, E., Shanske, S., Kranz-Eble, P., Jain, K.D., Bluestone, D.L., De Vivo, D.C. & DiMauro., S. (1996). Clinical heterogeneity associated with the mitochondrial DNA T8993C point mutation. Pediatr. Res. 39, 914–917.
Scharf, R.E. & Harker, L.A. (1987). Thrombosis and atherosclerosis: regulatory role of interactions among blood components and endothelium. Blut 55, 131–144.
Scheuner, M.T., Raffel, L.J. & Rotter, J.I. (1997). Genetics of diabetes. In: International Textbook of Diabetes Mellitus (Alberti, K.G.M.M., Zimmet, P., DeFronzo, R.A. & Keen, H., eds.), pp. 37–88. John Wiley & Sons, West Sussex.
Schievink, W.I. (1997a). Genetics of intracranial aneurysms. Neurosurgery 40, 651–662.
Schievink, W.I. (1997b). Intracranial aneurysms. N. Engl. J. Med. 336, 28–40.
Schievink, W.I., Limburg, M., Oorthuys, J.W., Fleury, P. & Pope, F.M. (1990). Cerebrovascular disease in Ehlers-Danlos syndrome type IV. Stroke 21, 626–632.
Schievink, W.I., Michels, V.V. & Piepgras, D.G.. (1994). Neurovascular manifestations of heritable connective tissue disorders: a review. Stroke 25, 889–903.
Schievink, W.I., Parisi, J.E. & Piepgras, D.G. (1997a). Familial intracranial aneurysms: an autopsy study. Neurosurgery 41, 1247–1251.
Schievink, W.I., Parisi, J.E., Piepgras, D.G. & Michels, V.V. (1997b). Intracranial aneurysms in Marfan's syndrome: an autopsy study. Neurosurgery 41, 866–870.
Schoenberg, B.S., Anderson, B.W. & Haerer, A.F. (1986). Racial differentials in the prevalence of stroke. Copiah County, Mississippi. Arch. Neurol. 43, 565–568.
Schroder, J.M., Sellhaus, B. & Jorg, J. (1995). Identification of the characteristic vascular changes in a sural nerve biopsy of a case with cerebral autosomal dominant arteriopathy with subcortical infarcts and leukoencephalopathy (CADASIL). Acta Neuropathol. (Berl.) 89, 116–121.
Schuster, H., Wienker, T.F., Bahring, S., Bilginturan, N., Toka, H.R., Neitzel, E.J., Toka, A., Gilbert, D., Lowe, A., Ott, J., Haller, H. & Luft, F.C. (1996). Severe autosomal dominant hypertension and brachydactyly in a unique Turkish kindred maps to human chromosome 12. Nat. Genet. 13, 98–100.
Selhub, J., Jacques, P.F., Bostom, A.G., D'Agostino, R.B., Wilson, P.W., Belanger, A.J., O'Leary, D.H., Wolf, P.A., Schaefer, E.J. & Rosenberg, I.H. (1995). Association between plasma homocysteine concentrations and extracranial carotid-artery stenosis. N. Engl. J. Med. 332, 286–291.
Seller, A., Kennedy, C.R., Temple, I.K. & Brown, G.K. (1997). Leigh syndrome resulting from de novo mutation at position 8993 of mitochondrial DNA. J. Inherit. Metab. Dis. 20, 102–103.

Shimkets, R., Warnock, D.G., Bositis, C.M., Nelson-Williams, C., Hansson, J.H., Schambelan, M., Gill, J.R., Ulick, S., Milora, R.V., Findling, J.W., Canessa, C.M., Rossier, B.C. & Lifton, R.P. (1994). Liddle's syndrome: heritable human hypertension caused by mutations in the beta subunit of the epithelial sodium channel. Cell 79, 407–414.

Simioni, P., de Ronde, H., Prandoni, P., Saladini, M., Bertina, R.M. & Girolami, A. (1995). Ischemic stroke in young patients with activated protein C resistance. A report of three cases belonging to three different kindreds. Stroke 26, 885–890.

Sonninen, V. & Savontaus, M.L. (1987). Hereditary multi-infarct dementia. Eur. Neurol. 27, 209–215.

Sourander, P. & Walinder, J. (1977). Hereditary multi-infarct dementia. Acta Neuropathol. 39, 247–254.

Stoy, N.S. (1997). Stroke and cholesterol: "enigma variations'? J. R. Coll. Physicians. Lond. 31, 521–526.

Sutherland, D.J., Ruse, J.L. & Laidlaw, J.C. (1966). Hypertension, increased aldosterone secretion, and low plasma renin activity relieved by dexamethasone. Can. Med. Assoc. J. 95, 1109–1119.

Tarter, R.E. (1995). Genetics and primary prevention of drug and alcohol abuse. Int. J. Addict. 30, 1479–1484.

Tatlisumak, T. & Fisher, M. (1996). Hematologic disorders associated with ischemic stroke. J. Neurol. Sci. 140, 1–11.

Thomas, G.R., Thibodeaux, H., Errett, C.J., Badillo, J.M., Keyt, B.A., Refino, C.J., Zivin, J.A. & Bennett, W.F. (1994). A long-half-life and fibrin-specific form of tissue plasminogen activator in rabbit models of embolic stroke and peripheral bleeding. Stroke 25, 2072–2078.

Thomas, T., Thomas, G., McLendon, C., Sutton, T. & Mullan, M. (1996). Beta-amyloid-mediated vasoactivity and vascular endothelial damage. Nature 380, 168–171.

Thorvaldsen, P., Asplund, K., Kuulasmaa, K., Rajakangas, A.-M. & Schroll, M. (1995). Stroke incidence, case fatality, and mortality in the WHO MONICA Project. Stroke 26, 361–367.

Tonstad, S., Joakimsen, O., Stensland-Bugge, E., Leren, T.P., Ose, L., Russell, D. & Bonaa, K.H. (1996). Risk factors related to carotid intima-media thickness and plaque in children with familial hypercholesterolemia and control subjects. Arterioscler. Thromb. Vasc. Biol. 16, 984–991.

Torack, R.M. (1975). Congophilic angiopathy complicated by surgery and massive hemorrhage. A light and electron microscopic study. Am. J. Pathol. 81, 349–365.

Torra, R., Badenas, C., Darnell, A., Nicolau, C., Volpini, V., Revert, L. & Estivill, X. (1996). Linkage, clinical features, and prognosis of autosomal dominant polycystic kidney disease types 1 and 2. J. Am. Soc. Nephrol. 7, 2142–2151.

Tournier-Lasserve, E., Iba-Zizen, M.T., Romero, N. & Bousser, M.G. (1991). Autosomal dominant syndrome with strokelike episodes and leukoencephalopathy. Stroke 22, 1297–1302.

Tournier-Lasserve, E., Joutel, A., Melki, J., Weissenbach, J., Lathrop, G.M., Chabriat, H., Mas, J.L., Cabanis, E.A., Baudrimont, M. & Maciazek, J. (1993). Cerebral autosomal dominant arteriopathy with subcortical infarcts and leukoencephalopathy maps to chromosome 19q12. Nat. Genet. 3, 256–259.

Tsai, M.Y., Garg, U., Key, N.S., Hanson, N.Q., Suh, A. & Schwichtenberg, K. (1996). Molecular and biochemical approaches in the identification of heterozygotes for homocystinuria. Atherosclerosis 122, 69–77.

Tsuruda, T., Nagatomo, Y., Fujimoto, S., Yamamoto, Y. & Eto, T. (1996). Polycystic disease associated with intracranial arteriovenous malformation: Another extrarenal manifestation? [letter]. Clin. Nephrol. 46, 147–148.

Tuite, P., Ahmad, F., Grant, I., Stewart, J.D., Carpenter, S. & Ethier, R. (1993). Cerebral vein thrombosis due to hereditary antithrombin III deficiency. Can. J. Neurol. Sci. 20, 158–161.

Tulinius, M.H., Houshmand, M., Larsson, N.G., Holme, E., Oldfors, A., Holmberg, E. & Wahlstrom, J. (1995). De novo mutation in the mitochondrial ATP synthase subunit 6 gene (T8993G) with rapid segregation resulting in Leigh syndrome in the offspring. Hum. Genet. 96, 290–294.

Ueland, P.M. & Refsum, H. (1989). Plasma homocysteine, a risk factor for vascular disease: plasma levels in health, disease, and drug therapy. J. Lab. Clin. Med. 114, 473–501.

Utsumi, K., Yamamoto, N., Kase, R., Takata, T., Okumiya, T., Saito, H., Suzuki, T., Uyama, E. & Sakuraba, H. (1997). High incidence of thrombosis in Fabry's disease. Intern. Med. 36, 327–329.

van den Berg, J.S., Limburg, M. & Hennekam, R.C. (1996). Is Marfan syndrome associated with symptomatic intracranial aneurysms? Stroke 27, 10–12.

van der Bom, J.G., Bots, M.L., Haverkate, F., Slagboom, P.E., Meijer, P., de Jong, P.T., Hofman, A., Grobbee, D.E. & Kluft, C. (1996). Reduced response to activated protein C is associated with increased risk for cerebrovascular disease. Ann. Intern. Med. 125, 265–269.

van Dijk, M.A., Chang, P.C., Peters, D.J. & Breuning, M.H. (1995). Intracranial aneurysms in polycystic kidney disease linked to chromosome 4. J. Am. Soc. Nephrol. 6, 1670–1673.

Velho, G. & Froguel, P. (1997). Genetic determinants of non-insulin-dependent diabetes mellitus: strategies and recent results. Diabetes Metab. 23, 7–17.

Vinters, H.V. (1987). Cerebral amyloid angiopathy. A critical review. Stroke 18, 311–324.

Wall, R.T., Harlan, J.M. & Harker, L.A. (1980). Homocysteine-induced endothelial cell injury in vitro: a model for the study of vascular injury. Thromb. Res. 18, 113–121.

Wang, Z.Z., Jensson, O., Thorsteinsson, L., & Vinters, H.V. (1997). Microvascular degeneration in hereditary cystatin C amyloid angiopathy of the brain. A.P.M.I.S. 105, 41–47.

Wattendorff, A.R., Bots, G.T., Went, L.N. & Endtz, L.J. (1982). Familial cerebral amyloid angiopathy presenting as recurrent cerebral haemorrhage. J. Neurol. Sci. 55, 121–135.

Watts, G.F. & Burke, V. (1996). Lipid-lowering trials in the primary and secondary prevention of coronary heart disease: new evidence, implications and outstanding issues. Curr. Opin. Lipidol. 7, 341–355.

Weimann, B.J., Kuhn, H. & Baumgartner, H.R. (1980). Effect of homocysteine on cultured bovine and human endothelial cells. Experientia 36, 762.

Weir, B. (1987). Special aneurysms (nonsaccular and saccular). In: Aneurysms Affecting the Nervous System, pp. 134–208. Williams & Wilkens, Baltimore.

Wendelhag, I., Wiklund, O. & Wikstrand, J. (1995). Intima-media thickness after cholesterol lowering in familial hypercholesterolemia. A three-year ultrasound study of common carotid and femoral arteries. Atherosclerosis 117, 225–236.

Williams, G.H., Jagger, P.I. & Braunwald, E. (1980). Hypertensive vascular disease. In: Harrison's Principles of Internal Medicine (Isselbacher, K.J., Adams, R.D., Braunwald, E., Petersdorf, R.G. & Wilson, J.D., eds.), pp. 1167–1178. McGraw-Hill, New York.

Wilson, P.W., Hoeg, J.M., D'Agostino, R.B., Silbershatz, H., Belanger, A.M., Poehlmann, H., O'Leary, D. & Wolf, P.A. (1997). Cumulative effects of high cholesterol levels, high blood pressure, and cigarette smoking on carotid stenosis. N. Engl. J. Med. 337, 516–522.

Wintzen, A.R., Broekmans, A.W., Bertina, R.M., Briet, E., Briet, P.E., Zecha, A., Vielvoye, G.J. & Bots, G.T. (1985). Cerebral haemorrhagic infarction in young patients with hereditary protein C deficiency: evidence for "spontaneous" cerebral venous thrombosis. Br. Med. J. 290, 350–352.

Wisniewski, T. & Frangione, B. (1992). Molecular biology of Alzheimer's amyloid—Dutch variant. Mol. Neurobiol. 6, 75–86.

Wisniewski, T., Ghiso, J. & Frangione, B. (1991). Peptides homologous to the amyloid protein of Alzheimer's disease containing a glutamine for glutamic acid substitution have accelerated amyloid fibril formation. Biochem. Biophys. Res. Commun. 179, 1247–1254.

Wolf, P.A., D'Agostino, R.B., Belanger, A.F. & Kannel, W.B. (1991). Probability of stroke: a risk profile from the Framingham Study. Stroke 22, 312–318.

Zeviani, M., Bertagnolio, B. & Uziel, G. (1996). Neurological presentations of mitochondrial diseases. J. Inherit. Metab. Dis. 19, 504–520.

Zuber, M., Toulon, P., Marnet, L, & Mas, J.L. (1996). Factor V Leiden mutation in cerebral venous thrombosis. Stroke 27, 1721–1723.

Chapter 8

Genetic Susceptibility in Multiple Sclerosis

ROBERT B. BELL

INTRODUCTION

Multiple sclerosis (MS) is a chronic inflammatory disease affecting the white matter of the central nervous system (CNS). The mechanism of disease is controversial, but the weight of scientific evidence implicates an autoimmune process in which

Advances in Cell Aging and Gerontology
Volume 3, pages 287–311

ISBN: 0-7623-0405-7

abnormalities of immune regulation allow autoreactive T cells to generate an aberrant immune response to self antigens in the brain and spinal cord (Storch and Lassmann, 1997). This is supported by clinical and histological observations as well as cellular and molecular immunological studies in experimental allergic encephalomyelitis (EAE), an animal model of autoimmune demyelination. Despite our understanding of many of the immunological mechanisms that occur in MS, we as yet have an incomplete understanding of the factors that influence susceptibility to MS or that participate in the underlying disease etiology. There is, however, considerable evidence to support the role of both genetic and environmental influences and their interactions in the pathogenesis of this and other complex human autoimmune disorders.

GENETIC EPIDEMIOLOGY

Historically, epidemiological studies demonstrating a nonrandom geographical distribution of MS independent of genetic/racial factors and the widely different disease prevalence rates of individuals of the same ethnic derivation, but from widely separated geographical areas, supported a role for environmental factors in the pathogenesis of this disease (Kurtzke, 1985; Sadovnick and Ebers, 1993; Compston, 1994; Ebers and Sadovnick, 1994). This was further substantiated by studies demonstrating changes in disease risk secondary to migration of individual racial groups between high and low risk MS zones (Alter et al., 1966; Dean, 1967; Elian et al., 1990a).

It has been recognized that the geographical distribution of MS varies widely between regions. In some areas, such as the Shetland and Orkney Islands off of the northern coast of Scotland, there is a very high prevalence of disease (approximately 200 per 100,000), whereas areas such as the southern United States have a very low prevalence (10 per 100,000) (Sawcer et al., 1997). In general, however, there is a north/south gradient in the northern hemisphere and the reverse gradient in the southern hemisphere with prevalence rates of MS increasing in more temperate climates (Kurtzke, 1985; Sadovnick and Ebers, 1993; Compston, 1994; Ebers and Sadovnick, 1994). Although this was initially suggested to support an environmental influence, the latitudinal distribution in Northern America may, however, reflect patterns of immigration of different ethnic groups in Europe (Sadovnick and Ebers, 1993; Ebers and Sadovnick, 1994). This was suggested by Davenport and later supported by Ebers and Bulman who examined U.S. military veterans' data and demonstrated that the pattern of disease distribution parallels the proportion of the population of North American states with Northern European and particularly Scandinavian heritage (Davenport, 1922; Kurtzke et al., 1979; Ebers and Bulman, 1986). It is evident, however, from epidemiological studies in Australia and New Zealand, where there is a relative uniform ethnic background, that rates of MS vary dramatically with latitude (Hammond et al., 1987; Skegg et al., 1987).

The migration studies that have supported an environmental influence are based on the comparison of disease prevalence in populations migrating between areas of differing disease risk. Confounding these studies, however, is the question of whether the migrant group is representative of the original population and whether or not following migration these individuals have been randomly distributed. It is evident that many migrations have been induced by religious, ethnic, or sectarian conflicts, and there may be considerable selection bias based on such factors as economics, health, or age. In general, the migration studies have examined a small migrating population and, in many circumstances, the immigrating population has not had good access to medical care. This was exemplified by studies on the Vietnamese immigrants to France, where the majority had at least one French parent that made immigration possible, and clearly differentiated them from the normal Vietnamese population control group (Kurtzke and Bui, 1980). The accumulative evidence from several migration studies, however, lends support to environmental factors influencing disease risk (Dean, 1949; Alter et al., 1971; Dean et al., 1976; Kurtzke and Bui, 1980; Elian et al., 1990b).

Although the potential role of genetic factors influencing disease susceptibility and course has come under increasing scrutiny in the past decade, the role of inherited factors was first proposed almost a century ago when Eichhorst depicted MS as an "inherited, transmissible" disease (Eichorst, 1896). Implicit in these remarks was the notion that MS perhaps was an acquired infectious process, but over time this has yielded to the concept that other nonheritable, but as yet unknown factors, as well as genetic influences, are more likely responsible.

Epidemiological evidence supporting the influence of genetic factors upon MS susceptibility includes the presence of ethnic and racial susceptibility and resistance, family and twin studies, and the association of MS with a particular major histocompatibility complex haplotype (Ebers and Sadovnick, 1994; Dyment et al., 1997; McFarland et al., 1997; Oksenberg and Hauser, 1997; Sawcer et al., 1997; Ebers and Sadovnick, 1998). In Canada, the prevalence of MS is approximately 1 per 1,000 or 0.1% among Caucasians of Central and Northern European origin. The average age for MS onset is between 28 and 30 years, and there is a disease ratio of females to males of approximately 2 to 1, with females having their first MS symptoms slightly earlier in life than males. Family studies demonstrate that the lifetime risk of MS is considerably increased for relatives of MS patients compared with the general population, and this increased risk decreases progressively for second- and third-degree relatives (Sadovnick, 1993; Sadovnick and Ebers, 1993; Robertson et al., 1996). Although the recurrence risks for first degree relatives of MS patients are relatively low (3% to 5%), this represents a substantial increase over the 0.1% lifetime risk in the general population. Age-adjusted familial risks for MS are documented in Table 1. A large reduction in disease risk between first-degree and more remote relatives supports the concept of oligo/polygenic inheritance with epistatic interactions occurring between susceptibility loci. Attempting to model the relative risk values does not discriminate between several

Table 1. Age-Adjusted Empirical Risks for Multiple Sclerosis

	Male Index Cases		*Female Index Cases*	
Relationship to Index Case	*Proportion Affected*	*% Risk ± 95% Confidence Intervals*	*Proportion Affected*	*% Risk ± 95% Confidence Intervals*
Mother	7/184	3.84 ± 1.42%	14/38	3.71 ± 0.97%
Father	1/128	0.79 ± 0.79%	6/303	2.00 ± 0.18%
Parents	8/312	2.56 ± 0.90%	20/686	2.95 ± 0.65%
Daughter	2/22	5.13 ± 3.53%	5/386	4.96 ± 2.17%
Son	0/248	0.00 ± 1.49%	0/411	0.00 ± 0.90%
Children	2/471	2.47 ± 1.72%	5/797	2.58 ± 1.14%
Sister	9/340	3.46 ± 1.14%	25/608	5.65 ± 1.10%
Brother	10/326	4.15 ± 1.28%	10/612	2.27 ± 0.71%
Siblings	19/666	3.81 ± 0.86%	35/1220	3.97 ± 0.66%
Aunt/uncle	15/560	2.68 ± 0.68%	23/1491	1.59 ± 0.33%
Niece/nephew	3/1000	1.47 ± 0.84%	7/1789	1.83 ± 0.69%
First cousins	7/795	1.53 ± 0.57%	34/2347	2.37 ± 0.40%

Source: Sadvonick et al. (1988).

genetic models. Racial differences in the risk of disease have been identified with the greatest vulnerability in Northern European Caucasoids, particularly those of Fenno Scandinavian descent. Relative racial resistance to MS is evident in black Africans, Orientals, and Hispanics, as well as in several ethnically distinct groups residing in close proximity to populations with a high disease prevalence, for example, Central European Gypsies, Hutterites, Lapps, Inuits, Amerindians, and Maoris (Dean, 1949; Detels et al., 1977; Enstrom and Operskalski, 1978; Popov, 1983; Gronning and Mellgren, 1985; Hader et al., 1985; Kurtzke et al., 1985; Skegg et al., 1987; Bharucha et al., 1988; Yaqub and Daif, 1988; Yu et al., 1989; Palffy et al., 1993; Hader et al., 1996).

A classical epidemiological method for distinguishing between the relative importance of genetic and environmental factors in disease has been the study of concordance rates in monozygotic (genetically identical) and dyzygotic twins (genetically equivalent to nontwin siblings). Similarly, a comparison of concordance rates between dyzygotic pairs (who may be expected to share a more common environment than siblings) and nontwin siblings is deemed to be a measure of the environmental influence on disease risk. It should be noted that epidemiological studies involving twin surveys may be deceiving. For example, monozygotic concordance has been observed in diseases with a pure infective etiology (Ebers and Sadovnick, 1998). A major confounding methodological problem in twin studies is ascertainment bias (Ott, 1991). Because of the requirement for the relatively large number of twin pairs for the conduct of these studies, some investigators have relied upon solicitation of volunteers by public appeal (Sawcer et al., 1997; Ebers and Sadovnick, 1998). This tends to result in an increased representation of monozygotic twin pairs, females, and concordant pairs. To minimize this bias, the identification of twin pairs from a population-based sample, either of patients with MS or of twins, is preferable. Other confounding factors may include inaccurate assignment of zygosity and the failure to recognize subclinical forms of disease. The majority of twin studies in MS demonstrate a concordance rate in monozygotic twins between 21% and 50% compared with 0% and 17% for dyzygotic twins (Table 2) (Mackay and Myrianthopoulos, 1966; Bobowick et al., 1978; Williams et al., 1980; Anonymous, 1982; Currier and Eldridge, 1982; McFarland et al., 1984; Ebers et al., 1986; Kinnunen et al., 1988; Anonymous, 1992; Sadovnick et al., 1993; Mumford et al., 1994).

The larger studies are quite consistent and demonstrate monozygotic concordance of around 25%. Further studies using magnetic resonance imaging to look for evidence of subclinical disease demonstrate abnormalities consistent with MS in a further 14% of unaffected monozygotic co-twins (Ebers et al., 1986; Mumford et al., 1994). Two studies demonstrate disparate results. The first of these reported on seven twins, of whom four were monozygotic, and included no concordant pairs (Uitdehaag et al., 1989). Subsequent magnetic resonance imaging (MRI) data, however, revealed changes compatible with MS in 75% of the clinically unaffected monozygotic twins. A French study relied on patient ascertainment by public appeal and thus was not population based (French Research Group on Multiple Sclerosis, 1992). Among the 116 twin pairs that were identified, over half of these could not be fully evaluated and the zygosity status of more than 20 pairs could not be

Table 2. Twin Studies

	Concordance	
	Monozygotic	*Dizygotic*
MacKay and Myrianthopoulos[a] (1996)	6/39 (15%)	3/29 (10%)
Bobowick et al.[a] (1978)	1/5 (20%)	0/4 (0%)
Currier and Eldridge[a, b] (1982)	8/22 (36%)	3/26 (11%) 2 sets unlike sex
Williams et al.[a, b] (1980)	6/12 (50%)	2/12 (17%)
McFarland et al.[a, b] (1984)	10/14 (71%)	2/12 (17%)
Heltberg and Holm (1982)	8/23 (35%)	2/29 (7%)
Kinnunen et al. (1988)	2/8 (25%)	0/6 (0%)
Ebers et al. (1986)	7/25 (28%)	1/40 (2.5%)
Mumford et al. (1994)	11/44 (25%)	2/61 (3.3%)
French Twin Study Group[a] (1992)	1/17 (6%)	1/41 (2.4%)

Notes: [a]Study subject to ascertainment bias. (Ref. under "French Research Group on Multiple Sclerosis".)
[b]Three studies of overlapping patient cohort.

determined. Although it has been suggested that there may be a difference in the magnitude of the genetic influence between France and other geographical areas, the 90% confidence limits are not significantly different from those of the Canadian twin study (Sadovnick et al., 1993; Ebers and Sadovnick, 1998).

Overall, the twin studies provide considerable support for a genetic component to susceptibility and suggest that this is unlikely to be accounted for by a single dominant, or even a single recessive gene. The monozygotic:dizygotic concordance ratio of approximately 7:1 implies that multiple genetic influences are likely to be involved. It is important to recognize that the results of these studies also indicate that the majority of monozygotic twins remain discordant for disease even after age correction and radiological imaging, suggesting that environmental effects are influential. Similar concordance rates for dyzygotic twins and siblings suggests that the environmental influence is present at a population level and is related to global influences such as diet, common childhood infection, vaccination, or climate. The tendency for sibling pairs to manifest disease at the same age, rather than during the same year, and the random order of affected sibling pairs in birth sequence, also support this.

The study of nonbiological relatives in MS has been used to conclusively demonstrate the role of genetic factors. Recent screening of a population sample of approximately 15,000 MS patients identified 238 MS adoptees who came into the adopted family before the age of 1 year and who lived with their adoptive parents (Ebers et al., 1995). The frequency of MS among first-degree nonbiological relatives in this case was demonstrated to be similar to the prevalence rate in the general population from which these individuals were drawn, indicating that a shared environment is not sufficient to explain familial aggregation. A study of 1,839 half-siblings derived from 939 MS index cases demonstrated an age-adjusted risk for half-siblings of 1.32% and for full siblings of 3.46% (Sadovnick et al., 1996). Because of the failure to identify differences between half-siblings who live together and half-siblings who never lived together (1.17 vs. 1.47%), it is again evident that shared environment does not account for familial aggregation in this disease. Studies of half-siblings have also negated the potential role of vertically transmitted infection, genomic imprinting, and mitochondrial inheritance (Ebers and Sadovnick, 1998). These data, however, must not be used to dissuade from the fact that population risk is strongly influenced by the environment, as is evidenced in populations where there is little genetic diversity and where there is a multifold difference in risk associated with a difference in latitude. Therefore, it is evident that whatever environmental factors are at play appear to influence the population as a whole and are likely to be related to rather ubiquitous factors such as diet or climate.

There are limited reports of conjugal MS and only one reasonably sized population-based study has been conducted (Robertson et al., 1995; Sawcer et al., 1997). This study failed to find evidence of an increase in concordance rate, clustering at year of onset, or distortion of the expected pattern of age of onset in the second affected spouse. However, when recurrence risks in offspring of MS conjugal pairs were examined, it was determined that 5.8% were concordant for MS and a further 4.7% reported isolated episodes of neurological dysfunction. Two of these later individuals had MRI abnormalities consistent with the radiological criteria for the diagnosis of MS. The recurrence risk in these individuals was significantly higher than the population-based recurrence risk for offspring of single affected parents and this has been interpreted to suggest that MS is inherited in a biparental fashion, weighing against the presence of genetic heterogeneity.

Family study data, particularly those derived from the Canadian and United Kingdom studies, clearly exclude simple mendelian (autosomal dominant, autosomal recessive) inheritance (Ebers and Sadovnick, 1998). Autosomal recessive inheritance is rejected based on the frequency of parent-child concordance being similar to that for siblings. A recessive locus with a dominant modifier is less easily excluded. Autosomal dominant inheritance would necessitate a very low penetrance and does not explain either the differential parent-child concordance rates or the altered sex ratio among affected individuals. X-linked recessive inheritance is incompatible with both the sex ratio of affected cases and the parental concordance pattern. X-linked dominant inheritance would account for the almost 2:1 sex ratio among affected cases and the low level of parental concordance. However, in multiplex pedigrees, parental concordance is no less frequent through the unaffected father than the mother. The frequency of concordant father-daughter pairs is as expected when adjusted for the sex ratio excluding vertical (mitochondrial) transmission. A Y-linked resistance gene could also explain the altered sex ratio but does not explain the observed lack of concordance between father-son pairs. Parental imprinting has been suggested as a mode of potential inheritance. This results from a differential degree of methylation in at least some genes depending on which parent transmits the gene. Parental imprinting would, however, have to differ by sex of offspring to explain the family data. The accumulated evidence from epidemiological studies, then, supports the hypothesis that susceptibility to MS is, in part, conferred by the influence of multiple genes and that the population prevalence of MS is determined by the strength of the genetic effect.

Support for a polygenic (multiple genes) or multifactorial (multiple genes interacting with the environment) influence on MS susceptibility comes from the study of experimental models of autoimmunity. In virtually all animal models of spontaneous autoimmunity, inheritance of susceptibility is polygenic, and the genetic loci that are involved appear to exert their influence at multiple levels or steps in the initiation of the autoimmune process. In the nonobese mouse, susceptibility to diabetes in determined by greater than ten specific susceptibility genes and the putative role of the major histocompatibility complex (MHC) in this process has been partially explained at a molecular level (Risch et al., 1993). In spontaneous lupus in the NZB murine strain, at least six genes have been determined to influence disease phenotype (Morel et al., 1994; Drake et al., 1995). When this strain is crossed with a nonsusceptible NZW strain, a severe nephritis results that is not present in either parental strain. This is an important example of the probable role for background genes in the determination of autoimmune susceptibility. It also appears that susceptibility is influenced more by the number of susceptibility genes than by the actual function of the individual genes and that there may be considerable epistatic interactions which, depending upon the number of genes involved and their specific effects, may result in a broad spectrum of phenotypic expression. It is also evident from these models that even in genetically identical individuals disease incidence of less than 100% may occur, and that the variation in the clinical expression of disease may be the result of environmental factors.

METHODOLOGICAL APPROACHES IN COMPLEX DISEASE

A genetically complex trait or disease is defined as one that is not inherited in a typical mendelian pattern (Ott, 1991; Lander and Schork, 1994). Whereas the manner of inheritance for traits such as cystic fibrosis (autosomal recessive) or Duchenne muscular dystrophy (X-linked recessive) can be explained by the effect of mutations in a single major gene, the inheritance of traits such as multiple sclerosis, insulin-dependent diabetes, hypertension, Alzheimer's disease, and many other more common medical disorders do not fit a simple mendelian pattern of inheritance, thereby implicating a multilocus or polygenic mode of inheritance. In this circumstance, alterations in more than one gene acting either alone, or in conjunction with other loci, may either increase or decrease the risk of developing a particular trait or disease. In many circumstances these genetic alterations may be common polymorphisms.

Methodological approaches to the mapping of genetically complex traits must take into account the potential for clinical and genetic heterogeneity as well as phenocopies, gene-gene, and gene-environment interactions (Ott, 1991; Lander and Schork, 1994; Pericack-Vance and Haines, 1997). The approach involved in the mapping of complex traits may also be unique to a particular trait depending on the resources that are available. Critical in this process is the accurate determination of clinical parameters of disease inclusive of disease subtypes and age effects. The availability of surrogate markers of disease may prove particularly useful. Assignment of phenotype must be consistent and rigorously done as misassigned phenotypes may have major negative implications in the analysis, leading to either false negative or false positive results. For traits in which there is sufficient epidemiological evidence to support a genetic component in susceptibility, it is first necessary to address the relative risk and age-adjusted relative risk to determine sample size requirements before studies proceed. For most genetically complex traits the underlying genetic model is usually unknown limiting the value of simulation studies, which in this circumstance may be used only as a general guideline. The type of analysis that is eventually chosen will depend on the nature of the trait and the types of families collected as well as the underlying genetic model, taking into account the power of the dataset. There are several methods of analysis that each have their relative pros and cons. In general, these include either association studies, affected relative pair analysis, sib pair analysis, or genetic model dependent linkage analysis (Ott 1991; Pericack-Vance and Haines, 1997). It must be noted, however, that these methodologies are not mutually exclusive approaches and it is likely that the successful identification of genes contributing to susceptibility in complex disease will require the use of multiple techniques, each relevant to different stages of the mapping process.

It is often of value to perform a genome wide screen or exclusion map to identify genetic regions that harbor susceptibility genes and to exclude regions that are unlikely to do so. Regions of interest identified through this process need to be confirmed, and this is usually accomplished using a second independent dataset and by genotyping additional markers. For genetically complex traits, the identification of recombinants is not useful because of the complexity of the underlying genetic model. Fine mapping of regions of interest, therefore, depends either on the identification of candidate genes that may have a biological function related to the trait, or saturation genotyping of the region with multiple markers utilizing both association and linkage analysis. It must be noted that a positive association could imply either direct action of the polymorphism identified or the presence of linkage disequilibrium with a functional polymorphism. The identification of genes within a critical region may take advantage of expressed sequence tags (ESTs), which are available in public databases, and in the future will undoubtedly rely more heavily on genetic sequence developed within the human genome initiative. It must be recognized that although genetic sequence will become increasingly available, the function of many of these loci and the molecules they encode are still unknown. For complex traits, polymorphisms responsible for susceptibility may be common in the population and may not have any evident functional effect on gene expression or function. Ultimately, the responsible genetic variation will therefore need to be studied in biological systems such as cultured cells, animal models, or human trials to conclusively demonstrate the importance of a specific polymorphism. The use of genetic manipulation using transgenic or knockout mice is now commonly used for this purpose.

Many complex diseases are considered qualitative traits that are a representation of a threshold affect where an underlying quantitative variable due to environmental or other genetic influences reaches a critical level or threshold that leads to disease expression. This is the case for MS as well as for insulin-dependent diabetes, lupus, and numerous other relatively common diseases. The methodologies that have proved useful in the determination of susceptibility genes in these diseases include both parametric (model-dependent) and nonparametric (model-independent) analysis, as well as allelic association. In linkage analysis, co-segregation of two or more genes (traits) is examined in the family unit to determine if they segregate independently according to Mendel's law or they do not segregate because of their close physical proximity (Ott, 1991). The method for detecting linkage which confers the greatest power is the parametric approach, which involves making an assumption of a genetic model as well as specifying certain parameters, such as mode of inheritance, disease and marker allele frequencies, mutation rate, and penetrance (Ott 1991; Pericack-Vance and Haines, 1997). The advantage of the parametric approach is that it provides an estimate of the recombination fraction and that it makes it possible to test data statistically for evidence of genetic heterogeneity. This approach is also referred to as lod score analysis. The method remains powerful, providing that the specification of the genetic model parameters are correct. However, power will be decreased significantly when wrong assumptions are made. The two most common problems in parametric analysis are mis-specification of the genetic model of mode of inheritance of the trait locus and mis-specification of the marker allele frequencies.

In nonparametric or model independent analysis, linkage is measured by allele sharing between a known marker and a disease where the mode of inheritance is unknown. The common nonparametric analysis methods are affected sib pair (ASP) analysis and affected relative pair (ARP) analysis. Affected sib pair analysis examines whether siblings in a pedigree inherit an identical allele or genetic marker (identical by descent, or IBD), meaning that the locus comes from the same parent. The alternative method, affected relative pair analysis, tests for deviation from the expected distribution of identity by state (IBS) relationships and is independent of parental information. Nonparametric analysis requires considerably more data than parametric analysis to provide potentially informative meioses. However, when a probable inheritance model cannot be assumed these methodologies offer a practical approach (Weeks and Hardy, 1995).

Allelic association refers to the increased or decreased frequency of occurrence of a marker allele with a disease trait representing a deviation from the random occurrence of the alleles with respect to disease phenotype. This can be secondary to either linkage disequilibrium or a true functional association. Association studies represent, therefore, another nonparametric approach to mapping gene loci. This approach is subject to methodological problems, particularly in relation to the choice of a wrong or inadequately matched control population. Family-based association studies using parenteral genotypes as controls may avoid this source of error. In diseases of mid-life to late-life onset, however, parental DNA may not always be available. Because linkage disequilibrium extends only approximately 1 cM from the susceptibility locus, the identification of allelic association that results from linkage disequilibrium encompasses only a limited genetic distance and restricts its use to the study of candidate genes. This is further confounded in complex disease by potential genetic heterogeneity or gene-gene interactions. However, the power of using linkage disequilibrium for fine mapping can be taken advantage of in genetically isolated populations which are perhaps more homogenous in terms of disease origin and in which perhaps fewer genes contribute toward a disease trait. Problematic in this approach is the observation that although the use of isolated populations may increase power, the entire spectrum of susceptibility genes may not be manifest in that population (Bergkvist, 1996).

GENOME SCREENS

Four separate groups have completed multistage genome screens in multiplex families that were collected in Canada, the United Kingdom, United States, and Finland (Ebers et al., 1996; Haines et al., 1996; Sawcer et al., 1996; Kuokkanen et al., 1997). These scans have used linkage or affected sib pair analysis of a large number of markers relatively evenly distributed throughout the chromosomes in an attempt to identify genomic regions potentially harboring susceptibility genes for MS. The U.S. study sampled 52 multiplex MS families, with many of these having between three and eight affected members (Haines et al., 1996). Both parametric and nonparametric linkage analysis strategies were used to identify 19 regions potentially encoding MS susceptibility. The regions of most significance included 5q13-23, 7q21-22, 19q13, and the MHC on 6p21. The later two of these regions were subsequently confirmed by genotyping an additional dataset. The U.K. study examined 129 families and identified 19 potentially significant regions (Sawcer et al., 1996). Six of these were then further evaluated in a second group of 98 families. There were two major regions of linkage identified 17q22 and 6p21 (MHC). In the Canadian dataset, the potential effect of the MHC loci was only determined after linkage disequilibrium studies had been conducted (Ebers et al., 1996). A further locus with a modest effect was also identified on chromosome 5p. The Finnish study focused on an isolated population that originates from a limited number of founders (Kuokkanen et al., 1997). Use of candidate gene analysis previously supported the presence of three distinct chromosomal regions potentially conferring MS susceptibility in this population. These regions include genes encoding myelin basic protein (MBP) on 18q22-q23, the MHC on 6p21, and a region on 5p14-p12 (Tienari et al., 1992, 1993; Kuokkanen et al., 1996; Tienari et al., 1998). The genome wide screen was performed in a selection of 16 pedigrees and found evidence for an additional predisposing locus on chromosome 17q22-q24, which was also revealed in the U.K. MS genome screen (Kuokkanen et al., 1997).

Despite the large number of families screened and the large number of genetic markers used, these studies did not define conclusively areas of genetic susceptibility but have suggested regions for further study. It is interesting to note the failure of these studies to demonstrate between them complete replication of regions of interest. This may be partially the result of deficiencies in the statistical approaches that continue to be explored and developed for complex diseases, the presence of etiological and genetic heterogeneity in MS, or an insufficient number of families to reveal true linkage for genes of modest effect. In this regard, an editorial by Risch and Merikangas (1996) suggested that the current approach of linkage analysis is likely to be ineffectual in identifying genes of modest effect in complex diseases because of limitations in the number of families required to identify such loci. As an alternative, they have proposed the use of very large-scale association studies encompassing most of the genes in the human genome. Although this would require considerable improvement in our technological resources, it would reduce the number of families necessary to identify genes conferring susceptibility and is likely a methodology that will become increasingly used as the human genome project provides us with more sequence data. It is also important to recognize the potential for genetic influence, not only in the development of the disease, but also in disease progression. It will likely take much larger datasets that are stratified for clinical variables to detect these.

CANDIDATE GENE ANALYSIS

A large number of studies have been performed on specific candidate genes that have been chosen for their biological relevance to the pathogenesis and pathophysiology of MS. These studies have used both association and linkage studies but have failed conclusively to detect and replicate genomic regions contributing to susceptibility in all ethnic backgrounds. This again may reflect the genetic and etiological heterogeneity of the disease in different populations.

MHC

The human MHC or HLA genes located on chromosome 6 are important determinants in the ability of an individual to respond to foreign or self antigens. The two groups of proteins of most interest in MS are the MHC class I and class II molecules whose role is to present unique amino acid sequences to T cells. MHC class I molecules (HLA-A, B, and C) are made up of polymorphic α chains combined with a conserved β2 microglobulin. They are expressed ubiquitously on cells and are responsible for the presentation of antigens to CD8 expressing class I restricted T cells. MHC class II molecules (HLA-DR, DQ, DP) are made up of a conserved α chain and a more polymorphic β chain. These are expressed primarily on antigen-presenting cells and are able to present antigen to CD4-expressing class II restricted T cells. There has been considerable change in the nomenclature of this region as a result of the development of increasingly sensitive and specific typing techniques. Originally, genotypes within this region were determined by serological techniques with additional specificities being identified through cellular typing. In more recent years, genotyping has utilized molecular methods which detect specific changes in nucleic acid and amino acid sequence. For each class II haplotype, one DQ and one DP α/β chain complex is expressed on cell surfaces. For DR the α chain can combine with alternative β chains and thus for each haplotype two DR molecules may be expressed. For the majority of DR genotypes, one of the expressed β chains is polymorphic while the second is more conserved. However, for the MS-associated DR haplotype DR2, both β chains are polymorphic and this results in two molecules DR2a (DRα and DRβ*0101) and DR2b (DRα and DRβ*1501), both of which are implicated in antigen presentation and which may differ in terms of the peptide epitopes that they present. An individual who inherits two different MHC haplotypes will express four DR molecules, two DQ molecules, and two DP molecules (Apple and Erlich, 1996; McFarland et al., 1997).

There are a number of potential mechanisms by which HLA class II molecules, in particular, may encode susceptibility to autoimmune disease (Oksenberg and Hauser, 1997). These include (1) the ability of particular MHC molecules to present specific epitopes of myelin peptides to encephalitogenic CD4 T cells, (2) the potential role of expressed MHC molecules in the development of immune tolerance within the embryonic thymus and the failure to negatively select or delete autoreactive encephalitogenic T cells, and (3) the ability of MHC molecules to present exogenous antigenic peptides with sequence or conformational similarities to CNS self proteins initiating the immune response through molecular mimicry.

The majority of MHC association studies in MS have been performed on Caucasians of Northern European descent. In this population, susceptibility to MS has been reproducibly associated with the DRB1*1501–DQA1*0102–DQB1*0602 extended haplotype. By serological techniques this was known as DR2. Accumulated data from multiple studies have failed, however, to conclusively localize the susceptibility genes encoded within the DR or DQ regions. The extensive linkage disequilibrium across the DR/DQ region and the fact that DRB1*1501 and DQB1*0602 are found exclusively on these haplotypes in Northern Europeans has prevented a clear resolution of the relative contributions of each gene (Oksenberg and Hauser, 1997). It is likely that combinations of specific alleles at DR and DQ loci or transcomplementation may be influential. Several studies in non-European populations have not supported the DR2 association. This suggests perhaps a degree of etiological heterogeneity with different susceptibility genes determining disease risk in different populations. This is supported by studies in the French Canadian, Japanese, and Sardinian populations (Marrosu et al., 1988; Kira 1996; Storch and Lassmann, 1997).

Despite the large number of association studies that have supported the role of the MHC in MS susceptibility, linkage analysis studies have been less conclusive (Ebers et al., 1982; Tienari et al., 1993; Kellar-Wood et al., 1995; Voskuhl et al., 1996). Several investigators have provided weak evidence of linkage, but some of these have been called into question as a result of flaws in the methodological approach (McFarland, et al., 1997; Ebers and Sadovnick, 1998). Sib pair analysis has failed to demonstrate an increase in DR2 (DR15) sharing by affected siblings that reaches more than of questionable significance (McFarland et al., 1997). In isolated populations such as sibling pairs derived from Finnish families, linkage analysis has supported a role for the MHC and also for the myelin basic protein locus in disease susceptibility (Tienari et al., 1993, 1998). The relative paucity of evidence in support of linkage in the presence of consistent associations between DR or DQ, or both, in MS may reflect differences between different geographical and ethnic populations and may also suggest perhaps that individual genes influencing susceptibility may not be necessary for disease development.

TNF-α

There is considerable evidence to support a potential role for tumor necrosis factor-α (TNF-α) in the pathophysiology of MS. TNF-α is a proinflammatory cytokine that has been demonstrated to upregulate MHC class II molecule expression in the central nervous system as well as to be myelinotoxic (Ledeen and Chakraborty, 1998), although TNF-α has also been shown to have beneficial effects on neurons (Bruce et al., 1996). Genetic variation at TNF-α has been correlated with increased production of this cytokine by HLA-DR2$^+$ T cells, suggesting a functional basis by which polymorphisms within the TNF gene may contribute to disease susceptibility (Zipp et al., 1995). The genetic region encoding TNF-α is located within the MHC on chromosome 6 and is in linkage disequilibrium with DR. This has made it difficult to separate the potential influence of this region from the effect on susceptibility contributed by class II molecules. Several association and linkage studies have been published in relation to this region, but the results have been conflicting (Fugger et al., 1990a; He et al., 1995; Vandevyver, et al., 1994b; Sandberg-Wollheim et al., 1995; Garcia-Merino, et al., 1996; Huizinga et al., 1997; Kirk et al., 1997; Schrijver et al., 1997; Weinshenker et al., 1997). Additional studies, perhaps in populations that do not demonstrate a strong class II association with disease, will be needed to further elucidate the potential relevance of this locus to MS susceptibility.

TAP1 and 2

Also encoded within the MHC are genes that are involved in the transport of peptides from the cytoplasm to the endoplasmic reticulum destined for cell surface expression by MHC class I molecules. These genes are located in the class II region between genes encoding DQ and DP. TAP genes have been demonstrated to be polymorphic and several studies have been conducted in different MS populations using somewhat different methodologies (Liblau et al., 1993; Kellar-Wood et al., 1994b; Middleton et al., 1994; Spurkland et al., 1994; Vandevyver et al., 1994c; Bennetts et al., 1995; Moins-Teisserenc et al., 1995). Again, the results have been inconclusive. It is also not clear that defined polymorphisms in this region are of functional significance.

T-cell Receptor

The potential importance of the T-cell receptor region in conferring MS susceptibility is suggested by the role of the receptor in antigen recognition and by the demonstrated relevance of the region in experimental allergic encephalomyelitis (EAE). In several strains of mice and rats, encephalitogenic T cells specific for peptide epitopes of myelin basic protein or proteolytic protein have a restricted T-cell receptor expression. Although this does not necessarily reflect polymorphic differences in the germline, it serves to substantiate the potential relevance of this region in disease development.

There has been little support for the role of T-cell receptor polymorphisms in relapsing, remitting MS. However, recent studies have demonstrated a potential role for polymorphisms within this region in determining a relapsing progressive disease course (Beall et al., 1989; Oksenberg et al., 1989; Seboun et al., 1989; Fugger et al., 1990b; Charmley et al., 1991a, 1991b; Hillert et al., 1991; Lynch et al., 1991; Hashimoto et al., 1992; Hillert et al., 1992; Hillert and Olerup 1992; Lynch et al., 1992; Robinson and Kindt, 1992; Usuku et al., 1992; Beall et al., 1993; Briant et al., 1993; Martinez-Naves et al., 1993; Eoli et al., 1994b; Vandevyver et al., 1994a; Wood et al., 1995b; Droogan et al., 1996; Hockertz et al., 1998). This will need to be substantiated in other populations. It is also evident that the putative influence of T-cell receptor gene polymorphisms on MS susceptibility is likely to be small. Because of the extent of linkage disequilibrium across this region, the study of polymorphic markers within each linkage group will likely be necessary to further confirm or exclude an effect upon disease susceptibility or disease course.

Immunoglobulin Genes

Both positive and negative association studies have been reported for genes encoding highly polymorphic regions of the constant region of the immunoglobulin heavy chain known as the GM locus (Pandey et al., 1981; Propert et al., 1982; Francis et al., 1986; Gaiser, et al., 1987; Francis et al., 1988). Other studies have identified a different GM phenotype as conferring a protective effect (Steinman, 1992). The failure to confirm the influence of this region may reflect variations in GM frequencies in different ethnic populations and potentially reflect the presence of population admixture in these studies. A more recent study has demonstrated an association with MS and a polymorphism within the immunoglobulin heavy chain variable region (IGVH-2). However, linkage to disease in families could not be identified (Tienari et al., 1993). A single study of patients from the United Kingdom has reported lod scores suggestive of linkage (Wood et al., 1995a). The accumulative results of studies within the immunoglobulin region suggest a weak association between immunoglobulin variable region polymorphisms and MS. These will need further substantiation in different ethnic backgrounds and with larger datasets.

Myelin Basic Protein

Because of the potential relevance of myelin basic protein (MBP) peptide epitopes as the autoantigenic target in MS, the genetic region encoding this protein has been extensively studied. Polymorphisms have been demonstrated in the upstream regulatory region of MBP, and several studies have demonstrated both linkage and association between a polymorphism in the 5′ end and MS (Boylan et al., 1990; Tienari et al., 1992, 1994; Fuerini, et al., 1997; He et al., 1998). Other studies, however, have failed to confirm this, again reiterating the likely weak influence of this region in terms of disease risk (Graham et al., 1993; Rose et al., 1993; Eoli et al., 1994a; Wood et al., 1994). Other myelin antigens including myelin oligodendrocyte glycoprotein (MOG), 2-3-cyclicnucleotide-3-phosphodiesterase (CNPase), and proteolipid protein are also potential antigenic targets and must be considered as candidate genetic regions. Studies of CNPase and MOG polymorphisms to date have not identified any significant association with MS (Roth et al., 1995; Thompson et al., 1996).

Mitochondrial Genes

Interest in the influence of mitochondrial DNA polymorphisms has come from the demonstration of a mutation in mitochondrial DNA in Leber's hereditary optic neuropathy (LHON), which is transmitted by females and closely resembles optic neuritis. There are a small number of patients with an illness that resembles both LHON and MS in whom a mutation in mitochondrial DNA has been identified. These patients have manifest bilateral optic neuritis in addition to a variety of other symptoms that are strongly suggestive of MS (Harding et al., 1992; Bet et al., 1994; Hanefeld et al., 1994). In addition, several of these patients demonstrated MRI abnormalities and had a family history LHON. Several large studies have subsequently examined a random sample of MS patients looking for evidence of mutations in mitochondrial DNA but have failed to identify such (Kellar-Wood et al., 1994a; Chalmers et al., 1996). It appears, however, that there may be a subset of patients with a disease mimicking MS, characterized by bilateral visual loss, in which mutations of mitochondrial DNA may be present. This reflects the potential degree of disease heterogeneity that exists and supports the development of large databases that will enable the examination of different disease phenotypes in terms of the potential for genetic influence upon disease course.

Other Genes

There have been a number of other candidate regions for which conflicting data have been published. These include heat shock protein-70, complement 3, interferon-γ, interleukin-1 receptor antagonist, interleukin-1β, complement 3, and α-1 antitrypsin (McCombe et al., 1985; Francis et al., 1988; Bulman et al., 1991; Crusius, et al., 1995; Ramachandran and Bell, 1995; Bergkvist et al., 1996; Arroyo et al., 1997; Crusius et al., 1997; de la Concha et al., 1997; Hillert et al., 1997; Wansen et al., 1997). A genetic region surrounding α-lipoprotein C2 on chromosome 19 has been implicated by linkage analysis (Oksenberg et al., 1997). There has been a single positive association study with intracellular adhesion molecule 1 and a negative association study with CTLA4 (Hillert et al., 1997).

CONCLUDING REMARKS

It is clear that the study of complex genetic diseases is more complicated than the study of simple mendelian traits, but the identification of susceptibility genes conferring disease risk to common disease is likely to give us insight into the pathogenesis of environmentally induced disease, improve our ability to identify risk factors be they either genetic or environmental, enable us to develop an understanding of gene and environmental interactions, and perhaps allow for successful therapeutic intervention based on the function of susceptibility loci and the genotype of the affected individual.

The development of large collaborative groups exploring the genetic basis of MS has fostered the development of large well-documented patient databases inclusive of families with several affected members. This, along with the increasing availability of genetic sequence data provided by the Human Genome Project and further advances in statistical approaches for the study of complex disease, is likely to allow the conclusive identification of genes conferring susceptibility to MS. The determination of the functional basis of these genetic variations in turn may provide important new insights into the underlying pathogenesis of this disease.

SUMMARY

Multiple sclerosis (MS) is a relatively common chronic central nervous system disease affecting individuals in young adult life. It is believed to result from an autoimmune attack on myelin proteins. Epidemiological studies have demonstrated that familial risk to this disease is genetically determined, while population risk is influenced by environmental factors. MS is considered a complex trait in which susceptibility and disease course are determined by the influences of multiple genetic loci. Several methodological approaches may be useful for the identification of genes conferring disease risk in complex traits. The eventual determination of the functional basis of these genetic alterations is likely to provide important insights into the underlying pathogenesis of this disease.

REFERENCES

Alter, M., Leibowitz, U. & Spear, J. (1966). Risk of multiple sclerosis related to age of migration to Israel. Arch. Neurol. 110–113.

Alter, M., Kahana, E. & Loewenson, R. (1971). Migration and risk of multiple sclerosis. Neurology 28, 1089–1093.

Apple, R.J. & Erlich, H.A. (1996). In: HLA and MHC: Genes, Molecules and Function (Browning, M. & McMichael, A., eds.), pp.97–109, Bios Scientific Publishers, Oxford, England.

Anonymous. (1982). Concordance in twins and recurrence in sibships in multiple sclerosis [letter]. Lancet 1, 1068.

Anonymous. (1992). Multiple sclerosis in 54 twinships: concordance rate is independent of zygosity. French Research Group on Multiple Sclerosis [see comments]. Ann. Neurol. 32, 724–727.

Arroyo, R., Fernandez-Arquero, M., Crusius, J.B.A., Vidal, F., Martin, C., de Siejas, E.V., Pena, A.S. & de la Concha, E.G. (1997). Combined effect of HLA-DRB1*1501 and interleukin-1 receptor antagonist gene allele 2 in susceptibility to relapsing/remitting multiple sclerosis. Mult. Scler. 3, 323–323.

Beall, S.S., Concannon, P., Charmley, P., McFarland, H.F., Gatti, R. A., Hood, L.E., McFarlin, D.E. & Biddison, W.E. (1989). The germline repertoire of T cell receptor beta-chain genes in patients with chronic progressive multiple sclerosis. J. Neuroimmunol. 21, 59–66.

Beall, S.S., Biddison, W.E., McFarlin, D.E., McFarland, H.F. & Hood, L.E. (1993). Susceptibility for multiple sclerosis is determined, in part, by inheritance of a 175-kb region of the TcR V beta chain locus and HLA class II genes. J. Neuroimmunol. 45, 53–60.

Bennetts, B.H., Teutsch, S. M., Heard, R.N., Dunckley, H. & Stewart, G.J. (1995). TAP2 polymorphisms in Australian multiple sclerosis patients. J. Neuroimmunol. 59, 113–121.

Bergkvist, M., Martinsson, T., Aman, P. & Sandberg-Wollheim, M. (1996). No genetic linkage between multiple sclerosis and the interferon alpha/beta locus. J. Neuroimmunol. 65, 163–165.

Bet, L., Moggio, M., Comi, G.P., Mariani, C., Prelle, A., Checcarelli, N., Bordoni, A., Bresolin, N., Scarpini, E. & Scarlato, G. (1994). Multiple sclerosis and mitochondrial myopathy: an unusual combination of diseases. J. Neurol. 241, 511–516.

Bharucha, N.E., Bharucha, E.P., Wadia, N.H., Singhal, B.S., Bharucha, A.E., Bhise, A.V., Kurtzke, J.F. & Schoenberg, B.S. (1988). Prevalence of multiple sclerosis in the Parsis of Bombay. Neurology 38, 727–729.

Bobowick, A.R., Kurtzke, J.F., Brody, J. A., Hrubec, Z. & Gillespie, M. (1978).Twin study of multiple sclerosis: an epidemiologic inquiry. Neurology 28, 978–987.

Boylan, K.B., Takahashi, N., Paty, D.W., Sadovnick, A.D., Diamond, M., Hood, L.E. & Prusiner, S.B. (1990). DNA length polymorphism 5′ to the myelin basic protein gene is associated with multiple sclerosis. Ann. Neurol. 27, 291–297.

Briant, L., Avoustin, P., Clayton, J., McDermott, M., Clanet, M. & Cambon-Thomsen, A. (1993). Multiple sclerosis susceptibility: population and twin study of polymorphisms in the T-cell receptor beta and gamma genes region. French Group on Multiple Sclerosis. Autoimmunity 15, 67–73.

Bruce, A.J., Boling, W., Kindy, M.S., Peschon, J., Kraemer, P.J., Carpenter, M.K., Holtsberg, F.W. & Mattson, M.P. (1996) Altered neuronal and microglial responses to brain injury in mice lacking TNF receptors. Nat. Med. 2, 788–794.

Bulman, D.E., Armstrong, H. & Ebers, G.C. (1991). Allele frequencies of the third component of complement (C3) in MS patients. J. Neurol. Neurosurg. Psychiatry 54, 554–555.

Chalmers, R.M., Robertson, N., Compston, D.S. & Harding, A.E. (1996). Sequence of mitochondrial DNA in patients with multiple sclerosis. Ann. Neurol. 40, 239–243.

Charmley, P., Beall, S.S., Concannon, P., Hood, L. & Gatti, R.A. (1991a). Further localization of a multiple sclerosis susceptibility gene on chromosome 7q using a new T cell receptor beta-chain DNA polymorphism. J. Neuroimmunol. 32, 231–240.

Charmley, P., Beall, S.S., Concannon, P., Hood, L. & Gatti, R.A. (1991b). Further localization of a multiple sclerosis susceptibility gene on chromosome 7q using a new T cell receptor beta-chain DNA polymorphism. J. Neuroimmunol. 32, 231–240.

Compston, A. (1994). The epidemiology of multiple sclerosis: principles, achievements and recommendations. Ann. Neurol. S211–S217.

Crusius, J.B., Pena, A.S., Van, O.B., Bioque, G., Garcia, A., Dijkstra, C.D. & Polman, C.H. (1995). Interleukin-1 receptor antagonist gene polymorphism and multiple sclerosis [letter] [see comments]. Lancet 346, 979.

Crusius, J.B.A., Schrijver, H.M., Garcia-Gonzalez, A., Swiers, A., Kostense, P.J., Uitdehaag, B.M.J., Pena, A.S. & Polman, C.H. (1997). The polymorphic IL-1B/IL-1RA gene cluster in multiple sclerosis. Mult. Scler. 3, 322–322.

Currier, R.D. & Eldridge, R. (1982). Possible risk factors in multiple sclerosis as found in a national twin study. Arch. Neurol. 39, 140–144.

Davenport, C.B. (1922). Multiple sclerosis from the standpoint of geographic distribution and race. Arch. Neurol. Psychiatry 8, 51–58.

de la Concha, E.G., Arroyo, R., Crusius, J., Campillo, J.A., Martin, C., Varella de Siejas, E., Pena, A.S., Claveria, L.E. & Fernandez-Arquero, M. (1997). Combined effect of HLA-DRB1*1501 and interleukin-1 receptor antagonist gene allele 2 in susceptibility to relapsing/remitting multiple sclerosis. J. Neuroimmunol. 80, 172–178.

Dean, G. (1949). Disseminated sclerosis in South Africa. Br. Med. J. 1, 842–845.

Dean, G. (1967). Annual incidence, prevalence and mortality of MS in white south African-born and white immigrants to South Africa. Br. Med. J. 724–726.

Dean, G., McLoughlin, H., Brady, R., et al. (1976). Multiple sclerosis among immigrants in Greater London. Br. Med. J. 1, 861–864.

Detels, R., Visscher, B.R., Malmgren, R.M., et al. (1977). Evidence for lower susceptibility to multiple sclerosis in Japanese-Americans. Am. J. Epidemiol. 105, 303–310.

Drake, C.G., Rozzo, S.J., Hirschfeld, H.F., Smarnworawong, N.P., Palmer, E. & Kotzin, B.L. (1995). Analysis of the New Zealand Black contribution to lupus-like renal disease. Multiple genes that operate in a threshold manner. J. Immunol.154, 2441–2447.

Droogan, A.G., Kirk, C. W., Hawkins, S.A., McMillan, S.A., Nevin, N.C. & Graham, C.A. (1996). T-cell receptor alpha, beta, gamma, and delta chain gene microsatellites show no association with multiple sclerosis. Neurology 47, 1049–1053.

Dyment, D.A., Sadnovich, A.D. & Ebers, G.C. (1997). Genetics of multiple sclerosis. Hum. Mol. Genet. 6, 1693–1698.

Ebers, G.C. & Bulman, D. (1986). The geography of MS reflects genetic susceptibility. Neurology 36, 108–108.

Ebers, G.C. & Sadovnick, A.D. (1994). The role of genetic factors in multiple sclerosis susceptibility. J. Neuroimmunol. 1–17.

Ebers, G.C. & Sadovnick, A.D. (1998). In: Multiple Sclerosis (Paty, D. & Ebers, GC., Eds.), pp. 29–47, F.A. Davis Co, Philadelphia.

Ebers, G.C., Paty, D.W., Stiller, C.R., Nelson, R.F., Seland, T.P. & Larsen, B. (1982). HLA-typing in multiple sclerosis sibling pairs. Lancet 2, 88–90.

Ebers, G.C., Bulman, D.E., Sadovnick, A.D., Paty, D.W., Warren, S., Hader, W., Murray, T.J., Seland, T.P., Duquette, P. & Grey, T. (1986). A population-based study of multiple sclerosis in twins. N. Engl. J. Med. 315, 1638–1642.

Ebers, G.C., Sadovnick, A.D. & Risch, N.J. (1995). A genetic basis for familial aggregation in multiple sclerosis. Canadian Collaborative Study Group [see comments]. Nature 377, 150–151.

Ebers, G.C., Kukay, K., Bulman, D.E., Sadovnick, A.D., Rice, G., Anderson, C., Armstrong, H., Cousin, K., Bell, R.B., Hader, W., Paty, D.W., Hashimoto, S., Oger, J., Duquette, P., Warren, S., Gray, T., O'Connor, P., Nath, A., Auty, A., Metz, L., Francis, G., Paulseth, J.E., Murray, T.J., Pryse-Phillips, W. & Risch, N. (1996). A full genome search in multiple sclerosis (see comments). Nat. Genet. 13, 472–476.

Eichorst, H. (1896). Uber infantile und hereditare multiple sclerose. Virchows Arch. 146, 173–193.

Elian, M., Nightingale, S. & Dean, G. (1990a). Multiple sclerosis among the United Kingdom born children of immigrants from the Indian Subcontinent. J. Neurol. Neurosurg. Psychiatry 906– 911.

Elian, M., Nightingale, S. & Dean, G. (1990b). Multiple sclerosis among United Kingdom-born children of immigrants from the Indian subcontinent, Africa and the West Indies [see comments]. J. Neurol. Neurosurg. Psychiatry 53, 906–911.

Enstrom, J.R. & Operskalski, EA. (1978). Multiple sclerosis among Spanish-surnamed Californians. Neurology 28, 434–438.

Eoli, M., Pandolfo, M., Milanese, C., Gasparini, P., Salmaggi, A. & Zeviani, M. (1994a). The myelin basic protein gene is not a major susceptibility locus for multiple sclerosis in Italian patients. J. Neurol. 241, 615–619.

Eoli, M., Wood, N.W., Kellar-Wood, H.F., Holmans, P., Clayton, D. & Compston, D.A. (1994b). No linkage between multiple sclerosis and the T cell receptor alpha chain locus. J. Neurol. Sci. 124, 32–37.

Francis, D.A., Brazier, D.M., Batchelor, J.R., McDonald, W.I., Downie, A.W. & Hern, J.E. (1986). Gm allotypes in multiple sclerosis: influence susceptibility in HLA-DQwl-positive patients from the North-East of Scotland. Clin. Immunol. Immunopathol. 41, 409–416.

Francis, D.A., Klouda, P.T., Brazier, D.M., Batchelor, J.R., McDonald, W.I. & Hern, J.E. (1988). Alpha-1-antitrypsin (Pi) types in multiple sclerosis and lack of interaction with immunoglobulin (Gm) markers. J. Immunogenet. 15, 251–255.

French Research Group on Multiple Sclerosis. (1992). Multiple sclerosis in 54 twinships: concordance rate is independent of zygosity. Ann. Neurol. 32, 724–727.

Fuerini, F.R., Losciale, L., Caputo, D., Boiko, A., Favorova, O., Gusev, E. & Ferrante, P. (1997). Association of MBP gene polymorphism and HLA DR2 antigen in Italian and Russian multiple sclerosis patients. Mult. Scler. 3, 321–321.

Fugger, L., Morling, N., Sandberg-Wollheim, M., Ryder, L.P. & Svejgaard, A. (1990a). Tumor necrosis factor alpha gene polymorphism in multiple sclerosis and optic neuritis. J. Neuroimmunol. 27, 85–88.

Fugger, L., Sandberg-Wollheim, M., Morling, N., Ryder, L.P. & Svejgaard, A. (1990b). The germline repertoire of T-cell receptor beta chain genes in patients with relapsing/remitting multiple sclerosis or optic neuritis. Immunogenetics 31, 278–280.

Gaiser, C.N., Johnson, M.J., de, L.G., Rassenti, L., Cavalli-Sforza, L.L. & Steinman, L. (1987). Susceptibility to multiple sclerosis associated with an immunoglobulin gamma 3 restriction fragment length polymorphism. J. Clin. Invest. 79, 309–313.

Garcia-Merino, A., Alper, C.A., Usuku, K., Marcus-Bagley, D., Lincoln, R., Awdeh, Z., Yunis, E.J., Eisenbarth, G.S., Brink, S.J. & Hauser, S.L. (1996). Tumor necrosis factor (TNF) microsatellite haplotypes in relation to extended haplotypes, susceptibility to diseases associated with the major histocompatibility complex and TNF secretion. Hum. Immunol. 50, 11–21.

Graham, C.A., Kirk, C.W., Nevin, N.C., Droogan, A.G., Hawkins, S.A., McMillan, S.A. & McNeill, T.A. (1993). Lack of association between myelin basic protein gene microsatellite and multiple sclerosis [letter; comment]. Lancet 341, 1596.

Gronning, M. & Mellgren, SI. (1985). Multiple sclerosis in the two northernmost counties of Norway. Acta. Neurol. Scand. 72, 321–327.

Hader, W.J., Feasby, T.E., Noseworth, J.H., Rice, G.P. & Ebers, G.C. (1985). Multiple Sclerosis in Canadian Native people. Neurology 35, 300.

Hader, W.J., Seland, T.P., Hader, M.B., Harris, C.J. & Dietrich, D.W. (1996). The occurrence of multiple sclerosis in Hutterites of North America. Can. J. Neurol. Sci. 23, 291–295.

Haines, J.L., Ter-Minassian, M., Bazyk, A., Gusella, J.F., Kim, D. J., Terwedow, H., Pericak-Vance, M.A., Rimmler, J.B., Haynes, C.S., Roses, A.D., Lee, A., Shaner, B., Menold, M., Seboun, E., Fitoussi, R.P., Gartioux, C., Reyes, C., Ribierre, F., Gyapay, G., Weissenbach, J., Hauser, S.L., Goodkin, D.E., Lincoln, R., Usuku, K. & Oksenberg, J.R. (1996). A complete genomic screen for multiple sclerosis underscores a role for the major histocompatability complex. The Multiple Sclerosis Genetics Group [see comments]. Nat. Genet. 13, 469–471.

Hammond, S.R., McLeod, J.G., Millingen, K.S., et al. (1987). The epidemiology of multiple sclerosis in three Australian cities: Perth, Newcastle and Hobart. Brain 111, 1–25.

Hanefeld, F.A., Ernst, B.P., Wilichowski, E. & Christen, H.J. (1994). Leber's hereditary optic neuropathy mitochondrial DNA mutations in childhood multiple sclerosis [letter]. Neuropediatrics 25, 331.

Harding, A.E., Sweeney, M.G., Miller, D.H., Mumford, C.J., Kellar-Wood, H., Menard, D., McDonald, W.I. & Compston, D.A. (1992). Occurrence of a multiple sclerosis-like illness in women who have a Leber's hereditary optic neuropathy mitochondrial DNA mutation. Brain 115, 979–989.

Hashimoto, L.L., Mak, T.W. & Ebers, G.C. (1992). T cell receptor alpha chain polymorphisms in multiple sclerosis. J. Neuroimmunol. 40, 41–48.

He, B., Navikas, V., Lundahl, J., Soderstrom, M. & Hillert, J. (1995). Tumor necrosis factor alpha-308 alleles in multiple sclerosis and optic neuritis. J. Neuroimmunol. 63, 143–147.

He, B., Yang, B., Lundahl, J., Fredrikson, S. & Hillert, J. (1998).The myelin basic protein gene in multiple sclerosis: identification of discrete alleles of a 1.3 kb tetranucleotide repeat sequence. Acta. Neurol. Scand. 97, 46–51.

Heltberg, A. & Holm, N.V. (1982). Concordance in twins and recurrence in sibships in multiple sclerosis. Lancet 1, 1068.

Hillert, J. & Olerup, O. (1992). Germ-line polymorphism of TCR genes and disease susceptibility—fact or hypothesis? (see comments) Immunol. Today. 13, 47–49.

Hillert, J., Leng, C. & Olerup, O. (1991). No association with germline T cell receptor beta-chain gene alleles or haplotypes in Swedish patients with multiple sclerosis. J. Neuroimmunol. 32, 141–147.

Hillert, J., Leng, C. & Olerup, O. (1992). T-cell receptor alpha chain germline gene polymorphisms in multiple sclerosis [see comments]. Neurology 42, 80–84.

Hillert, J., He, B., Ligers, A., Xu, C., Fogdell, A. & Olerup, O. (1997). Candidate gene analysis in multiple sclerosis: CTLA4, TGFB2, yIFN and TCRB. Mult. Scler. 3, 270–270.

Hockertz, M.K., Paty, D.W. & Beall, S.S. (1998). Susceptibility to relapsing-progressive multiple sclerosis is associated with inheritance of genes linked to the variable region of the TcR beta locus: use of affected family-based controls. Am J Hum Genet. 62, 373–385.

Huizinga, T.W., Westendorp, R.G., Bollen, E.L., Keijsers, V., Brinkman, B.M., Langermans, J.A., Breedveld, F.C., Verweij, C.L., van, d.G., Dams, L., Crusius, J.B., Garcia-Gonzalez, A., van, O.B., Polman, C.H. & Pena, A.S. (1997). TNF-alpha promoter polymorphisms, production and susceptibility to multiple sclerosis in different groups of patients. J. Neuroimmunol. 72, 149–153.

Kellar-Wood, H., Robertson, N., Govan, G.G., Compston, D.A. & Harding, A.E. (1994a). Leber's hereditary optic neuropathy mitochondrial DNA mutations in multiple sclerosis. Ann. Neurol. 36, 109–112.

Kellar-Wood, H.F., Powis, S.H., Gray, J. & Compston, D.A. (1994b). MHC-encoded TAP1 and TAP2 dimorphisms in multiple sclerosis. Tissue Antigens 43, 129–132.

Kellar-Wood, H.F., Wood, N.W., Holmans, P., Clayton, D., Robertson, N. & Compston, D.A. (1995). Multiple sclerosis and the HLA-D region: linkage and association studies. J. Neuroimmunol. 58, 183–190.

Kinnunen, E., Juntunen, J., Ketonen, L., Koskimies, S., Konttinen, Y.T., Salmi, T., Koskenvuo, M. & Kaprio, J. (1988). Genetic susceptibility to multiple sclerosis. A co-twin study of a nationwide series. Arch. Neurol. 45, 1108–1111.

Kira, J., Kanai, T., Nishimura, Y., Yamasaki, K., Matsushita, S., Kawano, Y., Hasuo, K., Tobimatsu, S. & Kobayashi, T. (1996). Western versus Asian types of multiple sclerosis: immunogenetically and clinically distinct disorders. Ann. Neurol. 40, 569–574.

Kirk, C.W., Droogan, A.G., Hawkins, S.A., McMillan, S.A., Nevin, N.C. & Graham, C.A. (1997). Tumour necrosis factor microsatellites show association with multiple sclerosis [published erratum appears in J. Neurol. Sci. 1997 Sept, 10;150(2):183]. J Neurol Sci. 147, 21–25.

Kuokkanen, S., Sundvall, M., Terwilliger, J.D., Tienari, P.J., Wikstrom, J., Holmdahl, R., Pettersson, U. & Peltonen, L. (1996). A putative vulnerability locus to multiple sclerosis maps to 5p14–p12 in a region syntenic to the murine locus Eae2 [see comments]. Nat. Genet. 13, 477–480.

Kuokkanen, S., Gschwend, M., Rioux, J.D., Daly, M.J., Terwilliger, J.D., Tienari, P.J., Wikstrom, J., Palo, J., Stein, L.D., Hudson, T.J., Lander, E.S. & Peltonen, L. (1997). Genomewide scan of multiple sclerosis in Finnish multiplex families. Am. J. Hum. Genet. 61, 1379–1387.

Kurtzke, J.F. (1985). In: Demyelinating Diseases (J.C. Koetsier, eds.), pp. 259–287, Elsevier, Amsterdam.

Kurtzke, J.F. & Bui, Q.U. (1980). Multiple sclerosis in a migrant population. Part II. Half Orientals immigrating in childhood. Neurology 8, 256–260.

Kurtzke, J.F., Beebe, G.W. & Norman, J.E.J. (1979). Epidemiology of multiple sclerosis in U.S. veterans: 1. Race, sex, and geographic distribution. Neurology 29, 1228–1235.

Kurtzke, J.F., Beebe, J.W. & Norman, J.E. (1985). Epidemiology of multiple sclerosis in US veterans. III. Migration and the risk of MS. Neurology 35, 672–678.

Lander, E.S. & Schork, N.J. (1994). Genetic dissection of complex traits [published erratum appears in Science (1994) Oct, 21;266 (5184):353]. Science 265, 2037–2048.

Ledeen, R.W. & Chakraborty, G. (1998) Cytokines, signal transduction, and inflammatory demyelination: review and hypothesis. Neurochem. Res. 23, 277–289.

Liblau, R., van, E.P., Sandberg-Wollheim, M., Patel, S.D., Lopez, M.T., Land, S., Fugger, L. & McDevitt, H.O. (1993). Antigen processing gene polymorphisms in HLA-DR2 multiple sclerosis. Neurology 43, 1192–1197.

Lynch, S.G., Rose, J.W., Petajan, J.H., Stauffer, D., Kamerath, C. & Leppert, M. (1991). Discordance of T-cell receptor beta-chain genes in familial multiple sclerosis. Ann. Neurol. 30, 402–410.

Lynch, S.G., Rose, J.W., Petajan, J.H. & Leppert, M. (1992). Discordance of the T-cell receptor alpha-chain gene in familial multiple sclerosis. Neurology 42, 839–844.

MacKay, R.P. & Myrianthopoulos, N. C. (1966). Multiple sclerosis in twins and their relatives. Arch. Neurol. 15, 449–462.

Marrosu, M.G., Muntoni, F., Murru, M.R., Spinicci, G., Pischedda, M.P., Goddi, F., Cossu, P. & Pirastu, M. (1988). Sardinian multiple sclerosis is associated with HLA-DR4: a serologic and molecular analysis. Neurology 38, 1749–1753.

Martinez-Naves, E., Victoria-Gutierrez, M., Uria, D.F. & Lopez-Larrea, C. (1993). The germline repertoire of T cell receptor beta-chain genes in multiple sclerosis patients from Spain. J. Neuroimmunol. 47, 9–13.

McCombe, P.A., Clark, P., Frith, J.A., Hammond, S.R., Stewart, G.J., Pollard, J.D. & McLeod, J.G. (1985). Alpha-1 antitrypsin phenotypes in demyelinating disease: an association between demyelinating disease and the allele PiM3. Ann. Neurol. 18, 514–516.

McFarland, H.F., Greenstein, J., McFarlin, D.E., Eldridge, R., Xu, X.H. & Krebs, H. (1984). Family and twin studies in multiple sclerosis. Ann. N. Y. Acad. Sci. 436, 118–124.

McFarland, H.F., Martin, R. & McFarlin, D.E. (1997). In: Multiple Sclerosis: Clinical and Pathogenetic Basis (Raine, C.S., McFarland, H.F., & Tourtellote, W.W., Eds.), pp. 205–219, Chapman and Hall, London.

Middleton, D., Megaw, G., Cullen, C., Hawkins, S., Darke, C. & Savage, D.A. (1994). TAP1 and TAP2 polymorphism in multiple sclerosis patients. Hum. Immunol. 40, 131–134.

Moins-Teisserenc, H., Semana, G., Alizadeh, M., Loiseau, P., Bobrynina, V., Deschamps, I., Edan, G., Birebent, B., Genetet, B. & Sabouraud, O. (1995). TAP2 gene polymorphism contributes to genetic susceptibility to multiple sclerosis. Hum. Immunol. 42, 195–202.

Morel, L., Rudofsky, U.H., Longmate, J.A., Schiffenbauer, J. & Wakeland, E.K. (1994). Polygenic control of susceptibility to murine systemic lupus erythematosus. Immunity 1, 219–229.

Mumford, C.J., Wood, N.W., Kellar-Wood, H., Thorpe, J.W., Miller, D.H. & Compston, D.A. (1994). The British Isles survey of multiple sclerosis in twins. Neurology 44, 11–15.

Oksenberg, J.R. & Hauser, SL. (1997). New insights into the immunogenetics of multiple sclerosis. Curr. Opin. Neurol. 10, 181–185.

Oksenberg, J.R., Sherritt, M., Begovich, A.B., Erlich, H.A., Bernard, C.C., Cavalli-Sforza, L.L. & Steinman, L. (1989). T-cell receptor V alpha and C alpha alleles associated with multiple and myasthenia gravis. Proc. Natl. Acad. Sci. U. S. A. 86, 988–992.

Oksenberg, J.R., Rimmler, J.B., Lincoln, R., Hauser, S.L., Agawral, R., Terwedow, H.A., Pericak-Vance, M.A. & Haines, J.L. (1997). The potential role of a locus on chromosome 19 in multiple sclerosis (MS). Am. J. Hum. Genet. 61, A289–A289.

Ott, J. (1991). In: Analysis of Human Genetic Linkage (Ott, J., Ed.), pp. 54–80, The Johns Hopkins University Press, Baltimore.

Palffy, G., Czopf, J. & Gyodi, E. (1993). In: Multiple Sclerosis in Europe (Finhaber, W. & Lauer, K., Eds.), pp. 274–278, Leuchtturm-Verlag LtV Press, Darmstadt, Germany.

Pandey, J.P., Goust, J.M., Salier, J.P. & Fudenberg, H.H. (1981). Immunoglobulin G heavy chain (Gm) allotypes in multiple sclerosis. J. Clin. Invest. 67, 1797–1800.

Pericack-Vance, M.A. & Haines, J.L. (1997). Mendel's pea soup: dissecting the genetics of common disease. American Academy of Neurology Annual Meeting—Neurogenetics Course Syllabi.

Popov, V.S. (1983). Clinical picture and epidemiology of disseminated sclerosis. Zh. Nevropatol. Psikhiatr. 83, 1330–1334.

Propert, D.N., Bernard, C.C. & Simons, M.J. (1982). Gm allotypes and multiple sclerosis. J. Immunogenet. 9, 359–361.

Ramachandran, S. & Bell, R.B. (1995). Heat shock protein 70 gene polymorphisms and multiple sclerosis. Tissue Antigens 46, 140–141.

Risch, N. & Merikangas, K. (1996). The future of genetic studies of complex human diseases (see comments). Science 273, 1516–1517.

Risch, N., Ghosh, S. & Todd, J.A. (1993). Statistical evaluation of multiple-locus linkage data in experimental species and its relevance to human studies: application to nonobese diabetic (NOD) mouse and human insulin-dependent diabetes mellitus (IDDM). Am. J. Hum. Genet. 53, 702–714.

Robertson, N., O'Riordan, J., Clayton, D., et al. (1995). Conjugal multiple sclerosis. J. Neurol. 242, S8–S8.

Robertson, N.P., Fraser, M., Deans, J., Clayton, D., Walker, N. & Compston, D.A. (1996). Age-adjusted recurrence risks for relatives of patients with multiple sclerosis. Brain 119, 449–455.

Robinson, M.A. & Kindt, T.J. (1992). Linkage between T cell receptor genes and susceptibility to multiple sclerosis: a complex issue. Res. Immunol. 4, 274–283.

Rose, J., Gerken, S., Lynch, S., Pisani, P., Varvil, T., Otterud, B. & Leppert, M. (1993). Genetic susceptibility in familial multiple sclerosis not linked to the myelin basic protein gene [see comments]. Lancet 341, 1179–1181.

Roth, M.P., Dolbois, L., Borot, N., Pontarotti, P., Clanet, M. & Coppin, H. (1995). Myelin oligodendrocyte glycoprotein (MOG) gene polymorphisms and multiple sclerosis: no evidence of disease association with MOG. J. Neuroimmunol. 61, 117–122.

Sadovnick, A.D. (1993). Familial recurrence risks and inheritance of multiple sclerosis. Curr. Opin. Neurol. Neurosurg. 6, 189–194.

Sadovnick, A.D. & Ebers, G.C. (1993). Epidemiology of multiple sclerosis: a critical overview. Can. J. Neurol. Sci. 20, 17–29.

Sadvonick, A.D., Baird P.A. & Ward, R.H. (1988). Multiple sclerosis: updated risks for relatives. Am. J. Med. Genet. 29, 553–511.

Sadovnick, A.D., Armstrong, H., Rice, G.P., Bulman, D., Hashimoto, L., Paty, D.W., Hashimoto, S.A., Warren, S., Hader, W. & Murray, T.J. (1993). A population-based study of multiple sclerosis in twins: update. Ann. Neurol. 33, 281–285.

Sadovnick, A.D., Ebers, G.C., Dyment, D.A. & Risch, N.J. (1996). Evidence for genetic basis of multiple sclerosis. The Canadian Collaborative Study Group. Lancet 347, 1728–1730.

Sandberg-Wollheim, M., Ciusani, E., Salmaggi, A. & Pociot, F. (1995). An evaluation of tumor necrosis factor microsatellite alleles in genetic susceptibility to multiple sclerosis. Mult. Scler. 1, 181–185.

Sawcer, S., Jones, H.B., Feakes, R., Gray, J., Smaldon, N., Chataway, J., Robertson, N., Clayton, D., Goodfellow, P.N. & Compston, A. (1996). A genome screen in multiple sclerosis reveals susceptibility loci on chromosome 6p21 and 17q22 [see comments]. Nat. Genet. 13, 464–468.

Sawcer, S., Robertson, N. & Compston, A. (1997). In: Multiple Sclerosis: Clinical Challenges and Controversies (Thompson, A.J., Polman, C. & Hohlfeld, R., eds.), pp. 13–34, Martin Dunitz Ltd., London.

Schrijver, H.M., Crusius, J.B.A., Garcia-Bonzalez, A., Swiers, A., Kostense, P.J., Uitdchaag, B.M.J., Pena, A.S., and Polman, C.H. (1997). TNF/LT haplotypes in subgroups of patients with multiple sclerosis. Mult. Scler. 3, 27127–271.

Seboun, E., Robinson, M.A., Doolittle, T.H., Ciulla, T.A., Kindt, T.J. & Hauser, S.L. (1989). A susceptibility locus for multiple sclerosis is linked to the T cell receptor beta chain complex. Cell 57, 1095–1100.

Skegg, D., Corwin, P.A., Craven, R., Malloch, J.A. & Pollack, M. (1987). Occurrence of multiple sclerosis in the north and south of New Zealand. J. Neurol. Neurosurg. Psychiatry 50, 134–139.

Spurkland, A., Knutsen, I., Undlien, D.E. & Vartdal, F. (1994). No association of multiple sclerosis to alleles at the TAP2 locus. Hum. Immunol. 39, 299–301.

Steinman, L. (1992). Autoimmune disease and the nervous system [editorial]. West. J. Med. 156, 664–666.

Storch, M. & Lassmann, H. (1997). Pathology and pathogenesis of demyelinating diseases. Curr. Opin. Neurol. 10, 186–192.

Thompson, R.J., Mason, C.R., Douglas, A.J., Hinks, L.J., Dwarswaard, A. & Price, S.E. (1996). Analysis of polymorphisms of the 2′,3′-cyclic nucleotide-3′-phosphodiesterase gene in patients with multiple sclerosis. Mult. Scler. 2, 215–221.

Tienari, P.J., Wikstrom, J., Sajantila, A., Palo, J. & Peltonen, L. (1992). Genetic susceptibility to multiple sclerosis linked to myelin basic protein gene (see comments). Lancet. 340, 987–991.

Tienari, P.J., Wikstrom, J., Koskimies, S., Partanen, J., Palo, J. & Peltonen, L. (1993). Reappraisal of HLA in multiple sclerosis: close linkage in multiplex families. Eur. J. Hum. Genet. 1, 257–268.

Tienari, P.J., Terwilliger, J.D., Ott, J., Palo, J. & Peltonen, L. (1994). Two-locus linkage analysis in multiple sclerosis (MS). Genomics 19, 320–325.

Tienari, P.J., Kuokkanen, S., Pastinen, T., Wikstrom, J., Sajantila, A., Sandberg-Wollheim, M., Palo, J. & Peltonen, L. (1998). Golli-MBP gene in multiple sclerosis susceptibility. J. Neuroimmunol. 81, 158–167.

Uitdehaag, B., Polman, C.H., Valk, J., Koetsier, J.C. & Lucas, C.J. (1989). Magnetic resonance imaging studies in multiple sclerosis twins. J. Neurol. Neurosurg. Psychiatry. 52, 1417–1419.

Usuku, K., Joshi, N. & Hauser, S.L. (1992). T-cell receptors: germline polymorphism and patterns of usage in demyelinating diseases. Crit. Rev. Immunol. 11, 381–393.

Vandevyver, C., Buyse, I., Philippaerts, L., Ghabanbasani, Z., Medaer, R., Carton, H., Cassiman, J.J. & Raus, J. (1994a). HLA and T-cell receptor polymorphisms in Belgian multiple sclerosis patients: no evidence for disease association with the T-cell receptor. J. Neuroimmunol. 52, 25–32.

Vandevyver, C., Raus, P., Stinissen, P., Philippaerts, L., Cassiman, J.J. & Raus, J. (1994b). Polymorphism of the tumour necrosis factor beta gene in multiple sclerosis and rheumatoid arthritis. Eur. J. Immunogenet. 21, 377–382.

Vandevyver, C., Stinissen, P., Cassiman, J.J. & Raus, J. (1994c). TAP 1 and TAP 2 transporter gene polymorphisms in multiple sclerosis: no evidence for disease association with TAP. J. Neuroimmunol. 54, 35–40.

Voskuhl, R.R., Goldstein, A.M., Simonis, T., Davey, R.J. & McFarland, H.F. (1996). DR2/DQw1 inheritance and haplotype sharing in affected siblings from multiple sclerosis families. Ann. Neurol. 39, 804–807.

Wansen, K., Pastinen, T., Kuokkanen, S., Wikstrom, J., Palo, J., Peltonen, L. & Tienari, P.J. (1997). Immune system genes in multiple sclerosis: genetic association and linkage analyses on TCR beta, IGH, IFN-gamma and IL-1ra/IL-1 beta loci. J. Neuroimmunol. 79, 29–36.

Weeks, D.E. & Hardy, L.D. (1995). The affected-pedigree-member method: power to detect linkage. Hum. Hered. 45, 13–24.

Weinshenker, B.G., Wingerchuk, D.M., Liu, Q., Bissonet, A.S., Schaid, D.J. & Sommer, S.S. (1997). Genetic variation in the tumor necrosis factor alpha gene and the outcome of multiple sclerosis. Neurology 49, 378–385.

Williams, A., Eldridge, R., McFarland, H., Houff, S., Krebs, H. & McFarlin, D. (1980). Multiple sclerosis in twins. Neurology 30, 1139–1147.

Wood, N.W., Holmans, P., Clayton, D., Robertson, N. & Compston, D.A. (1994). No linkage or association between multiple sclerosis and the myelin basic protein gene in affected sibling pairs. J. Neurol. Neurosurg. Psychiatry 57, 1191–1194.

Wood, N.W., Sawcer, S.J., Kellar-Wood, H.F., Holmans, P., Clayton, D., Robertson, N. & Compston, D.A. (1995a). Susceptibility to multiple sclerosis and the immunoglobulin heavy chain variable region. J. Neurol. 242, 677–682.

Wood, N.W., Sawcer, S.J., Kellar-Wood, H.F., Holmans, P., Clayton, D., Robertson, N. & Compston, D.A. (1995b). The T-cell receptor beta locus and susceptibility to multiple sclerosis. Neurology 45, 1859–1863.

Yaqub, B. & Daif, A.K. (1988). Multiple sclerosis in Saudi Arabia. Neurology 38, 621–623.

Yu, Y.L., Woo, E., Hawkins, B.R., Ho, H.C., Huang, C.Y. (1989). Multiple sclerosis amongst Chinese in Hong Kong. Brain 112, 1445–1467.

Zipp, F., Weber, F., Huber, S., Sotgiu, S., Czlonkowska, A., Holler, E., Albert, E., Weiss, E.H., Wekerle, H. & Hohlfeld, R. (1995). Genetic control of multiple sclerosis: increased production of lymphotoxin and tumor necrosis factor-alpha by HLA-DR2+ T cells [see comments]. Ann. Neurol. 38, 723–730.

Chapter 9

The Role of Mitochondrial Genome Mutations in Neurodegenerative Disease

GORDON W. GLAZNER

Advances in Cell Aging and Gerontology
Volume 3, pages 313–354

ISBN: 0-7623-0405-7

INTRODUCTION

From current theory, life has existed on Earth for a period of nearly 4 billion years, and for the first 3 billion or so, it was exclusively prokaryotic. One hypothesis states that then an event that was to change the face of life occurred, the symbiosis of two previously separate cells. According to this scenario, a large, relatively complex but energetically sluggish cell engulfed a smaller, more metabolically active one, an event which must have occurred uncounted times in the preceding eons. In this specific instance, however, the phagocytosed cell was not digested, but became a symbiote. The larger cell within which it lived provided the smaller with protection and nourishment while the smaller cell provided abundant energy to the partnership. The small symbiote became the organelle we call the mitochondria, and this relationship may have been a key in the incredible explosion of multicellular life that followed. After a billion years, mitochondria still carry the legacy of their history as an independent entity in the form of their own functional genome, mitochondrial DNA (mtDNA). Genetic analysis of mtDNA supports this hypothesis, showing significant homology between all eukaryotes. This separate genetic code still bears the mark of its bacterial origins, being circular, unprotected by histones, lacking introns, and free within the lumen of the mitochondria. While mtDNA no longer codes for structural proteins, it does code for critical components of the electron transport chain (ETC), the mechanism whereby the great majority of energy is supplied to the cell. The existence of a separate mitochondrial genome clearly confers important advantages to cells, but because of the locally high oxidizing environment in the mitochondria, the relative exposure of mtDNA, and the poor mitochondrial DNA repair mechanisms, mtDNA is much more susceptible to damage and mutation than is nuclear DNA. This damage decreases the efficiency of adenosine triphosphate (ATP) synthesis, and increases production of toxic reactive waste products. Whereas all cells may suffer by these events, neurons are particularly vulnerable, and a number of neurological diseases have mtDNA mutations either as a cause or contributing factor. This chapter discusses the causes of the increased vulnerability of mitochondria in neurons to mtDNA mutations, the reasons neurons are at special risk from mitochondrial dysfunction, and some of what is known of mtDNA mutations in relation to neuropathology.

MITOCHONDRIAL STRUCTURE

Organellar Structure

Mitochondria appear as small oblong or strand-like organelles under light microscopy, the name mitochondrion meaning "thread-like body." The number of mitochondria in a given cell can vary from as few as two to some thousands, and generally correlates to the energy requirements of that particular cell. There are several thousand mitochondria per hepatocyte, only a dozen or so in resting lymphocytes, and none at all in mature erythrocytes. Mitochondria are mobile within the cell, have been observed to constantly change morphology and position, and seem to accumulate within the cell in regions that have high energy requirements, such as cilia. This movement seems in many cases to be directional and linear, raising speculation that they may be transported along cytoskeletal elements within the cell. Mitochondria have two membranes, which have distinct functions and correspondingly differing physical properties, separated by an intermembrane space termed the mitochondrial matrix. The smooth outer membrane is freely permeable to low molecular weight molecules and ions, allowing passage of material into and out of the organelle. In addition, there are transmembrane pores (porins), which allow passage of larger proteins and other molecules through the

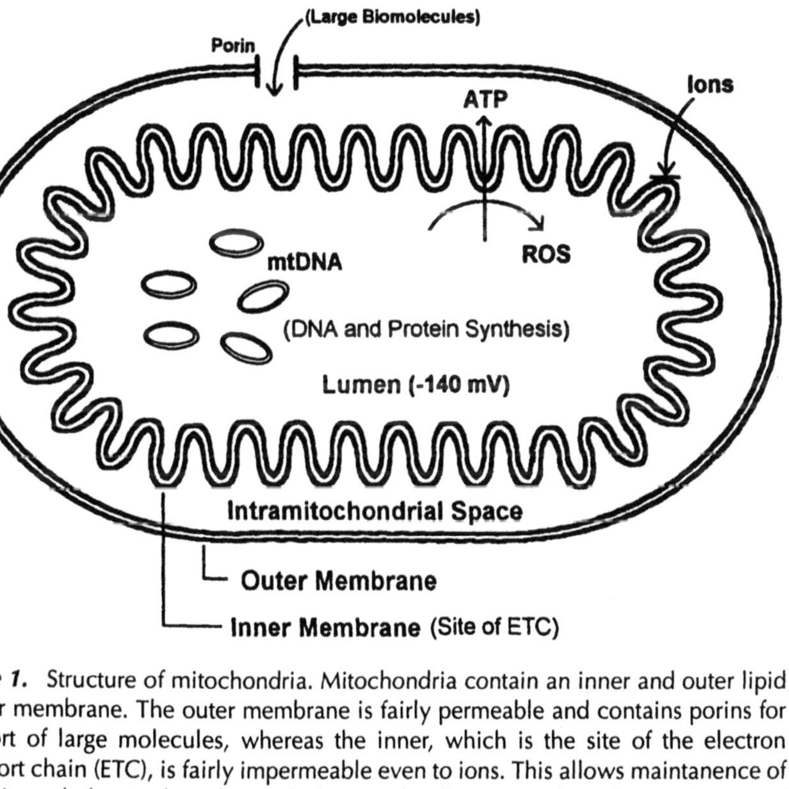

Figure 1. Structure of mitochondria. Mitochondria contain an inner and outer lipid bilayer membrane. The outer membrane is fairly permeable and contains porins for tranport of large molecules, whereas the inner, which is the site of the electron transport chain (ETC), is fairly impermeable even to ions. This allows maintanence of an ionic and electrical gradient which provides the motive force for production of adensosine triphosphate (ATP). The inner chamber, called the lumen, is the site of mtDNA, and DNA and protein synthesis.

outer membrane for maintenance of structural integrity and transport of the large proteins required for the ETC. The highly folded inner membrane, on the other hand, is relatively impermeable, especially to ions. This characteristic contributes to the development of an ionic, or proton, gradient across the membrane, which is critical for ATP production in the oxidative phosphorylation pathway. The inner mitochondrial space contains the circular mitochondrial genome (mtDNA) and is the site of mitochondrial protein and DNA synthesis (Figure 1).

Mitochondrial Genome

The first evidence of mitochondrial genomics came nearly 50 years ago, when Ephrussi et al. (1949) discovered a strain of yeast that grew in smaller than normal colonies in the presence of low glucose, which he termed "petite" mutants. These petite mutants were found to have a respiratory deficiency, which was inherited in a nonmendalian fashion. Ephrussi hypothesized from these studies that the deficiency was passed on by an extra-chromosomal, cytoplasmic element that he termed the ρ factor. In electron microscopy studies, Nass and Nass (1963a, 1963b) observed thread-like structures in the mitochondria of chick embryonic cells, which were destroyed by deoxyribonuclease but not by ribonuclease. Soon thereafter, DNA was detected and quantified in purified yeast mitochondria (Schatz and Klima, 1964). The ρ factor first detected by Ephrussi 20 years before was shown to be a unique feature of mitochondria among organelles; the presence of a genetic system independent of the nucleus. Some years later, Linnane et al. (1976) published the first circular genetic map of yeast mitochondria isolated from petite mutants. The first complete 16,569 base pair (bp) sequencing of human mtDNA was reported by Anderson et al. in 1981, and has been referred to as the most efficiently packed genome found in nature. This relatively short DNA sequence encodes 13 genes for proteins of the respiratory chain (seven subunits of complex I, one subunit of complex III, three subunits of complex IV, and two subunits of complex V) as well as 2 ribosomal RNA (rRNA) and 22 transfer RNA (tRNA) genes (Figure 2).

In addition to the abovementioned genes, there is a control region, which contains two adjacent promoter regions, (one for each strand), the replication origin (OrH) for the guanine-rich heavy (H) strand, and the associated displacement (D) loop. Binding of promoters generates continuous polycistronic transcripts, which are then processed into mature m, t, and rRNAs, necessitating expression of the entire mitochondrial genome for function. Replication of the genome is asynchronous in that the origin for the H strand is in the control region, whereas that for the cytosine-rich light (L) strand is two-thirds around the genome. Therefore, H strand replication begins at OrH of the L strand and proceeds two-thirds around the genome until it exposes the replication origin of the L-strand. At this point, L-strand replication is initiated and proceeds back along the displaced H-strand template. The D loop mentioned above is a triple-stranded region of the genome produced by synthesis of a short piece of H strand.

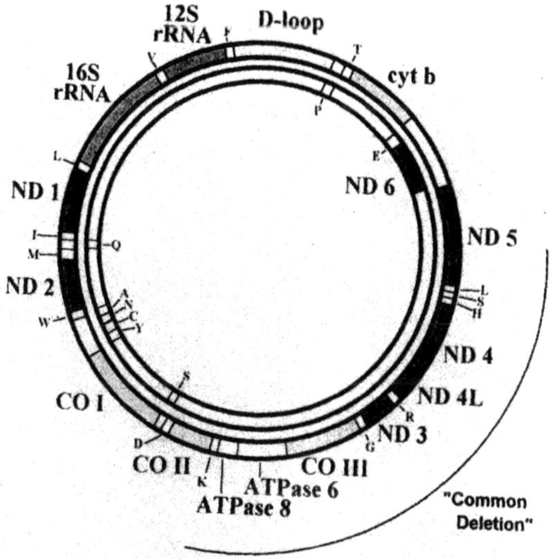

Figure 2. Structure of mtDNA. mtDNA is a closed, circular molecule containing no introns. This molecule codes for seven subunits of the electron transport chain (ETC) complex I (NADH ubiquinone oxidoreductase, ND 1, 2, 3, 4, 4L, 5, and 6), three subunits of complex IV (cytochrome c oxidase, CO I, II, and III), two subunits of complex V (ATPsynthase, ATPase 6 and 8), and one subunit of complex III (cytochrome c oxidoreductase, cytochrome b). The two mitochondrial rRNAs (16S and 12S) and 22 tRNAs (small letters) allow transcription and translation of these genes to occur in the mitochondria. In addition to encoding m, t, and rRNAs, there is a triple-stranded "D" loop, site of the control region and replication origin.

One of the most important tools for examination of the effects of mtDNA mutations is the ρ0 or null mutant cell. In 1986, Desjardins et al. treated an avian fibroblast cell line with low concentrations of ethidium bromide, effectively destroying the mtDNA without harming the nuclear genome. These cells were found to be auxotrophic for uridine, but continued to divide and grow in in the absence of functional mitochondria. Cells fused with cytoplasts derived from anucleated mitochondrial donor cells to produce transmitochondrial cybrids (cytoplasmic hybrids) were used to examine mitochondrial properties (Wallace et al., 1975). The first human cybrids were produced by King and Attardi (1989), and were found to require pyruvate in addition to uridine. As discussed later in the chapter, this technique allows one to isolate defects in cellular function caused by mtDNA mutations, as well as examine mitochondrial function with aging (Figure 3).

The structure of mtDNA, and the fact of an independent genetics, confers many advantages to the cell. Among these are rapid response to local conditions in the mitochondria, more direct control of expression, and an additional layer of control

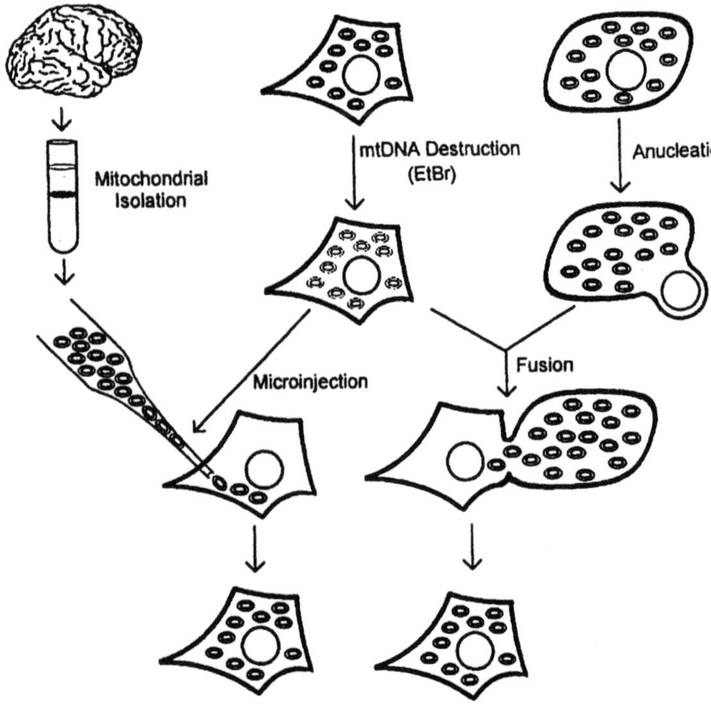

Figure 3. Production of cybrid cells. Cells containing mitochondria exclusively from another cell type or tissue are produced primarily by two methods. The acceptor cell is first treated with a compound that selectively destroys mtDNA, such as ethidium bromide (EtBr). These cells then become auxotrophic for pyruvate and uridine, but function normally otherwise. These cells can then be fused with anucleated cells by chemical means or electric pulse. The other method involves isolating purified mitochondria from tissue such as brain, and microinjecting them into the acceptor cell.

not found in nuclear genes. However, these advantages do not come without a price, and it is often very steep.

Properties of mtDNA

A number of features of mtDNA biology are of particular clinical importance. In humans, mtDNA is essentially maternally inherited (Anderson et al., 1981). At fertilization, the vast majority of cytoplasm, mitochondria, and thus mtDNA are derived from the ovum, as only the spermatozoa head enters the ovum and almost all the paternal mitochondria are located in the neck or tail of the spermatozoan.

Repeated back-crossing of mice has suggested that a few paternal mtDNA molecules may pass to the zygote (Gyllensten et al., 1991), but this has not been demonstrated in humans. Secondly, mtDNA transcription is not closely restricted by the cell cycle. Thus, in each cell cycle the nuclear DNA divides only once and is divided equally between daughter nuclei, whereas mitochondrial DNA may replicate several times during the cell cycle or not at all. Of interest in neurology, mtDNA continues dividing in postmitotic cells such as neurons. Thirdly, and most importantly from a disease standpoint, mtDNA is far more susceptible to mutations than is nuclear DNA for a number of reasons (Richter et al., 1988). As is discussed at greater length later, oxidative phosphorylation creates as a by-product reactive oxygen species (ROS). mtDNA is not protected by histones and is free in the mitochondrial matrix, and is therefore exposed to the elevated levels of these free radicals, which react with and damage mtDNA, causing increased mutations. Repair of DNA damage is also poor in mtDNA compared with nuclear DNA. Of the five types of eukaryotic polymerases known, only DNA polymerase γ is present in mitochondria. This specific polymerase is much less efficient at repairing mismatches than the other four nuclear enzymes (Kunkel and Loeb, 1979, 1981). Lastly, mtDNA synthesis and replication generally occurs with greater frequency than nuclear division, especially in postmitotic cells such as neurons, increasing the likelihood of passing on mismatches. Since each mitochondrion has between 2 and perhaps 20 copies of the genome, and there are anywhere from a few dozen to thousands of mitochondria per cell, the situation arises in a given cell that when mutations occur, there will be multiple populations of mtDNA (wild-type and the various mutant forms), a condition termed heteroplasmy. Because mitochondria segregate unevenly to daughter cells during mitosis, the rate of heteroplasmy between tissues and even neighboring cells may vary widely. If the new mutation confers a replicative advantage and is not pathological, it may become the dominant form of mtDNA within the cell, and even replace the original wild-type form, returning the cell to a homoplasmic state. If this occurs in the female germ line, the new mtDNA will be passed on to offspring. Current theory holds that this mechanism is responsible for the established polymorphism observed in different human subpopulations (Wallace, 1994a).

Although highly pathological mutations are unlikely to be passed on, owing to cell death, mild to moderately deleterious mutations can persist in heteroplasmy. This state is the most commonly seen in human mitochondrial disease. Probably because of differential segregation and rates of ROS production in different cells; the ratio of mutant to wild-type mtDNA often varies greatly between different tissues in the same individual, and even more so between offspring of a mother with a mitochondrial mutation. Rapidly dividing cells such as leukocytes often have low levels of heteroplasmy, whereas postmitotic cells such as neurons have higher ratios of mutant mtDNA. This may explain, in part, why mitochondrial disorders generally have a neuropathological component.

MITOCHONDRIAL FUNCTION

Oxidative Phosphorylation and Reactive Byproducts

The most universal and critical function of mitochondria, and the one most widely known, is oxidative phosphorylation. Essentially, this system transfers energy by means of the ETC derived from metabolism, and in the presence of oxygen produces ATP. This is accomplished through discrete metabolic reactions that produce reduced pyridine nucleotide (NADH, NADPH) and/or flavoproteins that transfer electrons from hydrogen to the ETC, ultimately reducing oxygen to water. This reaction in turn produces a proton gradient across the mitochondrial membrane, analogous to a battery in that it provides the necessary potential for synthesis of ATP from adenosine diphosphate (ADP) and phosphate, by the membrane complex ATP synthase (ATPase).

The oxidative phosphorylation system includes five large multienzyme assemblages, called ETC complexes I through IV, and complex V (ATP synthase). These complexes are inserted into the inner mitochondrial membrane, and consist of 77 separate proteins. Complex I (NADH-ubiquinone oxidoreductase) reduces ubiquinone, and is the largest complex, being composed of 44 proteins. Ubiquinone can also be reduced by either of two flavoproteins; complex II (succinate-ubiquinone oxidoreductase), or electron transfer flavoprotein, which is integral to fatty and amino acid oxidation. Complex III (cytochrome c oxidoreductase) transfers electrons from ubiquinone to reduce cytochrome c. The final step in the ETC is complex IV (cytochrome c oxidase) which reduces oxygen to form H_2O (Figure 4).

Over 90% of the oxygen in most cells is consumed in the electron transport system of the mitochondria. Among the many products of oxidative phosphorylation are ROS, which include the superoxide anion radical ($O_2\cdot^-$) and hydroxyl radical (OH·), both of which are highly reactive and destructive to many biomolecules, including DNA. When the first electron transfer occurs, the superoxide radical is formed. With the acceptance of the second electron, hydrogen peroxide, the protonated form of the peroxide radical is produced. Hydrogen peroxide is not truly a free radical, but it is highly reactive, and interacts with other substances to form free radicals. The third electron transfer results in formation of the highly toxic and reactive hydroxyl radical. The acceptance of the fourth electron generates a molecule of H_2O. Although $O_2\cdot^-$ is reactive, the proximate mediator of oxyradical toxicity is likely not $O_2\cdot^-$ but rather an ensuing flux of hydroxyl radical formed by either Fenton reactions that involve homolytic H_2O cleavage catalyzed by Fe^{2+} and other transition metals, or by decomposition of the peroxynitrite that is formed in reactions between $O_2\cdot^-$ and nitric oxide (reviewed by Gutteridge and Halliwell, 1989). Hydroxyl radicals are the most oxidizing radical species found in biological systems, with reaction constants so great they react within five molecular radii of their production. Considering the ubiquity of both transition metals and NO

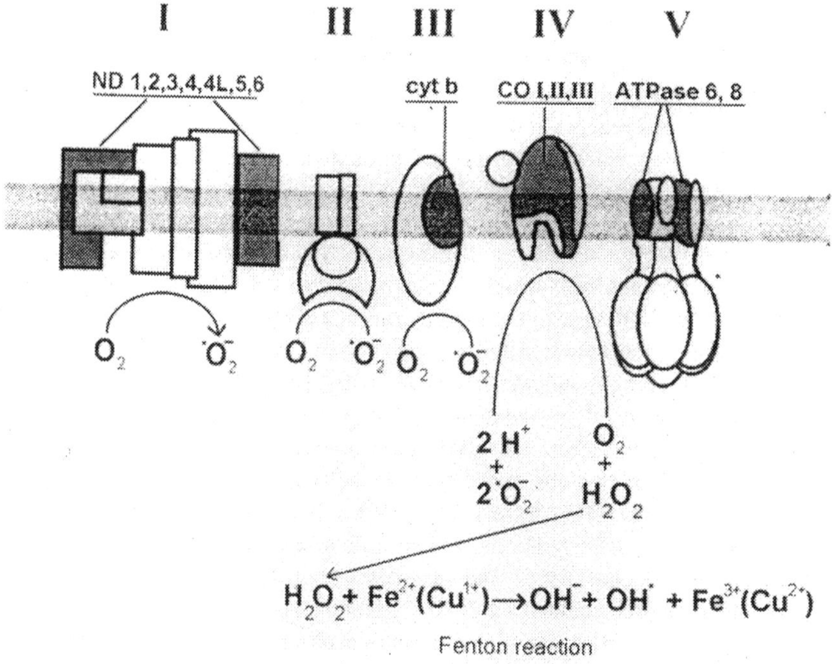

Figure 4. Subunits of oxidative phosphorylation. The electron transport chain (ETC) is composed of five complexes (I–V) inserted into the inner mitochondrial membrane. Each complex is composed of multiple polypeptide subunits encoded by either nuclear (white) or mitochondrial (gray) genes. As shown, mitochondria produce seven segments of complex I (ND 1, 2, 3, 4, 4L, 5, 6), three segments of complex IV (CO I, II, III), two segments of complex V (ATPase 6, 8), one segment of complex III (cyt b), and none of complex II. From this figure, one can see the greater risk to complex I and IV function from mtDNA mutations. The electron transfers which occur in complexes I, II, and III produce ROS in the form of superoxide ($O\cdot^-$). During the last transfer at complex IV, superoxide can react with H^+ to produce hydrogen peroxide (H_2O_2). H_2O_2 in the presence of transition metals such as Cu^+ or Fe^{2+} can form the highly reactive hydroxyl radical ($OH\cdot$).

synthetase in tissues including brain, $OH^{\cdot}$ is likely to be responsible for most oxyradical damage.

The free radicals produced in the normal efficient transfer of electrons are prevented from damaging the cellular components by being bound to the ETC enzymes during the reaction (Gutteridge and Halliwell, 1989). It is evident from this that a decrease in efficiency of one or more of the components, or mutations affecting the ability of the enzyme to bind the radical, or both, could have dire consequences for surrounding biomolecules.

Apoptosis

Early on it was observed that there are predictable phases of cell death in tissues as part of a normal process. It was further observed that this programmed cell death is often associated with characteristic morphological and biochemical changes, and the phenomenon was termed apoptosis (Warner, 1972; Wyllie et al., 1980). The central features of apoptosis include characteristic membrane alterations (blebbing), cell shrinkage, chromatin condensation, DNA cleavage, and a subsequent fragmentation of the cell, which is then phagocytosed. This particular form of cellular suicide is thought to be important in that apoptotic death does not release potentially toxic substances that may damage nearby cells (Searle et al., 1982). This phenomenon serves as a major mechanism for regulating cell numbers (Raff, 1992), by removing unwanted or potentially pathological cells such as tumor cells (Williams, 1992) or cells infected with viruses (Vaux et al., 1994). However, apoptosis may also contribute to a number of diseases including neurodegenerative diseases and ischemic stroke (Raff, et al., 1993). What is clear to date is that apoptosis requires energy, protein synthesis, and specific activation of a family of proteases called caspases (Cohen, 1997). Apoptosis is arbitrarily divided into three phases: (1) an activation phase in which various signaling pathways impact upon the central death-triggering mechanism; (2) an execution phase in which the now-activated components act upon various cellular targets; and (3) the destruction phase, in which the cell and its nucleus and DNA are broken down.

There is emerging literature indicating that mitochondria play a key role in activation of the apoptotic pathway in a number of ways. As discussed earlier, the mitochondrial membrane potential is key to oxidative phosphorylation in mitochondria, providing the ionic and electrical gradient necessary for ATP production. A drastic decrease in this potential is seen prior to DNA fragmentation during the apoptosis pathway (Vayssiere et al., 1994; Petit et al., 1995; Zamzami et al., 1995). This drop is responsible for the uncoupling of oxidative phosphorylation (Vayssiere et al., 1994) and occurs during apoptosis induced by both physiological (i.e., growth factor withdrawal, glucocorticoids) (Petit et al., 1995) and nonphysiological (i.e., irradiation, chemotherapy) (Marchetti et al., 1996) inducers of cell death, and therefore seems to be a universal apoptotic event. The central mechanism involved in this loss of potential seems to be the opening of pores in the inner membrane of the mitochondria, causing what is termed the permeability transition (PT). Opening of these pores causes a loss of the ionic and respiratory gradients leading to the arrest of ATP synthesis (Bernardi and Petronelli, 1992). In paradigms in which PT is blocked by addition of cyclosporin A, apoptosis is halted (Zamzami et al., 1996), whereas induction of PT induces DNA fragmentation (Chandrasekaran et al., 1992; Kroemer, 1996). It is not known how PT causes apoptosis, but recent evidence points to release of mitochondrial components as being crucial to induction. Cytochrome c, as well as an unidentified protein termed AIF (apoptosis initiating factor; Susin et al., 1996) are released during PT, and activate critical components of the apoptotic machinery. This process is critical to the following discussion of mitochondrial diseases because there is an increased susceptibility of PT in mitochondria that harbor specific mtDNA mutations (Wong and Cortopassi, 1997).

Calcium Regulation

Another function of mitochondria that has gained attention recently is regulation of intracellular Ca^{2+}. Recent studies examining the rates of change of cytoplasmic intracellular Ca^{2+} $[Ca^{2+}]_i$ after selective blockade of known mechanisms for cellular Ca^{2+} influx and extrusion showed that mitochondrial uptake accounts for as much as 70% of cytosolic Ca^{2+} removal during the initial rapid phase of recovery from large imposed Ca^{2+} loads in chromaffin cells (Herrington et al., 1996). These organelles therefore act as safety devices against potentially toxic increases of cytosolic Ca^{2+} (reviewed by Richter and Kass, 1991; Richter, 1997). This system of calcium regulation appears to be tightly controlled, as the organelles take up and release Ca^{2+} by separate mechanism, resulting in a dynamic process in which Ca^{2+} is cycled across the inner membrane (Carafoli, 1994).

It is thought that the major system by which mitochondria take up Ca^{2+} is via a Ca^{2+} uniporter, found in all vertebrate mitochondria examined (Gunter and Gunter, 1994). As mentioned previously, there is a large membrane potential as well as proton gradient across the inner membrane of the mitochondrion, which is key to ATP production. This potential is approximately 140 to 180 mV negative internal, and uptake of Ca^{2+} into mitochondria is therefore energetically favorable. In isolated liver mitochondria exposed to pulsatile challenge with extramitochondrial Ca^{2+}, this component of calcium uptake was essentially too rapid to be measured (Tang and Zucker, 1997). This function in particular is a critical one for neuronal survival. During neuronal depolarization of high intensity, mitochondria are important in buffering internal Ca^{2+} levels, rapidly sequestering Ca^{2+} after evoked influx (Friel and Tsien, 1994; Wang and Theyer, 1996) and regulating Ca^{2+} for some time thereafter via a Na^+/Ca^{2+} exchange mechanism (White and Reynolds, 1995). This regulation is more sensitive and complex than mere uptake, as shown by studies in which Ca^{2+} efflux after sequestration contributes perhaps the lions share of persistent Ca^{2+} elevation after neuronal depolarization (Friel and Tsien, 1994; White and Reynolds, 1995; Wang and Theyer, 1996; White and Reynolds, 1997). This theory is supported by observations that mitochondria are concentrated in close apposition to membrane domains with high levels of Ca^{2+} channels, perhaps permitting even more rapid buffering by mitochondria in these critical areas (Bindokas and Miller, 1995). Therefore, mitochondria buffer cells against excess Ca^{2+} influx, as well as regulate Ca^{2+} homeostasis during rest and recovery periods following Ca^{2+} load.

In addition to the regulation mitochondria have on Ca^{2+}, Ca^{2+} itself can regulate mitochondrial function, by activating several important enzymes within the ETC as well as ATP synthase (Gunter, 1994; Mildaziene et al., 1995). Increases in Ca^{2+}

that do not exceed a critical threshold therefore stimulate increased ATP production, which appears logical given the increased ATP demands upon the cell to remove Ca^{2+}. Increased mitochondrial uptake of Ca^{2+} in smooth muscle cells increases ATP production, therefore meeting the demands of both ATP-dependent ion pumps and actin/myosin ATPases (Drummond and Fay, 1996). This phenomenon of increased ATP production in the presence of Ca^{2+} has also been seen in isolated mitochondria (Li et al., 1995).

Functional Interaction

Although the functions of mitochondria have been presented separately, they are far from unrelated, and there are many feedback loops and direct forms of interaction between them. As mentioned, Ca^{2+} influx into the mitochondria below a certain level can increase production of ATP, but above this threshold, it has quite the opposite effect. Under greater loads of Ca^{2+}, mitochondria sequester this ion at the expense of the proton motive force, depolarizing the inner membrane and thus uncoupling ATP synthesis from respiration. Always to be kept in mind when discussing mtDNA mutations, the primary function of mitochondria is production of ATP, and in this specific context any decrease of ATP production may lead to increases in intracytoplasmic, and therefore intramitochondrial, Ca^{2+} levels. Mitochondrial calcium uptake appears to play a pivotal role in both apoptosis and necrosis, because ruthenium red (a specific inhibitor of mitochondrial calcium uptake) protects cells against both types of death (Kruman et al., 1998) (Figure 5).

Therefore, the two functions of ATP synthesis and Ca^{2+} sequestration are intricately linked at more than one level. In addition, it appears that influx of Ca^{2+} is a critical mediator of the opening of PT pores (Bernardi and Petronilli, 1992; Wang and Thayer, 1996; White and Reynolds, 1996; Kristal and Dubinsky, 1997), which are mediators of the apoptotic pathway. Therefore, the amount of Ca^{2+} which mitochondria can tolerate before depolarization and PT pore formation is crucial to cellular and mitochondrial function, and this parameter is known to be fairly plastic. The antiapoptotic proto-oncogene Bcl-2 protects against most forms of apoptosis, and at least part of its effect is within mitochondria (Henkart and Grinstein, 1996). Overexpression of Bcl-2 increases the amount of Ca^{2+} mitochondria can sequester by two- to fourfold (Murphy et al., 1996), while still maintaining respiratory function.

There are also significant interactions between ROS production and the various mitochondrial functions. Free radical generation is increased by inhibition of the ETC (Boveris and Chance, 1973). In yet another feedback system, oxygen radicals can directly inhibit oxidative phosphorylation by damaging both mtDNA and ETC components. In addition, reduced ATP levels result in increased free radical generation by cytochrome b566 (Nohl and Jordan, 1986). Complex I is particularly vulnerable to attack by both hydroxyl radical and superoxide anion, the inhibition of which causes still greater ROS production (Zhang et al., 1996). In vivo studies

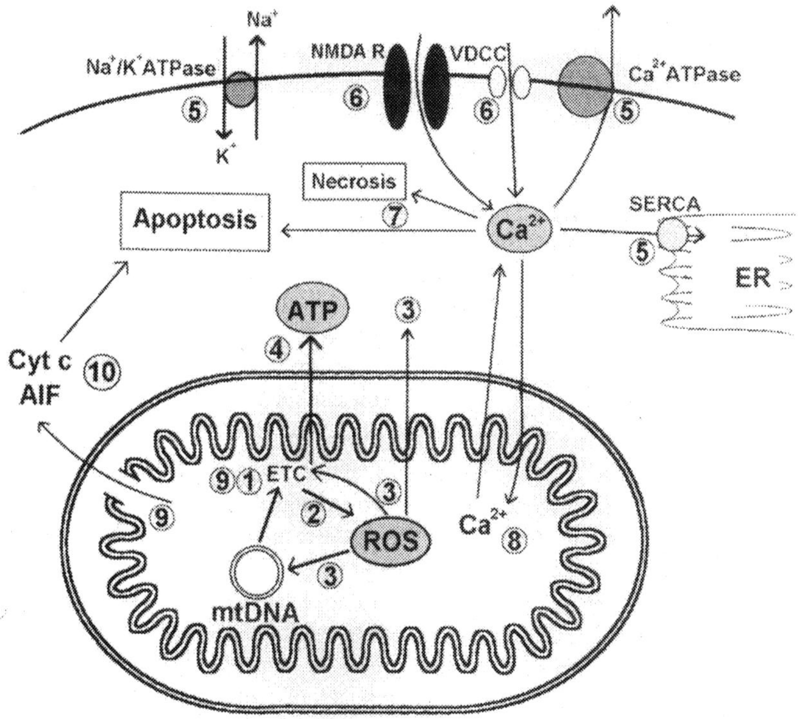

Figure 5. Pathways of neuronal death from mtDNA mutations. Mutations in mtDNA lead to production of aberrant subunits of the electron transport chain (ETC) (1), thus decreasing efficiency and increasing production of reactive oxygen species (ROS) (2). These ROS can then damage cellular biomolecules, including ETC subunits and mtDNA (3), further exacerbating ROS production and decreasing ATP production (4). Decreased availability of ATP to membrane Na^+/K^+ ATPase, Ca^{2+} ATPase and smooth endoplasmic reticulum ATPase (SERCA) (5) causes increased levels of intracellular Ca^{2+} and membrane depolarization. This depolarization in turn activates voltage-dependent Ca^{2+} channels (VDCC) and allows activation of NMDA receptors (6), further increasing $[Ca^{2+}]_i$, which can induce cellular necrosis or apoptosis (7). Mitochondria containing damaged mtDNA and inefficient ETC have a decreased Ca^{2+}-buffering ability (8), leading to a drop in mitochondrial membrane potential, which results in decoupling of the ETC, membrane permeability transition, and opening of pores in the inner membrane (9). Proteins that may induce apoptosis (10), such as cytochrome c and AIF, are then released through these openings.

have shown both complex I and IV are vulnerable to peroxynitrite, with complex IV being the most effected (Benzi et al., 1991, 1992). Although these effects may be direct, one should bear in mind the fact that the ETC complexes are membrane bound and very sensitive to the lipid environment (Fry and Green, 1981), and therefore the possibility exists that lipid peroxidation may play a role.

The other component of mitochondria vulnerable to attack by reactive compounds is the mtDNA. Because mtDNA codes for nothing but ETC components and the rRNA and tRNA necessary to express them, any pathological mutation of mtDNA will, regardless of other effects, decrease the efficiency of the ETC. DNA in the nucleus is compartmentalized and arranged in tightly bound helix, most of which is protected by histones, so the chance of free radical damage is minimized. In addition the repair processes in the nucleus are very efficient, and there is little evidence of substantial free radical–mediated damage to DNA in the nucleus (Beckman and Ames, 1997). Mitochondrial DNA has none of these advantages; it is free within the lumen, has no histones to guard it against radical attack, is in close proximity to the site of greatest cellular free radical generation, and of perhaps greatest importance, the repair mechanisms in the mitochondria are much less efficient than those found in the nucleus. When mtDNA damage does occur, there is an increasing order in which the various components are susceptible; pyrimidines have the greatest risk, followed by purines, and then the deoxyribose moiety. The superoxide radical in particular is involved in triggering strand breaks in DNA (Birnboim 1988), and deletions may occur as the result of free radical reaction with the sugar-phosphate backbone (Freeman and Crapo, 1982).

Even normal mitochondria exposed to oxidative stress become deadly producers of ROS. ρ0 cells (those containing no mtDNA and thus no ETC) are much more resistant to death caused by exposure to high O_2 concentrations than those cells containing functional mitochondria and mtDNA (Tanaka et al., 1996). Thus is set a theme to appear throughout the chapter; the vicious feedback cycle. In this case, mtDNA damage can lead to a decreased efficiency of the ETC, which in turn generates greater levels of free radicals, which then further damages the mtDNA. As discussed later, this certainly occurs in aging, and is thought to underlie the increasing severity with age of inherited mitochondrial disorders.

NEURONAL VULNERABILITY TO mtDNA MUTATION

Almost all diseases with mtDNA mutations as the causal factor have at least some neuropathy associated with them, and in many cases this is the primary symptom. Neurons are more vulnerable to mitochondrial defect than other cells, owing both to special metabolic requirements of these cells, as well as a special vulnerability to disregulation of ion balance. In addition, unlike most other cells, neurons that are lost cannot be replaced.

Postmitotic State and mtDNA Load

In the majority of mitochondrial disorders there is heteroplasmic distribution of mitochondria with mutated DNA. In dividing cells, there is generally a trend toward increasing the proportion of wild-type mitochondria as dysfunction in the mitochondria with mutated mtDNA tends to decrease the viability of the cell within which they reside. This is not the case in cells such as neurons, in which the proportion of dysfunctional mitochondria the cell has when it enters its postmitotic state generally does not decrease. In addition, because mtDNA is synthesized continuously, there is a high probability of spontaneous mutations, as discussed earlier, leading to an increased proportion of mitochondria carrying potentially damaging mtDNA.

Metabolic Properties

Neurons are among the few cells in the body that are totally dependent upon oxidative phosphorylation to supply energy needs and, as stated earlier, even a short period of ATP depletion leads to ionic imbalance and neuronal damage. In addition, neurons are very metabolically active. As an organ, the brain uses 20% of the total O_2 inspired, and yet accounts for only 2% of the weight of a normal individual. This high rate of O_2 usage in the brain leads to greater production of free radicals. An additional factor that puts the CNS at risk is that it is highly enriched in polyunsaturated fatty acids, which are susceptible to lipid peroxidation (Rice-Evans and Burdon, 1993). Cerebral spinal fluid contains relatively low levels of iron and copper binding compounds, which means that increased amounts of free Cu^+ and Fe^{2+} may be available to participate in generation of the hydroxyl radical via the Fenton reaction. Furthermore, neural tissue itself contains high levels of nonheme iron (Harman, et al., 1976). Therefore, the dependence upon mitochondria for energy, the high energy demands, and the relatively increased oxidative environment of the neuron all increase the vulnerability of these cells to mitochondrial dysfunction.

Calcium Homeostasis and Excitotoxicity

The ability of neurons to maintain low rest levels of intracellular free calcium, and modulate Ca^{2+} rapidly (spatially, temporally, and quantitatively) is conferred by an intricate set of membrane-associated and cytosolic calcium-regulating proteins (reviewed by Mattson and Mark, 1996; Carafoli, 1992, 1994). The concentration of calcium outside of nerve cells is typically in the 1- to 2-mM range, whereas $[Ca^{2+}]_i$ is in the range of 50- to 200-nM. This large transmembrane calcium concentration gradient is maintained largely by the plasma membrane calcium ATPase (PMCA), and is the central reason neuronal cells are so much more metabolically active than other types of cells, which do not require regeneration of this gradient following depolarization.

When neurons are stimulated by excitatory neurotransmitters or during an action potential, the $[Ca^{2+}]_i$ can rise within seconds to concentrations well over 10 μM. The $[Ca^{2+}]_i$ is then removed rapidly from the cytoplasm by at least four mechanisms. Perhaps the most important is the activity of the Na^+/Ca^{2+} exchanger in the plasma membrane, which has a very high capacity. A second mechanism involves activity of a smooth endoplasmic reticulum Ca^{2+}- ATPase (SERCA), which pumps Ca^{2+}

into the lumen of the endoplasmic reticulum. A third mechanism involves cytoplasmic Ca^{2+} binding proteins that may sequester Ca^{2+} or promote activation of membrane Ca^{2+} removal, or both. The fourth mechanism, sequestration of large amounts of calcium into the mitochondria, is discussed earlier.

Membrane depolarization in neurons requires constant regeneration of this and other ionic gradients, and therefore the importance of ATP produced by the mitochondria is even more critical, as both the plasma membrane ion-motive ATPases and the SERCA are each important for maintaining rest $[Ca^{2+}]_i$ (Thastrup, 1992). Mitochondria are of central importance for maintenance of intracellular Ca^{2+} homeostasis for a number of reasons. They act as perhaps the most significant rapid reservoir for Ca^{2+} after influx. They provide the ATP required for Ca^{2+} ATPases as well as that used to maintain the resting potential of the neuron through the activities of Na^+/K^+ ATPases.

The reason for this focus on regulation of Ca^{2+} homeostasis and the role of mitochondria therein is due to the compelling evidence that increased $[Ca^{2+}]_i$ precedes cell death in neurons (reviewed by Mattson and Mark, 1996). When this occurs due to exposure to excitatory neurotransmitters, it is referred to as excitotoxicity. Excitotoxicity has been shown to contribute to most forms of pathological neuronal death, including acute insults such as hypoxia, hypoglycemia and seizures (Olney and de Gubareff, 1978; Rothman et al., 1986; Mattson and Scheff, 1994), and chronic neurodegenerative disorders such as Parkinson's, Alzheimer's, and Huntington's diseases (Choi, 1988; Siesjo et al., 1989; McDonald and Johnston, 1990; Mattson et al., 1993).

The most important excitotoxic neurotransmitter is glutamate, first shown to be toxic to retinal neurons by Lucas and Newhouse in 1957. Glutamatergic excitotoxicity is primarily mediated through two different ionotropic glutamate receptors, the N-methyl-D-aspartate (NMDA) receptor and the AMPA/kainate receptor (reviewed by Gasic and Hollman, 1992; Seeburg, 1993). NMDA receptors are found throughout the brain, and all neurons shown to be vulnerable to NMDA receptor-mediated excitotoxicity also express the AMPA/kainate receptor as well. Moreover, it has been demonstrated that both NMDA and AMPA/kainate receptors localize to individual synapses in hippocampal neurons (Bekkers and Stevens, 1989). The NMDA protein forms a calcium-permeable channel when glutamate binds, thus causing an increase in intracellular Ca^{2+}. There is compelling evidence that the toxic effects of glutamate are caused by this influx of calcium. When extracellular calcium is removed from the medium, cultured neurons are protected against glutamate toxicity (Choi, 1987; Mattson et al., 1988, 1989), and a strong correlation between sustained elevation of $[Ca^{2+}]_i$ and neuronal death in vitro following glutamate exposure has been reported (Ogura et al., 1988). Under ischemic conditions, where ATP levels are suppressed, it becomes difficult for this gradient to be maintained. This hypothesis is supported by in vitro studies in which ATP depletion leads to increased $[Ca^{2+}]_i$, causing enhanced glutamate release and activation of NMDA receptors, further elevating $[Ca^{2+}]_i$ and leading to neuronal damage.

The demands for ATP and available energy are therefore increased after neuronal excitation in order to remove excess calcium and return the cell to the ionic resting state. Decreases in mitochondrial function would thus increase the vulnerability of the neuron to excitotoxicity. Indeed, in vitro studies have shown that when intracellular ATP levels are reduced in cultured neurons, they become more vulnerable to excitotoxicity largely mediated by NMDA receptors (Novelli et al., 1988). Cell culture studies in which neurons are deprived of glucose or oxygen, or both, have provided considerable insight into mechanisms of excitotoxic neuronal injury. Depleting ATP levels alone without adding exogenous glutamate induces cell injury by an NMDA receptor-mediated mechanism, because antagonists to this receptor decrease death in glucose-deprived neocortical (Moyner et al., 1990), and hippocampal (Cheng and Mattson, 1992) cell cultures. Studies in which levels of ATP and $[Ca^{2+}]_i$ were correlated with neuronal injury induced by glucose deprivation showed that ATP levels dropped to 20% of control levels within 8 hours, and $[Ca^{2+}]_i$ slowly increased to twice the normal level within 12 hours, followed by a fairly rapid increase of four to eightfold after 16 hours (Mattson et al., 1993). In these studies the mitochondria were healthy, but became significantly impaired with $[Ca^{2+}]_i$ increase.

As discussed earlier, increased Ca^{2+} levels can significantly decrease mitochondrial function, and the maintenance of Ca^{2+} homeostasis requires efficient mitochondrial production of ATP. Therefore, yet another pathogenic feedback cycle emerges with mtDNA damage, one in which mtDNA mutations lead to decreased availability of ATP for Ca^{2+} homeostasis, causing a rise in $[Ca^{2+}]_i$ to a level in which mitochondrial function itself is impaired.

The ways in which increased Ca^{2+} damages and kills neurons are unclear, but many lines of evidence indicate that Ca^{2+} leads to damage by increasing free radical production in neurons. In cultured hippocampal neurons, exposure to glutamate increased cellular oxidation (Mattson et al., 1995). This same effect was seen in synaptosomes isolated from cerebral cortex (Bondy and Lee, 1993; Oyama et al., 1993). The Ca^{2+} influx induced by glutamate receptor activation is critical for the increased ROS production seen during excitotoxicity. There are a number of ways Ca^{2+} can increase free radical production, and recent evidence has shown that direct effects on mitochondria are included. When exposed to elevated Ca^{2+} concentrations equivalent to those occurring during excitotoxic receptor stimulation (Murphy et al., 1987), isolated mitochondria from muscle produce a variety of radicals.

Oxyradicals can disrupt calcium homeostasis, whereas calcium elevations can induce ROS production. Because mitochondria are involved in both calcium and oxyradical homeostasis, they are likely to play a pivotal role in such processes. For example, mitochondria are extremely vulnerable to oxyradical mediated alteration, including decreased oxidative phosphorylation, increased mitochondrial calcium uptake, and reduced ATP production. Physiological (submicromolar) concentrations of nitric oxide may reversibly mediate such effects via transient binding to cytochrome oxidase. However, nitric oxide may combine with mitochondrial superoxide to form highly reactive peroxynitrite, which may irreversibly damage mitochondria.

MITOCHONDRIA AND DISEASES OF AGING

Mitochondrial Function in Normal Aging

Ames et al. (1995) have outlined a hypothetical series of events relating to the role of mitochondria and aging, drawing on the accumulating evidence that mtDNA mutations play a critical role in age-related diseases. This hypothesis states that cumulative mtDNA damage, mediated by oxidative stress, is a major contributor to the decrease in viability with age (also reviewed by Shigenaga, et al., 1994; Wallace et al., 1995). Analysis of mitochondrial oxidative phosphorylation enzymes in postmitotic tissues has revealed that the specific activities of the respiratory complexes decline with age (Trounce et al., 1989; Yen et al., 1989). This decline has been correlated with the progressive accumulation of somatic mtDNA mutations in postmitotic tissues. For example, the content of 8-OH-dG (8-hydroxydeoxyguanosine, a hydroxyl-radical adduct of deoxyguanosine) in mtDNA increases exponentially with age (Hayakawa et al., 1991, 1992, 1993). In studies of a 19-year-old patient with myocardial myopathy the levels of 8-OH-dG were comparable to that of a 78-year-old normal subject (Hayakawa et al., 1991). This patient also exhibited greatly increased mtDNA deletions, suggesting that germ-line mutations cause further damage to mtDNA. In addition, quantitation of a 5-kb mtDNA deletion (the so-called common deletion) in normal brain has revealed that the cerebellum accumulates relatively little mtDNA damage with age. In contrast, various cortical regions of the brain accumulate appreciable mtDNA damage after age 75, and the basal ganglia accumulates the highest level, with over 10% of the basal ganglia mtDNAs containing the 5-kb deletion by age 80 (Corral-Debrinski et al., 1992; Soong et al., 1992).

There is a large amount of evidence now indicating that these mutations during normal aging have functional consequences. In one study, mitochondria from muscle samples taken during surgical operation from 16- to 92-year-old patients (Trounce et al., 1989) were assayed for respiration rates with pyruvate/malate, glutamate/malate, and succinate as substrates. The results showed that there was a significant negative correlation between respiration rate and age with all substrates. This correlation was not found with the nuclear-encoded mitochondrial monoamine. These observations suggested a substantial decrease in respiratory capacity in aged muscle, correlating with observed decreases in exercise tolerance and aerobic ability with age, and that these changes seem to be the result of mtDNA encoded enzymes. Experiments done with liver obtained during surgery show similar findings (Yen et al., 1989). In that study, respiration rates of state 3 (activated) and state 4 (rested), the respiratory control ratio, and ADP/O were measured for subjects 35 to 76 years of age. Again, there was a significant negative correlation between age and respiratory ability and ADP/O ratios.

Although the number of fully functional mitochondria in cerebellum decreases with advancing age (Fattoretti et al., 1996), the underlying defects are not known (Craig and Hood, 1997). One hypothesis states that there is decreased efficiency of the ETC, which results in increased production of ROS (Hagen et al., 1997), as outlined earlier. Studies in a number of labs have shown that damage created by these radicals mimic many of the degenerative changes seen during aging (Cortopassi and Wang, 1995; Kalous and Drahota, 1996). This view is supported by the established relationship between dietary restriction, retarded age-associated increases in mitochondrial free radical production, and reduced oxidative damage within cells (Wachsman, 1996). Interestingly, dietary restriction results in decreased damage to striatal neurons, and improved behavioral outcome following administration of the mitochondrial toxin 3-nitropropionic acid to adult rats (Bruce-Keller et al., 1998). The accumulation of lipofuscin, which is typical of aged human nerve cells, has been associated with a age-related decrease in mitochondrial "detoxification" mechanisms (Chen and Yu, 1996). In addition, the amount of lipid peroxides measured as malondialdehyde and the activity of manganese-superoxide dismutase in mitochondria exhibit an age-dependent increase in different human tissues (Wei et al., 1996a, 1996b).

More direct evidence for an important role for mtDNA damage in the aging process comes from studies using experimental animals. When the various complexes of the ETC were examined in diaphragmatic mitochondria of rats aged 7 to 55 weeks, there was a significant decrease in both complex I and complex IV activity by 55 weeks (Torii et al., 1992). Recall that both of these complexes are composed largely of proteins encoded by mtDNA. In contrast there was no change in the activity of complex II, which has no proteins encoded by mtDNA, nor complex III, which has only one mtDNA-encoded unit. In an observation with clear implications for later discussion on encephalopathy, there was no discernible difference with age in any of the complexes in mitochondria isolated from liver. This indicates that mitochondria are more vulnerable in some tissues than in others, and that the complexes themselves differ in damage, with complex I being the most susceptible. One might expect tissues that were relatively more dependent on oxidative phosphorylation for energy, and those that are more metabolically active, to be at greater risk of oxidative damage to mitochondrial DNA. In addition, since these tissues depend so heavily upon mitochondrial ATP production, loss of ETC efficiency with mtDNA damage would have a relatively greater impact.

It has been suggested that mitochondria with reduced respiratory function, due to mutations of mtDNA affecting the respiratory chain, suffer less frequent lysosomal degradation because they inflict free radical damage more slowly on their own membranes (de Grey, 1997). Therefore, a mutation in a mitochondrion that inhibits its respiratory capacity in such a way as to reduce both ATP and free radical production, without affecting the replicative ability of the mtDNA, will actually have a replicative advantage over normally functioning mitochondria. If this occurs in a postmitotic cell, such as a neuron, this defective organelle eventually will become the predominate mitochondrial type, and destroy the respiratory capacity of that neuron.

If there were a threshold effect, in which the relative accumulation of mutations is the precipitating factor in mtDNA diseases, then the age of symptom onset and relative severity of the disease would be determined in part by the initial bioenergetic capacity at birth (Graeber and Muller, 1998). According to this mitochondrial theory of aging, those with higher mitochondrial capacity would require greater somatic mtDNA damage in order to exhibit pathology, and would thus remain more highly functional for a longer duration, this being true both for carriers of pathogenic mtDNA mutations, as well as normal individuals. In contrast, individuals who inherit significantly deleterious mtDNA mutations would start at a lower initial capacity and require fewer somatic mtDNA mutations to have the same effect, and would develop symptoms earlier.

Although probably not having mtDNA mutations as a primary etiology, a number of neurological diseases associated with aging may have mitochondrial dysfunction as a contributing factor to the age of onset and progression of the disease. The most common neuropathies associated with aging are Alzheimer's and Parkinson's diseases, and the role of mitochondrial dysfunction, as well as possible contributions of, and correlations with, mtDNA mutations have been studied.

Alzheimer's Disease

Alzheimer's disease (AD) is a progressive and always fatal neurodegenerative disorder characterized by the death of neurons in brain regions involved in learning and memory processes. Considerable data implicate accumulations of insoluble fibrillar aggregates of a protein called amyloid β-peptide (Aβ) in the pathogenesis of AD (see Mattson, 1997 for review). Aβ is associated with degenerating neurons in AD brain, and mutations for the amyloid precursor protein (the source of Aβ) cause a small percentage of cases of inherited familial AD (reviewed by Mullan and Crawford, 1994). In addition, mice genetically engineered to produce mutated human β-amyloid precursor protein exhibit Aβ deposition, neuronal degeneration, and cognitive impairments (Games et al., 1995; Hsiao et al., 1996). Aβ can damage and kill cultured neurons by a mechanism that is dependent upon increases in cytosolic calcium and ROS. Mitochondrial dysfunction resulting in increased ROS production may play a role in Aβ toxicity, as it has been observed that amyloidogenic amyloid precursor protein fragments aggregate in the presence of oxidation systems, a phenomenon that is inhibited by antioxidants (Dyrks et al., 1992, 1993). Amyloid itself may generate free radicals (Hensley et al., 1994) and inhibit mitochondrial function (Shearman et al., 1994; Keller et al., 1997), thus leading to another destructive feedback cycle, which theoretically could begin with slightly dyfunctional mitochondria resulting from mtDNA mutations. Other genes implicated in early-onset AD are the presenilins (PS-1 and PS-2), mutations that account for the majority of autosomal dominant inherited forms of this disease (Levy-Lahad et al., 1995; Sherrington et al., 1995). The proteins coded for by these genes are localized to the membrane of the endoplasmic reticulum (Kovacs et al., 1996) and are hypothesized to be involved in Ca^{2+} regulation of the endoplasmic reticulum (Guo et al, 1996, 1997, 1998). Vulnerability to this mutation may involve mitochondria, as it has been shown that there is an increased sensitivity to mitochondrial toxin-induced apoptosis in neural cells expressing mutant presenilin-1, and greater induction of ROS in these cells than control cells (Keller et al., 1998).

Among the factors found to be associated with AD are deficiencies in mitochondrial cytochrome c oxidase (complex IV) activity (Parker 1991; Chandrasekaran et al., 1992; Kish et al., 1992; Mutisya et al., 1994; Parker et al., 1994; Davis et al., 1997; Sheehan et al., 1997). The experimental and clinical correlations of complex IV activity and the symptomology and progression of AD are many, and exemplify the critical role of mitochondria in neuronal survival and function. Disruption of cytochrome c oxidase activity leads to neurodegenerative events discussed earlier, including accumulation of ROS, decreased ATP production, and disruption of Ca^{2+} homeostasis. Inhibition of cytochrome c oxidase with azide leads to defects in learning and memory, as well as altering hippocampal potentiation, a component of memory and learning (Bennett et al., 1992, 1996). The possibility that mtDNA mutations play a role is supported by a study in which there was a significant increase in oxidative damage in the form of 8-OH-dG in mtDNA in parietal cortex of AD patients compared with controls (Mecocci et al., 1994). Recent repots have shown that impairment of complex IV leads to increased ROS production. In studies in houseflies, complex IV activity declines with increasing age, and this decrease in activity is associated with increased H_2O_2 production (Sohal, 1993). Experimental inhibition of complex IV also has been shown to increase production of superoxide (Partridge et al., 1994).

More direct evidence that mtDNA mutations play a role in AD comes from a series of experiments using ρ cells that were repopulated with mitochondria from AD victims (Sheehan et al., 1997). In that study, clear differences were seen in key components of Ca^{2+} regulation and cell vulnerability. AD cybrids showed a significantly increased $[Ca^{2+}]_i$, which as discussed earlier is associated with both excitotoxicity and increased ROS. The most striking effect of transformation of cybrids with AD mitochondria was the delay in calcium recovery rate after stimulation of the IP3 pathway. This observation brings mtDNA into the picture as the possible explanation of other data, including the observations that AD fibroblasts have decreased Ca^{2+} uptake and increased $[Ca^{2+}]_i$, enhanced influx of Ca^{2+} after IP3 stimulation, and reduced Ca^{2+} uptake into mitochondria.

Bowling and Beal (1995) have reported preliminary finding of an increase in mutations in AD-PD at position 4336, in a sequence encoding tRNA Gln, as well as other mtDNA mutations in complex I, the 16S rRNA gene, and the 12S rRNA

Table 1. Characteristics of mtDNA Diseases

Disease	*mtDNA Mutation*	*ETC Defect*	*CNS Damage*	*Inheritance*
Alzheimer's disease	Unknown	Complex IV	Cortex Hippocampus	Sporadic
Parkinson's disease	Unknown	Complex I	Substantia nigra	Sporadic
KSS	Deletion 3271, 10423; multiple other deletions and duplications	Nonspecific/ unknown	Cerebellum Retina	Sporadic
MELAS	Point mutations in $tRNA^{Leucine}$	Nonspecific/ unknown	Cortex Basal ganglia Brain stem	Maternal
MERRF	Point mutations in $tRNA^{Lysine}$	Nonspecific/ unknown	Cerebellum Spinal cord	Maternal
LHON	Multiple point mutations in ETC mRNAs	Complex I	Optic nerve Basal ganglia	Maternal
Leigh's Syndrome	Point mutations in ATP6 gene	Complex V	Retina Basal ganglia	Maternal

Notes: ETC, electron transport chain; KSS, Kearns-Sayre syndrome; LHON, Leber's hereditary optic neuropathy; MELAS, mitochondrial encephalopathy, lactic acidosis, and stroke-like episodes; MERRF, myclonus epilepsy with ragged red fibers.

gene. In addition, mtDNA from AD temporal cortex shows a significant increase in the 5-kb mtDNA deletion associated with Kearns-Sayre syndrome, indicating that mutated DNA is preferentially accumulated in AD versus normal aged brain (Hamblet and Castora, 1997).

Therefore, a evidentiary chain can be constructed, from complex IV defects in AD leading to increased ROS production, further mtDNA damage, lose of $[Ca^{2+}]_i$ homeostasis and neuronal death.

Parkinson's Disease

Parkinson's disease (PD) is a very common disorder of movement characterized by neuronal cell death mainly affecting dopaminergic neurons of the substantia nigra, striking primarily the elderly. Pathologically, PD is characterized by degeneration of the substantia nigra and the locus ceruleus, and appearance of Lewy bodies in the neuronal cytoplasm. The substantia nigra sends dopaminergic fibers to the striatum; therefore, both dopamine content and tyrosine hydroxylase activity are greatly reduced in this target. In 1983, it was observed that the fungal neurotoxin 1-methyl-4-phenyl-1, 2, 3, 6-tetrahydropyridine (MPTP) caused a pathological syndrome remarkably similar to PD (Langston et al., 1983). Later studies showed the target of this toxin to be complex I of the mitochondrial ETC (McNaught et al., 1996). A great deal of interest has been generated of late by these findings, as it has been reported by several groups that there is a regional and disease-specific reduction in mitochondrial complex I activity in the substantia nigra of victims of PD (Schapira et al., 1989; Janetzky et al., 1994). This observation alone, however, still leaves the root cause of the disease a mystery.

As is the case for many neurodegenerative diseases, there are likely a number of different factors which eventually terminate at the same end-point. For example, there have been several nuclear genetic defects shown to be associated with this disease (Graeber et al., 1992; Furukawa et al., 1996; Nussbaum, 1997; Polymeropoulos et al., 1997), but the vast majority of cases do not have a family history of PD. The information indicating a complex I deficit has therefore focused some attention on possible mitochondrial DNA mutations, both inherited and spontaneous (Swerdlow et al., 1996; Graeber and Muller, 1998). Shoffner (1995) has reported increased incidence of an A-to-G mutation in the mitochondrial gene for tRNA (5.3% in PD versus 0.7% in control). Because so much of the mtDNA is devoted to coding for complex I products, this part of the ETC is the most likely to suffer damage from mitochondrial tRNA mutations, oxidative stress, and other insults, and is the most likely to be inherited. However, systematic sequencing studies of the complex I portion of mtDNA have been done by a number of groups (Ikebe et al., 1995; Kapsa et al., 1996; Kosel et al., 1997) and have led to the conclusion that over 90% of sporadic PD cannot be explained exclusively on this basis (Kosel et al., 1997). In studies of this type, one must keep in mind that most mitochondria in brain samples are derived from non-neuronal cells, which, because of reasons already discussed, are likely to be less at risk for development of mtDNA mutations during aging.

The possibility remains that mtDNA mutations may be responsible for a subset of PD, and for the severity and progression of those PD cases that do not have mtDNA mutations as the proximal cause of the disease. Direct evidence for this comes from studies in vitro in which the relative contributions of environmental toxins, complex I nuclear DNA mutations, and mitochondrial mutations were analyzed (Swerdlow et al., 1996). From these studies it was concluded that a portion of the complex I defect found in PD arises from mtDNA mutations. More supporting evidence comes from studies of a rare form of familial dystonia, in which a single point mutation of mtDNA complex I causes a basal ganglia dysfunction with characteristics of PD (Jun et al., 1994; Shoffner, 1995). Miller et al. (1996) recently produced a ρ cell line starting with the human SH-SY5Y neuroblastoma cells. Exogenous mitochondria were introduced via fusion of human platelets from either PD patients or nonsufferers of the disease with the ρ0 cells (Swerdlow et al., 1996). Following fusion, complex IV ETC activity was comparable in the parent SH-SY5Y cell line and in the PD and control cybrids. However, complex I activity was reduced by 20% in the PD cybrids as compared with the controls.

There is significant evidence that overproduction of the enzyme monoamine oxidase B (MAO B) is a primary cause of PD, and that the damage done to the cell is mediated both by neurotransmitter depletion and by production of ROS. Excessive oxidation of dopamine, the principal target of MAO B, leads to a depletion of this neurotransmitter, and an increase in levels of H_2O_2 and other ROS (Riederer et al., 1989). The weight of evidence points to some role of ROS in PD, and it is clear that both complex I and mtDNA are particularly vulnerable to damage by these oxyradicals (Jesberger and Richardson, 1990). It is hypothesized that the basal ganglion dopaminergic neurons are more at risk from oxidative damage because the deamination of dopamine by monoamine oxidase B increases with aging (Benedetti and Dostert, 1994) and results in increased formation of H_2O_2 and other toxic byproducts such as free radicals, 6- hydroxydopamine and quinones.

Again, a feedback cycle can be envisioned, in which some primary etiology, such as overproduction of MAO B, leads to radical production, inhibiting complex I either directly or by damage to mtDNA. There is a large variability in mtDNA heteroplasmy amongst individuals, and relative efficiencies of complex I vary between both individuals and mitochondria due to random or inherited mutations, suggesting that such variability could play a role in determining the severity and progression of the disease. In addition, as discussed earlier, mtDNA mutations during aging play a key role in the activity of complex I, which decreases over time. Some groups have reported that PD patients exhibit increased levels of the mtDNA "common mutation," which would further erode mitochondrial activity and increase ROS production, whereas others see no change between control and PD tissues. As this mutation is a 5-kb deletion between the 13-bp direct repeat sequence encompassing genes for ATPase 6/8, CO 3, ND3, ND4, ND4L, this mutation would cause a loss of complex I. Regardless of the possible increased occurrence of the common mutation in PD patients, it is known to accumulate normally in aging, and may account at least in part to the age of onset of PD.

Though evidence suggests that mitochondrial mutations are not the proximal cause of PD (reviewed by Singer et al., 1995), there are a number of lines of evidence indicting that the relative load of mitochondrial mutations carried by an individual may predispose to the disease, and perhaps more importantly, that the disease itself induces damaging mtDNA mutations which further exacerbate the progression.

HERITABLE MITOCHONDRIAL DISEASES

The first disease described having a mitochondrial defect as the primary etiology, the myopathy Luft's syndrome, was described less than 40 years ago (Luft et al., 1962). Since then well over 100 diseases have been discovered and described containing primary defects in mitochondrial function. A large proportion of these have as the defect mutations in the mtDNA that lead to a wide range of pathological conditions. This field has made spectacular progress since the first mtDNA mutation-linked disorders were first described only 10 years ago (Holt et al., 1988; Wallace et al., 1988). Because of the properties of neurons discussed earlier, these cells are more vulnerable to mitochondrial dysfunction, and nearly all mitochondrial diseases described to date have at least some neuropathology. Although over 100 different disease states have been ascribed to mtDNA mutations, the great majority of these are found in very few patients, and often in only a single individual. Of the well-studied mtDNA neuropathies, the following five account for the majority of cases (Table 1).

Kearns-Sayre Syndrome

Kearns and Sayre (1958) described two patients who presented with retinitis pigmentosa, external ophthalmoplegia, and complete heart block, and put their name to the syndrome (reviewed by Brown and Squier, 1996). Later the criteria for this disease were defined as onset before age 20, progressive external ophthalmoplegia, retinitis pigmentosa, and at least one of the following: complete heart block, cerebellar dysfunction, and elevated levels of protein in CSF (Pavlakis et al., 1984). The causes of Kearns-Sayre syndrome (KSS), multiple mtDNA rearrangements, were the first pathological mtDNA mutations described (Holt et al., 1988). There are a host of different mutations that cause KSS; both mtDNA deletions and mtDNA duplication mutations have been described (Holt et al., 1988; Poulton et al., 1989, 1993; Brockington et al., 1995) and more are likely to be found. The deletion mutations described consist of loss of a section of mtDNA that includes contiguous tRNA and oxidative phosphorylation polypeptide genes, but rarely rRNA genes. As is common in mtDNA diseases, there is significant heteroplasmy, ranging from 9% to 50% of the mitochondrial genome, which likely accounts for the variation seen between cases and tissues. The smallest mutation reported that can cause KSS is a single nucleotide deletion in the $tRNA^{Leucine\ (UUR)}$ at position 3271, a point at which A-to-G substitution can cause MELAS, as discussed later. This specific mutation causes Fahr's disease, a condition marked by extensive cerebral calcification and neurodegeneration. The duplication mutation produces two tandemly arranged mtDNA molecules consisting of a length of 16.6 kb coupled to a mtDNA deletion mutation. Approximately 80% of patients with KSS have mtDNA rearrangements (Holt et al., 1989; Moraes et al., 1989a, 1989b), which usually arise spontaneously after oocyte fertilization, with mtDNA duplication mutations having the greatest chance of maternal transmission. Sequence analysis of the region adjacent to the deletion in a KSS patient suggested a possible topoisomerase II site (Blok et al., 1995). The 5-kb "common deletion," which is present in such high amounts in KSS, is detectable in human oocytes (Chen et al., 1995), supporting the theory that pathological mutations can be transmitted in the female germ line.

As with most mtDNA diseases, the severity depends on the relative degree of heteroplasmy, and the nature of the mutation. It has also been observed that there is a reduction in absolute amount of wild-type mtDNA, which may also be a factor in the pathogenicity (Moraes et al., 1995).

MELAS

Perhaps the most widely studied of the mitochondrial diseases is mitochondrial encephalopathy, lactic acidosis, and stroke-like episodes (MELAS), was first reported by Garder-Medwin in 1975. This disorder is characterized by large or small vessel stroke, often associated with a migraine headache or seizures, and accompanied by one or more of the following: lactic acidosis, ataxia, cardiomyopathy, diabetes mellitus, renal tube defects, and retinitis pigmentosa (for reviews see Hilton, 1995; Shoffner, 1996). It may also be accompanied by sensorimotor hearing loss, ophthalmoplagia, dementia, and cardiomyopathy (de Vries et al., 1994; van den Ouweland et al., 1994). The cerebellar ataxia is often the first symptom, and may precede stroke events by years. The symptomology and severity of this, as with most mtDNA diseases, shows wide variability, and genetic studies are usually necessary for diagnosis (de Vries et al., 1997). The cause of MELAS is a mtDNA point mutation concentrated in the $tRNA^{Leucine\ (UUR)}$ (Goto et al., 1981), usually an A-to-G mutation at position 3243, found in about 80% of MELAS cases. This specific mutation is associated with increased risk of diabetes mellitus and is found in approximately 1% of adult-onset diabetic patients tested (Otabe et al., 1994). The other more infrequent mutations of the $tRNA^{Leucine\ (UUR)}$ that produce MELAS are at positions 3252 (Morten et al., 1993), 3256 (Sato et al., 1994), 3271 (Goto et al., 1991), and 3291 (Goto et al., 1994). Mutations other than those at the $tRNA^{Leucine\ (UUR)}$ associated with this disorder are a point mutation in the $tRNA^{ser(UCN)}$ gene (Nakamura et al., 1995) and a mutation in the $tRNA^{valine}$ gene (Taylor et al., 1996).

These mutations result in abnormalities in protein synthesis (Chomyn et al., 1992; King et al., 1992), ETC efficiency (Goto et al., 1992; Muller-Hocker et al., 1993), mitochondrial membrane potential (Moudy et al., 1995; James et al., 1996), and calcium sequestration (Moudy et al., 1995). Thus, these mutations have been shown to disrupt all of the major mitochondrial functions. As discussed, the functions of mitochondria are closely linked, as shown in an elegant study by Wong and Cortopassi (1997). In this study, ρ cells were tranfected with mitochondria from patients with MELAS, LHON (Leber's hereditary optic neuropathy), and MERRF (myoclonus epilepsy with ragged red fibers), and the relative sensitivity to oxidative stress measured. All three mtDNA mutant cell lines were more sensitive to H_2O_2 treatment than nonmutant parent lines, with the MELAS-transformed cells being the most vulnerable. Showing the linkage between ROS, Ca^{2+}, and PT, both depletion of Ca^{2+} in the extracellular medium, as well as treatment with cyclosporin A, protected mtDNA mutant transformed ρ cells against H_2O_2. Therefore, mtDNA mutations not only increase production of ROS, but also increase cell vulnerability to ROS. The mechanism of this sensitivity appears to be decreased capacity of mitochondria to sequester calcium before mitochondrial inner membrane depolarization and PT.

As discussed, the progressive nature of mtDNA diseases, and other diseases associated with aging, may be caused by a gradual increase in the proportion of mutated mtDNA molecules. Kovalenko et al. (1996) showed that the proportion of various non-$tRNA^{Leucine\ (UUR)}$ mutations in mtDNA from a MELAS patient was much greater than that in corresponding controls, supporting the hypothesis that mutations secondary to the primary mutation may be critical in progression of this, and by extension other, mitochondrial disorders. In cybrids containing both mitochondria with a heteroplasmic MELAS mtDNA and those with normal mtDNA, those mitochondria containing MELAS sequences had a replicative advantage, leading rapidly to homoplasmic MELAS cells (Attardi et al., 1995). This finding raises the possibility that mitochondrial division plays a critical role in at least this form of mtDNA disease.

MERRF

Myoclonus epilepsy with ragged red fibers (MERRF) is characterized by myoclonus epilepsy, cerebellar ataxia, and myopathy with ragged red fibers. Additional symptoms include seizures, cardiomyopathy, hearing loss, and dementia. The mutation most associated with MERRF is an A-to-G transition in the $tRNA^{Lys}$ gene at position 8344 (A834G) (Shoffner et al., 1990, 1991). There is also a mutation in position 3243 that is associated with symptoms similar to MELAS.

Although the particular pathogenesis of the A8344G is unknown, tRNA mutations can effect any one of the ETC polypeptides, as well as transcription. In studies in human myoblast cultures heterogenous for the mutation, no transcriptional defects were observed, but there was a decrease in protein synthesis correlating to the proportion of mutant mtDNA present (Hanna et al., 1995). Of interest, there was evidence that the different ETC subunits were differentially affected. The degree of pathogenicity in different organs does not appear to be correlated with the amount of mutant mtDNA present (Huang et al., 1995; Ozawa et al., 1995), indicating that there are different thresholds of disease expression amongst the different tissues (Tanno et al., 1993). One explanation for this may be the relative sensitivities of the tissues to oxidative damage. Cortopassi and Wang (1995) have shown that a decrease in complex I causes an increased production of ROS associated with greater electron movement through complex II, and sensitivity to oxidative damage varies widely among tissues. In support of this, a study of MERRF patients has shown that there is a high proportion of the A8344G mutation in all tissues, but few tissues outside the brain were affected (Oldfors et al., 1995). As in other mtDNA diseases, there appears to be other factors than the mtDNA mutation that lead to disease states, as Munscher et al. (1993) has observed the A8344G mutation in some healthy people at levels near those found in people exhibiting the disease.

LHON

Leber's hereditary optic neuropathy (LHON) was first described over 100 years ago (Leber, 1871). This disease accounts for approximately 3% of blindness in young adult males. LHON is clinically manifested by adult onset, bilateral, acute or subacute, central vision loss and eventually optic nerve atrophy (for review, see Newman, 1993; Wallace 1994a; Schapira 1998).

There have been 19 different mutations reported to be associated with LHON; however, it appears that only 5 play a primary role in causing the disease (Brown et al., 1992a; 1992b; Wallace, 1994b). These mutations are a G-to-A transition at np 14459 in the ND6 gene (Jun et al., 1994), a G-to-A transition at np 11778 in the ND4 gene (Wallace et al., 1988), a G-to-A transition at np 3640 in the ND1 gene (Huoponen et al., 1991, Johns, 1992; Johns et al., 1992a), a T-to-C transition at np14484 in the ND6 gene (Johns et al., 1992b), and a G-to-A transition at np 15257 in the cytochrome B gene (Johns and Neufeld, 1991).

As in any discussion of diseases with a primarily mitochondrial etiology, there is evidence of additional genetic or environmental factors, or both, which may influence the expression and severity of the disease. For example, the preponderance of LHON cases are male; 4:1 in the case of the 11778 and 3460 mutation, and over 7:1 for the 14484 mutation (Ortiz et al., 1993; Harding et al., 1995). This observed ratio is difficult to explain if the disorder were caused entirely by inherited mutant mitochondria. The possibility of an X-linked factor is hypothesized, but recent studies suggest this is not the case (Chalmers et al., 1996; Oostra et al., 1996; Pegoraro et al., 1996). There is also the curious phenomenon that one sibling who carries the same mutation as an affected sibling can remain totally asymptomatic. Since many of these patients are homoplasmic for the mutations, its unlikely to be explained by relative mutant load specific to the optic nerves.

Although all five of the common mutations can result in the same phenotype, they differ in their severity, as revealed by their ability to cause additional neurological symptoms, their propensity for causing the disease in the absence of other contributing factors, the frequency with which they are heteroplasmic, and their tendency for visual recovery. The 14459 mutation is by far the most severe of the forms, and provides a good example of the heterogeneity of disease progression and symptomology when discussing mitochondrial disorders. This mutation was first discovered in a large family in which there were two strikingly different phenotypic manifestations. One of the phenotypes was typical of LHON, but the other involved more severe neurological symptoms that mimicked those of PD in many ways (Jun et al., 1994). These symptoms included early-onset generalized dystonia (a movement disorder involving progressive rigidity associated with basal ganglia degeneration or bilateral striatal necrosis) together with pseudobulbar degeneration, short stature, and impaired cognition. Of the 43 maternal relatives in the family, 19% had LHON, 31% dystonia, and 2% both. Whereas this was the first documented case of disease arising from this specific mutation, a subsequent survey of dystonia and LHON revealed two addition positive families: an isolated child with dystonia and a mother and daughter, with both families being heteroplasmic (Novotny et al., 1986). Given the observation that one family was Caucasian while the other was of African descent, and the original observed case was Hispanic, it is believed that the np 14459 mutation has arisen in each of these families independently and can result in LHON, dystonia, or both.

Although the 14459 mutation is the most severe form, the 11778 mutation is the most common, accounting for over half of cases of LHON; it is primarily homoplasmic, and rarely shows visual recovery (Stone et al., 1992). The other three mutations previously mentioned tend to be much less common, as well as less severe.

Recent studies using cybrids have been performed examining the role of the specific mutations. Cybrids containing the 14459 mutation seem to have altered coenzyme Q binding to complex I (Jun et al., 1996). Two other mutations, at 11696 and 14596, are also associated with decreased complex I activity in a family with LHON. In studies of LHON tissues, de Vries et al. (1994) found biopsied muscle had a severe complex I deficiency and Yen et al. (1996) has shown an elevation in complex II, which is hypothesized to be a nuclear compensation mechanism. Cells from LHON patients exhibit clear deficiencies in oxygen consumption and complex I activity (James et al., 1996).

Leigh Syndrome

There are two separate mutations in the mtDNA ATP6 gene that have been shown to cause neurodegenerative disease. Both occur in codon 156 and are highly pathogenic, suggesting that most if not all families are independent mutations. The 8993G mutation converts a highly conserved leucine to an arginine (Holt et al., 1990) whereas the 8993C converts a leucine to a proline (de Vries et al., 1993). The two mutations are always heteroplasmic, and as they segregate they generate a wide range of neurological symptoms from mild retinitus pigmentosa, through macular degeneration, mental retardation, and olivopontocerebellar atrophy to Leigh syndrome (Tatuch et al., 1992; Ortiz et al., 1993). Leigh syndrome is a frequently lethal childhood disease characterized by the progressive degeneration of the basal ganglia. There is wide variability in the progression of the disease, presumably reflecting the different replicative segregation of the heteroplasmic mutations.

The biochemical abnormality of the 8993G mutation was shown to be a defect in the proton channel of the ATP synthase. Lymphoblastoid cell lines harboring the mutations were prepared from patients, and the mitochondria were analyzed for oxidative phosphorylation defects by respiration studies. Respiratory complexes I, II, and IV of the ETC were shown to be normal, as the patient's mitochondria had the same maximum respiration rate as normal mitochondria, when respiring in the presence of the uncoupler nitrophenol. However, respiration coupled to ATP synthesis and the ADP/0 ratio were both reduced 30% to 40% in the patient mitochondria, suggesting a defect in the proton channel of the ATP synthase. These defects were then linked to the 8993G mutation using cybrid transfer, and the resulting cybrids proved to be either homoplasmic mutant, homoplasmic normal, or heteroplasmic. Respiration studies of mitochondria isolated from homoplasmic normal cybrids were normal, while mitochondria from homoplasmic mutant cybrids showed the same 30% to 40% reduction in ADP-stimulated respiration and ADP/0 ratio as the parental patient cell lines. Hence, the ATP synthase defect and the variable clinical phenotypes are both the result of the heteroplasmic T-to-G transversion at np 8993 of the ATP6 gene (Trounce et al., 1994).

MITOCHONDRIAL MEDICINE

Our current understanding of mtDNA disorders indicates that the neuropathologies are determined primarily by three interrelated effects of the mutations: a decrease in available ATP to neurons, an increase in damaging ROS, and increased excitotoxicity caused by the lack of Ca^{2+} regulation. With this knowledge, a number of clinical treatment strategies have been published regarding patients suffering from these diseases. It is important to bear in mind, however, that most reports have been of single cases, and are therefore somewhat anecdotal (reviewed in Luft et al., 1996).

The most widely reported treatments showing promise are those using substrates and coenzymes of the ETC. Coenzyme Q (CoQ) and succinate treatment has shown benefits in a patient with complex I defects and KSS (Shoffner and Wallace, 1994) and another with a primary complex IV defect (Jinnai, 1990), and CoQ administration alone benefited a patient with ocular myopathy and five patients with KSS (Ogasahara, 1986). The rationale for this particular treatment rests on observations of low levels of CoQ in serum and skeletal muscle biopsies (Ogasahara et al., 1986), which leads to defective energy supply (Mortensen, 1993). To address oxidative damage, antioxidant therapy has been tried with some success in a patient with severe complex III deficiency (Eleff et al., 1984).

Recently, attention has been focused on the use of molecular biological means to treat the cause of the disorders, loss of functional expression of mtDNA-coded genes. The most promising strategy rests on attempts to relocate normal mitochondrial genes to the nucleus of the patient's cells in hopes that the gene products will be delivered to the organelle from the cytoplasm. However, owing to the constraints imposed by the inner mitochondrial membrane, these attempts have met with limited success.

CONCLUDING REMARKS

Few fields in medicine have experienced the incredible explosion of information that has occurred since the discovery of the first disease state caused by mtDNA mutations only 10 years ago. Since that time, dozens of diseases with mtDNA mutation as primary etiology have been described, and evidence now implicates mtDNA mutations as a contributing factor in many others. The complete mapping of the mitochondrial genome, and availability of cybrid cell lines, has allowed researchers to isolate the cause of these diseases to a level of understanding achieved with few other inherited illnesses. However, treatment of these disorders remains primitive, and the relative contributions of mtDNA mutations to other neurodegenerative diseases is still poorly understood. Nevertheless, the rapid pace of work in this field and emerging molecular techniques hold out great promise for both understanding the causes and effects of mtDNA mutations, and development of treatments for diseases caused by them.

REFERENCES

Ames, B.N., Shigenaga, M.K. & Hagen, T.M. (1995). Mitochondrial decay in aging. Biochim. Biophys. Acta. 1271, 165–170.

Anderson, S., Bankier, A.T., Barrell, B.G., de Bruijn, M.H., Coulson, A.R., Drouin, J., Eperon, I.C., Nierlich, D.P., Roe, BA., Sanger, F., Schreier, PH., Smith, A.J., Staden R. & Young, I.G. (1981). Sequence and organization of the human mitochondrial genome. Nature 290, 457–465.

Attardi, G., Yoneda, M. & Chomyn, A. (1995). Complementation and segregation behavior of disease-causing mitochondrial DNA mutations in cellular model systems. Biochim. Biophys. Acta 1271, 241–248.

Beckman, K.B. & Ames, B.N. (1997). Oxidative decay of DNA. J. Biol. Chem. 272, 19633–19636.

Bekkers, J.M. & Stevens, C.F. (1989). NMDA and non-NMDA receptors are co-localized at individual excitatory synapses in cultured rat hippocampus. Nature 341, 230–233.

Benedetti, M.S. & Dostert, P. (1994). Contribution of amine oxidases to the metabolism of xenobiotics. Drug Metab. Rev. 26, 507–355.

Bennett, M.C., Diamond, D.M., Stryker, S.L., Parks, J.K., Parker, W.D., Jr., Drouin, J., Eperon, I.C., Nierlich, D.P., Roe, B.A., Sanger, Schreier, P.H., Smith, A.J., Staden, R. & Young, I.G. (1992). Cytochrome oxidase inhibition, a novel model of Alzheimer's disease. J. Geriatr. Psychiatry Neurol. 5, 93–101.

Bennett, M.C., Mlady, G.W., Fleshner, M. & Rose, G.M. (1996). Synergy between chronic corticosterone and sodium azide treatments in producing a spatial learning deficit and inhibiting cytochrome oxidase activity. PNAS. 93, 1330–1334.

Benzi, G., Curti, D., Pastoris, O., Marzatico, F., Villa, R.F. & Dagani, F. (1991). Sequential damage in mitochondrial complexes by peroxidative stress. Neurochem. Res. 16, 1295–1302.

Benzi, G., Pastoris, O., Marzatico, F., Villa, RF., Dagani, F. & Curti, D. (1992). The mitochondrial electron transfer alteration as a factor involved in the brain aging. Neurobiol. Aging 13, 361–368.

Bernardi, P. & Petronilli, V. (1992). The permeability transition pore as a mitochondrial calcium release channel, a critical appraisal. J. Bioenerg. Biomembr. 28, 131–138.

Bindokas, V.P. & Miller, R.J. (1995). Excitotoxic degeneration is initiated at non-random sites in cultured rat cerebellar neurons. J. Neurosci. 15, 6999–7011.

Birnboim, H.C. (1988). A superoxide anion induced DNA strand-break metabolic pathway in human leukocytes, effects of vanadate. Biochem. Cell Biol. 66, 374–381.

Blok, R.B., Thorburn, D.R., Thompson, G.N. & Dahl, H.H. (1995). A topoisomerase II cleavage site is associated with a novel mitochondrial DNA deletion. Hum. Genet. 95, 75–81.

Bondy, S.C. & Lee, D.K. (1993). Oxidative stress induced by glutamate receptor agonists. Brain Res. 610, 229–233.

Boveris, A. & Chance, B. (1973). The mitochondrial generation of hydrogen peroxide. General properties and effect of hyperbaric oxygen. Biochem. J. 134, 707–716.

Bowling, A.C. & Beal, M.F. (1995). Bioenergetic and oxidative stress in neurodegenerative diseases. Life Sci. 56, 1151–1171.

Brockington, M., Alsanjari, N., Sweeney, M.G., Morgan-Hughes, J.A., Scaravilli, F. & Harding, A.E. (1995). Kearns-Sayre syndrome associated with mitochondrial DNA deletion or duplication, a molecular genetic and pathological study. J. Neurol. Sci. 131, 78–87.

Brown, G. K. & Squier, M.V. (1996). Neuropathology and pathogenesis of mitochondrial diseases. J. Inherit Metab. Dis. 19, 553–572.

Brown, M.D., Voljavec, A.S., Lott, M.T., MacDonald, I. & Wallace, D.C. (1992a). Leber's hereditary optic neuropathy: a model for mitochondrial neurodegenerative diseases. FASEB J. 6, 2791–2799.

Brown, M.D., Yang, C.C., Trounce, I., Torroni, A., Lott, MT. & Wallace, D.C. (1992b). A mitochondrial DNA variant, identified in Leber hereditary optic neuropathy patients, which extends the amino acid sequence in cytochrome c oxidbase subunit I. Am. J. Hum. Genet. 51, 378–385.

Bruce-Keller, A.J., Begley, J.G., Fu, W., Butterfield, D.A., Bredesen, D.E., Hutchins, J.B., Hensley, K. & Mattson, M.P. (1998). Bcl-2 protects isolated plasma and mitochondrial membranes against lipid peroxidation induced by hydrogen peroxide and amyloid β-peptide. J. Neurochem. 70, 31–39.

Carafoli, E. (1992). The Ca^{2+} pump of the plasma membrane. J. Biol. Chem. 267, 2115–2118.

Carafoli, E. (1994). The signaling function of calcium and its regulation. J. Hypertens. Suppl. 12, 47–56.

Chalmers, R.M., Davis, M.B., Sweeney, M.G., Wood, N.W. & Harding, A.E. (1996). Evidence against an X-linked visual loss susceptibility locus in Leber hereditary optic neuropathy. Am. J. Hum. Genet. 59, 103–108.

Chandrasekaran, K., Stoll, J., Giordano, T., Atack, J.R., Matocha, M.F., Brady, D.R. & Rapoport, S.I. (1992). Differential expression of cytochrome oxidase. (COX). genes in different regions of monkey brain. J. Neurosci. Res. 32, 415–423.

Chen, J.J. & Yu, B.P. (1996). Detoxification of reactive aldehydes in mitochondria, effects of age and dietary restriction. Aging 8, 334–340.

Chen, X., Prosser, R., Simonetti, S., Sadlock, J., Jagiello, G. & Schon, E.A (1995). Rearranged mitochondrial genomes are present in human oocytes Am. J. Hum. Genet. 57, 239–247.

Cheng, B. & Mattson, M.P. (1992). IGF-I and IGF-II protect cultered hippocampal and septal neurons against calcium-mediated hypoglycemic damage. J. Neurosci. 12, 1558–1566.

Choi, D.W. (1987). Ionic dependence of glutamate neurotoxicity. J. Neurosci. 7, 369–379.

Choi, D.W. (1988). Glutamate neurotoxicity and diseases of the nervous system. Neuron 1, 623–634.

Chomyn, A., Martinuzzi, A., Yoneda, M., Daga, A., Hurko, O., Johns, D., Lai, S.T., Nonaka, I., Angelini, C. & Attardi, G. (1992). MELAS mutation in mtDNA binding site for transcription termination factor causes defects in protein synthesis and in respiration but no change in levels of upstream and downstream mature transcripts. PNAS. 89, 4221–4225.

Cohen, G.M. (1997). Caspases, the executioners of Apoptosis. Biochem. J. 326, 1–16.

Corral-Debrinski, M., Horton, T., Lott, M.T., Shoffner, J.M., Beal, M.F., & Wallace, D.C. (1992). Nat. Genet. 2, 324–329

Cortopassi, G. & Wang E. (1995). Modelling the effects of age-related mtDNA mutation accumulation, complex I deficiency, superoxide and cell death. Biochim Biophys. Acta 1271, 171–176.

Craig, E.E. & Hood, D.A. (1997). Influence of aging on protein import into cardiac mitochondria. Am. J. Physiol. 272, 2983–2988.

Davis, R.E., Miller, S., Herrnstadt, C., Ghosh, S.S., Fahy, E., Shinobu, L.A., Galasko, D., Thal, L.J., Beal, M.F., Howell, N. & Parker, W.D., Jr. (1997). Mutations in mitochondrial cytochrome c oxidase genes segregate with late-onset Alzheimer disease. PNAS. 94, 4526–4531.

de Grey, A.D. (1997). A proposed refinement of the mitochondrial free radical theory of aging. Bioessays 19, 161–166.

de Vries, D.D., van Engelen, B.G., Gabreelsm, F.J., Ruitenbeek, W. & van Oost, B. (1993). A second missense mutation in the mitochondrial ATPase 6 gene in Leigh's syndrome. Ann. Neurol. 34, 410–422.

de Vries, D.D., Went, L.N., Bruyn, G.W., Scholte, H.R., Hofstra, R.M., Bolhuis, P.A. & van Oost, B. (1994). Genetic and biochemical impairment of mitochondrial complex I activity in a family with Leber hereditary optic neuropathy and hereditary spastic dystonia. Am. J. Hum. Genet. 58, 703–711.

de Vries, D., de Wijs, I., Ruitenbeek, W., Begeer, J., Smit, P., Bentlage, H. & van Oost B. (1997). Extreme variability of clinical symptoms among sibs in a MELAS family correlated with heteroplasmy for the mitochondrial A3243G mutation. J. Neurol. Sci. 124, 77–82.

Desjardins, P., de Muys, J.M. & Morais, R. (1986). An established avian fibroblast cell line without mitochondrial DNA. Somat. Cell. Mol. Genet. 12, 133–139.

Drummond, R.M. & Fay, F.S. (1996). Mitochondria contribute to Ca^{2+} removal in smooth muscle cells. Pflugers Arch. 431, 473–482.

Dyrks, T., Dyrks, E., Hartmann, T., Masters, C. & Beyreuther, K. (1992). Amyloidogenicity of β A4 and β A4-bearing amyloid protein precursor fragments by metal-catalyzed oxidation. J. Biol. Chem. 267, 18210–18217.

Dyrks, T., Dyrks, E., Masters, C.L. & Beyreuther K. (1993) Amyloidogenicity of rodent and human β A4 sequences. FEBS Lett. 324, 231–236.

Eleff, S., Kennaway, N.G., Buist, N.R., Darley-Usmar, V.M., Capaldi, R.A., Bank, W. J. & Chance, B. (1984). ^{31}P NMR study of improvement in oxidative phosphorylation by vitamins K3 and C in a patient with a defect in electron transport at complex III in skeletal muscle. PNAS. 81, 3529–3533.

Ephrussi, B., Hottinguer, B.H. & Chimenes, A.M. (1949) Action de l'acriflavine sur les levures. I. La mutation "petite colonie." Ann. Inst. Pasteur 76, 351–367.

Fattoretti, P., Bertoni-Freddari, C., Caselli, U., Paoloni, R. & Meier-Ruge, W. (1996). Morphologic changes in cerebellar mitochondria during aging. Anal. Quant. Cytol. Histol. 18, 205–208.

Freeman, B.A. & Crapo, J.D. (1982). Biology of disease, free radicals and tissue injury. Lab. Invest. 47, 412–426.

Friel, D.D. & Tsien, R.W. (1994). An FCCP-sensitive Ca^{2+} store in bullfrog sympathetic neurons and its participation in stimulus-evoked changes in $[Ca^{2+}]_i$. J. Neurosci. 14, 4007–4024.

Fry, M. & Green, D.E. (1981). Cardiolipin requirement for electron transfer in complex I and III of the mitochondrial respiratory chain. J. Biol. Chem. 256, 1874–1880.

Furukawa, Y., Mizuno, Y. & Narabayashi, H. (1996). Early-onset parkinsonism with dystonia. Clinical and biochemical differences from hereditary progressive dystonia or DOPA-responsive dystonia. Adv. Neurol. 69, 327–337.

Games, D., Adams, D., Alessandrini, R., Barbour, R., Berthelette, P., Blackwell, C., Carr, T., Clemens, J., Donaldson, T., Gillespie, F., Guido, T., Hagopian, S., Johnson-Wood, K., Khan, K., Lee, M., Leibowitz, P., Lieberburg, I., Little, S., Masliah, E., McConlogue, L., Montoya-Zavala, M., Mucke, L., Paganini, L., Penniman, E., Power, M., Schenk, D., Seubert, P., Snyder, B., Soriano, F., Tan, H., Vitale, J., Wadsworth, S., Wolozin, B. & Zhao, J. (1995). Alzheimer-type neuropathology in transgenic mice overexpressing V717F β-amyloid precursor protein. Nature 373, 523–527.

Gardner-Medwin, D., Dale, G. & Parkin, J.M. (1975). Lactic acidosis with mitochondiral myopathy and recurrent coma [abstract]. First International Congress of Child Neurology No 47.

Gasic, G.P. & Hollmann, M. (1992). Molecular neurobiology of glutamate receptors. Annu. Rev. Physiol. 54, 507–536.

Goto, Y., Nonaka, I. & Horai, S. (1981). A mutation in the $tRNA^{Leucine\ (UUR)}$ gene associated with the MELAS subgroup of mitochondrial encephalomyopathies. Nature 348, 651–653.

Goto, Y., Nonaka, I. & Horai, S. (1991). A new mtDNA mutation associated with mitochondrial myaopathy, encephalopathy, lactic acidasis and stroke-like episodes. (MELAS). Biochem. Biophys. Acta 1097, 238–240.

Goto, Y., Horai, S., Matsuoka, T., Koga, Y., Nihei, K., Kobayashi, M. & Nonaka, I. (1992). Mitochondrial myopathy, encephalopathy, lactic acidosis, and stroke-like episodes (MELAS): a correlative study of the clinical features and mitochondrial DNA mutation. Neurology 42, 545–550.

Goto, Y., Tsugane, K., Tanabe, Y., Nonaka, I. & Horai, S. (1994). A new point mutation at nucleotide pair 3291 of the mitochondrial tRNA. ($tRNA^{Leucine\ (UUR)}$). gene in a patient with mitochondrial myopathy, encephalopathy, lactic acidosis, and stroke-like episodes (MELAS). Biochem. Biophys. Res. Commun. 202, 1624–1630.

Graeber, M.B. & Muller, U. (1998). Recent developments in the molecular genetics of mitochondrial disorders. J. Neurol. Sci. 153, 251–263.

Graeber, M.B., Kupke, K.G. & Muller, U. (1992). Delineation of the dystonia-parkinsonism syndrome locus in Xq13. PNAS. 89, 8245–8248.

Gunter, T.E. (1994). Cation transport by mitochondria. J. Bioenerg. Biomembr. 26, 465–469.

Gunter, K.K. & Gunter, T.E. (1994). Transport of calcium by mitochondria. J. Bioenerg. Biomembr. 26, 471–485.

Guo Q., Furukawa K., Sopher B.L., Pham D.G., Xie J., Robinson N., Martin G.M. & Mattson M.P. (1996). Alzheimer's PS-1 mutation perturbs calcium homeostasis and sensitizes PC12 cells to death induced by amyloid β-peptide. Neuroreport 8, 379–383.

Guo Q., Sopher B.L., Furukawa K., Pham D.G., Robinson N., Martin G.M. & Mattson M.P. (1997). Alzheimer's presenilin mutation sensitizes neural cells to apoptosis induced by trophic factor withdrawal and amyloid β-peptide: involvement of calcium and oxyradicals. J. Neurosci. 17, 4212–4222.

Guo Q., Robinson N. & Mattson M.P. (1998). Secreted β-amyloid precursor protein counteracts the proapoptotic action of mutant presenilin-1 by activation of NF-kappaB and stabilization of calcium homeostasis. J. Biol. Chem. 273, 12341–12351.

Gutteridge, J.M. & Halliwell, B. (1989). Iron toxicity and oxygen radicals. Baillieres Clin. Haematol. 2, 195–256.

Gyllensten, U., Wharton, D., Josefsson, A. & Wilson, A.C. (1991). Paternal inheritance of mitochondrial DNA in mice. Nature 352, 255–257.

Hagen, T.M., Yowe, D.L., Bartholomew, J.C., Wehr, C.M., Do, K.L., Park, J.Y. & Ames, B.N. (1997). Mitochondrial decay in hepatocytes from old rats, membrane potential declines, heterogeneity and oxidants increase. PNAS. 94, 3064–3069.

Hamblet, N.S. & Castora, F.J. (1997). Elevated levels of the Kearns-Sayre syndrome mitochondrial DNA deletion in temporal cortex of Alzheimer's patients. Mutat. Res. 379, 253–262.

Hanna, M.G., Nelson, I.P., Morgan-Hughes, J.A. & Harding, A.E. (1995). Impaired mitochondrial translation in human myoblasts harbouring the mitochondrial DNA $tRNA^{lysine}$ 8344 A→G (MERRF) mutation, relationship to proportion of mutant mitochondrial DNA. J. Neurol. Sci. 130, 154–160.

Harding, A.E., Sweeney, M.G., Govan, G.G. & Riordan-Eva, P. (1995). Pedigree analysis in Leber's hereditary optic neuropathy families with a pathogenic mtDNA mutation. Am. J. Hum. Genet 57, 77–86.

Harman, D., Eddy, D.E. & Noffsinger, J. (1976). Free radical theory of aging; inhibition of amyloidosis in mice by antioxidants, possible mechanism. J. Am. Geriatr. Soc. 24, 203–210.

Hayakawa, M., Tori, K., Sugiyama, S., Tanaka, M. & Ozawa, T. (1991). Age-associated accumulation of 8-hydroxydeoxyguanosine in mitochondrial DNA of human diaphragm. Biochem. Biophys. Res. Commun. 179, 1023–1029.

Hayakawa, M., Hattori, K., Sugiyama, S. & Ozawa, T. (1992). Age-associated oxygen damage and mutations in mitochondrial DNA in human hearts. Biochem. Biophys. Res. Commun. 189, 979–985.

Hayakawa, M., Sugiyama, S., Hattori, K., Takasawa, M. & Ozawa, T. (1993). Age-associated damage in mitochondrial DNA in human hearts. Mol. Cell. Biochem. 119, 95–103.

Henkart, P.A. & Grinstein, S. (1996). Apoptosis, mitochondria resurrected? J. Exp. Med. 183, 1293–1295.

Hensley, K., Carney, J.M., Mattson, M.P., Aksenova, M., Harris, M., Wu, J.F., Floyd, R.A. & Butterfield, D.A. (1994). A model for β-amyloid aggregation and neurotoxicity based on free radical generation by the peptide, relevance to Alzheimer disease. PNAS. 91, 3270–3274.

Herrington, J., Park, Y.B., Babcock, D.F. & Hille B. (1996). Dominant role of mitochondria in clearance of large Ca^{2+} loads from rat adrenal chromaffin cells. Neuron 16, 219–228.

Hilton, G. (1995). MELAS, a mitochondrial encephalomyopathy syndrome. J. Neurosci. Nurs. 27, 278–282.

Holt, I.J., Harding, A.E. & Morgan-Hughes, J.A. (1988). Mitochondrial DNA polymorphism in mitochondrial myopathy. Hum. Genet. 79, 53–57.

Holt, I.J., Harding, A.E. & Morgan-Hughes, J.A. (1989). Deletions of muscle mitochondrial DNA in mitochondrial myopathies, sequence analysis and possible mechanisms. Nucleic Acids Res. 17, 4465–4469.

Holt, I.J., Harding, A.E., Petty, R.K. & Morgan-Hughes. (1990). A new mitochondrial disease associated with mitochondrial DNA heteroplasmy. Am. J. Hum. Genet. 46, 428–433.

Hsiao, K., Chapman, P., Nilsen, S., Eckman, C., Harigaya, Y., Younkin, S., Yang, F. & Cole, G. (1996). Correlative memory deficits, Aβ elevation, and amyloid plaques in transgenic mice. Science 274, 99–102.

Huoponen, K., Vilkki, J., Aula, P., Nikoskelainen, E.K. & Savontaus, M.L. (1991). A new mtDNA mutation associated with Leber hereditary optic neuroretinopathy. Am. J. Hum. Genet. 48, 1147–1153.

Huang, C.C., Chu, N.S., Shih, K.D., Pang, C.Y. & Wei, Y.H. (1995). Distribution and clinical expression of the tRNA. (Lys) mutation in mitochondrial DNA in MERRF syndrome. J. Formos. Med. Assoc. 94, 159–163.

Ikebe, S., Tanaka, M. & Ozawa, T. (1995). Point mutations of mitochondrial genome in Parkinson's disease. Mol. Brain. Res. 28, 281–295.

James, A.M., Wei, Y.H., Pang, C.Y. & Murphy, M.P. (1996). Altered mitochondrial function in fibroblasts containing MELAS or MERRF mitochondrial DNA mutations. Biochem. J. 318, 401–407.

Janetzky, B., Hauck, S., Youdim, M.B., Riederer, P., Jellinger, K., Pantucek, F., Zochling, R., Boissl K.W. & Reichmann, H. (1994). Unaltered aconitase activity, but decreased complex I activity in substantia nigra pars compacta of patients with Parkinson's disease. Neurosci. Lett. 169, 126–128.

Jesberger, J.A., Richardson, J.S. (1991). Oxygen free radicals and brain dysfunction. Int. J. Neurosci. 57, 1–17.

Jinnai K., Yamada H., Kanda F., Masui Y., Tanaka M., Ozawa T. & Fujita, T. (1990). A case of mitochondrial myopathy, encephalopathy and lactic acidosis due to cytochrome c oxidase deficiency with neurogenic muscular changes. Eur. Neurol. 30, 56–60.

Johns, D.R. (1992). Mitochondrial ND-I mutation in Leber hereditary optic neuropathy. Am. J. Hum. Genet. 50, 872–874.

Johns, D.R. & Neufeld, M.J. (1991). Cytochrome b mutations in Leber hereditary optic neuropathy. Biochem. Biophys. Res. Commun. 181, 1358–1364.

Johns, D.R., Smith, K.H. & Miller, N.R. (1992a). Leber's hereditary optic neuropathy. Clinical manifestations of the 3460 mutation. Arch. Ophthalmol. 110, 1577–1581.

Johns, D.R., Neufeld, M.J. & Park, R.D. (1992b). An ND-6 mitochondrial DNA mutation associated with Leber hereditary optic neuropathy. Biochem. Biophys. Res. Commun. 187, 1551–1557.

Jun, A.S., Brown, M.D. & Wallace, D.C. (1994). A mitochondrial DNA mutation at nucleotide pair 14459 of the NADH dehydrogenase subunit 6 gene associated with maternally inherited Leber hereditary optic neuropathy and dystonia. PNAS. 91, 6206–6210.

Jun, A.S., Trounce, I.A., Brown, M.D., Shoffner, J.M. & Wallace, D.C. (1996). Use of transmitochondrial cybrids to assign a complex I defect to the mitochondrial DNA-encoded NADH dehydrogenase subunit 6 gene mutation at nucleotide pair 14459 that causes Leber hereditary optic neuropathy and dystonia. Mol. Cell. Biol. 16, 771–777.

Kalous, M. & Drahota, Z. (1996). The role of mitochondria in aging. Physiol. Res. 45, 351–359.

Kapsa, R.M., Jean-Francois, M.J., Lertrit, P., Weng, S., Siregar, N., Ojaimi, J., Donnan, G., Masters, C. & Byrne. E. (1996). Mitochondrial DNA polymorphism in substantia nigra. J. Neurol. Sci. 144, 204–211.

Kearns, N.P. & Sayre, G.P. (1958). Retinitis pigmentosa, external ophthalmoplegia, and complete heart block. Arch. Ophthalmol. 60, 280–289.

Keller, J.N., Pang, Z., Geddes, J.W., Begley, J.G., Germeyer, A., Waeg, G. & Mattson, M.P. (1997). Impairment of glucose and glutamate transport and induction of mitochondrial oxidative stress and dysfunction in synaptosomes by amyloid β-peptide: role of the lipid peroxidation product 4-hydroxynonenal. J. Neurochem. 68, 273–284.

Keller, J.N., Guo, Q., Holtsberg, F.W., Bruce-Keller., A.J. & Mattson, M.P. (1998). Increased sensitivity to mitochondrial toxin-induced apoptosis in neural cells expressing mutant presenilin-1 is linked to perturbed calcium homeostasis and enhanced oxyradical production. J. Neurosci. 18, 4439–4450.

King, M.P. & Attardi, G. (1989). Human cells lacking mtDNA, repopulation with exogenous mitochondria by complementation. Science 246, 500–503.

King, M., Koga, Y., Davidson, M. & Schon, E. (1992). Defects in mitochondrial protein synthesis and respiratory chain activity segragate with the $tRNA^{Leucine (UUR)}$ mutation associated with mitochondrial myopathy, encephalopathy, lactic acidosis, and strokelike episodes. Mol. Cell. Biol. 12, 180–190.

Kish, S.J., Bergeron, C., Rajput, A., Dozic, S., Mastrogiacomo, F., Chang, L.J., Wilson, J.M., DiStefano, L.M. & Nobrega, J.N. (1992). Brain cytochrome oxidase in Alzheimer's disease. J. Neurochem. 59, 776–779.

Kosel, S., Egensperger, R., Schnopp, N.M. & Graeber, M.B. (1997). The 'common deletion' is not increased in parkinsonian substantia nigra as shown by competitive polymerase chain reaction. Mov. Disord. 12, 639–645.

Kovacs, D.M., Fausett, H.J., Page, K.J., Kim, T.W., Moir, R.D., Merriam, D.E., Hollister, R.D., Hallmark, O.G., Mancini, R., Felsenstein, K.M., Hyman, B.T., Tanzi, R.E. & Wasco, W. (1996) Alzheimer-associated presenilins 1 and 2: neuronal expression in brain and localization to intracellular membranes in mammalian cells. Nat. Med. 2, 224–229.

Kovalenko, S.A., Tanaka, M., Yoneda, M., Iakovlev, A.F. & Ozawa T. (1996). Accumulation of somatic nucleotide substitutions in mitochondrial DNA associated with the 3243 A-to-G $tRNA^{Leucine (UUR)}$ mutation in encephalomyopathy and cardiomyopathy. Biochem. Biophys. Res. Commun. 222, 201–207.

Kristal, B.S. & Dubinsky, J.M. (1997). Mitochondrial permeability transition in the central nervous system, induction by calcium cycling-dependent and -independent pathways. J. Neurochem. 69, 524–538.

Kroemer, G. (1996). Mitochondrial permeability transition triggers lymphocyte apoptosis. J. Immunol. 157, 4830–4836.

Kruman, I., Guo, Q. & Mattson, M.P. (1998). Calcium and reactive oxygen species mediate staurosporine-induced mitochondrial dysfunction and apoptosis in PC12 cells. J. Neurosci. Res. 51, 293–308.

Kunkel, T.A. & Loeb, L.A. (1979). On the fidelity of DNA replication. Effect of divalent metal ion activators and deoxyribonucleoside triphosphate pools on *in vitro* mutagenesis. J. Biol. Chem. 254, 5718–5725.

Kunkel, T.A. & Loeb, L.A. (1981). Fidelity of mammalian DNA polymerases. Science 213, 765–767.

Langston, J.W., Ballard, P., Tetrud, J.W. & Irwin I. (1983). Chronic Parkinsonism in humans due to a product of meperidine-analog synthesis. Science 219, 979–980.

Leber, T. (1871). Ueber hereditare und congenital angelegte Sehnopsinlieden. Albrecht v. Graefes Arch. Ophthalmol. 17, 249–291.

Levy-Lahad, E., Wasco, W., Poorkaj, P., Romano, D.M., Oshima, J., Pettingell, W.H., Yu, C.E., Jondro, P.D., Schmidt, S.D., Wang, K., Crowly, A.C., Fu, Y-H., Guenette, S.Y., Galas, D., Nemens, Wijsman, E.M., Bird, T.D., Schellenberg, G.D. and Tanzi, R.E. (1995). Candidate gene for the chromosome 1 familial Alzheimer's disease l. Science 269, 973–7.

Li, L., Zheng, L.X. & Yang, Y. (1995) Effect of propensity of hexagonal II phase formation on the activity of mitochondrial ubiquinol-cytochrome c reductase and H.(+)-ATPase. Chem. Phys. Lipids 76, 135 144.

Linnane, A.W., Lukins, H.B., Molloy, P.L., Nagley, P., Rytka, J., Sriprakash, K.S. & Trembath, M.K. (1976). Biogenesis of mitochondria, molecular mapping of the mitochondrial genome of yeast. PNAS. 73, 2082–2085.

Lucas, D.R. & Newhouse, J.P. (1957). The toxic effect of sodium L-glutamate on the inner layers of the retina. Arch. Ophthalmol. 58, 193–201.

Luft, R. (1996). The development of mitochondrial medicine. PNAS. 91, 8731–8738.

Luft, R., Ikkos, D. & Palmieri, G. (1962) A case of severe hypermetabolism of nonthyroid origin with a defect in the maintenance of mitochondrial respiratory control; a correlated clinical, biochemical and morphological study. J. Clin. Invest. 41, 1776–1804.

Marchetti, P., Hirsch, T., Zamzami, N., Castedo, M., Decaudin, D., Susin, S.A., Masse, B. & Kroemer, G. (1996). Mitochondrial permeability transition triggers lymphocyte apoptosis. J. Immunol. 157, 4830–4836.

Mattson, M.P. (1997). Cellular actions of β-amyloid precursor protein and its soluble and fibrillogenic derivatives. Physiol. Rev. 77, 1081–1132.

Mattson, M.P. & Cheng, B. (1993). Growth factors protect neurons against excitotoxic/ischemic damage by stabilizing calcium homeostasis. Stroke 24, 136–140.

Mattson, M.P. & Mark, R.J. (1996). Excitotoxicity and excitoprotection *in vitro*. Adv. Neurol. 71, 1–30.

Mattson, M.P. & Scheff, S.W. (1994) Endogenous neuroprotection factors and traumatic brain injury: mechanisms of action and implications for therapy. J. Neurotrauma 11, 3–33.

Mattson, M.P., Dou, P. & Kater, S.B. (1988). Outgrowth-regulating actions of glutamate in isolated hippocampal pyramidal neurons. J. Neurosci. 8, 2087–2100.

Mattson, M.P., Guthrie, P.B. & Kater, S.B. (1989). A role for Na^+-dependent Ca^{2+} extrusion in protection against neuronal excitotoxicity. FASEB J. 3, 2519–2526.

Mattson, M.P., Barger, S.W., Cheng, B., Lieberburg, I., Smith-Swintosky, V.L. & Rydel R.E. (1993). β-Amyloid precursor protein metabolites and loss of neuronal Ca^{2+} homeostasis in Alzheimer's disease. Trends Neurosci. 16, 409–414.

Mattson, M.P., Lovell, M.A., Furukawa, K. & Markesbery, W.R. (1995). Neurotrophic factors attenuate glutamate-induced accumulation of peroxides, elevation of intracellular Ca^{2+} concentration, and neurotoxicity and increase antioxidant enzyme activities in hippocampal neurons. J. Neurochem. 65, 1740–1751.

McDonald, J.W. & Johnston, M.V. (1990). Physiological and pathophysiological roles of excitatory amino acids during central nervous system development. Brain Res. Rev. 15, 41–70.

McNaught, K.S., Thull, U., Carrupt, P.A., Altomare, C., Cellamare, S., Carotti, A., Testa, B., Jenner, P. & Marsden, C.D. (1996). Inhibition of [H3]dopamine uptake into striatal synaptosomes by isoquinoline derivatives structurally related to 1-methyl-4-phenyl-1, 2, 3, 6-tetrahydropyridine. Biochem. Pharmacol. 52, 29–34.

Mecocci, P., MacGarvey, U. & Beal, M.F. (1994). Oxidative damage to mitochondrial DNA is increased in Alzheimer's disease. Ann. Neurol. 36, 747–751.

Mildaziene, V., Baniene, R., Nauciene, Z., Bakker, B.M., Brown, G.C., Westerhoff, H.V. & Kholodenko, B.N. (1995). Calcium indirectly increases the control exerted by the adenine nucleotide translocator over 2-oxoglutarate oxidation in rat heart mitochondria. Arch. Biochem. Biophys. 324, 130–134.

Miller, S.W., Trimmer, P.A., Parker, W.D., Jr. & Davis, R.E. (1996). Creation and characterization of mitochondrial DNA-depleted cell lines with "neuronal-like" properties. J. Neurochem. 67, 1897–1907.

Moraes, C.T., DiMauro, S., Zeviani, M., Lombes, A., Shanske, S., Miranda, A.F., Nakase, H., Bonilla, E., Werneck, L.C. & Servidei, S. (1989a). Mitochondrial DNA deletions in progressive external ophthalmoplegia and Kearns-Sayre. N. Engl. J. Med. 320, 1293–1299.

Moraes, C.T., Schon, E.A., DiMauro, S. & Miranda, A.F. (1989b). Heteroplasmy of mitochondrial genomes in clonal cultures from patients with Kearns-Sayre syndrome. Biochem. Biophys. Res. Commun. 160, 765–771.

Moraes, C.T., Sciacco, M., Ricci, E., Tengan, C.H., Hao, H., Bonilla, E., Schon, E.A. & DiMauro, S. (1995). Phenotype-genotype correlations in skeletal muscle of patients with mtDNA deletions. Muscle Nerve 3, 150–153.

Morten, K.J., Cooper, J.M., Brown, G.K., Lake, B.D., Pike, D. & Poulton, J. (1993). A new point mutation associated with mitochondrial encephalomyopathy. Hum. Mol. Genet. 2, 2081–2087.

Mortensen, S.A. (1993). Perspectives on therapy of cardiovascular diseases with coenzyme Q10 (ubiquinone). Clin. Invest. 71, 116–123.

Moudy, A.M., Handran, S.D., Goldberg, M.P., Ruffin, N., Karl I. & Kranz-Eble, P. (1995). Abnormal calcium homeostasis and mitochondrial polarization in a human encephalomyopathy. Proc. Natl. Acad. Sci. U.S.A. 92, 729–733.

Moyner, H., Goldberg, M.P. & Choi, D.W. (1990). Glucose deprivation neuronal injury in cortical culture. Brain Res. 483, 347–354.

Mullan, M. & Crawford, F. (1994) The molecular genetics of Alzheimer's disease. Mol. Neurobiol. 9, 15–22.

Muller-Hocker, J., Hubner, G., Bise, K., Forster, C., Hauck, S., Paetzke, I., Pongratz, D. & Kadenbach, B. (1993). Generalized mitochondrial microangiopathy and vascular cytochrome c oxidase deficiency. Occurrence in a case of MELAS syndrome with mitochondrial cardiomyopathy-myopathy and combined complex I/IV deficiency. Arch. Pathol. Lab. Med. 117, 202–210.

Munscher, C., Rieger, T., Muller-Hocker, J. & Kadenbach, B. (1993). The point mutation of mitochondrial DNA characteristic for MERRF disease is found also in healthy people of different ages. FEBS Lett. 317, 27–30.

Murphy, A.N., Bredesen, D.E., Cortopassi, G., Wang, E. & Fiskum, G. (1996). Bcl-2 potentiates the maximal calcium uptake capacity of neural cell mitochondria. PNAS. 93, 9893–9898.

Murphy, J.G., Marsh, J.D. & Smith, T.W. (1987). The role of calcium in ischemic myocardial injury. Circulation 75, 15–24.

Mutisya, E.M., Bowling, A.C. & Beal, M.F. (1994). Cortical cytochrome oxidase activity is reduced in Alzheimer's disease. J. Neurochem. 63, 2179–2184.

Nakamura, M., Nakano, S., Goto, Y., Ozawa, M., Nagahama, Y., Fukuyama, H., Akiguchi, I., Kaji, R. & Kimura, J. (1995). A novel point mutation in the mitochondrial $tRNA^{ser(UCN)}$ gene detected in a family with MERRF/MELAS overlap syndrome. Biochem. Biophys. Res. Commun. 214, 86–93.

Nass, M.M. & Nass, S. (1963a). I. Fixation and electron staining reactions. J. Cell Biol. 19, 593–611.

Nass, S. & Nass, M.M. (1963b). Intramitochondral fibers with DNA characteristics. II. Enzymatic and other hydrolytic treatments. J. Cell. Biol. 19, 613–29.

Newman, N.J. (1993). Neuro-ophthalmology, the afferent visual system. Curr. Opin Neurol. 738–746.

Nohl, H. & Jordan, W. (1986). The mitochondrial site of superoxide formation. Biochem. Biophys. Res. Commun. 138, 533–539.

Novelli, A., Reilly, J.A., Lysko. P.G. & Henneberry, R.C. (1988). Glutamate becomes neurotoxic via the N-methyl-D-aspartate receptor when intracellular energy levels are reduced. Brain Res. 451, 205–212.

Novotny, E.J., Jr., Singh, G., Wallace, D.C., Dorfman, L.J., Louis, A., Sogg, R.L. & Steinman, L. (1986). Leber's disease and dystonia, a mitochondrial disease. Neurology 36, 1053–1060.

Nussbaum, R.L. (1997). Mutation in the alpha-synuclein gene identified in families with Parkinson's disease. Science 276, 2045–2047.

Ogasahara, S., Nishikawa, Y., Yorifuji, S., Soga, F., Nakamura, Y., Takahashi, M., Hashimoto, S., Kono, N. & Tarui, S. (1986). Treatment of Kearns-Sayre syndrome with coenzyme Q10. Neurology 36, 45–53.

Ogura, A., Miyamoto, M. & Kudo, Y. (1988). Neuronal death *in vitro*, parallelism between survivability of hippocampal neurones and sustained elevation of cytosolic Ca^{2+} after exposure to glutamate receptor agonist. Exp. Brain Res. 73, 447–458.

Oldfors, A., Moslemi, A.R., Fyhr, I.M., Holme, E., Larsson, N.G. & Lindberg, C. (1995). Mitochondrial DNA deletions in muscle fibers inclusion body myositis. J. Neuropathol. Exp. Neurol. 54, 581–587.

Olney, J.W. & de Gubareff, T. (1978). Glutamate neurotoxicity and Huntington's chorea Nature 271, 557–559.

Oostra, R.J., Kemp S., Bolhuis, P.A. & Bleeker-Wagemakers, E.M. (1996). No evidence for 'skewed' inactivation of the X-chromosome as cause of Leber's hereditary optic neuropathy in female carriers. Hum. Genet. 97, 500–505.

Ortiz, R.G., Newman, N.J., Shoffner, J.M., Kaufman, A.E., Koontz, D.A. & Wallace, D.C. (1993). Variable retinal and neurologic manifestations in patients harboring the mitochondrial DNA 8993 mutation. Arch. Ophthalmol. 111, 1525–1530.

Otabe, S., Sakura, H., Shimokawa, K., Mori Y., Kadowaki, H., Yasuda, K., Nonaka, K., Hagura, R., Akanuma, Y. & Yazaki, Y. (1994). The high prevalence of the diabetic patients with a mutation in the mitochondrial gene in Japan. J. Clin. Endocrinol. Metab. 79, 768–771.

Oyama, Y., Hayashi, A., Ueha, T., Chikahisa, L. & Furukawa, K. (1993). Fluorescent estimation on the effect of Ca^{2+} antagonists on the oxidative metabolism in dissociated mammalian brain neurons. Brain Res. 610, 172–175.

Ozawa, M., Goto, Y., Sakuta, R., Tanno, Y., Tsuji, S. & Nonaka, P. (1995). The 8,344 mutation in mitochondrial DNA, a comparison between the proportion of mutant DNA and clinico-pathologic findings. Neuromuscul. Disord. 5, 483–488.

Parker, W.D., Jr. (1991). Cytochrome oxidase deficiency in Alzheimer's disease. Ann. N.Y. Acad. Sci. 640, 59–64.

Parker, W.D., Jr., Mahr, N.J., Filley, C.M., Parks, J.K., Hughes, D., Young, D.A. & Cullum C.M. (1994). Reduced platelet cytochrome c oxidase activity in Alzheimer's disease. Neurology 44, 1086–1090.

Partridge, R.S., Monroe, S.M., Parks, J.K., Johnson, K., Parker, W.D., Jr., Eaton, G.R. & Eaton, S.S. (1994). Spin trapping of azidyl and hydroxyl radicals in azide-inhibited rat brain submitochondrial particles. Arch. Biochem. Biophys. 310, 210–217.

Pavlakis, S.G., Phillips, P.C., DiMauro, S., De Vivo, D.C. & Rowland, L.P. (1984). Mitochondrial myopathy, encephalopathy, lactic acidosis, and strokelike episodes, a distinctive clinical syndrome. Ann. Neurol. 16, 481–488.

Pegoraro, E., Carelli, V., Zeviani, M., Cortelli, P., Montagna, P., Barboni, P., Angelini, C. & Hoffman, E.P. (1996). X-inactivation patterns in female Leber's hereditary optic neuropathy patients do not support a strong X-linked determinant. Am. J. Med. Genet. 61, 356–362.

Petit, P.X., Lecoeur, H., Zorn, E., Dauguet, C., Mignotte, B. & Gougeon, M.L. (1995). Alterations in mitochondrial structure and function are early events of dexamethasone-induced thymocyte apoptosis. J. Cell Biol. 130, 157–167.

Polymeropoulos, M.H., Lavedan, C., Leroy, E., Ide, S.E., Dehejia, A., Dutra, A., Pike, B., Root, H., Rubenstein, J., Boyer, R., Stenroos, E.S., Chandrasekharappa, S., Athanassiadou, A., Papapetropoulos, T., Johnson, W.G., Lazzarini, A.M., Duvoisin, R.C., Di Iorio, G., Golbe, L.I. & Nussbaum, R.L. (1997). Mutation in the alpha-synuclein gene identified in families with Parkinson's disease. Science 276, 2045–2047.

Poulton, J., Deadman, M.E. & Gardiner, R.M. (1989). Tandem direct duplications of mitochondrial DNA in mitochondrial myopathy, analysis of nucleotide sequence and tissue distribution. Nucleic Acids Res. 17, 10223–10229.

Poulton, J., Deadman, M.E., Bindoff, L., Morten, K., Land, J. & Brown, G. (1993). Families of mtDNA re-arrangements can be detected in patients with mtDNA deletions, duplications may be a transient intermediate form. Hum. Mol. Genet. 2, 23–30.

Raff, M.C. (1992). Social controls on cell survival and cell death. Nature 356, 397–400.

Raff, M.C., Barres, B.A., Burne, J.F., Coles, H.S., Ishizaki, Y. & Jacobson, M.D. (1993). Programmed cell death and the control of cell survival, lessons from the nervous system. Science 262 (5134), 695–700.

Rice-Evans, C. & Burdon R. (1993). Free radical-lipid interactions and their pathological consequences. Prog. Lipid. Res. 32, 71–110.

Richter, C. (1997). Reactive oxygen and nitrogen species regulate mitochondrial Ca^{2+} homeostasis and respiration. Biosci. Rep. 17, 53–66.

Richter, C. & Kass, G.E. (1991). Oxidative stress in mitochondria, its relationship to cellular Ca^{2+} homeostasis, cell death, proliferation, and differentiation. Chem. Biol. Interact. 77, 1–23.

Richter, C., Park, J.W. & Ames, B.N. (1988). Normal oxidative damage to mitochondrial and nuclear DNA is extensive. PNAS. 85, 6465–6467.

Riederer, P., Konradi, C., Hebenstreit, G. & Youdim, M.B. (1989). Neurochemical perspectives to the function of monoamine oxidase. Acta Neurol. Scand. Suppl. 126, 41–45.

Rothman, S.M. (1986). Glutamate and anoxic neuronal death *in vitro*. Adv. Exp. Med. Biol. 203, 687–695.

Sato, W., Hayasaka, K., Shoji, Y., Takahashi, T., Takada, G., Saito, M., Fukawa, O. & Wachi, E. (1994). A mitochondrial $tRNA^{Leucine\ (UUR)}$ mutation at 3,256 associated with mitochondrial myopathy, encephalopathy, lactic acidosis, and stroke-like episodes. (MELAS). Biochem. Mol. Biol. Int. 33, 1055–1061.

Schapira, A.H. (1998). Human complex I defects in neurodegenerative diseases. Biochim. Biophys. Acta 1364, 261–270.

Schapira, A.H., Cooper, J.M., Dexter, D., Jenner, P., Clark, J.B. & Marsden, C.D. (1989). Mitochondrial complex I deficiency in Parkinson's disease. Lancet 1, 1269.

Schatz, G. & Klima, J. (1964). Triphosphopyridine nucleotide, cytochrome c reductase of saccharomyces cerevisiae, "microsomal" enzyme. Biochim. Biophys. Acta. 981, 448–461.

Searle, J., Kerr, J.F. & Bishop, C. (1982). Necrosis and apoptosis, distinct modes of cell death with fundamentally different significance. Pathol. Annu. 17, 229–259.

Seeburg, P.H. (1993). The TIPS/TINS lecture, the molecular biology of mammalian glutamate receptor channels. Trends Pharmacol. Sci. 14, 297–303.

Shearman, M.S., Ragan, C.I. & Iversen, L.L. (1994). Inhibition of PC12 cell redox activity is a specific, early indicator of the mechanism of β-amyloid-mediated cell death. Proc. Natl. Acad. Sci. U.S.A. 91, 1470–1474.

Sheehan, J.P., Swerdlow, R.H., Miller, S.W., Davis, R.E., Parks, J.K., Parker, W.D. & Tuttle, J.B. (1997). Calcium homeostasis and reactive oxygen species production in cells transformed by mitochondria from individuals with sporadic Alzheimer's disease. J. Neurosci. 17, 4612–4622.

Sheerington, R., Rogaev, E.I., Liang, Y., Rogaev, E.I., Levesque, G., Ikeda, M., Chi, H., Li, G., Holman, K., Tsuda, T., Mar, L., Foncin, J.F., Bruni, A.C., Montesi, M.P., Sorbi, S., Rainero, I., Pinessi, L., Nee, L., Chumakov, I., Pollen, D., Brookes, A., Sanseau, P., Polinsky, R.J., Wasco, W., DaSilva, H.A.R., Haines, J.L., Pericak-Vance, M.A., Tanzi, R.E., Roses, A.D., Fraser, P.E., Rommens, J.M. & St.George-Hyslop, P.H. (1995). Cloning of a gene bearing missense mutations in early-onset familial Alzheimer's disease. Nature 375, 754–60.

Shigenaga, M.K., Hagen, T.M. & Ames, B.N. (1994). Oxidative damage and mitochondrial decay in aging. PNAS. 91, 10771–10778.

Shoffner, J.M. (1995). Mitochondrial defects in basal ganglia diseases. Curr. Opin. Neurol. 8, 474–479.

Shoffner, J.M. (1996). Maternal inheritance and the evaluation of oxidative phosphorylation diseases. Lancet 348, 1283–1288.

Shoffner, J.M. & Wallace, D.C. (1994). Oxidative phosphorylation diseases and mitochondrial DNA mutations: diagnosis and treatment. Annu. Rev. Nutr. 14, 535–568.

Shoffner, J.M., Lott, M.T., Lezza, A.M.S., Seibel, P., Ballinger, S.W. & Wallace, D.C. (1990). Myoclonic epilepsy and tagged-red fiber disease (MERRF) is associated with a mitochondrial DNA tRNA (Lys) mutation. Cell 61, 931–937.

Shoffner, J.M., Lott, M.T. & Wallace, D.C. (1991). MERRF: a model disease for understanding the principles of mitochondrial genetics. Rev. Neurol. 147, 431–435.

Siesjo, B.K., Agardh, C.D. & Bengtsson, F. (1989). Free radicals and brain damage. Cerebrovasc. Brain Metab. Rev. 1, 165–211.

Singer, T.P., Ramsay, R.R. & Ackrell, B.A. (1995). Deficiencies of NADH and succinate dehydrogenases in degenerative diseases and myopathies. Biochim. Biophys. Acta 1271, 211–219.

Sohal, R.S. (1993). Aging, cytochrome oxidase activity, and hydrogen peroxide release by mitochondria. Free Radic. Biol. Med. 14, 583–588.

Soong, N.W., Hinton, D.R., Cortopassi, G., & Arnheim, N. (1992). Mosaicism for a specific somatic mitochondrial DNA mutation in adult human brain. Nat. Genet. 2, 318–323.

Stone, E.M., Newman, N.J., Miller, N.R., Johns, D.R., Lott, M.T. & Wallace, D.C. (1992). Visual recovery in patients with Leber's hereditary optic neuropathy and the 11778 mutation. J. Clin. Neuroophthalmol. 12, 10–14.

Susin, S.A., Zamzami, N., Castedo, M., Hirsch, T., Marchetti, P., Macho, A., Daugas, E., Geuskens, M. & Kroemer, G. (1996). Bcl-2 inhibits the mitochondrial release of an apoptogenic protease. J. Exp. Med. 184, 1331–1341.

Swerdlow, R.H., Parks, J.K., Miller, S.W., Tuttle, J.B., Trimmer, P.A., Sheehan, J.P., Bennett, J.P., Jr., Davis, R.E. & Parker, W.D., Jr. (1996). Origin and functional consequences of the complex I defect in Parkinson's disease. Ann. Neurol. 40, 663–671.

Tanaka, M., Kovalenko, S.A., Gong, J.S., Borgeld, H.J., Katsumata, K., Hayakawa, M., Yoneda, M. & Ozawa, T. (1996). Accumulation of deletions and point mutations in mitochondrial genome in degenerative diseases. Ann. N.Y. Acad. Sci. 786, 102–111.

Tang, Y. & Zucker, R.S. (1997). Mitochondrial involvement in post-tetanic potentiation of synaptic transmission. Neuron 18, 483–491.

Tanno, Y., Yoneda, M., Tanaka, K., Kondo, R., Hozumi, I., Wakabayashi, K., Yamada, M., Fukuhara, N., Ikuta, F. & Tsuji, S. (1993). Uniform tissue distribution of tRNA. (Lys). Mutation in mitochondrial DNA in MERRF patients. Neurology 43, 1198–1200.

Tatuch, Y., Christodoulou, J., Feigenbaum, A., Clarke, J.T., Wherret, J., Smith, C., Rudd, N., Petrova-Benedict, R. & Robinson, B.H. (1992). Heteroplasmic mtDNA mutation. (T-G) at 8993 can cause Leigh disease when the percentage of abnormal mtDNA is high. Am. J. Hum. Genet. 50, 852–858.

Taylor, R.W., Chinnery, P.F., Haldane, F., Morris, A.A., Bindoff, L.A., Wilson, J. & Turnbull, D.M. (1996). MELAS associated with a mutation in the valine transfer RNA gene of mitochondrial DNA. Ann. Neurol. 40, 459–462.

Thastrup, O. (1992). Role of Ca^{2+}-ATPases in regulation of cellular Ca^{2+} signalling, as studied with the selective microsomal Ca^{2+}-ATPases inhibitor, thapsigargin. Agents Actions 29, 8–15.

Torii, M., Ito, H. & Suzuki, T. (1992). Lipid peroxidation and myocardial vulnerability in hypertrophied SHR myocardium. Exp. Mol. Pathol. 57, 29–38.

Trounce, I., Byrne, E. & Marzuki, S. (1989). Decline in skeletal muscle mitochondrial respiratory chain function, possible factor in ageing. Lancet 1, 44–45.

Trounce, I., Neill, S. & Wallace, D.C. (1994). Cytoplasmic transfer of the mtDNA nt 8993 T→G. (ATP). point mutation associated with Leigh syndrome into mtDNA-less cells demonstrates cosegregation with a decrease in state III respiration and ADP/O ratio. PNAS. 91, 8334–8338.

van den Ouweland, J.M., Lemkes, H.H., Trembath, R.C., Ross, R., Velho, G., Cohen, D., Froguel, P. & Maassen, J.A. (1994). Maternally inherited diabetes and deafness is a distinct subtype of diabetes and associates with a single point mutation in the mitochondrial $tRNA^{Leucine (UUR)}$ gene. Diabetes 43, 746–751.

Vaux, D.L., Haecker, G. & Strasser, A. (1994). An evolutionary perspective on apoptosis. Cell 76, 777–779.

Vayssiere, J.L., Petit, P.X., Risler, Y. & Mignotte, B. (1994). Commitment to apoptosis is associated with changes in mitochondrial biogenesis and activity in cell lines conditionally immortalized with simian virus 40. PNAS. 91, 11752–11756.

Wachsman, J.T. (1996). The beneficial effects of dietary restriction, reduced oxidative damage and enhanced apoptosis. Mutat. Res. 350, 25–34.

Wallace, D.C. (1994a). Mitochondrial DNA sequence variation in human evolution and disease. PNAS. 91, 8739–8746.

Wallace, D.C. (1994b). Mitochondrial DNA mutations in diseases of energy metabolism. J. Bioenerg. Biomembr. 26, 241–250.

Wallace, D.C., Bunn, C.L. & Eisenstadt, J.M. (1975). Cytoplasmic transfer of chloramphenicol resistance in human tissue culture cells. J. Cell Biol. 67, 174–188.

Wallace, D.C., Singh, G., Lott, M.T., Hodge, J.A., Schurr, T.G., Lezza, A.M., Elsas, L.J. & Nikoskelainen, E.K. (1988). Mitochondrial DNA mutation associated with Leber's hereditary optic neuropathy. Science 242, 1427–1430.

Wallace, D.C., Shoffner, J.M., Trounce, I., Brown, M.D., Ballinger, S.W., Corral-Debrinski, M., Horton, T., Jun, A.S. & Lott, M.T. (1995). Mitochondrial DNA mutations in human degenerative diseases and aging. Biochim. Biophys. Acta 1271, 141–151.

Wang, G.J. & Thayer, S.A. (1996). Sequestration of glutamate-induced Ca^{2+} loads by mitochondria in cultured rat hippocampal neurons. J. Neurophysiol. 76, 1611–1621.

Warner, T.F. (1972). Apoptosis. Lancet 2, 1252.

Wei, Y.H., Kao, S.H. & Lee, H.C. (1996a). Simultaneous increase of mitochondrial DNA deletions and lipid peroxidation in human aging. Ann. N.Y. Acad. Sci. 786, 24–43.

Wei, Y.H., Pang, C.Y., You, B.J. & Lee, H.C. (1996b). Tandem duplications and large-scale deletions of mitochondrial DNA are early molecular events of human aging process. Ann. N.Y. Acad. Sci. 786, 82–101.

White, R.J. & Reynolds, I.J. (1995). Mitochondria and Na^+/Ca^{2+} exchange buffer glutamate-induced calcium loads in cultured cortical neurons. J. Neurosci. 15, 1318–1328.

White, R.J. & Reynolds, I.J. (1996). Mitochondrial depolarization in glutamate-stimulated neurons, an early signal specific to excitotoxin exposure. J. Neurosci. 16, 5688–5697.

White, R.J. & Reynolds, I.J. (1997). Mitochondria accumulate Ca^{2+} following intense glutamate stimulation of cultured rat forebrain neurones. J. Physiol. 498, 31–47.

Williams, G.T. (1992). Programmed cell death, apoptosis and oncogenesis. Cell 65, 1097–1098.

Wong, A. & Cortopassi, G. (1997). mtDNA mutations confer cellular sensitivity to oxidant stress that is partially rescued by calcium depletion and cyclosporin A. Biochem. Biophys. Res. Comm. 239, 139–145.

Wyllie, A.H., Kerr, J.F. & Currie, A.R. (1980). Cell death, the significance of apoptosis. Int. Rev. Cytol. 68, 251–306.

Yen, M.Y., Lee, H.C., Liu, J.H. & Wei, Y.H. (1996). Compensatory elevation of complex II activity in Leber's hereditary optic neuropathy. Br. J. Ophthalmol. 80, 78–81.

Yen, T.C., Chen, Y.S., King, K.L., Yeh, S.H. & Wei, Y.H. (1989). Liver mitochondrial respiratory functions decline with age. Biochem. Biophys. Res. Commun. 165, 944–1003.

Zamzami, N., Marchetti, P., Castedo, M., Decaudin, D., Macho, A., Hirsch, T., Susin, S.A., Petit, P.X., Mignotte, B. & Kroemer, G. (1995). Sequential reduction of mitochondrial transmembrane potential and generation of reactive oxygen species in early programmed cell death. J. Exp. Med. 182, 367–377.

Zamzami, N., Marchetti, P., Castedo, M., Hirsch, T., Susin, S.A., Masse, B. & Kroemer, G. (1996). Inhibitors of permeability transition interfere with the disruption of the mitochondrial transmembrane potential during apoptosis. FEBS Lett. 384, 53–57.

Zhang, Y., Marcillat, O., Giulivi, C., Ernster, L. & Davies, K.J. (1996). The oxidative inactivation of mitochondrial electron transport chain components and ATPase. J. Biol. Chem. 265, 16330–16336.

Chapter 10

Hereditary Disorders of Copper Metabolism

ZEYNEP TÜMER and NINA HORN

Advances in Cell Aging and Gerontology
Volume 3, pages 355–389

ISBN: 0-7623-0405-7

MENKES DISEASE

The X-linked recessive Menkes disease (MD) is a multisystemic lethal disorder of copper metabolism. MD was first recognized in 1962 as a new neurodegenerative disorder (Menkes et al., 1962) and involvement of copper metabolism was described 10 years later by Danks et al. (1972), showing low serum levels and defective intestinal absorption of the metal. At first, copper malabsorption was thought to be the primary cause, but later investigations demonstrating copper accumulation in extrahepatic tissues, except for the brain, indicated a multisystemic involvement (Heydorn et al., 1975; Horn et al., 1978). Most of the clinical features of MD are attributable to deficiency of one or more important copper-requiring enzymes such as cytochrome c oxidase (electron transport), superoxide dismutase (free radical detoxication), dopamine β hydroxylase (catecholamine production), lysyl oxidase (cross-linking of collagen and elastin), and peptidyl-glycine α-amidating monooxygenase (PAM; bioactivation of peptide hormones).

Clinical Features

MD shows clinical variability to an extent, but about 90% to 95% of the patients suffer from the severe lethal form (Table 1; Figure 1) (Horn et al., 1995). Besides a number of mildly affected patients with different degrees of nervous system or connective tissue involvement, occipital horn syndrome (OHS) represents a distinct subgroup. Clinical, pathological, and physiopathological aspects of the disease have been reviewed previously (Horn et al., 1992, 1995; Danks, 1995) and the reader is referred to these manuscripts for the original references.

In the classical and most common form of Menkes disease the clinical picture is characterized by severe progressive psychomotoric retardation with seizures, marked connective tissue symptoms, and peculiar hair (see Figure 1). Patients often have a characteristic facial appearance with frontal or occipital bossing, micrognatia, and pudgy cheeks.

Pregnancy is usually uncomplicated, but premature delivery is frequent and the birth weight is low. The developmental milestones are usually normal for the first few months, though hypothermia and subtle hair changes may be present already, and the appearance may be described as being odd. A genetic disease is rarely suspected at this stage.

From about 2 to 3 months of age therapy-resistant convulsions start. They may be generalized or focal, and myoclonic jerks may also occur. Additional initial symptoms are marked hypotonia, vomiting, diarrhea, and feeding problems. Developmental regression becomes obvious around 5 to 6 months of age; early skills such as head control or smiling are lost, and spontaneous movements become limited. Muscular tonus is often decreased in early life and is later replaced by spasticity of the extremities, which finally develops into paresis. The psychomotoric retardation shows progression, spontaneous movements become limited, and

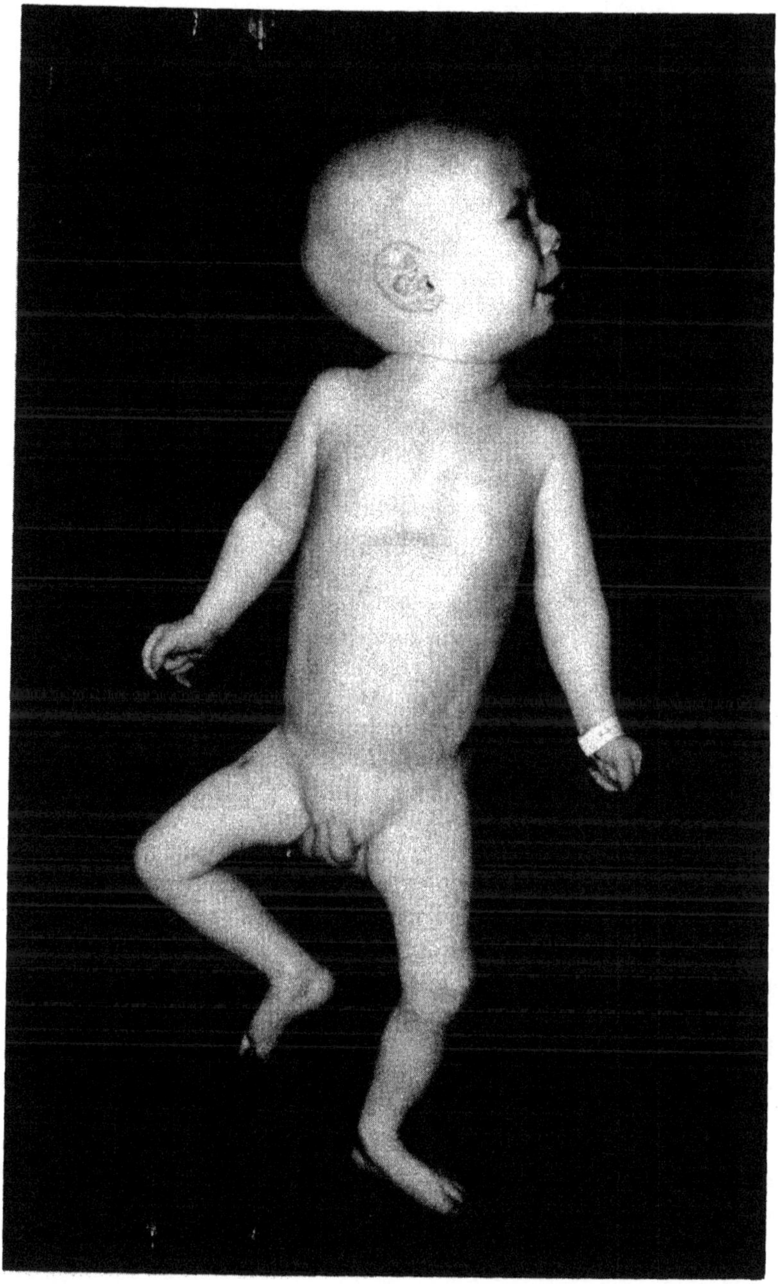

Figure 1. Clinical picture of a 9-month-old boy affected with Menkes disease.

Table 1. Menkes Disease versus Wilson Disease

	Menkes Disease	*Wilson Disease*
Inheritance	X-linked recessive	Autosomal recessive
Main clinical features	Neurodegeneration	Hepatic disease
	Connective tissue symptoms	Neurological dysfunction
Basic pathology	Impaired function of copper enzymes	Impaired biliary excretion of copper
		Impaired incorporation of copper into ceruloplasmin
Animal model	Mottled mouse	LEC rats
		Toxic milk mice
Predicted protein tissue expression	1,500 AA copper-ATPase	1,465 AA copper-ATPase
	All tissues	Mainly in liver and kidney
	Hardly detectable in liver	Low in other tissues
Cellular localization	Secretory pathway: Trans-Golgi network	Secretory pathway: Trans-Golgi network
Gene defect	Cytogenetical abnormalities	Small base pair changes
	Gross rearrangements	
	Small base pair change	

Note: AA, amino acid; LEC, Long-Evans-Cinnamon.

drowsiness and lethargy emerge. At this age, Menkes disease is often suspected owing to the peculiar hair. The kinky, colorless, and friable hair is a striking and pathognomonic feature of Menkes disease. Hair microscopy shows twisting about its own axis (pili torti), periodic narrowing of the shaft (monilethrix), and fragmentation at regular intervals (trichorrhexis nodosa). Significantly decreased serum copper and ceruloplasmin levels support the clinical diagnosis (see "Diagnosis").

Connective tissue symptoms are numerous. Vascular abnormalities such as tortuous vessels and vascular rupture, multiple bladder diverticula, and inguinal hernia are observed frequently. The joints are hyperextensive, and loose and dry skin may be observed very early. Patients have characteristic skeletal changes including pectus excavatum or pectus carrinatum. Skeletal X-ray films show widening of the flared metaphyses, diaphyseal periosteal reaction, and thickening. Rib fracture as a result of osteoporosis is a common finding.

Late manifestations of the disease are blindness, subdural hematomas, respiratory failure, and liability to respiratory infections. Most of the patients die before the third year of life because of infections or vascular complications.

Occipital Horn Syndrome

The occipital horn syndrome (OHS) originally described by Lazoff et al. (1975) is the mildest form of Menkes disease. The neurological symptoms in OHS are relatively few and include orthostatic hypotension and diarrhea. The intellectual capacity is described as low to borderline normal. The patients show numerous skeletal abnormalities including the occipital horns, and connective tissue manifestations, such as bladder diverticula, joint laxity, and cutis laxa. The facial appearance is often distinctive. Unusual features include a long thin face, high forehead, down-slanting eyes, hooked or prominent nose, long philtrum, high arched palate, and prominent large ears. The hair is usually not conspicuous. The life span is almost normal in OHS patients. Serum copper and ceruloplasmin are low, and cultured fibroblasts show increased ^{64}Cu incorporation as in MD. Tissue copper distribution in OHS also is similar to that observed in classical Menkes disease (Heydorn et al., 1995). MD and OHS have been suggested to be allelic owing to their biochemical and clinical resemblances, and their homology with the allelic forms of the mottled mouse (see "Mottled Mouse—The Animal Model"). However, confirmation of this allelism had to await the cloning of the gene defective in Menkes disease.

In OHS, the major effect is on lysyl oxidase, an essential enzyme in cross-linking of collagen and elastin, and cultured fibroblasts from OHS patients (as well as MD patients) show markedly low lysyl oxidase activity (Royce et al., 1980; Peltonen et al., 1983). The possibility of a primary defect in the lysyl oxidase gene has been excluded by the assignment of this gene to chromosome 5 (Hämäläinen et al., 1991). Following the isolation of the gene defective in Menkes disease, OHS patients were investigated for a defect within this gene and the finding of splicing mutations gave the molecular genetic evidence that these two disorders were allelic (Kaler et al., 1994; Das et al., 1995).

Pathology

The most pronounced pathological changes are seen in the central nervous system (CNS), but distinct systemic changes involving especially connective tissue are also observed. The pathological changes are widespread in the cerebral hemispheres and the cerebellum. The lesions may indicate both antenatal and postnatal disturbances aggravated by degenerative changes.

The brain weight is considerably reduced and the cerebral hemispheres show a symmetrical and generalized atrophy. The corpus callosum is thin and the cerebral cortex has varying thickness in all lobes. The ventricular system is enlarged. The cerebellum is small with narrow gyri and deep sulci. The brain stem and the spinal cord appear normal.

The cortical layer reveals diffuse and focal depopulation of neurons in varying degrees. In the temporal lobes the neurons may be totally absent and in cortex they may be changed into a coarse mesh work of astrocytes with scattered rod cells, filled with lipids and cholesterol crystals, and proliferating capillaries. The preserved nerve cells show unspecific changes and the number of mitochondria in the perikaryon is increased. The neuronal loss is most pronounced in the neocortex and cerebellum, and may be accompanied by calcification. Oligodendroglial cells are very few and satellitosis is not present.

The cerebral white matter reveals gross deficiency of myelin, especially in the temporal lobes. The few preserved myelin sheaths show severe degeneration. Severe gliosis is present, and neutral fat and cholesterol compounds are accumulated in glial cells, rod cells, and in the extracellular space. The well-preserved capsula interna forms a sharp border against the severely damaged centrum semiovale.

The changes in the cerebellum are particularly severe, with atrophy of the molecular and granular layer, and varying loss of Purkinje cells. The preserved Purkinje cells are displaced more or less deeply into the internal granular layer, and they frequently show somal sprouts, abnormal dendritic tree, and sometimes torpedoes. These changes have been considered unique to MD and similar changes have been found in mottled mice. A marked increase in the number of mitochondria is seen in the perikaryon of Purkinje cells. Basket cells are absent and fat products can be seen in all cortical layers, together with gliosis. The white substance shows degeneration and gliosis, but is relatively well preserved. In the late stages of the disease, brain infarction and hemorrhage secondary to arterial disease can be observed.

Changes outside the CNS are also typically observed. In skeletal muscle, mitochondrial changes with "ragged red" fibers and glycogen accumulation have been observed. In the eye, the number of retinal ganglion cells may be decreased and optic nerve atrophy is evident. The pigment epithelium shows hypoplasia and hypotrophia of melanin granules. The central retinal blood vessels are often tortuous. Eyes without pathological changes have also been described. There is profound evidence for a systemic circulatory failure in Menkes disease, with tortuosity and narrowing of the aorta and larger arteries, which have also lost their elasticity. Multiple diverticula of the urinary bladder are often present.

Physiopathology

Copper is crucial for the normal activity of numerous oxidative enzymes involved in vital metabolic processes in all living cells. In MD cellular copper uptake is normal, but owing to a defective intracellular transport copper cannot be exported from the cell, while copper requiring enzymes cannot receive copper necessary for their normal function. Dysfunction of these enzymes leads to the widespread manifestations of the disease. Copper is accumulated in extrahepatic tissues, except for the brain, implying a true copper deficiency in the brain, versus the functional deficiency present in other extrahepatic tissues.

In the brain, essentiality of copper is underlined by its role as a co-factor for several enzymes involved in the formation and metabolism of neurologically active substances; for example, dopamine β-hydroxylase (DBH) catalyzing the production of adrenaline and noradrenaline, and peptidyl-glycine α-amidating monooxygenase (PAM) necessary for the post-translational maturation of neuropeptides. Copper is also essential in cellular respiration, free radical defense, as well as blood vessel stability and tissue iron mobilization, being a co-factor for enzymes such as cytochrome c oxidase (COX), cytosolic and extracellular forms of Cu/Zn-superoxide dismutase (SOD), lysyl oxidase (LOX), and ceruloplasmin (Cp), respectively. Diamine oxidase and monoamine oxidase are probably also of importance for the metabolism of neurotransmitters.

Neurodegeneration in MD appear to be a result of a complex process and deficiency of several enzymes may play a synergistic role. Malfunction of DBH, COX, SOD, as well as LOX may all add to the neurological symptoms. In brain, besides a defective intracellular transport, the overall copper content is low, possibly as a result of impaired uptake. This may add to the severity of the malfunction of the copper enzymes in this organ.

DBH is crucial for the production of the catecholamines, adrenaline, and noradrenaline. A critical balance between the adrenergic and cholinergic system is required for the normal function of many important hypothalamic centers regulating appetite, thirst, sleep, blood pressure, cardiovascular activity, and body temperature. A disturbance of this balance may result in a relative deficiency of noradrenaline and lead to anorexia, dehydration, somnolence, low blood pressure, and hypothermia. Reduced noradrenergic fluorescence in the tegmental and hypothalamic regions (Uno et al., 1983), disturbed synthesis of catecholamine metabolites (Kaler, 1994), and abnormal catecholamine levels in blood and CSF (Grover et al., 1982; Kaler, 1994) document the disturbed balance in the central nervous system of patients with MD. Many of the early clinical features in MD may thus result from a hypothalamic disturbance. Orthostatic hypotension, which is a continuous problem in occipital horn syndrome patients, as well as in Menkes patients treated with copper histidine, may also be attributed to DBH malfunction (Christodoulou et al., 1998).

In the extrapyramidal system, an imbalance between dopaminergic and cholinergic systems may result in abnormalities of muscle coordination. Hypofunction of the dopaminergic system may therefore lead to ataxia, which is a characteristic symptom in milder cases and in treated Menkes patients. Ataxia may also be a result of cerebellar dysfunction. Nerve conduction is affected directly not only by the impaired production of the adrenergic neurotransmitters resulting from DBH deficiency, but also by a decreased COX activity, which reduces the energy supply. It is well established that genetic defects in the COX genes can cause serious neurological symptoms comprising ataxia, muscle weakness, hypotonia, and visual impairment in infants, and these symptoms are also observed in patients with MD.

A disturbed oxidative respiration caused by defective COX activity will also lead to an increase in free radical formation. Free radicals are superoxide anions (O_2^-), which are constantly produced in the body by partial reduction of molecular oxygen in several metabolic processes such as oxidative respiration. These highly reactive molecules are normally detoxified by SOD, and when this enzyme is defective the cells cannot be protected against the destructive effects of the free radicals. Unsaturated fatty acids will be oxidized to the saturated form, which will provoke degeneration of the myelin, leading to short circuits in nerve conduction. Lipid peroxidation is believed to play a major role in disruption of neuronal ion homeostasis and promotion of excitotoxic cascades in a variety of neurodegenerative disorders (Mattson, 1998). It is known that brain lipids are particularly sensitive to oxidative damage, and increased content of oxidized lipids in the brain and reduced amount of polyunsaturated glycolipids have been observed in MD patients (O'Brien and Sampson, 1966; French et al., 1972; Lou et al., 1974). Decreased function of SOD will therefore have a synergistic effect on the already disturbed nerve conduction. Clinically this may appear as hypertonicity, later replaced by spasticity.

In copper-deficient animals, primary changes of myelin have also been observed, and the lack of copper during development is reflected by a general impairment of myelination (Smith, 1981). The degree of neuronal degeneration present at birth indicates that Menkes fetuses are also copper deficient. Probably the normal cooperation between nerve cells, oligodendroglial cells, and astrocytes is disturbed causing a dysmyelination (Reske-Nielsen et al., 1987).

The role of PAM in MD is speculative. PAM is essential for full biological potency of important neuropeptides and several peptide hormones (Southan and Kruse, 1989). The well-demonstrated copper deficiency in the brains of MD patients may result in a deficient PAM (and DBH), severely affecting the neuropeptides that function in this organ.

Some of the degenerative changes in the brain may also be explained by vascular lesions resulting from reduced activity of LOX. The fragility of the vessels increases the tendency to intracranial hemorrhage, a symptom frequently observed in MD patients. Significant secondary mitochondrial dysfunction resulting from COX deficiency can explain some of the metabolic features (Barkovich et al., 1993), and deficient LOX and SOD may explain some of the structural and vascular features.

Lack of three enzymes, namely LOX, tyrosinase, and a yet unidentified enzyme ("cross-linkase") responsible for cross linking of the SH-bridges in the hair, may be responsible for the characteristic somatic appearance in MD. LOX deficiency has been well documented in several MD and OHS patients (Royce et al., 1980; Peltonen et al., 1983; Kemppainen et al, 1996). Deficiency of this enzyme may be responsible for all the connective tissue abnormalities observed. Depigmentation of the hair, skin, and retinal cells can be explained by malfunction of tyrosinase. Furthermore, copper is necessary for the formation of disulfide bonds in keratin, and in the hair of MD patients, free sulfhydryl groups are increased. Deficiency of "cross-linkase" may explain the hair abnormalities, as well as the dry and scaly skin.

Identification of the Defective Gene

The gene defective in MD was assigned to the X chromosome when it was first described in 1962 as a new neurological disorder (Menkes et al. 1962). The disease locus (MNK) was further localized to the proximal long arm of the X chromosome by linkage studies, exclusion mapping, and comparative gene mapping between human and mouse (Wieacker et al., 1983; Horn et al., 1984; Yang et al., 1990; Brockdorff et al., 1991; Davisson, 1991). The physical evidence for the localization of MNK came by finding of two patients with chromosome aberrations: a female patient carrying a balanced X;2 translocation (Kapur et al., 1987) and a male patient carrying a unique intrachromosomal rearrangement of the X chromosome (Tümer et al., 1992). In both patients, one of the X chromosome breakpoints was at q13.3, the candidate locus for MD (Verga et al., 1991; Tümer et al., 1992). These rearrangements suggested that the X chromosome break had occurred at, or very near, the disease gene directly affecting its function.

Finding of the chromosome breakpoints within the same candidate region in two MD patients enabled three groups, including ours, to isolate the gene defective in MD using positional cloning strategies (Chelly et al., 1993; Mercer et al., 1993; Vulpe et al., 1993). Positional cloning requires knowledge about the chromosomal position of the gene, and it is the strategy used when the basic defect is unknown and a candidate gene cannot be proposed. The MD gene encodes a 1,500 amino acid protein predicted to be a copper translocating P-type ATPase (designated as ATP7A), owing to its striking sequence homology to a bacterial copper binding protein (Figure 2) (Odermatt et al., 1993). The gene had been named *MNK* until prediction of the protein product and now it is designated as *ATP7A*. The 8.5-kb mRNA transcript of *ATP7A* is expressed in all tissues analyzed, though in trace amounts in liver, if any. This result is in line with the multisystemic involvement in MD.

ATP7A is organized in 23 exons spanning a genomic region of about 150 kb. The first exon is a leader exon containing only untranslated sequences, and the ATG translation start codon is in the second exon. The last exon contains a 274-bp translated sequence, the TAA translation termination site, the 3.8-kb 3′-untranslated region, and a polyadenylation site (Dierick et al., 1995; Tümer et al., 1995). The promoter region of *ATP7*A is also characterized and includes three tandem repeats of a 98-bp sequence (Levinson et al., 1996).

Underlying Genetic Defect

To date, a total of 192 mutations affecting *ATP7A* have been identified in unrelated MD patients with the classical severe form, or with one of the atypical phenotypes, and these mutations show great variety (Tümer et al., 1998).

A large proportion of patients (n = 150) have small base pair changes (point mutations). These mutations are missense (14%) or nonsense mutations (14%),

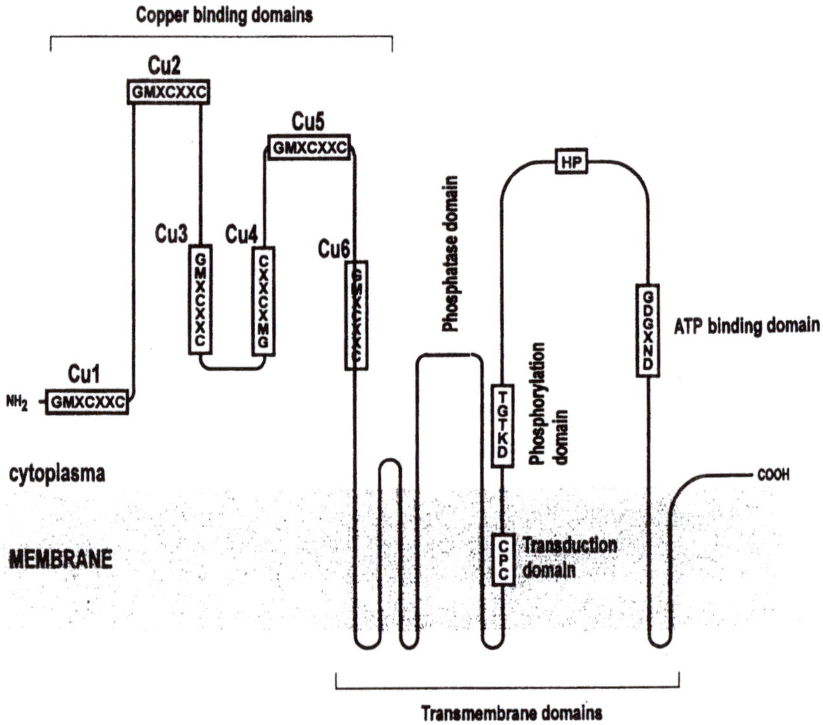

Figure 2. Predicted protein structure of ATP7A and ATP7B, with the conserved functional domains and motifs. The amino termini of ATP7A and ATP7B contains six repetitive domains. The consensis GMXCXXC motif is also found in other copper binding proteins. This motif with a pair of conserved cysteine residue is involved in the binding copper (Cu). Phosphatase domain contains the TGE motif, which may have a role in removing the phosphate from the phosphorylated aspartic acid (D) as part of the cation transport. CPC, is part of the predicted cation channel. The proline residue (P) is highly conserved among P-type ATPases, and it is proposed to participate in the transduction of energy from the phosphorylation site to cation transport. In copper P-type ATPases (except for CopB in bacteria), the proline is surrounded by cysteine residues (C), which provides specificity for heavy metals. The phosphorylation domain contains an invariant cytoplasmic DKTGT motif. The aspartate residue (D) is crucial for the enzyme activity, and it is phosphorylated with the terminal phosphate of ATP in the cation transport cycle. The HP motif is specific for all the copper transporting proteins, but function of this motif is yet unknown. The ATP binding domain is an extramembraneous segment, responsible for ATP binding in P-type ATPases, and it is one of the most conserved sites. A consensus motif within this domain is GDGXND. Transmembrane domains are a common structural feature of P-type ATPases and anchor the protein into the membrane. Both proteins are predicted to have eight transmembrane domains.

small deletions or insertions resulting in a frameshift and premature termination of the mRNA transcript (24%), and splice site mutations leading to skipping of one or more exons (22%). These changes are predicted to affect the protein product at variable degrees depending on the nature and localization of the mutation. Point mutations are detected throughout the gene except for exons 1, 2, 5, and the last exon. An accumulation of mutations can be recognized in the middle of the gene (exons 7 to 15) corresponding to the last copper binding domain, the first six transmembrane domains, the phosphatase domain, and the transduction domain of the protein (see Figure 2). Nineteen of these mutations have been published by other groups (Das et al., 1994, 1995; Kaler et al., 1994, 1995, 1996; Levinson et al., 1996; Ronce et al., 1997; Qi and Byers, 1998) and 44 by ours (Tümer et al., 1994, 1996, 1997). Seventeen point mutations were observed more than once in unrelated families, and six of these have been published previously (Das et al., 1994; Kaler et al., 1994).

A substantial number of patients (n = 35) do not have small base pair changes, but instead have partial gene deletions. Sizes and locations of these mutations are also variable (Tümer and Horn, in preparation).

Chromosome abnormalities affecting *ATP7A* have been detected in seven patients. One of these patients was a male with a unique chromosome abnormality, in which the segment Xq13.3-q21.1 was inserted into the short arm of the X chromosome (Tümer et al., 1992). One of the female patients was mosaic for the Turner karyotype (Barton et al., 1983) and the rest had X;autosome translocations (Barton et al., 1983; Kapur et al., 1987; Beck et al., 1994; M. Tsukahara, D. Wattana, and S. Mohammed, personal communication).

Diagnosis

Initial diagnosis of MD is suggested by the clinical features (especially the typical hair changes) and supported by the reduced levels of serum copper and ceruloplasmin. However, interpretation of these markers may be difficult in the first months of life, as serum copper and ceruloplasmin levels may also be low in normal infants in this period. A definitive biochemical diagnosis exists and is based on the intracellular accumulation of copper caused by impaired efflux. Accumulation is evaluated in cultured cells by measuring radioactive copper (^{64}Cu) retention after a 20-hour pulse, and impaired efflux is directly determined after a 24-hour pulse-chase. However, these analyses demand expertise and are carried out only in a few specialized centers in the world (Tümer and Horn, 1997).

Ultimate diagnostic proof of MD is the demonstration of the molecular defect in *ATP7A*. However, because of the large size of the gene and the variety of the mutations observed in different families, detection of the genetic defect in a given family may take time. This fact should be taken into consideration in at-risk pregnancies, where a prenatal diagnosis by molecular means can only be carried out if the mutation of the family has already been established. In cases where the mutation is unknown, biochemical analysis remains a possibility. First trimester biochemical diagnosis is carried out by measuring the total copper content of chorionic villi, and in the second trimester copper accumulation in amniotic fluid cells is evaluated. Though potential pitfalls also exist for these analyses, they have been carried out routinely at the John F. Kennedy Institute in Denmark since 1975 (Tümer and Horn, 1997).

Mottled Mouse—The Animal Model

Mottled mouse, now considered as the animal model of MD, arose spontaneously in 1953 (Fraser et al., 1953). Mutations at the X-linked mottled locus (*Mo*) were leading to a characteristic mottled pigmentation of the coat of the female heterozygotes. There are at least 23 different mottled mutants which show considerable variability of neurological and connective tissue abnormalities (Y. Boyd, personal communication; Green, 1989; Davisson et al., 1991).

In mottled mouse, copper metabolism defect was first demonstrated in 1974 (Hunt, 1974). Since then several similarities between the biochemical pathology and phenotypic features of the mottled mice and MD have been demonstrated (Danks, 1986; Green, 1989). Two of the most well-characterized mottled mutants, the mottled brindled (Mo^{br}) and the mottled blotchy (Mo^{blo}), are suggested as murine models for the classical forms of MD and OHS, respectively. Mo^{br} phenotype presents with severe neurological impairment and death at an early age (Hunt, 1974; Danks, 1986). On the other hand, Mo^{blo} phenotype resembles OHS showing predominantly connective tissue manifestations (Rowe et al., 1977).

Following the isolation of *ATP7A*, the mouse gene was cloned using the human sequences (Levinson et al., 1994; Mercer et al., 1994). The predicted protein product (atp7a) showed great sequence homology to ATP7A, and the tissue expression profile was also similar to the human counterpart. Later, identification of an *atp7a* mutation in one of the alleles confirmed the status of mottled mouse as the animal model for MD (Das et al., 1995). This allele was mottled blotchy, the suggested animal model for OHS (Rowe et al., 1977). Today *atp7a* mutations are known in eight mottled alleles (Das et al., 1995; Cecchi et al., 1997; Grimes et al., 1997; Levinson et al., 1997a, 1997b; Murata et al., 1997; Mori and Nishimura, 1997; Ohta et al., 1997; Reed and Boyd, 1997).

The exon structure of the mouse gene *atp7a* is almost identical to its human counterpart, indicative of the high conservation of these genes through evolution. The 8.0 kb transcribed sequence consists of 23 exons spanning about 120 kb, and all the introns interrupt the aligned coding regions of *ATP7A* and *atp7a* at the same positions (Cecchi and Avner, 1996). The sizes of the exons are thus identical, except exons 1, 2 and 5, which are shorter in the mouse corresponding to the nine amino acid difference in the respective proteins.

Treatment

In MD, copper uptake is normal, but a defective ATP7A disturbs the intracellular copper balance. Copper cannot be extruded from the cell and the copper pool shifts to metallothionein. As a result, copper is unavailable to enzymes requiring this metal for their normal function. The objective of a treatment is thus to provide extra copper to the tissues and the copper enzymes. The most reasonable way is to administer copper parenterally, to bypass the intestines, the first step defective in the overall copper metabolism. However, trials with various copper preparations (e.g., copper sulphate or copper-EDTA) did not produce substantial clinical improvement in MD patients (see Kaler, 1994, for review). On the other hand, the advent of copper-histidine gave favorable results in four unrelated patients (see Christodoulou et al., 1998, for review). These patients do not show the devastating neurological problems and, most interestingly, their clinical course resemble OHS. A critical point in this treatment is early administration of copper-histidine as the patients receiving treatment after the first few months of age do not benefit in the same way, although survival may be prolonged (Horn et al., 1995). Studies with the mottled mouse imply that there is a critical stage in brain development at which copper is essential (Danks, 1995). This suggests that the therapy should be initiated very early, before irreparable neurodegeneration occurs. However, it is very difficult to diagnose the sporadic cases in the neonatal period, and in familial cases the parents usually prefer to discontinue the pregnancy. Clinical trials for the early treatment of MD with copper-histidine will therefore remain very limited.

In the four previously mentioned patients, the main neurological symptoms have been improved, but features such as connective tissue abnormalities, orthostatic hypotension, and diarrhea persisted. These observations suggested that at least two enzymes, DBH and LOX, could not receive copper necessary for their normal function. In one of these patients, hypotension and diarrhea could be corrected by administration of L-DOPS (Christodoulou et al., 1998).

WILSON DISEASE

Clinical Features

Wilson disease (WD) is an autosomal recessive disorder of copper metabolism reflecting the toxic effects of copper, in contrast to MD, which mimics copper deficiency. WD is mainly characterized by different degrees of liver disease, neurological, or psychiatric symptoms. Kayser-Fleisher rings resulting from the accumulation of copper at the limbus of the cornea are a pathognomonic feature when present. WD shows differences in clinical expression and the age of onset of the symptoms is also variable (see Brewer and Yuzbasiyan-Gurkan, 1992, for review; Yarze et al., 1992).

The progression of the disease has been described in five stages (Deiss et al., 1971). Stage 1 begins at birth, but patients are asymptomatic while copper accumulates in the liver, and stage 1 may blend into stage 3. In stage 2, the liver is saturated and copper is released to the serum. It can debut with hemolytic anemia (2a) or liver failure (2b). Copper accumulates in extrahepatic tissues, especially brain and eyes (Kayser-Fleischer rings), in stage 3. In the fourth stage, neurological disease develops, and in the last stage therapy is initiated and the symptoms start to fade away.

In WD, the liver is the main target and a broad spectrum of acute and chronic liver disease is present. The symptoms are manifest usually between 8 and 16 years. Acute episodes of jaundice, vomiting, and malaise are quite frequent. Acute hemolysis may occur in acute hepatic episodes without obvious liver symptoms.

Neurological symptoms are very unusual before the age of 12 years. The initial symptoms include tremor, dysarthria, incoordination of especially fine movements, and ataxia. Late findings include dystonia, spasticity, rigidity, and seizures, but these symptoms are rarely observed owing to early diagnosis and treatment. Cognitive and sensory functions are infrequently affected.

Other clinical features include ocular symptoms in the form of Kayser-Fleischer rings, which are yellow-brown granular deposits of copper on the Descement membrane at the limbus of the cornea. Renal symptoms are caused by accumulation of copper in the renal cortex and manifest mainly as tubular dysfunction or Fanconi's syndrome. Bone and joint disorders include osteoporosis, reduction in the joint spaces in limbs and spine, osteophytes around large joints, and ligamentous laxity. Hemolytic crises, resulting from sudden release of large amounts of copper, may occur in WD.

Cloning of the Defective Gene

The locus (WND) for the gene defective in WD was assigned to chromosome 13, based on a close linkage to the locus for the red cell enzyme esterase-D (ESD) (Frydman et al., 1985). Extensive linkage studies localized *WND* to an approximately 2 cM region at 13q14.3 between the flanking markers D13S31 and D13S59, by genetic linkage analyses (Bowcock et al., 1987; 1988; Farrer et al., 1991), and microsatellites isolated from this region enabled identification of disease-specific haplotypes (Thomas et al., 1993).

In contrast to MD, visible cytogenetic rearrangements, which could lead to straightforward isolation of the gene, were not reported for WD. Identification of *ATP7A* encoding a putative copper transporting ATPase, and its hardly detectable expression in liver, led to the suggestion that WD might be caused by a defective liver specific copper transporter (Vulpe et al., 1993). Soon after, the gene defective in WD was cloned by three independent groups. Two of these groups isolated the gene using probes from the proposed copper binding domain encoded by *ATP7A* (Bull et al., 1993; Yamaguchi et al. 1993). The third group used a degenerate oligonucleotide corresponding to a novel heavy metal binding site within the Aβ region of the amyloid β-protein precursor to screen a brain cDNA library (Tanzi et al., 1993). The WD gene is now designated as *ATP7B*.

Northern blot analysis revealed a 7.5-kb mRNA transcript expressed predominantly in liver, kidney, and placenta, while the message was in minute amounts in heart, brain, lung, muscle, and pancreas. The putative protein product is also a copper binding P-type ATPase (designated ATP7B), highly homologous to ATP7A (57%). *ATP7B* has 22 exons (Petrukhin et al., 1994; Thomas et al., 1995a), and its genomic organization shows remarkable similarity to *ATP7A* and *atp7a* (Tümer et al., 1995; Cecchi and Avner, 1996). Starting from the exons coding for the fifth copper binding domain (exon 5 in *ATP7A* and exon 3 in *ATP7B)* the coding regions of both genes are organized into 19 exons, showing an almost identical structure (Tümer et al., 1995).

Pathogenesis

ATP7B is expressed mainly in liver, kidney, and placenta. It is involved in supplying copper to ceruloplasmin and in the biliary excretion of copper. A defect in the protein will result in deficient production of ceruloplasmin and copper will accumulate in the hepatocytes, resulting in toxic effects, which are characteristic for WD. Pathogenesis in WD is explained by tissue injury and is rather straightforward compared to the pathogenesis of MD, involving several copper enzymes.

Hepatotoxicity observed in WD may be explained by the oxidative injury or the lysosomal injury hypotheses, or both (Britton, 1996). According to the former hypothesis, excess copper results in the formation of free radicals which will damage cellular compartments. The latter hypothesis proposes that excessive accumulation of copper, which is normally sequestered within lysosomes, can impair lysosomal function and release of hydrolytic enzymes will cause cellular damage.

Neurological degeneration in WD appears to be associated with anatomical disruption of the basal ganglia, deep cerebral cortical layers, cerebellum, and to a lesser extent, brain stem. Lesions in thalamus, internal capsule, white matter, and dentate nucleus are also observed. Though excess copper is distributed throughout the brain, anatomical changes are rather site specific. Brain injury secondary to accumulation of copper is still poorly understood, but may be explained by the abovementioned hypotheses.

Copper accumulation in the brain of patients with WD is generally regarded as a result of overflow from liver, when the storage capacity is saturated in this organ. It is, however, noteworthy that ceruloplasmin synthesis also occurs in brain tissues such as the choroid plexus, and a membrane-bound form has recently been identified in astrocytes (Patel and David, 1997). The physiological role of this form of ceruloplasmin is yet unknown, but may contribute to the physiopathology when defective. Incorporation of copper into ceruloplasmin at these sites, therefore, remains to be investigated in both normal and abnormal conditions of copper metabolism.

Genetic Defect in *ATP7B*

In contrast to *ATP7A*, chromosome mutations or partial gene deletions have not been reported for *ATP7B*. The lack of chromosome aberrations disrupting *ATP7B* is not unexpected, as WD is inherited as an autosomal recessive trait and chromosome abnormalities are very rare in such disorders. The largest deletion observed in *ATP7B* was a 24-bp deletion (Figus et al., 1995; Orru et al., 1997), while in *ATP7A* an approximately 100-kb deletion removed the whole gene except for the first two exons, the leader exon, and exon 2 with the translation start site.

To our knowledge, at least 88 mutations have been identified in *ATP7B* and all these mutations are small base pair changes (Bull et al., 1993; Tanzi et al., 1993; Figus et al., 1995; Houwen et al., 1995; Shimizu et al., 1995; Thomas et al., 1995a, 1995b, 1995c; Chuang et al., 1996; Loudianos et al., 1996; Waldenström et al., 1996; Czlonkowska et al., 1997; Kemppainen et al., 1997; Nanji et al., 1997; Orru et al., 1997; Shah et al., 1997). Most of these mutations are missense mutations (49%) or small deletions/insertions (%35) and nonsense or splice mutations make up only a small percentage (8% each).

Animal Models

WD has two animal models, the LEC rat and the toxic milk mouse, which provide invaluable systems for studying the liver pathogenesis in WD and therapeutic possibilities, including gene therapy.

The Long-Evans Cinnamon (LEC) rat has clinical and biochemical features in common with WD (Li et al., 1991). LEC rats develop acute hepatitis a couple of months after birth and if they survive, they develop chronic hepatitis and hepatocellular carcinoma. Serum ceruloplasmin and copper levels are decreased and liver copper value is abnormally high. These findings are similar to those observed in WD patients. An important difference, however, is that LEC rats do not show neurological abnormalities.

The rat gene *(atp7b)* is isolated using DNA sequences corresponding to *ATP7B* and the predicted protein products of the human and rat genes show strong homology (Wu et al., 1994). Identification of a deletion in the coding sequences of *atp7b* in LEC rat demonstrated that it was a true animal model for WD.

Toxic milk *(tx)* is a recessive mutation in mice and results in the toxic accumulation of copper in liver. Phenotypical features of the toxic milk mouse resemble those found in LEC rat and WD patients. However, the milk produced by *tx* mouse is copper deficient and this feature together with the histological differences between the livers of *tx* mouse and WD patients have raised the question whether this mutant was a true model for WD.

Mouse *atp7b* gene was identified using DNA sequences from the rat *atp7b* gene (Theophilos et al., 1996). The predicted protein products of the mouse, rat, and human genes are all copper binding P-type ATPases, which show a high level of amino acid identity. A missense mutation within the mouse *atp7b* gene in *tx* mouse gave strong evidence that this mouse mutant was a true model for WD.

Treatment

WD is one of the few genetic disorders that can be treated to such an extent that the progression of the disease can be prevented almost completely. The objective of the treatment is to prevent the toxic accumulation of copper with anticopper agents such as D-penicillamine or zinc, and the treatment of WD has been reviewed elsewhere (Brewer and Yuzbasiyan-Gurkan, 1992; Yarze et al., 1992).

Establishment of LEC rat and toxic milk mouse as animal models for WD enables researchers to assess the possibility of a liver-directed genetic therapy. Correction of the metabolic defect in vitro in the hepatocytes of Watanabe heritable hyperlipidemic rabbit, an animal model for familial hypercholesterolemia, has been an initial step toward treatment of a liver disease by gene therapy (Wilson et al., 1988). Today, intensive research is under way into the gene therapy of organ-specific hereditary disorders.

COPPER BINDING P-TYPE ATPASES

Remarkable sequence similarity of ATP7A to bacterial P-type ATPases involved in copper translocation in *Enterococcus hirae* (Odermatt et al., 1993), led to the prediction of the first intracellular copper binding P-type ATPase in eukaryotes (see Figure 2) (Vulpe et al., 1993). Since then a number of related genes have been described in different species and their protein products now form a distinct subgroup of the large family of P-type ATPases. For a detailed review of Cu-ATPases, the reader is referred to a recent review by Solioz (1998).

The human members of copper translocating P-type ATPases are ATP7A and ATP7B. In mouse, atp7a and atp7b, and in rat, atp7b belong to this family. In *Saccharomyces cerevisiae* (baker's yeast) two Cu-ATPases (Pca1p and Ccc2p) are identified (Rad et al., 1994; Fu et al., 1995). Ccc2p is likely to be the yeast homolog of ATP7A/ATP7B (Fu et al., 1995) and Pca1p is suggested to have a role in copper extrusion from the cell (Rad et al., 1994). Among the bacterial members of this family there are four copper ATPases, CopA and CopB of *Enterococcus hirae* (Odermatt et al., 1993), and CtaA and PacS of *Synechococcus* (Kanamaru et al., 1994; Phung et al., 1994). All these proteins are involved in copper homeostasis. CopA is likely to be serving in the uptake and CopB in the extrusion of copper. CtaA is suggested to function as a cation pump similar to the yeast Ccc2p, and PacS is probably involved in intracellular distribution of copper (Solioz, 1998).

The eukaryotic and most of the prokaryotic cells have copper enzymes involved in electron transfer and therefore require well-balanced copper transport systems. These systems in bacteria and yeast provide an invaluable model for understanding the overall intracellular copper homeostasis in higher eukaryotes, and studies carried out in yeast have already provided important new information.

P-type ATPases are energy utilizing membrane proteins, functioning as cation pumps, for uptake, export, or exchange of cations through a membrane (see Figure 2). They are called "P-type" ATPases, as they form a phosphorylated intermediate during the transport of cations across a membrane. Ca^{2+}-ATPases and the Na^{+}-K^{+}-ATPases are two of the well studied members of this large protein family.

Independent of the metal they translocate, P-type ATPases have common functional domains with highly conserved motifs (see Figure 2). Phosphorylation domain contains an invariant cytoplasmic DKTGT motif and the aspartate residue (D) is phosphorylated with the terminal phosphate of ATP in the cation transport cycle. ATP binding domain with the (GDGXND) responsible for the binding of ATP is one of the most conserved sites. Phosphatase domain contains the consensus TGE sequence, which has a role in removing the phosphate from the phosphorylated aspartic acid (D) as part of the cation transport.

Besides the features common for all P-type ATPases, copper ATPases contain some motifs specific for this subgroup. Interestingly, these motifs are also conserved in P-type ATPases transporting other heavy metals such as mercury and cadmium. One of the most striking features of Cu-ATPases is presence of one or more copper binding domains with the consensus GMXCXXC motif at the amino terminal region. ATP7A, ATP7B, and mouse atp7a have six successive repeats of copper binding domains, whereas in mouse and rat atp7b the fourth repeat seems to have lost its function during evolution. In yeast Ccc2p protein, there are two copper domains while Pca1p and the bacterial proteins have only one domain. Recent studies suggest that the amino terminal region of ATP7A and ATP7B is directly involved in selective binding of heavy metals and the highest affinity was against copper followed by zinc, nickel, and cobalt (DiDonato et al., 1997; Lutsenko et al., 1997b). This domain is likely to bind six copper atoms, one atom per metal binding repeat (DiDonato et al., 1997; Lutsenko et al., 1997a). NMR studies with the fourth metal binding domain of ATP7A demonstrated that Ag(I) was bound to the two cysteine (C) residues of the conserved GMXCXXC motif (Gitschier et al., 1998). These cysteine residues are, therefore, essential in copper transport. However, site-directed mutagenesis and complementation studies carried out in yeast have suggested that each metal binding domain was not equally important for the normal function of the protein (Payne and Gitlin, 1998).

Another remarkable feature of Cu-ATPases is a characteristic transduction motif CPC unique for this subgroup (except for copB, which has CPH). In other heavy metal binding ATPases this motif is CPC, CPH, or CPS, and the name "CPX-type ATPases" has been proposed for this group (Solioz, 1998).

The HP motif situated in the cytoplasmic loop C-terminal to the aspartate residue (D) is another consensus motif in Cu-ATPases. The HP motif is conserved in all the known and putative copper transport ATPases characterized so far, but its specific function is yet unknown. A common missense mutation of the histidine residue (His1069Glu) observed in WD patients underscores the importance of this motif for the normal function of the human Cu-ATPases. This observation was also supported by site-directed mutagenesis and complementation studies carried out with *CCC2*-deficient yeast mutants using *ATP7A* cDNA (Payne and Gitlin, 1998). Another common feature of Cu-ATPases is the presence of only eight transmembrane domains, whereas other P-type ATPases have ten or more. This domain anchors the protein into the membrane.

COPPER METABOLISM

Copper is an essential trace element with profound influence on neurological development and function. Though required for normal activity of several important oxidative enzymes, excess copper has detrimental effects. Careful control of copper levels is therefore crucial.

Whole body steady levels of copper are maintained by a balance between intestinal absorption and biliary excretion (Figure 3). Copper absorbed from the intestines is transported by the portal route to the liver and from there redistributed to all tissues and organs. Copper filtered through the glomeruli is reabsorbed very efficiently and urine excretion of copper is usually negligible. Copper enters the brain probably via choroid plexus, as the blood-brain barrier represented by the microvascular endothelial cells restricts the passage of molecules from blood to the cerebrospinal fluid very efficiently (Janzer, 1993; Segal, 1993). Copper concentration is highest in liver, the central organ for copper homeostasis, and brain has the second highest value (Linder and Hazegh-Azam, 1996).

Our understanding of copper transport at the cellular level has increased significantly in the past few years. Following the cloning of the genes defective in MD and WD, a number of genes coding for proteins involved in intracellular copper transport have been identified. Most interestingly, these genes have homologs in yeast, indicating that copper (and iron) homeostasis in yeast and human is likely to be very similar. Our current knowledge about the intracellular copper metabolism in human is based mainly on the functional studies carried out in yeast. Biochemical studies performed on MD or WD patients and the animal models of these two disorders are other important sources for understanding the intracellular process. We discuss the cellular copper transport in human briefly, in the light of the recent findings (Figure 4).

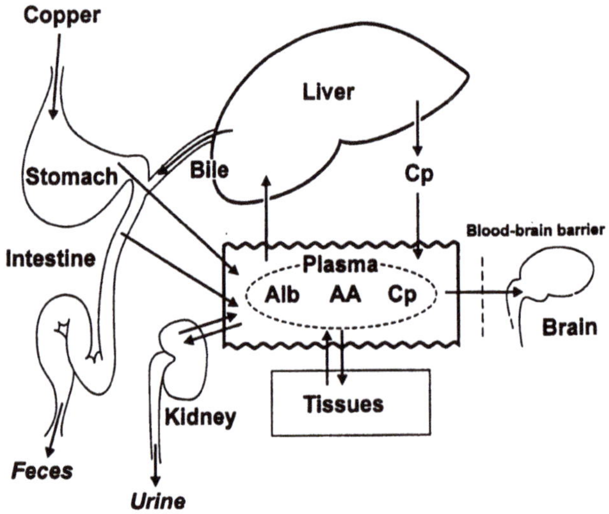

Figure 3. Copper metabolism. Copper homeostasis depends on a balance between intestinal absorption and biliary excretion. Copper absorbed from the intestine is attached to albumin (Alb) and amino acids (AA), and transported to the liver, the central organ of copper homeostasis, where storage and biliary excretion of copper, and ceruloplasmin (Cp) synthesis take place. Copper is secreted from the hepatocytes to the plasma bound to ceruloplasmin. In Wilson disease, the main target is the liver, where deficient incorporation of copper into ceruloplasmin and impaired biliary copper excretion lead to toxic accumulation of the metal. On the other hand, Menkes disease is a multisystemic disorder, where copper cannot be extruded from the cell, but cannot be utilized by copper requiring enzymes, either. The normal route of copper in the body is indicated by arrows.

Delivery of Copper to the Cells and Copper Uptake

In blood, the main copper carrying molecules are ceruloplasmin (Cp), albumin, and copper amino acid complexes, including histidine (Sarkar and Kruck, 1966). In the cerebrospinal fluid, albumin and ceruloplasmin levels are low compared to blood and this is not unexpected owing to the efficient blood-brain barrier (Janzer 1993; Segal, 1993). Copper appears to be taken up by cells using two different systems, a Cp-mediated transfer and a quantitatively more important uptake from the amino acid–bound pool. Important new knowledge about copper uptake and delivery of copper to cellular compartments has been accumulated during the last couple of years.

Cp constitutes the major part of the plasma pool, but the metal is bound in a not readily exchangeable form (Harris and Percival, 1989). Cp is believed to play an important role in delivery of copper to cells and potential Cp receptors have been

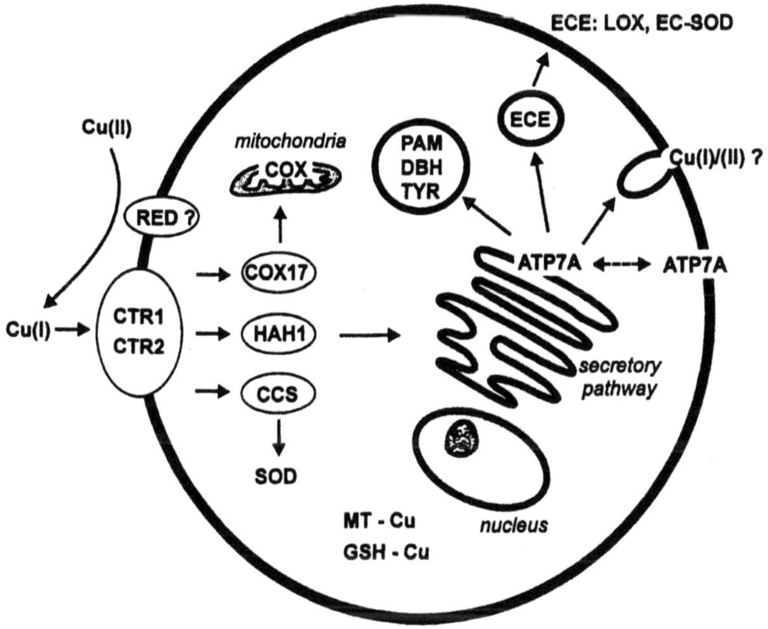

Figure 4. A proposed model for intracellular copper metabolism. CTR1 and CTR2 are the predicted copper uptake proteins. Copper is likely to be reduced by a reductase (RED) before being translocated through the plasma membrane. The reduced ion is then transported by one of the cytosolic copper proteins (COX17, HAH1, CCS) to different compartments and enzymes. Metallothionein and glutathione are likely to comprise an intracellular buffer system for copper. ATP7A is involved in delivering copper to some intracellular and extracellular enzymes and it is also involved in export of copper. Function and localization of ATP7B is similar to that of ATP7B. ATP7B is involved in delivery of copper to apoceruloplasmin in liver cells and may be in astrocytes in the brain. COX, cytochrome c oxidase; DBH, dopamine β-hydroxylase; ECE, extracellular enzymes; EC-SOD, extracellular superoxide dismutase; GSH, glutathione; LOX, lysyl oxidase; MT, metallothionein; PAM, peptidyl-glycine α-amidating monooxygenase; RED, a yet unknown reductase; SOD, superoxide dismutase; TYR, tyrosinase.

identified on several cell types and tissues. There is evidence to suggest that copper derived from Cp enters the cells via endosomes, but the protein itself is not endocytosed. However, Cp does not seem to be essential as a copper donor since copper deficiency symptoms were absent in aceruloplasminemia patients, who in turn, showed iron metabolism disturbances (Logan et al., 1994; Harris et al., 1995; Yoshida et al., 1995; Takahashi et al., 1996). Cp does not seem to play a major role in delivery of copper to cells within the nervous system, either. It does not transverse the blood-brain barrier and the concentration of the soluble protein is very low in the cerebrospinal fluid (Segal, 1993). Still, Cp may have a significant role in the brain, as indicated by the expression of a membrane-bound form in astrocytes, the cell type that comprises about 25% of the total volume of the brain (Patel and David, 1997).

The second route for delivery of copper into cells involves albumin or amino acid–bound copper. In blood the copper-albumin pool is in balance with the amino acid–bound fraction, especially copper-histidine. This suggests that albumin-histidine may act as a buffer system for the metal and, thus, exhibit a regulatory role in making the element available for cellular uptake (Ettinger, 1987). In the cerebrospinal fluid a different buffering system must be operating, as the albumin concentration is much lower than in blood (Segal, 1993). It is conceivable that the amino acids and low molecular weight peptides are the predominant form for copper delivery to the brain. There is evidence to suggest two separate mechanisms of copper uptake into brain tissue, a high-affinity but low-capacity process, and a low-affinity but high-capacity process (Hartter and Barnea, 1988).

Recently two genes, h*CTR1* and h*CTR2*, coding for putative membrane-bound copper transporters have been isolated (Zhou and Gitschier, 1997). h*CTR1* was cloned by functional complementation of a yeast mutant *(CTR1)* defective in high-affinity copper uptake (Dancis et al., 1994a). h*CTR2* was identified in a database search and its yeast homolog encodes a protein (Ctr2p) that is involved in low-affinity copper uptake in yeast. It is yet unknown whether hCTR1 (or hCTR2) transports copper in the reduced form (CuI), but as the cytoplasm is a reducing environment, copper is likely to be as Cu(I) once it is in the cell. In yeast, two plasma membrane proteins with Fe(III)/Cu(II) reductase activity have been identified (Fre1p, Fre2p), but similar enzymes have not yet been characterized in humans (Dancis et al., 1990, 1992; Anderson et al., 1992; Georgatsou and Alexandraki, 1994; Hasset and Kosman, 1995; Georgatsou et al., 1997).

In conclusion, copper is made available to the cell by one or more of the several serum copper carriers such as ceruloplasmin, albumin, and histidine. The metal is likely to be taken up as Cu(I), which suggests the presence of a reducing step before entering the cells. Cu(I) is then transported across the membrane by hCTR1 and hCTR2, which deliver copper to intracellular transporters to be carried to the final destinations.

Intracellular Copper Transport and Compartmentalization

Copper is a highly reactive transition element that will bind unspecifically to many biologically important molecules, and careful control of cellular copper levels is, therefore, crucial. All cells have developed highly specialized complexing proteins to recruit, deliver, and excrete copper, thereby assuring copper delivery to its appropriate biological destination. Until the identification of *ATP7A* (and *ATP7B)*, very little information existed on the mode of delivery of copper to subcellular compartments. Both proteins belong to a unique family of metal-transporting P-type ATPases, and the discovery of the strong evolutionary conservation of protein sites used for metal coordination constituted a major break through in identification of other specific carrier molecules. Now several different proteins involved in copper transport to specific compartments and enzymes have been identified.

Copper is associated with several intra- and extracellular enzymes that are involved in vital metabolic processes, including neurotransmitter biosynthesis, connective tissue cross-linking, cellular respiration, neuropeptide maturation, and antioxidant defense. The copper-containing proteins and enzymes are localized in various cellular compartments. Cu/Zn superoxide dismutase (Cu/Zn SOD, also known as SOD1) is found in the cytosol and another form of SOD encoded by a different gene is secreted from the cell (EC-SOD, extracellular-SOD). LOX is also secreted to the extracellular space. COX is localized in the inner mitochondrial membrane, PAM and DBH are stored in the secretory vesicles, and tyrosine in pigment granules.

Very recently three genes coding for cytosolic copper-transporting proteins have been identified using the respective yeast homolog: human *COX17*/yeast *COX17*, human *CCS1*/yeast *LYS7*, and human *HAH1*/yeast *ATX1*. All three human proteins were functional in yeast as demonstrated by complementation studies. These studies are based on expression of the human cDNA in yeast and when the corresponding yeast gene is mutated, the human protein replaces the function of the natural yeast protein. Functional similarities of the yeast and human proteins present a simple, but invaluable system to study intracellular copper metabolism in human. To explore the amino acid residues crucial for the human proteins, complementation studies can also be combined with site-directed mutagenesis, where selected mutations are inserted in the human cDNA, and its expression is studied in yeast.

Each of the three cytosolic copper proteins is involved in trafficking copper from the uptake protein CTR1 to a specific compartment. COX17 (yeast Cox17p) provides copper to mitochondria and hence to cytochrome c oxidase (Glerum et al., 1996; Amaravadi et al., 1997; Beers et al., 1997), and CCS1 (yeast Lys7p) delivers copper to the cytosolic Cu/Zn SOD (Horecka et al., 1995; Culotta et al., 1997). The third cytosolic protein, HAH1 (yeast Atx1p) (Lin and Culotta, 1995; Klomp et al., 1997), is likely to deliver copper from the cell surface to the secretory pathway where ATP7A and ATP7B (yeast Ccc2p) are localized (Petris et al., 1996; Yamaguchi et al., 1996; Dierick et al., 1997; Lin et al., 1997; Yuan et al., 1997).

This new class of metal binding cytosolic proteins which exhibit specific carrier functions can be considered as copper chaperones, as they guide the metal ion to the specific target compartments and enzymes, protecting the cell from damage, while the enzymes requiring copper as a co-factor are loaded. The chaperones are supposed to deliver their metal by a direct metal-induced protein-protein interaction with their natural recipient. The region crucial for this interaction has been identified for Atx1p (human HAH1) (Pufahl et al., 1997). This region includes the metal binding motif, MXCXXC, which is highly conserved among heavy metal binding proteins (Solioz, 1998). Site-directed mutagenesis studies with HAH1 demonstrated that the two cysteine residues were functioning as copper ligands (Hung et al., 1998). Studies with HAH1 and Atx1p also suggested that both proteins bind copper as Cu(I) (Pufahl et al., 1997; Hung et al., 1998).

Another intracellular protein to mention is metallothionein, a low molecular weight heavy metal binding protein, which plays an important role in copper homeostasis and detoxification (Hamer 1986; Kägi and Schäffer, 1988). Together with the reducing agent glutathione (GSH), metallothionein may constitute an intracellular buffering system for copper and other transition metals.

The two other proteins that play a crucial role in intracellular copper metabolism are ATP7A and ATP7B (yeast Ccc2p). These proteins are the only intracellular copper binding proteins that are known to be associated with a disease, namely MD and WD. ATP7A is involved in the biosynthesis of secreted enzymes as well as enzymes usually contained in vesicles. ATP7B supplies copper to apoceruloplasmin for the synthesis of ceruloplasmin (yeast Fet3p) (Yuan et al., 1995). Recent studies demonstrated that mutant yeast Ccc2p could be functionally complemented both by ATP7B and ATP7A, as indicated by the restoration of copper incorporation into Fet3p (human Cp) (Hung et al., 1997; Payne and Gitlin, 1998). These findings suggest that ATP7A and ATP7B have similar functions in different tissues.

ATP7A seems to have an important role in the regulation of intracellular copper homeostasis in several cell types, and this function seems to be overtaken by ATP7B in hepatocytes. When ATP7A is defective (as in MD), copper which cannot be utilized induces metallothionein transcription. Metallothionein binds copper, minimizing its toxic effects. While protecting the cell, metallotionein induction may prevent transport of copper by the cytosolic chaperones to the target enzymes, which leads to secondary copper deficiency symptoms. As an example, if COX17 cannot transport copper to mitochondria, COX function may be affected. Interestingly, when the intracellular copper level is high enough to saturate metallothionein, excess copper seems to be delivered to COX. However, this is not true for enzymes such as DBH and LOX, which probably receive their copper directly from ATP7A. This hypothesis may explain why some of the symptoms of MD can be corrected with copper-histidine administration, whereas others can not.

In conclusion, copper translocated from the plasma membrane is probably in the reduced Cu(I) form and might be complexed by GSH (Vulpe and Packman, 1995; Solioz, 1998). Copper is then delivered by the so-called HAH1, CCS, and COX17, to different intracellular compartments and enzymes.

Copper Export

ATP7A and ATP7B have been localized to the trans-Golgi network, implicating a role in copper loading of extracellular and vesicular enzymes (Petris et al., 1996; Yamaguchi et al., 1996; Dierick et al., 1997; Hung et al., 1997). When exposed to high copper levels, ATP7A is reversibly allocated to the plasma membrane, indicating that it also plays a direct role in exocytosis of copper. This occurs either by pumping of the metal into vesicles at the trans-Golgi network, which are subsequently sorted to the plasma membrane where they release their content to the extracellular space, or by direct efflux through ATP7A relocated at the plasma membrane. This copper-dependent relocalization appears to be a common mechanism for cellular extrusion of surplus copper and is probably of physiological significance (Petris et al., 1996; Ackland et al., 1997). Studies with ATP7B also revealed that an increase in the intracellular copper resulted in relocalization of the protein from the trans-Golgi network to a cytoplasmic vesicular compartment, which may be a step in extrusion of copper (Hung et al., 1997). However, these investigators could not detect ATP7B on the plasma membrane, and studies using different cell types are required to understand and evaluate the extent of this process. Recent studies also demonstrated a different localization (endoplasmic reticulum) of an abnormal variant of ATP7A in an occipital horn syndrome patient (Qi and Byers, 1998). Further studies investigating the localization of ATP7A (and ATP7B) in patients with different types of mutations will undoubtedly increase our knowledge about the behavior of these proteins.

Copper–Iron Connection

Iron is the most abundant metal in humans and it is required for several metabolic functions, because of its redox properties. As is the case with copper, iron can also be toxic in excess amounts. Iron imbalances have been observed in neurodegenerative disorders such as Alzheimer's disease (Connor et al., 1992), Parkinson's disease, and Hallervorden-Spatz syndrome (Gerlach et al., 1994); in the aging process; and in aceruloplasminemia (Harris et al., 1995; Yoshida et al., 1995; Takahashi et al., 1996). Iron and copper can promote membrane lipid peroxidation, which in neurons can lead to impairment of membrane ion-motive ATPases and glucose transporters, resulting in disruption of cellular ion homeostasis and neuronal degeneration (Mark et al., 1997a, 1997b). Recent studies demonstrated a relationship between copper and iron metabolisms both in humans and in yeast. For details the reader is referred to an extensive review by De Silva et al. (1996).

Apart from being a major copper binding component in serum, ceruloplasmin (Cp) is suggested to be involved in iron metabolism, especially in mobilization of iron. Cp has a ferroxidase activity that appears to be required for oxidation of Fe(II) (ferrous iron) to Fe(III) (ferric iron). This step is important in the release of iron from the intracellular iron storage protein ferritin and its attachment to transferrin, the protein that carries iron in the blood. Cp is synthesized in hepatocytes and a membrane-bound form has recently been identified in astrocytes (Patel and David, 1997).

In yeast, copper-dependent oxidase activity of Fet3p (human Cp) is required for high-affinity iron uptake (Askwith et al., 1994; Dancis et al., 1994b; De Silva et al.,

1995). At the plasma membrane, ferric iron is reduced to ferrous iron by two ferric reductases, Fre1p and Fre2p (Dancis et al., 1990, 1992; Anderson et al., 1992; Georgatsou and Alexandraki, 1994; Georgatsou et al., 1997). These reductases are regulated by a copper-sensing transcription factor, Mac1p (Georgatsou et al., 1997; Zhu et al., 1998). Two interesting points to mention are that Fre1p and Fre1p are also copper reductases (Hasset and Kosman, 1995), and Mac1p regulates not only iron but also copper uptake by direct binding to the promoter region of Ctr1p (human hCTR1) (Yamaguch-Iwai et al., 1997).

It is suggested that iron is translocated through the plasma membrane into the cell by a complex comprising of Fet3p (human Cp) and a permease, Ftr1 (Stearman et al., 1996). Fet3p and Ftr1p are likely to be assembled as a complex in a cellular compartment early in the secretory pathway, and this complex moves to a post- (or late) Golgi compartment where Ccc2 (human ATP7A/ATP7B) loads Fet3p (human Cp) with copper, which is required for its normal function (Yuan et al., 1997). Fet3p alone or as a complex then progresses to the plasma to be involved in the translocation of iron (Stearman et al., 1996). Fet3p (human Cp) is likely to reoxidize ferrous iron (FeII) to ferric (FeIII) iron, which is then transported in this form (Askwith et al., 1994 Dancis et al., 1994b).

Studies of the rare autosomal recessive disorder aceruloplasminemia have provided strong evidence for the involvement of Cp in iron metabolism and mobilization (Harris et al., 1995; Yoshida et al., 1995; Takahashi et al., 1996). Aceruloplasminemia patients have greatly decreased serum copper and Cp, and increased serum ferritin. Excessive iron deposition is observed in the brain, liver, and pancreas leading to severe tissue damage and clinical symptoms. The identification of Cp gene mutations in three unrelated aceruloplasminemia patients with systemic hemosiderosis demonstrated the importance of Cp in iron metabolism and supported the studies suggesting Cp as a ferro-oxidase (Harris et al., 1995; Yoshida et al., 1995; Takahashi et al., 1996). Surprisingly, these patients did not show any copper metabolism disturbances, indicating that the role of Cp as a copper transporter was dispensable. The important role of Cp in iron metabolism is also reflected by the correction of anemia in copper-deficient swine by Cp administration (Lee et al., 1968). Iron levels have not been well documented in MD or WD, but further studies in this field may yield interesting new knowledge.

In conclusion, data suggest that there is a strong interaction between iron and copper metabolisms in both yeast and human, as reflected by alterations in iron metabolism when the multicopper oxidases (yeast Fet3p and human Cp) are defective.

SUMMARY

Copper is an indispensable trace element with profound influence on neurological development and function through its role in the normal activity of several important enzymes, including DBH, COX, and SOD. On the other hand, copper has detrimental effects in excess. A fine regulation of copper levels is therefore essential for all biological systems from bacteria, to yeast, to human.

In the past half decade, tremendous knowledge about the intracellular copper metabolism has been accumulated. Today we are aware of eight proteins involved in intracellular copper transport and homeostasis. Apart from the genes coding for metallothioneins, all these genes were identified only very recently, following the cloning of the genes defective in Menkes and Wilson diseases in 1993. These two inherited disorders of copper metabolism reflect the contradictory effects of copper, being an essential, yet toxic element.

REFERENCES

Ackland, M.L., Cornish, E.J., Paynter, J.A., Grimes, A., Michalczyk. A. & Mercer, J.F.B. (1997). Expression of Menkes disease gene in mammary carcinoma cells. Biochem. J. 328, 237–243.

Amaravadi. R., Glerum, D.M. & Tzagoloff, A. (1997). Isolation of a cDNA encoding the human homologue of COX17, a yeast gene essential for mitochondrial copper recruitment. Hum. Genet. 99, 329–333.

Anderson, G.J., Lesuisse, E., Dancis, A., Roman, D.G., Labbe, P. & Klausner, R.D. (1992). Ferric iron reduction and iron assimilation in Saccharomyces cerevisiae. J. Inorg. Biochem. 47, 249–255.

Askwith, C., Eide, D., Van Ho, A., Bernard, P.S., Li. L., Davis-Kaplan, S., Sipe, D.M. & Kaplan, J. (1994). The FET3 gene of S. cerevisiae encodes a multicopper oxidase required for ferrous iron uptake. Cell 76, 403–410.

Barkovich, A.J., Good, W.V., Koch, T.K. & Berg, B.O. (1993). Mitochondrial disorders: analysis of their clinical and imaging characteristics. J. Neuroradiol. 14, 1119–1137.

Barton, N.W., Dambrosia, J.M. & Barranger, J.A. (1983). Menkes' kinky-hair syndrome: report of a case in a female infant. Neurology 33 (suppl 2), 154.

Beck, J., Enders, H., Schliephacke, M., Buchwald-Saal, M. & Tümer, Z .(1994). X;1 translocation in a female Menkes patient: characterization by fluorescence in situ hybridization. Clin. Genet. 46, 295–298.

Beers, J., Glerum, D.M. & Tzagoloff, A. (1997). Purification, characterization, and localization of yeast cox17p, a mitochondrial copper shuttle [In Process Citation]. J. Biol. Chem. 272, 33191–33196.

Bowcock, A.M. (1988). Eight closely linked loci place the Wilson disease locus within 13q14-q21. Am. J. Hum. Genet. 43, 664–674.

Bowcock, A.M., Farrer, L.A., Cavalli-Sforza, L.L., Hebert, J.M., Kidd, K.K., Frydman, M. & Bonné-Tamir, B. (1987). Mapping the Wilson disease locus to a cluster of linked polymorphic markers on chromosome 13. Am. J. Hum. Genet. 41, 27–35.

Brewer, G.J. & Yuzbasiyan-Gurkan, V. (1992). Wilson disease. Medicine 71, 139–164.

Britton, R.S. (1996). Metal-induced hepatotoxicity. Semin. Liver Dis. 16, 3–12.

Brockdorff, N., Kay, G., Smith, S., Keer, J.T., Hamvas, R.M.J., Brown, S.D.M. & Rastan, S. (1991). High-density molecular map of the central span of the mouse X chromosome. Genomics 10, 17–22.

Bull, P.C., Thomas, G.R., Rommens, J.M., Forbes, J.R. & Cox, D.W. (1993). The Wilson disease gene is a putative copper transporting P-type ATPase similar to the Menkes gene. Nat. Genet. 5, 327–337.

Cecchi, C. & Avner, P. (1996). Genomic organization of the mottled gene, the mouse homologue of the human Menkes disease gene. Genomics 37, 96–104.

Cecchi, C., Biasotto, M., Tosi, M. & Avner, P. (1997). The mottled mouse as a model for human Menkes disease: identification of mutations in the Atp7a gene. Hum. Mol. Genet. 6, 829.

Chelly, J., Tümer, Z., Tønnesen, T., Petterson, A., Ishikawa-Brush, Y., Tommerup, N., Horn, N. & Monaco, A.P. (1993). Isolation of a candidate gene for Menkes disease that encodes a potential heavy metal binding protein. Nat. Genet. 3, 14–19.

Christodoulou, J., Danks, D.M., Sarkar, B., Baerlocher, K.E., Casey, R., Horn, N., Tümer, Z.& Clarke, J.T. (1998). Early treatment of Menkes disease with parenteral copper-histidine: long-term follow-up of four treated patients. Am. J. Med. Genet. 76, 154–164.

Chuang, L.M., Wu, H.P., Jang, M.H., Wang, T.R., Sue, W.C., Lin, B.J., Cox, D.W. & Tai, T.Y. (1996). High frequency of two mutations in codon 778 in exon 8 of the ATP7B gene in Taiwanese families with Wilson disease. J. Med. Genet. 33, 521–523.

Connor, J.R., Menzies, S.L., St. Martin, S.M. & Mufson, E.J. (1992) A histochemical study of iron, transferrin, and ferritin in Alzheimer's disease brains. J. Neurosci. Res. 31, 75-83.

Culotta, V.C., Klomp, L.W., Strain, J., Casareno, R.L., Krems, B. & Gitlin, J.D. (1997). The copper chaperone for superoxide dismutase. J. Biol. Chem. 272, 23469–23472.

Czlonkowska, A., Rodo, M., Gajda, J., Ploos van Amstel, H.K., Juyn, J. & Houwen, R.H. (1997). Very high frequency of the His1069Gln mutation in Polish Wilson disease patients. J. Neurol. 244, 591–592.

Dancis, A., Klausner, R.D., Hinnebusch, A.G. & Barriocanal, J.G. (1990). Genetic evidence that ferric reductase is required for iron uptake in Saccharomyces cerevisiae. Mol. Cell. Biol. 10, 2294–2301.

Dancis, A,. Roman, D.G., Anderson, G.J,. Hinnebusch, A.G.& Klausner, R.D. (1992). Ferric reductase of Saccharomyces cerevisiae: molecular characterization, role in iron uptake, and transcriptional control by iron. Proc. Natl. Acad. Sci. U. S. A. 89, 3869–3873.

Dancis, A., Haile, D., Yuan, D.S. & Klausner, R.D. (1994a). The Saccharomyces cerevisiae copper transport protein (Ctr1p). Biochemical characterization, regulation by copper, and physiologic role in copper uptake. J. Biol. Chem. 269, 25660–25667.

Dancis, A., Yuan, D.S., Haile, D., Askwith, C., Eide, D., Moehle, C., Kaplan, J. & Klausner, R.D. (1994b). Molecular characterization of a copper transport protein in S. cerevisiae: An unexpected role of copper in iron transport. Cell 76, 393–402.

Danks, D.M. (1986). Of mice and men, metals and mutations. J. Med. Genet. 23, 99–106.

Danks, D.M. (1995). Disorders of copper transport. In: The Metabolic Basis of Inherited Disease, 7th ed. (Scriver, J.R., Beaudet, A.L., Sly, W.S. & Valle, D., eds.), pp. 2211–2235. McGraw-Hill, New York.

Danks, D.M., Stevens, B.J., Campbell, P.E., Gillespie, J.M., Walker-Smith, J., Blomfield, J. & Turner, B. (1972). Menkes' kinky-hair syndrome. Lancet 1, 1100–1102.

Das, S., Levinson, B., Whitney, S., Vulpe, C., Packman, S. & Gitschier, J. (1994). Diverse mutations in patients with Menkes disease often lead to exon skipping. Am. J. Hum. Genet. 55, 883–889.

Das, S., Levinson, B., Vulpe, C., Whitney, S., Gitschier, J. & Packman, S. (1995). Similar splicing mutations of the Menkes/Mottled copper transporting ATPase gene in occipital horn syndrome and the blotchy mouse. Am .J. Hum. Genet. 56, 570–576.

Davisson, M.T., Lalley, P.A., Peters, J., Doolittle, D.P., Hillyard, A.L. & Searle, A.G. (1991). Report of the comparative committee for human, mouse and other rodents. Cytogenet. Cell Genet. 58, 1152–1189.

De Silva, D.M., Askwith. C.C., Eide, D.& Kaplan, J. (1995). The *FET3* gene product required for high affinity iron transport in yeast is a cell surface ferroxidase. J. Biol. Chem. 270,1098–1101.

De Silva, D.M., Askwith, C.C.& Kaplan, J. (1996). Molecular mechanisms of iron uptake in eukaryotes. Physiol. Rev. 76,31–47.

Deiss A., Lynch, R.E., Lee, G.R. & Cartwright, G.E. (1971) Long-term therapy of Wilson's disease. Ann. Intern. Med. 75, 57–65.

DiDonato, M., Narindrasorasak, S., Forbes, J.R., Cox, D.W. & Sarkar, B. (1997). Expression, purification, and metal binding properties of the N-terminal domain from the Wilson disease putative copper-transporting ATPase (ATP7B). J. Biol. Chem. 272, 33279–33282.

Dierick, H.A., Ambrosini, L., Spencer, J., Glover, T.W. & Mercer, J.F.B. (1995). Molecular structure of the Menkes disease gene (*ATP7A*). Genomics 28, 462–469.

Dierick, H.A., Adam, A.N., Escara-Wilke, J.F. & Glover, T.W. (1997). Immunocytochemical localization of the Menkes copper transport protein (ATP7A) to the trans-Golgi network. Hum. Mol. Genet. 6, 409–416.

Ettinger, M.J. (1987). Cellular biochemistry of copper and inherited defects of copper metabolism. Life Chem. Rep. 5, 169–186.

Farrer, L.A., Bowcock, A.M., Hebert, J.M., Bonné-Tamir, B., Sternlieb, I., Giagheddu, M., St.George-Hyslop, P., Frydman, M., Lössner, J., Demelia, L., Carcassi, C., Lee, R., Beker, R., Bale, A.E., Donis-Keller, H., Scheinberg, I.H. & Cavalli-Sforza, L.L. (1991). Predictive testing for Wilson's disease using tightly linked and flanking DNA markers. Neurology 41, 992–999.

Figus, A., Angius, A., Loudianos, G., Bertini, C., Dessi, V., Loi, A., Deliana, M., Lovicu, M., Olla, N., Sole, G., De Virgiliis, S., Lilliu, F., Farci, A.M.G., Nurchi, A., Giacchino, R., Barabino, A., Marazzi, M., Zancan, L., Greggio, N.A., Marcellini, M., Solinas, A., Deplano, A., Barbera, C., Devoto, M., Ozsoylu, S., Kocak, N., Akar, N., Karayalcin, S., Mokini, V., Cullufi, P., Balestrieri, A,, Cao, A. & Pirastu, M. (1995). Molecular pathology and haplotype analysis of Wilson disease in Mediterranean populations. Am. J. Hum. Genet. 57, 1318–1324.

Fraser, A.S., Sobey, S. & Spicer, C.C. (1953). Mottled, a sex-modified lethal in the house mouse. J. Genet. 51, 217–221.

French, J.H., Sherard, S.H., Lubell, H., Brotz, M. & Moore, C.L. (1972). Trichopoliodystrophy I. Report of a case and biochemical studies. Acta Neurol. 26, 229–244.

Frydman, M., Bonné-Tamir, B., Farrer, L.A., Conneally, P.M., Magazanik, A., Ashbel, S. & Goldwitch, Z. (1985). Assignment of the gene for Wilson disease to chromosome 13: linkage to the esterase D locus. Proc. Natl. Acad. Sci. U. S. A. 82, 1819–1821.

Fu, D., Beeler, T.J. & Dunn, T.M. (1995). Sequence, mapping and disruption of CCC2, a gene that cross-complements the Ca2+-sensitive phenotype of csg1 mutants and encodes a P-type ATPase belonging to the Cu2+-ATPase subfamily. Yeast 11, 283–292.

Georgatsou, E. & Alexandraki, D. (1994). Two distinctly regulated genes are required for ferric reduction, the first step of iron uptake in Saccharomyces cerevisiae. Mol. Cell. Biol. 14, 3065–3073.

Georgatsou, E., Mavrogiannis, L.A,. Fragiadakis, G.S. & Alexandraki, D. (1997). The yeast Fre1p/Fre2p cupric reductases facilitate copper uptake and are regulated by the copper-modulated Mac1p activator. J. Biol. Chem. 272, 13786–13792.

Gerlach, M., Ben-Shachar. D., Riederer, P. & Youdim, M.B. (1994). Altered brain metabolism of iron as a cause of neurodegenerative diseases? J. Neurochem. 63, 793–807.

Gitschier, J., Moffat, B., Reilly, D., Wood, W.I. & Fairbrother, W.J. (1998). Solution structure of the fourth metal-binding domain from the Menkes copper-transporting ATPase. Nat. Struct. Biol. 5, 47–54.

Glerum, D.M., Shtanko, A. &Tzagoloff, A. (1996). Characterization of *COX17*, a yeast gene involved in copper metabolism and assembly of cytochrome oxidase. J. Biol. Chem. 271, 14504–14509.

Green, M.C. (1989). Catalog of mutant genes and polymorphic loci. In: Genetic Variants and Strains of the Laboratory Mouse, 2nd ed. (Lyon, M.F. & Searle, A.G., eds.), pp. 241–244. Oxford University Press, Oxford.

Grimes, A., Hearn, C.J., Lockhart, P., Newgreen, D.F. & Mercer, J.F. (1997). Molecular basis of the brindled mouse mutant (Mo(br)): a murine model of Menkes disease. Hum. Mol. Genet. 6, 1037–1042.

Grover, W.D., Henkin, R.I., Schwartz, M., Brodsky, N., Hobdell, E. & Stolk, J.M. (1982). A defect in catecholamine metabolism in kinky hair disease. Ann. Neurol. 12, 263.

Hämäläinen, E-R., Jones, T.A., Sheer, D., Taskinen, K., Pihlajaniemi, T. & Kivirikko, K.I. (1991). Molecular cloning of human lysyl oxidase and assignment of the gene to chromosome 5q23.3-31.2. Genomics 11, 508–516.

Hamer, D.H. (1986). Metallothionein. Ann. Rev. Biochem. 55, 913–951.

Harris, E.D. & Percival, S.S. (1989). Copper transport: insights into a ceruloplasmin-based delivery system. Adv. Exp. Med. Biol. 258, 95–102.

Harris, Z.L., Takahashi, Y., Miyajima, H., Serizawa, M., MacGillivray, R.T.A. & Gitlin, J.D. (1995). Aceruloplasminemia: molecular characterization of this disorder of iron metabolism. Proc. Natl. Acad. Sci. U. S. A. 92, 2539–2543.

Hartter, D.E. & Barnea, A. (1988). Brain tissue accumulates ^{67}copper by two ligand-dependent saturable processes: a high affinity, low capacity and a low affinity, high capacity process. J. Biol. Chem. 263, 799–805.

Hassett, R. & Kosman, D.J. (1995). Evidence for Cu(II) reduction as a component of copper uptake by Saccharomyces cerevisiae. J. Biol. Chem. 270, 128–134.

Heydorn, K., Damsgaard, E., Horn, N., Mikkelsen, M., Tygstrup, I,, Vestermark, S. & Weber, J. (1975). Extra-hepatic storage of copper. A male fetus suspected of Menkes' disease. Humangenetik. 29, 171–175.

Heydorn, K., Damsgaard, E., Horn, N., Tønnesen, T., Mussalo-Rauhamaa, H. & Kaitila, I. (1995). Comparison of trace element distributions in occipital horn syndrome and Menkes disease with normal subjects by neutron activation analysis. J. Trace Elem. Exp. Med. 8, 241–247.

Horecka, J., Kinsey, P.T. & Sprague, G.F. (1995). Cloning and characterization of the Saccharomyces cerevisiae LYS7 gene: evidence for function outside of lysine biosynthesis. Gene 162, 87–92.

Horn, N., Heydorn, K., Damsgaard, E., Tygstrup, I. & Vestermark, S. (1978). Is Menkes syndrome a copper storage disorder? Clin. Genet. 14, 186–187.

Horn, N., Stene, J., Møllerkaer, A.-M. & Friedrich, U. (1984). Linkage studies in Menkes' disease. The Xg blood group system and C-banding of the X chromosome. Ann. Hum. Genet. 48, 161–172.

Horn, N., Tønnesen, T. & Tümer, Z. (1992). Menkes disease: an X-linked neurological disorder of the copper metabolism. Brain Pathol. 2, 351–362.

Horn, N., Tønnesen, T. & Tümer, Z. (1995). Variability in clinical expression of an X-linked copper disturbance, Menkes disease. In: Genetic Response to Metals (Sarkar, B., ed.), pp 285–303. Marcel Dekker, New York.

Houwen, R.H.J., Juyn, J., Hoogenraad, T.U., Ploos van Amstel, J.K. & Berger, R. (1995). H714Q mutation in Wilson disease is associated with late, neurological presentation. J. Med. Genet. 32, 180–182.

Hung, I.H., Suzuki, M., Yamaguchi, Y., Yuan, D.S., Klausner, R.D. & Gitlin, J.D. (1997). Biochemical characterization of the Wilson disease protein and functional expression in the yeast Saccharomyces cerevisiae. J. Biol. Chem. 272, 21461–21466.

Hung, I.H., Casareno, R.L., Labesse, G., Mathews, F.S. & Gitlin, J.D. (1998). HAH1 is a copper-binding protein with distinct amino acid residues mediating copper homeostasis and antioxidant defense. J. Biol. Chem. 273, 1749–1754.

Hunt, D.M. (1974). Primary defect in copper transport underlies mottled mutants in the mouse. Nature 249, 852–854.

Janzer, R.C. (1993). The blood-brain barrier: cellular basis. J. Inherit. Metab. Dis. 16, 639–647.

Kägi, J.H.R. & Schäffer, A. (1988). Biochemistry of Metallothionein. Biochem. J. 27, 8509–8515.

Kaler, S.G. (1994). Menkes disease. Adv. Pediatr. 41, 263–304.

Kaler, S.G., Gallo, L.K., Proud, V.K., Percy, A.K., Mark, Y., Segal, N.A., Goldstein, D.S., Holmes, C.S. & Gahl, W.A. (1994). Occipital horn syndrome and a mild Menkes phenotype associated with splice site mutations at the MNK locus. Nat. Genet. 8, 195–202.

Kaler, S.G., Buist, N.R.M., Holmes, C.S., Goldstein, D.S., Miller, R.C. & Gahl, W.A. (1995). Early copper therapy in classic Menkes disease patients with a novel splicing mutation. Ann. Neurol. 38, 921–928.

Kaler, S.G., Das, S., Levinson, B., Goldstein, D.S., Holmes, C.S., Patronas, N.J., Packman, S. & Gahl, W.A. (1996). Successful early copper therapy in Menkes disease associated with a mutant transcript containing a small in-frame deletion. Biochem. Mol. Med. 57, 37–46.

Kanamaru, K., Kashiwagi, S. & Mizuno, T. (1994). A copper-transporting P-type ATPase found in the thylakoid membrane of the cyanobacterium *Synechococcus* species PCC7942. Mol. Microbiol. 13, 369–377.

Kapur, S., Higgins, J.V., Delp, K. & Rogers, B. (1987). Menkes syndrome in a girl with X-autosome translocation. Am. J. Med. Genet. 26, 503–510.

Kemppainen, R., Hämäläinen, E.R., Kuivaniemi, H., Tromp, G., Pihlajaniemi, T. & Kivirikko, K.I. (1996). Expression of mRNAs for lysyl oxidase and type III procollagen in cultured fibroblasts from patients with the Menkes and occipital horn syndromes as determined by quantitative polymerase chain reaction. Arch. Biochem. Biophys. 328, 101–106.

Kemppainen, R., Palatsi, R., Kallioinen, M. & Oikarinen, A .(1997). A homozygous nonsense mutation and a combination of two mutations of the Wilson disease gene in patients with different lysyl oxidase activities in cultured fibroblasts. J. Invest. Dermatol. 108, 35–39.

Klomp, L.W., Lin, S.J., Yuan, D.S., Klausner, R.D., Culotta, V.C. & Gitlin, J.D. (1997). Identification and functional expression of HAH1, a novel human gene involved in copper homeostasis. J. Biol. Chem. 272, 9221–9226.

Lazoff, S.G., Rybak, J.J., Parker, B.R. & Luzzatti, L. (1975). Skeletal dysplasia, occipital horns, diarrhea and obstructive uropathy—a new hereditary syndrome. Birth Defects 11, 71–74.

Lee, G.R., Williams, D.M. & Cartwright, G.E. (1968). Iron metabolism in copper-deficient swine. J. Clin. Invest. 47, 2058–2069.

Levinson, B., Vulpe, C., Elder, B., Martin, C., Verley, F., Packman, S. & Gitschier, J. (1994). The mottled gene is the mouse homologue of the Menkes disease gene. Nat .Genet. 6, 369–373.

Levinson, B., Conant, R., Schnur, R., Das, S., Packman, S. & Gitschier, J. (1996). A repeated element in the regulatory region of the *MNK* gene and its deletion in a patient with occipital horn syndrome. Hum. Mol. Genet. 5, 1737–1742.

Levinson, B., Packman, S. & Gitschier, J. (1997a). Mutation analysis of mottled pewter. Mouse Genome 95, 163–165.

Levinson, B., Packman, S. & Gitschier, J. (1997b). Deletion of the promoter region of the *Atp7a* gene of the *mottled dappled* mouse. Nat. Genet. 16, 223–224.

Li ,Y., Togashi, Y., Sato, S., Emoto, T., Kang, J-H., Takeichi, N., Kobayashi, H., Kojima, Y., Une, Y. & Uchino, J. (1991). Spontaneous hepatic copper accumulation in Long-Evans Cinnamon rats with hereditary hepatitis. J. Clin. Invest. 87, 1858–1861.

Lin, S. & Culotta, V.C. (1995). The *ATX1* gene of *Saccharomyces cerevisiae* encodes a small metal homeostasis factor that protects cells against reactive oxygen toxicity. Proc. Natl. Acad. Sci. U. S. A. 92, 3784–3788.

Lin, S.J., Pufahl, R.A., Dancis, A., O'Halloran, T.V. & Culotta, V.C. (1997). A role for the Saccharomyces cerevisiae ATX1 gene in copper trafficking and iron transport. J. Biol. Chem. 272, 9215–9220.

Linder, M.C. & Hazegh-Azam, M. (1996). Copper biochemistry and molecular biology. Am. J. Clin. Nutr. 63, 797S–811S.

Logan, J.I., Harveyson, K.B., Wisdom, G.B., Hughes, A.E. & Archbold, G.P.R. (1994). Hereditary ceruloplasmin deficiency, dementia and diabetes mellitus. Q. J. Med. 87, 663–670.

Lou, H.C., Hølmer, G.K., Reske-Nielsen, E. & Vagn-Hansen, P. (1974). Lipid composition of the brain in Menkes' disease. J. Neurochem. 22, 377–381.

Loudianos, G., Dessi, V., Angius, A., Lovicu, M., Loi, A., Deiana, M., Akar, N., Vajro, P., Figus, A., Cao, A. & Pirastu, M. (1996). Wilson disease mutations associated with uncommon haplotypes in Mediterranean patients. Hum. Genet. 98, 640–642.

Lutsenko, S., Petrukhin, K., Cooper, M.J., Gilliam, C.T. & Kaplan, J.H. (1997a). N-terminal domains of human copper-transporting adenosine triphosphatases (the Wilson's and Menkes disease proteins) bind copper selectively in vivo and in vitro with stoichiometry of one copper per metal-binding repeat. J. Biol. Chem. 272, 18939–18944.

Lutsenko, S., Petrukhin, K., Gilliam ,T.C. & Kaplan, J.H. (1997b). Heterologous expression of the metal-binding domains of human copper-transporting ATPases (P1-ATPases). Ann. N.Y. Acad. Sci. 834, 155–157.

Mark, R.J., Lovell, M.A., Markesbery, W.R., Uchida, K. & Mattson, M.P. (1997a). A role for 4-hydroxynonenal in disruption of ion homeostasis and neuronal death induced by amyloid β-peptide. J. Neurochem. 68, 255–264.

Mark, R.J., Pang, Z., Geddes, J.W. & Mattson, M.P. (1997b). Amyloid β-peptide impairs glucose uptake in hippocampal and cortical neurons: involvement of membrane lipid peroxidation. J. Neurosci. 17, 1046–1054.

Mattson, M. P. (1998). Modification of ion homeostasis by lipid peroxidation: roles in neuronal degeneration and adaptive plasticity. Trends Neurosci. 21, 53–57.

Menkes, J.H., Alter, M., Steigleder, G., Weakley, D.R. & Sung, J.H. (1962). A sex-linked recessive disorder with retardation of growth, peculiar hair and focal cerebral and cerebellar degeneration. Pediatrics 29, 764–779.

Mercer, J.F.B., Livingston, J., Hall, B., Paynter, J.A., Begy, C., Chandrasekharappa, S., Lockhart, P., Grimes, A., Bhave, M., Siemieniak, D. & Glover, T.W. (1993). Isolation of a partial candidate gene for Menkes disease by positional cloning. Nat. Genet. 3, 20–25.

Mercer, J.F.B., Grimes, A., Ambrosini, L., Lockhart, P., Paynter, J.A., Dierick, H.A. & Glover, T.W. (1994). Mutations in the murine homologue of the Menkes gene in dappled and blotchy mice. Nat. Genet. 6, 374–378.

Mori, M. & Nishimura, M. (1997). A serine-to-proline mutation in the copper-transporting P-type ATPase gene of the macular mouse. Mamm. Genome 8, 407–410.

Murata, Y., Kodama, H., Abe, T., Ishida, N., Nishimura, M., Levinson, B., Gitschier, J. & Packman, S. (1997). Mutation analysis and expression of the mottled gene in the macular mouse model of Menkes disease [In Process Citation]. Pediatr. Res. 42, 436–442.

Nanji, M.S., Nguyen, V.T., Kawasoe, J.H., Inui, K., Endo, F., Nakajima, T., Anezaki, T. & Cox, D.W. (1997). Haplotype and mutation analysis in Japanese patients with Wilson disease. Am. J. Hum. Genet. 60, 1423–1429.

O'Brien, J.S. & Sampson, E.L. (1966). Kinky hair disease II. biochemical studies. J. Neuropath. Exp. Neurol. 5, 523–530.

Odermatt, A,, Suter, H., Krapf, R. & Solioz, M. (1993). Primary structure of two P-type ATPases involved in copper homeostasis in Enterococcus hirae. J. Biol. Chem. 268,12775–12779.

Ohta, Y., Shiraishi, N. & Nishikimi, M. (1997). Occurrence of two missense mutations in Cu-ATPase of the macular mouse, a Menkes disease model. Biochem. Mol. Biol. Int. 43, 913–918.

Orru, S., Thomas, G., Cox, D.W. & Contu, L. (1997). 24 bp deletion and Ala_{1278} to Val mutation of the ATP7B gene in a Sardinian family with Wilson disease. Am. J. Hum. Genet. 10, 84–85.

Patel, B.N. & David, S. (1997). A novel glycosylphosphatidylinositol-anchored form of ceruloplasmin is expressed by mammalian astrocytes. J. Biol. Chem. 272, 20185–20190.

Payne, A.S. & Gitlin, J.D. (1998). Functional expression of the Menkes disease protein reveals common biochemical mechanisms among the copper-transporting P-type ATPases. J. Biol. Chem. 273, 3765–3770.

Peltonen. L., Kuivaniemi, H., Palotie, A., Horn, N., Kaitila, I. & Kivirikko, K.I. (1983). Alterations in copper and collagen metabolism in the Menkes syndrome and a new subtype of the Ehlers-Danlos syndrome. Biochem. J. 22, 6156–6163,

Petris, M.J., Mercer, J.F.B., Culvenor, J.G., Lockhart, P, Gleeson, P.A. & Camakaris, J. (1996). Ligand-regulated transport of the Menkes copper P-type ATPase efflux pump from the Golgi apparatus to the plasma membrane: a novel mechanism of regulated trafficking. EMBO. J. 15, 6084–6095.

Petrukhin, K., Lutsenko, S., Chernov, I., Ross, B.M., Kaplan, J.H. & Gilliam, T.C. (1994). Characterization of the Wilson disease gene encoding a P-type copper transporting ATPase: Genomic organization, alternative splicing, and structure/function predictions. Hum. Molec. Genet. 3, 1647–1656.

Phung, L.T., Ajlani, G. & Haselkorn, R. (1994). P-type ATPase from the cyanobacterium Synechococcus 7942 related to the human Menkes and Wilson disease gene products. Proc. Natl. Acad. Sci. U. S. A. 91, 9651–9654.

Pufahl, R.A., Singer, C.P., Peariso, K.L., Lin, S., Schmidt, P.J., Fahrni, C.J., Culotta, V.C. & Penner-Hahn, J.E. (1997). Metal ion chaperone function of the soluble Cu(I) receptor Atx1. Science 278, 853–856.

Qi, M. & Byers, P.H. (1998). Constitutive skipping of alternatively spliced exon 10 in the ATP7A gene abolishes Golgi localization of the Menkes protein and produces the occipital horn syndrome. Hum. Mol. Genet. 7, 465–469.

Rad, M.R., Kirchrath, L. & Hollenberg, C.P. (1994). A putative P-type Cu2+-transporting ATPase gene on chromosome II of Saccharomyces cerevisiae. Yeast 10, 1217–1225.

Reske-Nielsen, E., Gregersen, M. & Lund, E. (1987). Astrocytes in the postnatal nervous system. From birth to 14 years of age. An Immunohistochemical study of paraffin-embedded material. Acta. Path. Microbiol. Immunol. Scand. 95, 347–356.

Reed, V. & Boyd, Y. (1997). Mutation analysis provides additional proof that mottled is the mouse homologue of Menkes' disease. Hum. Mol. Genet. 6, 417–423.

Ronce, N., Moizard, M.P., Robb, L., Toutain, A., Villard, L. & Moraine, C. (1997). A C2055T transition in exon 8 of the ATP7A gene is associated with exon skipping in an occipital horn syndrome family. Am. J. Hum. Genet. 61, 233–238.

Rowe, D.W., McGoodwin, E.B., Martin, G.R. & Grahn, D, (1977). Decreased lysyl oxidase activity in the aneurysm-prone, mottled mouse. J. Biol. Chem. 252, 939–942.

Royce, P.M., Camakaris, J. & Danks, D.M. (1980). Reduced lysyl oxidase activity in skin fibroblasts from patients with Menkes' syndrome. Biochem. J. 192, 579–586.

Sarkar, B. & Kruck, T.P.A .(1966). Copper-amino acid complexes in human serum. In: The Biochemistry of Copper (Peisach, J., Aisen, P. & Blumberg, W.E., eds.), pp. 183–196. Academic press, New York.

Segal, M.B. (1993). Extracellular and cerebrospinal fluids. J. Inher. Metab. Dis. 16, 617–638.

Shah, A.B., Chernov, I., Zhang, H.T., Ross, B.M., Das, K., Lutsenko, S, Parano, E., Pavone, L., Evgrafov, O., Ivanova-Smolenskaya, I.A., Anneren, G., Westermark, K., Urrutia, F.H., Penchaszadeh, G.K., Sternlieb, I., Scheinberg, I.H., Gilliam, T.C. & Petrukhin, K. (1997). Identification and analysis of mutations in the Wilson disease gene (ATP7B): population frequencies, genotype-phenotype correlation, and functional analyses. Am. J. Hum. Genet. 61, 317–328.

Shimizu, N., Kawase, C., Nakazono, H., Hemmi, H., Shimatake, H. & Aoki, T. (1995). A novel RNA splicing mutation in Japanese patients with Wilson disease. Biochem. Biophys. Res. Comm. 217, 16–20.

Smith, R.M. (1981). Menkes' disease and the role of copper in brain development. In: Fetal Brain Disorders—Recent Approaches to the Problem of Mental Deficiency (Hetzel, B.S.& Smith, R.M., eds), pp.365–388. Elsevier/North-Holland.

Solioz, M. (1998). Copper homeostasis by CPX-type ATPases: the new class of heavy metal P-type ATPases. Adv. Mol. Cell. Biol. 23A, 167–203.

Southan, C. & Kruse, L.I. (1989). Sequence similarity between dopamine β hydroxylase and peptide a-amidating enzyme: evidence for a conserved catalytic domain. FEBS Lett.255, 116–120.

Stearman, R., Yuan, D.S., Yamaguchi-Iwai, Y., Klausner, R.D. & Dancis, A. (1996). A permease-oxidase complex involved in high-affinity iron uptake in yeast. Science 271, 1552–1557.

Takahashi, Y., Miyajima, H., Shirake, T., Suenaga, A. & Gitlin, J.D. (1996). A novel mutation resulting in aceruloplasminemia. Hum. Mol. Genet. 5, 81–84.

Tanzi, R.E., Petrukhin, K., Chernov, I., Pellequer, J.L., Wasco, W., Ross, B., Romano, D.M., Parano, E., Pavone, L., Brzustowicz, L.M., Devoto, M., Peppercorn, J., Bush, A.I., Sternlieb, I., Pirastu, M., Gusella, J.F., Evgrafov, O., Penchaszadeh, G.K., Honig, B., Edelman, I.S., Soares, M.B., Scheinberg, I.H. & Gilliam, T.C. (1993). The Wilson disease gene is a copper transporting ATPase with homology to the Menkes disease gene. Nat. Genet. 5, 344–357.

Theophilos, M.B., Cox, D.W. & Mercer, J.F.B. (1996). The toxic milk mouse is a murine model of Wilson disease. Hum. Mol. Genet. 5, 1619–1624.

Thomas, G.R., Roberts, E.A., Rosales, T.O., Moroz, S.P., Lambert, M.A,. Wong, L.T.K. & Cox D.W. (1993). Allelic association and linkage studies in Wilson disease. Hum. Mol. Genet. 2, 1404–1405.

Thomas, G.R., Forbes, J.R., Roberts, E.A., Walshe, J.M. & Cox, D.W. (1995a). The Wilson disease gene: spectrum of mutations and their consequences. Nat. Genet. 9, 210–217.

Thomas, G.R., Jensson, O., Gudmundsson, G., Thorsteinsson, L. & Cox, D.W. (1995b). Wilson disease in Iceland: a clinical and genetic study. Am. J. Hum. Genet. 56, 1140–1146.

Thomas, G.R., Roberts, E.A., Walshe, J.M. & Cox D.W. (1995c). Haplotypes and mutations in Wilson disease. Am. J. Hum. Genet. 56, 1315–1319.

Tümer, Z. & Horn, N. (1997). Menkes disease: recent advances and new aspects. J. Med. Genet. 34, 265–274.

Tümer, Z. & Horn, N. (1998). Deletion spectrum of Menkes disease (in preparation).

Tümer, Z., Tommerup, N., Tønnesen, T., Kreuder, J., Craig, I.W. & Horn, N. (1992). Mapping of the Menkes locus to Xq13.3 distal to the X-inactivation center by an intrachromosomal insertion of the segment Xq13.3-q21.2. Hum. Genet. 88,668–672.

Tümer, Z., Tønnesen, T., Böhmann, J., Marg, W. & Horn, N. (1994). First trimester prenatal diagnosis of Menkes disease by DNA analysis. J. Med. Genet. 31, 615–617.

Tümer, Z., Vural, B., Tønnesen, T., Chelly, J., Monaco, A.P. & Horn, N. (1995). Characterization of the exon structure of the Menkes disease gene using Vectorette PCR. Genomics 26, 437–442.

Tümer, Z., Horn, N., Tønnesen, T., Christodoulou, J., Clarke, J.T.R. & Sarkar, B. (1996). Early copper-histidine treatment for Menkes disease. Nat. Genet. 12, 11–13.

Tümer, Z., Lund, C., Tolshave, J., Vural, B., Tønnesen, T. & Horn, N. (1997). Identification of point mutations in 41 unrelated patients affected with Menkes disease. Am. J. Hum. Genet. 60, 63–71.

Tümer, Z., Møller, L.B. & Horn, N. (1999). Mutation spectrum of *ATP7A*, the gene defective in Menkes disease. In: Copper Transport and its Disorders: Molecular and Cellular Aspects, Advances in Experimental Medicine and Biology (Leone, A.& Mercer, J.,eds). Plenum Press (in press).

Uno, H., Arya, S., Laxova, R. & Gilbert, E.F. (1983). Menkes' syndrome with vascular and adrenergic nerve abnormalities. Arch. Pathol. Lab. Med. 107, 286–289.

Verga, V., Hall, B.K., Wang, S., Johnson, S., Higgins, J.V. & Glover, T.W. (1991). Localization of the translocation breakpoint in a female with Menkes syndrome to Xq13.2-q13.3 proximal to PGK-1. Am. J. Hum. Genet. 48, 1133–1138.

Vulpe, C. & Packman, S. (1995). Cellular copper transport. Annu. Rev. Nutr. 15, 293–322.

Vulpe, C., Levinson, B., Whitney, S., Packman, S. & Gitschier, J. (1993). Isolation of a candidate gene for Menkes disease and evidence that it encodes a copper-transporting ATPase. Nat. Genet. 3, 7–13.

Waldenström, E., Lagerkvist, A., Dahlman, T., Westermark, K. & Landegren, U. (1996). Efficient detection of mutations in Wilson disease by manifold sequencing. Genomics 37, 303–309.

Wieacker, P., Horn, N., Pearson, P., Wienker, T.F., McKay, E., Ropers, H.H. (1983). Menkes kinky hair disease: a search for closely linked restriction fragment length polymorphism. Hum. Genet. 64, 139–142.

Wilson, J.M., Johnston, D.E., Jefferson, D.M. & Mulligan, R.C. (1988). Correction of the genetic defect in hepatocytes from the Watanabe heritable hyperlipidemic rabbit. Proc. Natl. Acad. Sci. U. S. A. 85, 4421–4425.

Wu, J., Forbes, J.R., Chen, H.S. & Cox, D.W. (1994). The LEC rat has a deletion in the copper transporting ATPase gene homologous to the Wilson disease gene. Nat. Genet. 7, 541–545.

Yamaguchi, Y., Heiny, M.E. & Gitlin, J.D. (1993). Isolation and characterization of a human liver cDNA as a candidate gene for Wilson disease. Biochem. Biophys. Res. Comm. 197, 271–277.

Yamaguchi, Y., Heiny, M.E., Suzuki, M. & Gitlin, J.D. (1996). Biochemical characterization and intracellular localization of the Menkes disease protein. Proc. Natl. Acad. Sci. U. S. A. 93,14030–14035.

Yamaguchi-Iwai, Y., Serpe, M., Haile, D., Yang, W., Kosman, D.J., Klausner, R.D. & Dancis, A. (1997). Homeostatic regulation of copper uptake in yeast via direct binding of MAC1 protein to upstream regulatory sequences of FRE1 and CTR1. J. Biol. Chem. 272, 17711–17718.

Yang, H-M., Lund, T., Niebuhr, E., Nørby, S., Schwartz, M. & Shen, L. (1990). Exclusion mapping of 12 X-linked disease loci and 10 DNA probes from the long arm of the X-chromosome. Clin. Genet. 38, 94–104.

Yarze, J.C., Martin, P., Muñoz, S.J., Friedman, L.S. (1992). Wilson disease: current status. Am. J. Med. 92, 643–654.

Yoshida, D.S., Furihata, K., Takeda, S.S., Nakamura, A., Yamamoto, K., Morita, H., Hiyamuta, S., Ikeda, S., Shimizu, N. & Yanagisawa, N. (1995). A mutation in the ceruloplasmin gene is associated with systemic homodiserosis in humans. Nat. Genet. 9, 267–272.

Yuan, D.S., Stearman, R., Dancis, A., Dunn, T., Beeler, T. & Klausner, R.D. (1995). The Menkes/Wilson disease gene homologue in yeast provides copper to a ceruloplasmin-like oxidase required for iron uptake. Proc. Natl. Acad. Sci. U. S. A. 92, 2632–2636.

Yuan, D.S., Dancis, A. & Klausner, R.D. (1997). Restriction of copper export in saccharomyces cerevisiae to a late golgi or post-golgi compartment in the secretory pathway. J. Biol. Chem. 272, 25787–25793.

Zhou, B. & Gitschier, J. (1997). hCTR1: a human gene for copper uptake identified by complementation in yeast. Proc. Natl. Acad. Sci. U. S. A. 94, 7481–7486.

Zhu, Z,. Labbe, S., Pena, M.M. & Thiele, D.J. (1998). Copper differentially regulates the activity and degradation of yeast Mac1 transcription factor. J. Biol. Chem. 273, 1277–1280.

Chapter 11

The Neuronal Ceroid-lipofuscinoses (Batten Disease)

R.D. JOLLY, A. KOHLSCHÜTTER, D.N. PALMER, and S.U. WALKLEY

INTRODUCTION

The neuronal ceroid-lipofuscinoses are an enigmatic group of inherited neurodegenerative diseases of humans and animals that present as lysosomal storage

Advances in Cell Aging and Gerontology
Volume 3, pages 391–420

ISBN: 0-7623-0405-7

Table 1. Synopsis of the Major Forms of Human Neuronal Ceroid-lipofuscinosis (Batten Disease)

Disease	*Infantile NCL*	*Late Infantile NCL*	*V. Late Infantile NCL—Finnish*	*V. Late Infantile NCL*	*Juvenile NCL*	*V. Juvenile NCL (GRODs)*	*Adult NCL*
Eponym	Santavuori-Haltia	Jansky-Bielschowsky			Spielmeyer-Voght-Batten		Kufs
Genetic locus definition	CLN1	CLN2	CLN5	CLN6	CLN3	V.CLN1	
McKusick Cat. No.	256730	204500	256731	601780	204200	600680	204300
Lead clinical symptoms							
Visual failure [onset and progression]	12–22 months	2–3 years	4–5 years	Findings as for CLN2	4–9 years	Findings as for CLN3	---
Onset of seizures	14–24 months	2.5–3.5 years	5–6 years		8–16 years		30 years
Onset dementia	6–18 months	2.5–3.5 years	5–9 years		7–9 years		
Inability to walk	12–18 months	3.5–6 years	10–20 years		10–20 years		
Motor disorders	Chorreoathetosis	Myoclonus	Myoclonus		Myoclonus/rigor Slurred speech		Myoclonus Psychosis
Death by –	< 14 years	8–15 years	10–30 years		20–40 years		
Retinal degeneration							
Macular degeneration	+++	+++			+++	+	---
Pigment aggregation	---	+++	---		+++	---	---
ERG absent by –	12 months	3–4 years	4–5 years		5–7 years	variable	---

Electrophysiology							
EEG: helpful findings	Flat. No photic response	Photic spikes [early]	Photic spikes [late]				
Visual evoked potential	↓, later absent	↑↑ [early]	↑↑ [8–10 years]		↓		↓
Neuroimaging							
Brain atrophy	+++	+	+		+		+
Diagnostic pathology							
Lymphocyte vacuoles	---	---	---	---	+++	---	---
EM of storage bodies	GRODs	CB	CB & FP	CB & FP	FP mainly, CB	GRODs	GRODs & FP
Preferred tissue	Skin/lymphocyte/ rectal	Skin/lymphocyte/ rectal	Skin/lymphocyte/ rectal	Skin/lymphocyte/ rectal	Skin/lymphocyte/ rectal	Skin/lymphocyte/ rectal	Skin/muscle
Biochemistry							
Main storage material	SAPs A & D.	Subunit c	Subunit c				Subunit c?
Protein deficiency	Palmitoyl-protein thioesterase	Lysosomal peptidase			438 as protein	Palmitoyl-protein thioesterase	
Chromosome location	1p32	11p15	13q22	15q21–23	16p12	1p32	

Notes: AA, amino acids; CB, curvilinear bodies; EM, electron microscopy; FP, finger print bodies; GRODs, granular osmiophilic bodies; Subunit c, Subunit c of Mitochondrial ATP synthase; V., variant form of disease; +, degree present; ---, absent; ↑, elevated; ↓, decreased. The table follows the format introduced by Kohlschütter et al. (1993).

diseases. The first succinct clinical description was that of Stengel (1826) and a number of later more detailed reports gave rise to several eponyms of limited use today. For simplicity in communication, the term *Batten disease,* originally used among others for the juvenile form of disease, is now used in a generic sense for the whole group (Rider and Rider, 1988).

The term *neuronal ceroid-lipofuscinosis* (NCL or CLN) was introduced to cover this subgroup of amaurotic familial idiocies in which the storage material had tinctorial, histochemical, fluorescent, and ultrastructural similarities to the lipopigments ceroid and lipofuscin (Zeman and Dyken, 1969). This name is misleading in that accumulation of lipopigment occurs in cell types other than neurons, and the nature of lipopigment is now known to differ from that of these two classical pigments. The descriptor *neuronal* is useful from a clinical point of view as clinical signs reflect only those of neurodegeneration.

Several disease entities are included within the ceroid-lipofuscinoses (Batten disease), and these are presently differentiated in terms of age of onset, sometimes with other qualifying terms. Classical entities are the infantile, late infantile, juvenile, and adult forms, but variant diseases are also recognized (Dyken and Wisniewski, 1995) (Table 1). As they are defined by positional cloning and biochemical defect, classification by age of onset becomes less satisfactory as more than one genetically distinct disease may exist within an age classification. Disease entities are now more precisely defined by a numerical schema in which the different forms are designated CLN1 to CLN6 (Table 1). This also has limitations because at any gene locus there may be more than one mutation resulting in more than one form of disease.

Despite genetic heterogeneity, there are a number of unifying features in the various diseases that suggest certain common features in pathogenesis and in secondary pathways of neurodegeneration. As this is not yet understood for any form of disease, a combined and comparative approach is taken here in trying to rationalize mechanisms of disease.

FEATURES THAT UNIFY THE CEROID-LIPOFUSCINOSES

The neuronal ceroid-lipofuscinoses are characterized by progressive neurodegeneration leading in human patients to blindness, dementia, myoclonus, seizures, and premature death. Similar disease patterns are seen in affected animals, although species differences in manifestation of clinical signs occur. Pathologically, the hallmarks of disease are brain atrophy and the accumulation of a fluorescent lipopigment in neurons and many other cell types throughout the body. With the exception of one extended family suffering an adult-onset dominantly inherited disease (Boehme et al., 1971; Arnold et al., 1987), all forms are inherited as recessive traits.

The fluorescent storage material is not soluble in solvents used in the preparation of paraffin or epoxy resin embedded tissue sections, so pigment is observed in situ in both types of tissue section. This characteristic differentiates the ceroid-lipofuscinoses from most, but not all, other types of lysosomal storage disease, where much of the material is dissolved in either lipid or aqueous solvents. In paraffin sections, storage material invariably stains with Sudan black (Figure 1A and B) and usually with periodic-acid Schiff (PAS) and Luxol fast blue stains. It has a yellow fluoresence under ultraviolet or blue light fluorescence microscopy. The perikarya may be distended (ballooned) by lipopigment and particularly in pyramidal cells this accumulation extends into the axon hillock (Figure 1B) forming nonspiny meganeurites (Figure 2A) (Williams et al.,1977). This contrasts with

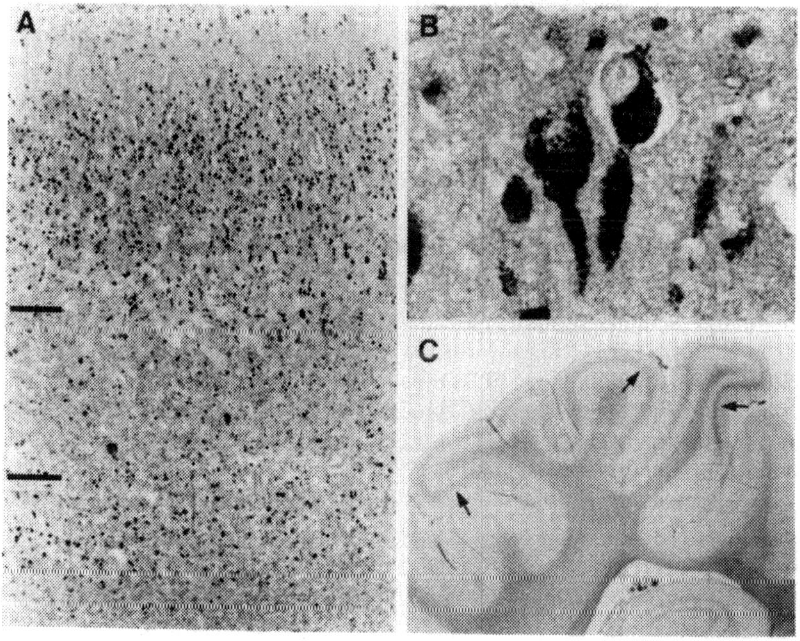

Figure 1. (**A**) Low power (× 50) of Sudan black stained paraffin section of the neocortex of a 24-year-old patient with juvenile NCL shows a marked loss of pigmented neurons in the lower layers (i.e. between bars). (**B**) Sudan black stained storage inclusions in the perikaryon of neurons in the neocortex of a patient with late infantile NCL. Note that they extend into the axon hillock to form a "meganeurite" (× 320). (**C**) Paraffin section of cerebral cortex of a 5-month-old Southhampshire lamb with ceroid-lipofuscinosis immunostained with antibody against glial fibrillary acidic protein (GFAP). The dark line (*arrows*) indicates a reactive astrocytosis where there has been selective death of neurons (× 2.5). (**B**) and (**C**) are republished in modified form from Jolly et al., 1989 and Jolly and Palmer, 1995, respectively, with the permission of the editor.

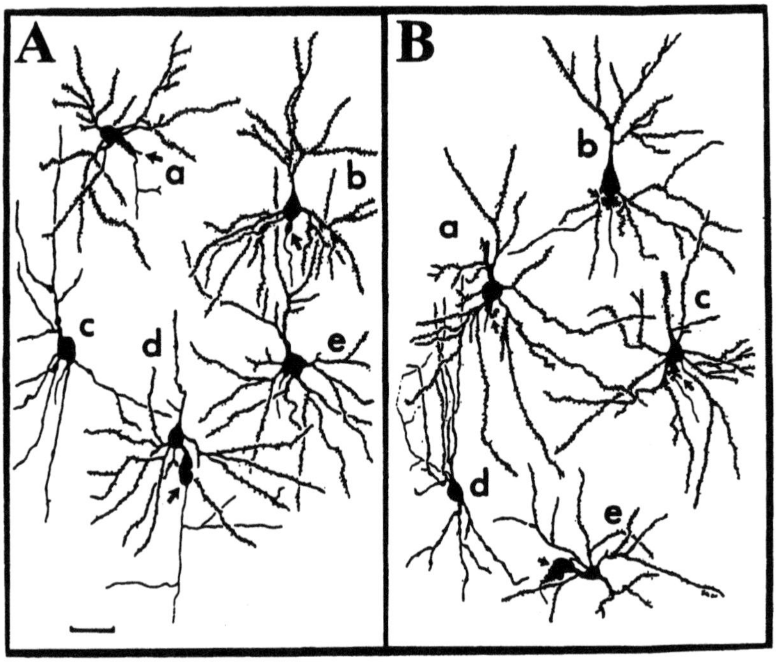

Figure 2. (**A**) Camera lucida drawing of Golgi-impregnated cortical neurons of an 18-month-old Southhampshire lamb with ceroid-lipofuscinosis in which many pyramidal neurons exhibited meganeurites (a, b, d). With rare exceptions (e.g., a), these structures were aspiny in form. (**B**) Similar meganeurites are shown in ovine GM_1 gangliosidosis. Unlike those in ceroid-lipofuscinosis, cortical neurons exhibit exuberant ectopic dendritogenesis as spine and neurite covered meganeurites and axon hillocks. (a, b, c, e); bar equals 50 m.

spiny meganeurites (Figure 2B), as seen in Golgi-stained neurons in other types of storage diseases (Purpura and Suzuki, 1976; Ahern-Rindell et al., 1988; Walkley and Wurzelmann, 1995; Walkley et al., 1996). In the central nervous system, lipopigment also accumulates in astrocytes and macrophages. The latter are often observed in or about the perivascular spaces, suggesting an excretory route for phagocytosed pigment.

Ultrastructurally, lipopigment is electron dense and either shows a variety of lamellar patterns known as curvilinear, rectilinear, fingerprint, and multilamellar, or is granular in appearance. This latter type is sometimes described as granular osmiophilic deposits (GRODs). These patterns have nosological significance (see Table 1) (Wisniewski et al., 1992; Lake, 1997), particularly as confusion between variants is clarified by positional cloning.

Intraneuronal storage of lipopigment is widespread, affecting most if not all neurons, in all forms of neuronal ceroid-lipofuscinosis. In contrast, significant neuronal cell death and accompanying atrophy is more selective, affecting the cerebral cortex and in some forms of disease, the cerebellum and retina. Neurons in other regions such as the basal ganglia, thalamus, and other diencephalic structures, brain stem, spinal cord, and visceral ganglia, may exhibit heavy pigment storage, but do not die. The degree of atrophy varies with the form of disease, with infantile patients showing the greatest degree of neuronal loss and adult patients the least. Late infantile and juvenile forms of ceroid-lipofuscinosis exhibit intermediate degrees of atrophy (Lake, 1997).

With infantile NCL, intraneuronal accumulation of pigment, cortical neuron loss, astrogliosis, and macrophage invasion is well advanced at 1 year of age. By terminal disease, the cerebral cortex is greatly diminished with only an occasional neuron within an area of reactive gliosis (Haltia, 1982). The cerebellum may be similarly affected. At this stage the brain may weigh as little as 250 to 400 g. In contrast, a cortical biopsy in a 4-year-old-girl with late infantile NCL showed minimal gliosis and neuronal loss, but pyramidal neurons were noted to exhibit marked axon hillock swellings (nonspiny meganeurites) and a diminution of type II (inhibitory) synapses over their soma-dendrite domains (Williams et al., 1977). By terminal disease, there is rarefaction of cortex, particularly in the region of laminae IV and V (Jolly, 1995).

Postmortem studies of the cerebral cortex from a 17-year-old patient with juvenile NCL showed evidence of loss of neurons with rarefaction of layers II to III, IV, and V (Braak and Goebel 1978, 1979). Loss of small nonpyramidal type neurons (likely inhibitory interneurons) predominated in the upper cortical layers. Surviving pyramidal neurons again showed the presence of meganeurites. An example of severe neuron loss, particularly in the deeper layers, is shown for a 24-year-old patient with the 1.02-Kb deletion of the CLN3 gene (see Figure 1A). Similar changes occur in adult-onset NCL (Goebel et al., 1982), but neuron loss is proportionally less pronounced (Berkovic et al., 1988).

In affected Southhampshire lambs, brain weights are normal at birth with normal growth to 4 months of age; thereafter, there is a decline and by terminal disease brain weight is reduced to approximately half that of normal individuals. Atrophy is associated with an initial lamina loss of neurons noted from 10 weeks of age (Jolly et al., 1989). This becomes more extensive and diffuse with increasing age and is accompanied by an astrocytosis (see Figure 1C). Similar changes occur in affected English setter dogs (see Figure 3D and C) (Walkley 1998), and by terminal disease, brains are 60% to 70% of normal weight (Koppang 1992). In contrast, neuron loss in the analogous disease in the *mnd/mnd* mouse (Faust et al., 1994) primarily affects the spinal cord (Bronson et al., 1993).

Retinal degeneration is a feature of all but the adult onset human forms of ceroid-lipofuscinosis; it occurs likewise in several analogous animal diseases (Jolly and Palmer, 1995). The essential lesion is loss of photoreceptor cells, but other layers may be secondarily affected. In time-course studies on affected Southhampshire sheep, rod cells were more susceptible than cone cells and remaining photoreceptors had short "dystrophic" outer segments (Graydon and Jolly, 1984;

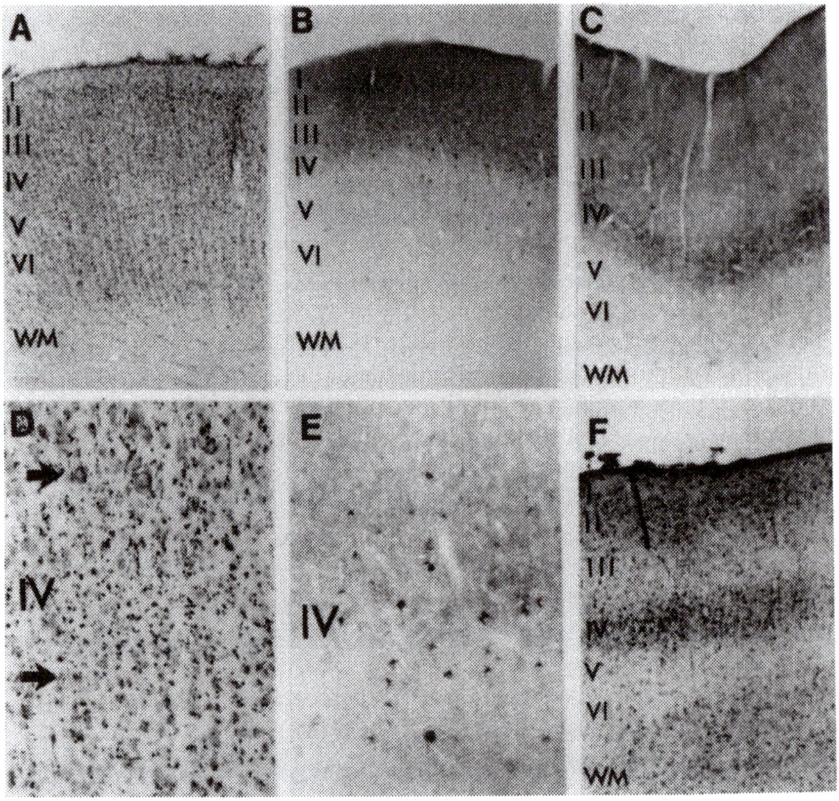

Figure 3. Variously stained vibratome sections of the cerebral cortex of a 20-month-old English setter dog with ceroid-lipofuscinosis (**A**, **B**, **D**, **E**, **F**) and a normal dog (**C**). (**A**) Nissl stain of neocortex (× 15) showing layers I through VI and the underlying white matter (WM). A higher magnification (× 60) is shown in (**D**) and reveals an abnormal number of small nuclei in nonneuronal-like cells. (**B**, **C**) Cytochrome oxidase histochemical stain of cortical sections (× 15) shows an absence of the cytochrome oxidase–positive band at layer IV in (**B**) which is present in normal cortex (**C**). (**E**) Higher magnification (× 60) of layer IV in (**B**) reveals the presence of scattered cytochrome oxidase–enriched neurons. Many of these cells contain abnormally enlarged cytochrome oxidase–positive mitochondria (see Walkley et al., 1995). (**F**) Another cortical section similar to those shown in A and B immunostained for glial fibrillary acidic protein (GFAP) shows the gliotic response is specific to lamina IV (× 15).

Mayhew et al., 1985). In the canine species, retinal degeneration is not a feature (Jolly et al., 1994) except in NCL of miniature Schnauzer dogs (Smith et al., 1996).

CLINICAL AND GENETIC ASPECTS OF THE MAJOR FORMS OF CEROID-LIPOFUSCINOSIS

Ceroid-lipofuscinosis (Batten Disease) in Humans

In the human ceroid-lipofuscinoses, degeneration of neuronal structures leads to blindness, dementia, and some form of epileptic seizures, with most types manifesting themselves during childhood or adolescence. A few cases of congenital NCL have been described but they are poorly defined. The better understood types are listed in Table 1, together with their defining features. Other atypical diseases are discussed by Dyken and Wisniewski (1995).

In infantile NCL (CLN1) the patients develop normally during the first 6 months of life (Santavuori, 1973). Arrest of psychomotor development between 6 and 18 months is the first symptom of disease and the ability to walk is lost between 12 and 14 months. Choreoathetosis and peculiar stereotype hand movements appear. Visual impairment is noted before the age of 2 years, leading rapidly to blindness. Myoclonic jerks, several types of convulsions, and spastic tetraparesis occur. The full syndrome may be observed after less than 1 year of disease; thereafter the condition of the child may become clinically stable for several years. In the final vegetative state there is coma-like unresponsiveness, myoclonus, and generalized tonic seizures. Death occurs between 5 and 13 years of age.

Electroencephalogram (EEG) changes are dramatic and consist of a marked diminution in amplitude before finally becoming isoelectric by the age of 2 or 3 years, while the electroretinogram (ERG) may be strikingly reduced well before impairment of vision becomes evident. Cranial computerized topography (CT) and magnetic resonance imaging (MRI) show severe brain atrophy most pronounced supratentorially and in the brain stem. Diagnosis may be confirmed by electron microscopy of peripheral lymphocytes or skin biopsy and noting the GRODs characteristic of the disease (Haynes et al., 1979; Ikeda and Goebel, 1979). Infantile NCL (CLN1) maps to chromosome 1p32 (Järvelä, 1991).

The late infantile NCL (CLN2) usually starts with major motor or myoclonic seizures at 21/2 to 3 years of age (Boustany, 1992). Concomitantly, there is loss in intellectual capacity; speech becomes slurred and eventually lost. Motor deterioration is characterized by successive development of hypotonia, spasticity, flexor spasms, and contractures. Retinal degeneration leads to blindness by about 6 years of age. Difficulty with swallowing and clearing secretions sets in and survival then depends on artificial feeding and bronchial toilet. Death occurs between about age 7 to 15 years.

The EEG shows high amplitude spikes or polyspikes after single flash photic stimulation and the ERG is lost early in the course of disease (Pampiglione and Harden, 1977). CT and MRI show a diffuse cerebral atrophy. A suspected diagnosis can be confirmed by electron microscopy of peripheral lymphocytes or skin biopsy and demonstration of curvilinear type storage material. CLN2 maps to chromosome 11p15 (Sharp et al., 1997). A similar disease, but with both fingerprint and curvilinear deposits, maps to chromosome 15q21-23 and is classified as CLN6 (Sharp et al., 1997).

In Western Finland, another variant form of late infantile NCL (CLN5) occurs with a relatively high local incidence (Varilo et al., 1996). The clinical features resemble those of late infantile disease (CLN2) but onset may be slightly later (Santavuori et al., 1982). It maps to chromosome 13q 22 (Klockars et al., 1996). A similar disease called early juvenile NCL has been described in a few families outside of Finland (Lake and Cavanagh, 1978), but may not be the same entity (Lake, 1997).

Juvenile NCL (CLN3) has visual failure as the leading symptom at about 4 to 7 years of age, failure being due to a pigmentary retinal degeneration (Hansen, 1979). Progression is rapid and leads to blindness in a further 2 to 4 years. Dementia and decay of speech become evident only after the onset of visual problems, but progression is slow (Kohlschütter et al., 1988). The severity of the seizure component and loss of motor capacities are variable (Boustany, 1992). There is unrest, depression, and psychotic symptoms (hallucinations) which can be profound problems, but patients may survive to their third or fourth decade.

In juvenile NCL the EEG shows spike-and-slow-wave complexes and the ERG is extinguished early. Progressive brain atrophy can be demonstrated neuroradiologically after the age of 9 years and is particularly pronounced in the cerebellum (Raininko et al., 1990). A suspected diagnosis can be confirmed by demonstration of vacuolated peripheral blood lymphocytes and electron microscopic demonstration of fingerprint patterned storage material in lymphocytes or skin biopsy. Juvenile NCL (CLN3) maps to chromosome 16p 12 (International Batten Disease Consortium, 1995). A variant juvenile NCL has been termed *juvenile NCL with GRODs* because of the electron microscopic appearance of storage material. It maps to chromosome 1 and may be an allelic variant of CLN1 (Mitchision et al., 1998).

Adult NCL or Kufs disease may reflect a heterogeneous group of neuronal storage disorders, with dementia, seizures, and extrapyramidal motor deterioration starting in adulthood, but without visual involvement. Although not defined genetically, it is referred to as CLN4. The development of abnormal behavior and a decline in intellectual capacity at around 30 years of age may first suggest a psychiatric condition (Augustine et al., 1993). Seizures of a myoclonic or grand mal type and extrapyramidal symptoms draw attention to organic brain disease. Diagnosis during life has been achieved by brain biopsy (Augustine et al., 1993) and skin biopsy (Vercruyssen et al., 1982). Ultrastructurally, storage material has both granular and fingerprint appearance.

A dominantly inherited form of adult NCL, with clinical signs of cerebellar syndrome from 31 years of age and with major fits and myoclonic jerks, has been described in an extended family (Boehme et al., 1971; Arnold et al., 1987).

Ceroid-lipofuscinosis in Animals

Ceroid-lipofuscinosis has been described in most common species of domestic animals (Jolly and Palmer, 1995); those in dogs have been reviewed (Jolly et al., 1994). A number of entities are mere case reports, others have been of veterinary importance, and yet others have been studied extensively as animal models of the analogous disease of human beings. The latter forms of NCL are in English setter dogs originally found in Norway (Koppang, 1973/74, 1992) and in Southhampshire sheep in New Zealand (Jolly et al., 1982).

Affected English setters develop normally from birth to 12 to 14 months of age. Thereafter, mental dullness and reduced vision become obvious but some vision remains even in terminal disease. There is progressive deterioration and dogs begin to stagger and show a stiffening of their extremities. There are also behavioral changes such as loss of learned behavior and apprehension. Convulsions start from about 17 to 20 months of age, and death usually occurs between 16 and 24 months of age. No affected animal has lived longer than 26 months. As with most other canine forms of NCL, the retina is essentially intact.

Clinical onset of the Southhampshire ovine disease is at 8 to 9 months of age when partial loss of vision is first noted and attributable to atrophy of the occipital cortex. However, as the disease progresses there is a retinal component to developing blindness associated with retinal degeneration (Mayhew et al., 1985). There is progressive abnormal behavior and intermittent episodes of tremor, affecting lips, eyelids, ears, face, head, neck, and occasionally thoracic limb musculature, which have been interpreted as partial seizures that did not become generalized (Mayhew et al., 1985). Affected sheep may live to 24-26 months of age. This form of ceroid-lipofuscinosis maps to ovine chromosome 7q13-15, an area syntenic to15q14-23 in humans (Broom et al., 1998). It is thus a probable model of CLN6.

A recently described adult onset form of NCL in miniature Schnauzer dogs (Smith et al., 1996; Palmer et al., 1997b) and a congenital form in Swedish Landrace lambs (Järplid and Haltia 1993; Haltia et al., 1996) have chemical and ultrastructural similarities of storage material to that in infantile NCL (CLN1). In contrast to other canine forms of disease, that in the miniature Schnauzer is characterized by severe retinal atrophy (Smith et al., 1996).

BIOCHEMICAL ASPECTS OF LIPOPIGMENT FORMATION

Nature of Lipopigment

The term *ceroid-lipofuscinosis* is derived from the histochemical and fluorescent similarities of the storage material to the lipopigments ceroid and lipofuscin (Zeman and Dyken, 1969). At that time, a theory developed that these pigments were the product of lipid peroxidation, with the production of malondialdehyde which formed insoluble Schiff base polymers with proteins and phospholipids. In early analytical studies of "ceroid" isolated from cases of neuronal ceroid-lipofuscinosis, approximately half was described as an "acidic lipid polymer" (Siakotas et al., 1972). It was postulated and concluded that the neuronal ceroid-lipofuscinoses were caused by a disorder of peroxidation of polyunsaturated fatty acids (Zeman 1974, 1976). Further support for this came from the finding of an apparent deficiency of a leukocyte peroxidase (Armstrong et al., 1973; Armstrong, 1982), a finding reaffirmed in some additional studies but questioned in others. The lipid peroxidation theory was also parsimonius with data suggesting a central role for membrane lipid peroxidation in the pathogenesis of neuronal degeneration in several different age-related disorders, including Alzheimer's disease (Mark et al., 1997; Mattson, 1998) and amyotrophic lateral sclerosis (Pedersen et al., 1998). An assumption that the ceroid-lipofuscinoses were associated with a primary defect involving abnormal lipid peroxidation held sway for more than a decade but has not stood the tests of time and experimentation.

Characterization of the Southhampshire ovine model was pivotal to present understanding of the biochemical nature of the storage material and denial of the primary lipid peroxidation theory. Isolation and analyses of storage bodies showed that the dominant chemical species in this lipopigment was protein (64% to 76%) with smaller amounts of phospholipids (11%), neutral lipids (11%), dolichyl pyrophosphate linked–oligosaccharides (< 2%), and a small amount of metals (Palmer et al., 1986 a, 1986b, 1988, 1990; Hall et al., 1989). Phospholipids were those found in normal mammalian membranes and included phosphatidylcholine, phosphatidylethanolamine, phosphatidlyinositol, and an additional lysosomal marker, bis(monoacylglycero)phosphate. Fatty acids showed a full range of unsaturated species. Neutral lipids were retinol, ubiquinone, dolichol, dolichyl esters, free fatty acids, and cholesterol. Isolated bodies retained their fluorescent properties observed *in situ*, but fluorescence was lost on separation into various components (Palmer et al., 1986b; Palmer et al., 1993). Extracted lipids from storage bodies contained a number of weak fluorophors, but none of these was a major component that equated with a putative Schiff base polymer consequent to lipid peroxidation. It was concluded that the fluorescence emitted from protein in its unique lipid environment.

The protein separated on SDS-PAGE as discrete species, which included a major band with apparent molecular weight of 3500. This proved to be subunit c of mitochondrial adenosine triphosphate (ATP) synthase (Palmer et al., 1986a, 1989; Fearnley et al., 1990), a very hydrophobic protein that separated in chloroform/methanol (2:1) with lipids and whose molecular weight calculated from the amino acid sequence is 7608. The term proteolipid has been used for proteins soluble in lipid solvents, but in the case of subunit c there is no implication of covalent bonding to lipid. Its hydrophobicity explains many of the difficulties of dealing with it *in vitro* that were met in its characterization, including the need for the unconventional technique of applying impure preparations to the protein sequencer. Being of low molecular weight and the dominant protein present, an identifiable sequence was obtained. Furthermore, by knowing the amount of total protein loaded by quantitative amino acid analysis of an aliquot, it was possible to calculate the percentage that was subunit c. For storage material from the ovine pancreas, subunit c accounted for 50% of total components. This is a minimum estimate, as initial coupling in the sequencer is not quantitative (Palmer et al., 1989). Later studies showed that the molecular weight of subunit c from storage bodies is 7650, indicating a post-translational modification. Although this change is not clearly identified, it is consistent with acetylation or trimethylation of lysine-43 (Buzy et al., 1996). This modification is also present on subunit c from normal mitochondria, so it is not disease specific.

More limited studies on isolated storage material likewise showed accumulation of subunit c in cases of late infantile (approximately 90% of accumulated protein), juvenile (approximately 45% of accumulated protein) (Palmer et al.,1990, 1992), and the Finnish late infantile variant (Tyynelä et al., 1997) forms of NCL. By implication it probably accumulates in another late infantile variant disease CLN6, as linkage analysis has shown this to be syntenic to the disease in Southhampshire sheep (Broom et al., 1998). An association of Kufs disease (adult-onset NCL) with mitochondrial subunit c accumulation is based on immunocytochemistry (Westlake et al., 1995), but this is not quantitative.

Subunit c also accumulates in three forms of NCL of English setter, Border collie, and Tibetan terrier dogs (Jolly et al., 1994; Palmer et al., 1997a), in NCL of Devon cattle (Martinus et al., 1991), and in the *mnd/mnd* mouse (Faust et al., 1994). In the latter, a related protein of higher molecular weight, subunit c of vacuolar ATPase, also accumulates. This protein occurs in smaller amounts in storage material from Southhampshire sheep and an intermediate amount in the Border collie, English setter, and Tibetan terrier dogs (Palmer et al., 1997a).

The diseases characterized by accumulation of subunit c of mitochondrial ATP synthase have a variety of lamellar patterns when viewed with an electron microscope (see earlier discussion and Table 1). It is likely that these are associated with the presence of polar phospholipids in the storage material.

Subunit c of mitochondrial ATP synthase does not accumulate in infantile NCL (CLN1) (Palmer et al., 1990, 1992). Compositional analyses of storage material from brains of patients with this disease revealed 43% of its dry weight was protein and 37% lipids (Tyynelä et al., 1993). Dominant proteins were sphingolipid activating proteins (SAPs) A and D. Their migration rates on PAGE were consistent with the complete and normal proteins. Saps B and C were not detected. In the lipid fractions, vitamin E, dolichols, cholesterol, and ubiquinone were identified and considered nonspecific as they are also found in the Southhampshire ovine subunit c storage disease. Two glycosphingolipids, globotriaosylceramide and fucosylneolactoteraosylceramide, were also present in minor amounts (1% to 2%) (Tyynelä, 1997). Phospholipids are not recorded as a significant component and this paucity of polar lipids may explain the granular rather than multilamellar patterns of storage material observed by electron microscopy.

There are three other forms of ceroid-lipofuscinosis that resemble CLN1, these being early juvenile variant with GRODs, which shares the same gene locus (O'Rawe et al., 1997), the adult-onset form in miniature Schnauzer dogs (Smith et al., 1996), and a congenital disease in Swedish landrace sheep (Järplid and Haltia, 1993). These latter animal diseases are also characterized ultrastructurally by GRODs, the absence of subunit c accumulation, and the presence of SAPS A and D (Haltia et al., 1996; Palmer et al., 1997b).

Putative Pathogenic Mechanisms of Lipopigment Formation

Whereas there are a number of gross and histopathological features common to the ceroid-lipofuscinoses that lead them to be considered together, ultrastructure and chemical analyses of storage bodies indicate at least two distinct groups, as follows: (1) Those that have multilamellar profiles and accumulation of subunit c of mitochondrial ATP synthase, of which the Southhampshire ovine disease is the prototype, and (2) those that have granular appearing inclusions (GRODs) and do not accumulate subunit c, of which the human infantile disease (CLN1) is the prototype.

NCL Characterized by the Accumulation of Subunit c of Mitochondrial ATP Synthase

Freeze-fracture electron micrographs, powder x-ray defraction studies, and chemical analyses of isolated storage material from affected Southhampshire sheep (Palmer et al., 1986a, 1986b; Jolly et al., 1988) were contrary to expectations arising from the lipid peroxidation hypothesis of pigmentogenesis (Zeman, 1974, 1976).

The dominant metabolite that accumulates in ovine ceroid-lipofuscinosis is the normal post-translationally modified subunit c of mitochondrial ATP synthase (Palmer et al., 1990, 1992; Fearnley et al., 1990; Buzy et al., 1996). Two expressed genes (P1 and P2) for subunit c encoded the same mature protein in both normal sheep and those with ceroid-lipofuscinosis. The mitochondrial import sequences for P1 and P2 genes, which are cleaved off to yield the mature protein, differed, but were the same for each gene in normal and affected sheep. The levels of expression of P1 and P2 genes were similar in normal and diseased individuals, so the protein is unlikely to accumulate because of excessive transcription of either gene (Palmer et al., 1990, 1992; Medd et al., 1993). These findings led to the hypothesis that ovine ceroid-lipofuscinosis should reflect a defect in the specific degradation pathway of subunit c (Palmer et al., 1989, 1990, 1992; Fearnley et al., 1990).

Tissue culture studies on fibroblasts from patients with late infantile Batten disease confirmed normal incorporation of subunit c into mitochondria (Ezaki et al., 1995a) and also indicated a defect in proteolysis (Ezaki et al., 1995b, 1996). A recent report that late infantile NCL (CLN2) is associated with a deficiency of a novel pepstatin insensitive peptidase activity (Sleat et al., 1997), implies a primary lysosomal defect for this disease.

With inborn errors of metabolism and catabolism, there is an expectation that the dominantly accumulated chemical species would reflect the substrate of a missing enzyme. The fact that there are at least four, and possibly more, genetic forms of human Batten disease that accumulate subunit c, suggests a more complicated course of events, particularly as there are a myriad of proteins to be degraded and lysosomal proteases have overlapping activities rather than being protein specific.

Subunit c occurs in multiple copies in the F_0 component of ATP synthase, where it helps form a transmembrane proton channel within the inner membrane. The normal turnover pathway of subunit c is not known nor is there much knowledge concerning the turnover of other mitochondrial membrane proteins. Proteins in the inner membrane have half-lives of several days (Grisolía et al., 1985) and this would appear to be so for subunit c (Ezaki et al., 1995a). It is considered unlikely that inner membrane proteins are transported out of mitochondria because of the complexity of transport mechanisms and the energy requirements that would be needed to free them from, or transport them across, membranes (Jolly, 1997). Effete mitochondria are believed to enter the lysosomal system by the process of autophagy, and there is evidence that 70% of mitochondrial synthesized proteins of long half-life are degraded by lysosomes (Grisolía et al., 1985). In the rat liver, at least 50% of mitochondrial protein is degraded by other than autophagy and it is likely some of these are short-lived proteins degraded by mitochondrial proteases. A defect in the catabolic pathway of subunit c could thus reside either in mitochondria or lysosomes.

Although bilayer membranes are fluid structures and molecules may move freely within each layer, lipid molecules adjacent to proteins are partially immobilized. This is known as boundary lipid immobilization. Lipid molecules between closely oriented proteins may exhibit strong perturbation of lipid bilayer fluidity (Marsh et al., 1978). The inner mitochondrial membrane contains 76% protein, much of it in hydrophobic complexes such as the transmembrane protein channel, which is rich in subunit c. The lipid:protein ratio lies in the range for which strong perturbations may be expected. The F_0 complex should thus be regarded in relation to its partially immobilized lipids with which it forms an F_0 complex domain (Jolly, 1997). It was postulated that disassembly of this F_0 complex domain would be the initial step in the catabolic pathway, which could be expected to be under some form of control involving several steps of intermembrane signaling or enzymic activity. As the complex is disassembled, subunit c may then be exposed to catabolism by proteases. This could occur within mitochondria, or alternatively, within lysosomes or late endosomes following autophagy of effete mitochondria. Perturbation of the disassembly process could result in abnormal accumulations of subunit c and associated lipids, that could rearrange to form paracrystalline structures that characterize the storage material (Jolly et al., 1988). Complexed in this way, subunit c was resistant to catabolism by proteases in at least four different types of normal cell, including macrophages maintained and observed in tissue culture for 7 to 21 days (Elleder et al., 1995). Subunit c accumulation would thus be an accident of its extreme hydrophobicity as well as a result of the biochemical defect. The hypothesis helps accommodate the different genetic forms of disease in which subunit c accumulates and is not excluded by the reported deficiency of a lysosomal peptidase in late infantile disease (CLN2) (Sleat et al., 1997).

The defect, however, could be more general than that previously described and also affect the turnover of other membrane systems or complexes. If this were so, the apparent specificity implied by subunit c accumulation would again be a result of its extreme hydrophobicity and propensity to form complexes with lipids; other less hydrophobic proteins would not accumulate because they did not form similar complexes and were catabolized. This extended hypothesis is in accord with the co-accumulation of varying amounts of vacuolar ATPase subunit c in various forms of the disease (Faust et al.,1994 ; Palmer et al., 1997a).

Dolichyl pyrophosphate–linked oligosaccharides are another stigma of storage material in the ceroid-lipofuscinoses and at one time were considered to directly reflect the biochemical anomaly (Hall et al., 1989; Pullarkat et al., 1992). However, in the ovine model, quantitative measurements showed them to be minor components and hence unlikely to be causally important (Hall et al., 1989; Palmer et al., 1990, 1992). Dolichol is a transmembrane molecule and dolichyl pyrophosphate–linked oligosaccahrides function in the endoplasmic reticulum. They could enter the lysosomal apparatus during the process of autophagy as part of the autophagic membrane, react with the hydrophobic subunit c, and contribute to the storage complex.

NCL Characterized by GRODs and Lack of Accumulation of Mitochondrial ATP Synthase Subunit c

The second group of ceroid-lipofuscinoses is composed of the infantile human form (CLN1), the juvenile variant with GRODs, the congenital form in Swedish landrace lambs, and adult-onset disease in miniature Schnauzer dogs. In this group, subunit c of mitochondrial ATP synthase does not accumulate. Instead, the dominant proteins recognized are sphingolipid activator proteins SAPs A and D (Tyynelä et al., 1993; Haltia et al., 1996; Palmer et al., 1997a).

The infantile human disease (CLN1) is now known to be caused by a deficiency of palmitoyl-protein thioesterase (Hofmann et al., 1997), a lysosomal enzyme that removes fatty acyl groups esterified to cysteines of proteins (Verkruyse and Hofmann, 1996). In immortalized lymphocytes from CLN1 patients, the catabolism of a distinct chloroform/methanol soluble cysteine thioester is blocked (Lu et al., 1996; Hofmann et al., 1997). However, such compounds have not been described as major components of storage material. The relationship of this deficiency and the accumulation of SAPs A and D, the major stored components, is not known. Hofmann et al. (1997) suggest that thioester compounds may be "toxic to lysosomal proteases, many of which are thiol proteases that may be sensitive to high energy thiol compounds." An alternative explanation may relate to the enzyme's neutral pH optimum, indicating that it may function in the endosomal compartment. Processing of SAPs A, B, C, and D begins in the endosome and is completed in lysosomes (Vielhaber et al., 1996; Hiraiwa et al., 1997). An abnormality in processing could lead to abnormal accumulation in lysosomes.

PUTATIVE MECHANISMS OF NEURODEGENERATION

NCL Characterized by Accumulation of Subunit c of Mitochondrial ATP Synthase

From positional cloning studies there are at least three genetic forms of subunit c storage diseases in humans associated with mutations at 11p15, 13q22, and 16p12 (see Table 1). A fourth form is implied through synteny of CLN6 with the Southhampshire ovine disease (Broom et al., 1998). However, the similarities of storage material, neuropathology, and clinical manifestations in human diseases and those in animals suggests commonality in underlying disease processes, as discussed earlier. This is tentatively assumed in developing hypotheses concerning the degeneration of nervous tissue.

The ability to carry out time-course experimental studies on analogous animal forms of NCL has helped elucidate the nature of the changes observed. In affected lambs, loss of vision was attributed to both central and retinal lesions (Graydon and Jolly, 1984; Mayhew et al., 1985). The latter were associated with loss of photoreceptors, but also with short "dystrophic" outer segments on those that remained. This latter lesion implies an inability to maintain membrane integrity of outer segments. The central lesion would be associated with the atrophy of the visual cortex and a similar lesion is the likely cause of loss of vision in affected dogs, as retinal atrophy is not a feature of ceroid-lipofuscinosis in that species. In human patients, electrophysiology and pathology likewise implicate both central and retinal components in loss of vision. The degree of retinal degeneration at end-stage disease in affected human patients is severe. Although comparisons with the time-course series of changes seen in the ovine model are not possible, similarities in pathology suggest a similar course of events.

Neuronal death in Batten disease, like that of most neurodegenerative diseases (Alzheimer's, Parkinson's, etc.), is largely restricted to select regions of the brain despite the fact that intracellular storage of lipopigment is near universal in neurons and non-neuronal cell types. Exactly what imparts this selective vulnerability in cerebral cortex, cerebellum, and retina is unclear. That this may involve mitochondria was initially signaled by the intralysosomal accumulation of a mitochondrial protein, subunit c of mitochondrial ATP synthase. Recent experiments have also suggested altered ATP synthase regulation, at least in some forms of the disease. In fibroblasts from patients with CLN2 and CLN3, there was a decrease in basal ATP synthase activity relative to that in normal cells. In both types of disease, activity was downregulated in response to anoxia or when functionally uncoupled, but upregulated in response to supraphysiological calcium concentration in incubation media (Das and Kohlschütter, 1996). In contrast, fibroblasts from affected Southhampshire sheep had increased basal activity. As with CLN2 and CLN3 fibroblasts, those from NCL sheep were downregulated by anoxia or uncoupling, but in contrast to the human diseases were also downregulated (55%) in response to high calcium (Das, et al., 1998). Respiratory chain enzyme activities of complex I + III and II + III and cytochrome c-oxidase did not differ significantly between fibroblasts from normal and affected sheep and humans. Fibroblast studies from patients with CLN2 and CLN3 (but not CLN1) have also shown oxidation defects affecting three types of fatty acid, suggesting a mitochondrial membrane defect (Dawson et al., 1996). Mitochondrial ATP synthase is responsible for most of the ATP production in aerobic tissues such as the brain. If defects found in fibroblasts are also reflected in neurons, this could lead to cellular ATP depletion, resulting in potential cell dysfunction and eventually death of neurons.

In the cerebral cortex of human Batten disease patients, atrophy has been associated with loss of specific types of neurons as discussed earlier (Williams et al., 1977; Braak and Goebel 1978, 1979). A similar loss of neurons is observed in affected Southhampshire sheep and English setter dogs with a laminar pattern (Jolly et al., 1989; Walkley, 1998). Early changes in cerebral cortex are evident in lamina IV and, with time, spread from this area to other laminae, particularly the supragranular laminae. This phenomenon has been studied most extensively in these animal models, being documented through a variety of methods. These include histochemistry for cytochrome oxidase activity and immunocytochemistry with glial fibrillary acidic protein (GFAP) and SMI32 antibodies for astroglial and cortical pyramidal cell labeling, respectively (Campbell and Morrison, 1989). In the canine (Walkley, 1998) and ovine (Walkley and Jolly, unpublished) models, it has been found that the early and consistent features of cortical atrophy are reduced cytochrome oxidase activity, neuron loss, and a marked astrocytic response, with these phenomena centered in mid-level cortex between pyramidal cell layers III and V (Figure 3). The affected layer (IV) in normal brain represents the major receiving zone in the cortex for excitatory thalamocortical inputs. This layer in normal animals demonstrates the greatest endogenous cytochrome oxidase staining, with this believed to be indicative of high sustained metabolic activity (Wong-Riley, 1989). The death of neurons and subsequent gliosis that predominate in this area in the ceroid-lipofuscinoses is thus likely to be related to this high metabolic and, hence, mitochondrial activity. In addition, a subpopulation of gamma-aminobutyric acid-ergic (GABAergic) neurons scattered throughout all cortical laminae, in the canine and murine disease models, exhibits greatly enlarged mitochondria, and this may be further evidence of mitochondrial involvement (March et al., 1995; Walkley et al., 1995). Enlarged mitochondria are less conspicuous in the ovine model, but GABAergic terminals on perikarya of surviving deep pyramidal neurons (lamina

V) in late disease are dramatically enlarged and contain conspicuous numbers of mitochondria (Walkley et al., 1995).

Chronic excitotoxicity has been postulated to account for selective neuronal vulnerability in neurodegenerative diseases (Choi, 1988; Beal, 1992; Mattson et al., 1993a). Accordingly, the neurotransmitter glutamate is believed to be deleterious to specific types of neurons, either because it is overabundant, or because these neurons have become vulnerable to its effects due to receptor changes or to metabolic defects within the cell. This latter view, the so-called weakened target cell model (Horowski et al., 1994), is based on impaired energy production within neurons possessing glutamate receptors. Therefore, suboptimal mitochondrial function in neurons receiving abundant excitatory glutamatergic input would be anticipated to lead to excessive calcium influx, free radical formation, and cell death (Novelli et al., 1988). This scenario conceivably would express itself differently in different types of neuron since metabolic requirements and types of synaptic inputs vary widely. Neurons with higher metabolic rates or with substantial glutamatergic input, or both, would be anticipated to be most vulnerable if compromised by a defect in ATP synthase regulation. Neural elements in lamina IV (Wong-Riley, 1989), cortical GABAergic neurons (Houser et al., 1984), and basket cells of the cerebellum (Luo et al., 1989) have all been proposed to have relatively high metabolic rates.

Studies of mitochondrial function of neurons in Batten disease are presently lacking. Given the wide heterogeneity of mitochondrial function in different types of cells (Lai, 1992) and indeed, in different parts of individual cells like neurons (Wong-Riley, 1989), the issues here are likely to be quite complex. Nonetheless, an initial gliotic response and putative neuron loss occurring specifically in cortical lamina IV is consistent with excitotoxicity playing a dominant role as a pathogenic mechanism. It could be argued also that loss of a growth factor secondary to loss of thalamic inputs to this layer might be responsible (Mattson et al., 1993b; Dunn et al., 1994). However, at present, there is no evidence that thalamocortical afferents themselves are degenerating in the ceroid-lipofuscinoses, and the thalamus itself is not a region displaying significant neuron death. Thus, it is more likely that changes observed in lamina IV are the result of a functional insult to these neurons, with neurodegenerative changes being secondary to metabolic compromise and sustained thalamocortical input.

The presence of enlarged mitochondria in select GABAergic neurons in canine NCL may represent a compensatory response. Alternatively, the altered mitochondrial structure in these neurons may be the direct result of the primary metabolic defect, possibly associated with altered trafficking of subunit c of miotochondrial ATP synthase, and thus represent a pathological change. If GABAergic neurons are lost in the process of cortical atrophy, pathogenic mechanisms associated with excitotoxicity would be anticipated to be enhanced. However, any such progression of events would be difficult to predict with certainty since disinhibition is prominent in cerebral cortex and presumably would be lost secondary to GABAergic neuron loss (Walkley et al., 1995).

Cerebellar events similar to those occurring in the cortex also have been reported in the English setter model of Batten disease. Significant loss of basket cells and basket cell terminals, which supply inhibitory inputs to Purkinje cells, occurs before Purkinje cell loss and cerebellar atrophy (March et al., 1995). The basket cell population of the cerebellum, which is believed characterized by high metabolic activity similar to that of cortical inhibitory neurons Luo, et al., (1989), also displays similarly enlarged and abnormal appearing mitochondria. There is also a loss of cytochrome oxidase activity in the cerebellum as noted by histochemical staining. Similar findings have recently been reported in Leigh's disease (subacute necrotizing encephalopathy), in which there is a primary mitochondrial compromise (Cavanagh, 1994). An excitotoxicity mechanism of Purkinje cell loss has likewise been proposed to explain the neurodegenerative features of this disorder.

There is a close correlation between the brain regions showing loss of neurons and clinical neurological disease exhibited by both animals and humans in subunit c–accumulating forms of ceroid-lipofuscinosis. These findings of dementia, seizures, ataxia, and blindness have also been reported in patients with primary defects in mitochondrial ATP synthase (Clark et al., 1984; Holt et al., 1990), suggesting that select types of neurons are indeed highly susceptible to compromise in energy production by mitochondria.

The preceding model of neurodegeneration could extend to include the photoreceptors in the retina, but does not adequately explain the chronic abnormalities in outer segments that are associated with otherwise healthy looking inner segments (Graydon and Jolly 1984; Mayhew et al., 1985). These suggest a primary defect affecting membranes. Such a defect could also underly the suboptimal mitochondrial function that has been argued contributes to neuron death. The mechanism of neuron death in Batten disease may involve apoptosis as reported in human cases and the ovine model (Lane et al., 1996), or cell necrosis. The presence of a macrophage and astrocytic reaction implies that the latter may also occur. If apoptosis is a major mechanism of cell death, then knowing the types of cell that are vulnerable to this change might provide important insights into the pathogenesis of disease and suggest ways to slow cell death. Likewise, if excitotoxicity is a major contributor to the pathogenic cascade, pharmacological intervention (Horowski et al., 1994) might aid in slowing the rate of neuron death and ameliorate the neurological progression.

NCL Characterized by GRODs and Nonaccumulation of Mitochondrial ATP Synthase

Cerebral atrophy is so severe in CLN1 that no pattern of degeneration has been observed that can be exploited observationally, or experimentally, to help understand its pathogenesis, as in the subunit c–accumulating diseases. It is difficult to rationalize a mechanism for catastrophic loss of neurons associated with the passive accumulation of storage material in lysosomes as a result of a deficiency of palmitoyl-protein esterase. There are many palmitoylated proteins in the nervous system (Bizzozero, 1997) including a Na-channel (α subunit) and seven G-protein–coupled neurotransmitter receptors. A defect involving deacylation of one or more of these could provide ample opportunity for abnormal function that could lead to neuron death. The localization of palmitoyl-protein esterase in endosomes or lysosomes makes its involvement in the depalmitoylation of proteins at other subcelluar locations an unlikely possibility (Bizzozero, 1997). Therefore, deacylation of these proteins as part of the catabolic process is likely to occur within lysosomes after their entry into the lysosomal apparatus by autopaghy of effete organelles.

It is possible that the thioesters are transferred from lysosomes and accumulate in other compartments, where they are toxic or cause an excitotoxicity similar to that previously described for the subunit c–accumulating diseases. Cysteine sulfinate and a number of other structurally related sulfur-containing amino acids are known to act in an excitatory fashion on neurons and, indeed, have been proposed as possible excitatory neurotransmitters (Griffiths, 1990). Compounds like the thio-ester, S-sulfocysteine bear strong structural resemblance to known excitatory amino acids such as glutamate and aspartate. Accumulation of this particular compound has been implicated as a neurotoxic agent in rare genetic diseases characterized by sulfite oxidase deficiency (Olney et al., 1975). In this disorder and that of the related molybdenum co-factor deficiency (Johnson and Wadman, 1995), absence of sulfite oxidase leads to elevated levels of S-sulfocysteine in tissues, plasma, and urine. There is consequent progressive neuron death in cerebral cortex and other regions, leading to severe brain atrophy. Affected children survive only a few years and exhibit seizures that are refractory to anticonvulsants, hypotonia, visual loss, myoclonus, and psychomotor retardation. Experimentally, cysteine-S-sulfate injected into animals has been found to cause a glutamate-like excitotoxic lesion (Olney et al., 1975). If accumulating thioesters in CLN1 include compounds such as S-sulfocysteine or related metabolites even in minute amounts, their presence could lead to a progressive cascade of neuron death affecting a variety of brain regions, particularly those with prominent excitatory amino acid neurotransmitter receptors such as in the cerebral cortex.

The preceding comments are speculative, but it is logical to postulate that neurodegeneration is related in some way to the high-energy thioesters that are expected to accumulate from a deficiency of a palmitoyl-protein thioesterase.

DISCUSSION

Progress has been made in understanding the pathogeneses of the various diseases that make up the Batten disease group, but they are still incompletely understood. The enigma lies in the number of genetically distinct entities that share many common pathological and clinical features and the difficulty in relating the severe and selective neurodegeneration with what appears to be passive accumulation of storage material within lysosomes. Of these, CLN1 appears to be quite different from at least four distinct entities that accumulate subunit c of mitochondrial ATP synthase (i.e., CLN2, CLN3, CLN5, CLN6, and probably CLN4). Of these latter, the biochemical defect is defined only for CLN2 as a lysosomal pepstatin insensitive peptidase deficiency (Sleat et al., 1997).

The gene defect for CLN3 is defined but the protein defect is not. It is thought to involve a large transmembrane protein (Mitchison et al., 1997). This has been variously located by immunochemistry in mitochondria of Müller cells and photoreceptors of the retina (Katz et al., 1997), and in lysosomes by confocal immunofluorescence microscopy (Järvelä et al., 1998).

The very hydrophobic nature of subunit c and a propensity to form poorly soluble protease resistant complexes with lipids may also be a factor in its accumulation. Thus, an indirect biochemical defect affecting disassembly of the F_o *complex domain*, or of proteolysis in general, might also perturb the normal and orderly catabolic pathway of this protein and favor formation of these complexes.

Despite advances in understanding the nature of the storage products and in some cases the biochemical defect, the relationships between storage disease and the severe neurodegeneration that characterizes these diseases is far from clear. As with defining the storage material, animal models have been useful in exploring the pathogenesis of neurodegeneration, at least for the subunit c–accumulating diseases. There is good evidence, as previously discussed, to invoke suboptimal mitochondrial function and chronic excitotoxicity in the neurodegenerative process. Neurons with high metabolic rates or with substantial glutamatergic input, or both, would be the most vulnerable.

Although evidence for apoptosis as a means of cell death is reported (Lane et al., 1996), inflammatory and reactive components to pathology of these diseases may indicate that necrosis is also involved. These various elements of pathogenesis need to be understood before therapeutic or ameliorating strategies can be developed in a logical manner. Well-defined experimental models such as the English setter dog and Southhampshire sheep will still be needed for this aspect of research, even when each human form of NCL is defined by enzyme deficiency and molecular pathology.

A similar situation exists for CLN1 inasmuch as a deficiency of palmitoyl-protein esterase neither explains the dominant components of storage material nor the catastrophic neurodegeneration that occurs early in the disease process. The expected storage metabolites are energy-rich thioesters which may be unstable and toxic. Their extension beyond the lysosomal compartment could induce instability in membrane systems. An argument that they, or related metabolites, might also exert excitotoxicity was also advanced. A precedence for such extension of a metabolite beyond lysosomes exists in several other lysosomal storage diseases, where GM_2 ganglioside accumulates directly or indirectly as a result of a lysosomal enzyme defect. In this situation, it is implicated as a pathogenic mechanism underlying the growth of ectopic dendrites on neuronal meganeurites (see Figure 2B) (Walkley, 1995). These may form new synapses which have been proposed as an underlying cause of brain dysfunction in some lysosomal storage diseases (Walkley et al., 1991; Walkley and Wurzelmann 1995).

The variant juvenile ceroid-lipofuscinosis with GRODs is presently thought to be analogous to infantile ceroid-lipofuscinosis (CLN1), but with later onset (Mitchison et al., 1998). This could be expected if the product of the mutant gene had some residual activity. Two potential models for infantile ceroid-lipofuscinosis exist in Swedish landrace lambs and miniature Schnauzer dogs, but neither is yet developed nor shown to reflect the analogous gene.

SUMMARY

The neuronal ceroid-lipofuscinoses are a group of inherited neurodegenerative diseases also characterized by accumulation of a fluorescent lipopigment in neurons and many other cell types. There is severe atrophy of the brain and retina. Ultrastructurally, storage bodies are electron dense and have either a multilamellar or granular appearance. In diseases characterized by multilamellar inclusions, the dominant accumulating material is subunit c of mitochondrial ATP synthase, but this very hydrophobic protein does not accumulate in those with granular inclusions

In humans there are at least four genetically distinct subunit c accumulating diseases, these being CLN2, CLN3, CLN5, and by implication CLN6. Similar diseases occur in domestic animals and those in the Southhampshire sheep and English setter dog have been studied extensively as models of the analogous human diseases. The pathogeneses of accumulation of subunit c are obscure, although CLN2 is associated with a deficiency of a lysosomal protein with peptidase activity. It is postulated that hydrophobicity and a propensity to form poorly soluble, protease-resistant complexes, may be an additional factor in accumulation of this protein. Neurodegeneration may be associated with excitotoxicity of select populations of neurons in cerebral cortex and cerebellum secondary to suboptimal mitochondrial function. The relationship of this to lysosomal storage of protein and lipid is not clear.

Infantile ceroid-lipofuscinosis (CLN1) is associated with a deficiency of palmitoyl-protein thioesterase but the relationship between this and the main storage material, sphingolipid activating proteins A and D, is also obscure. The theoretical enzyme substrates are energy rich thioesters. It is possible that they become distributed beyond the lysosomal compartment and are toxic either to membrane systems or induce excitotoxicity of neurons with consequent catastrophic neurodegeneration early in the course of the disease.

A full understanding of the pathogeneses of these diseases is necessary to allow specific therapeutic strategies to be developed. The animal models will remain critical to this even after the biochemical and molecular lesions are determined for each form of the disease.

ACKNOWLEDGMENTS

For work attributable to the authors we wish to acknowledge U.S. Public Health Service grants NS 11238 (RDJ), NS 32348 (DNP), NS 18804, and NS 30163 (SUW) and financial support from NCL Deutschland e.V.(AK).

NOTE ADDED IN PROOF

Two further genetically distinct forms of ceroid-lipofuscinosis have recently been described. These are a variant late infantile onset disease designated CLN7 (Wheeler et al., 1999) and an adult form designated CLN8. This latter maps to chromosome 8p23 and also accumulates subunit c of ATP synthase (Haltia et al., 1999).

REFERENCES

Ahern-Rindell, A.J., Prieur, D.J., Murnane, R.D., Raghaven, S.S., Daniel, P.F., McCleur, R.H., Walkley, S.U. & Parish, S.M. (1988). Inherited lysosomal storage disease associated with deficiencies of β-galactosidase and α-neuraminidase in sheep. Am. J. Med. Genet. 31, 39–56.

Armstrong, D. (1982). Peroxidase activity in human and canine ceroid-lipofuscinosis, with special reference to the eye. In: Ceroid-lipofuscinosis (Batten's Disease) (Armstrong, D., Koppang, N. & Rider, J.A., eds.), pp.247–270. Elsevier Biomedical Press, Amsterdam.

Armstrong, D., Dimmitt, S., Grider, L, van Wormer, D. & Austin, J. (1973). Deficient leukocyte peroxidase activity in Batten-Spielmeyer-Vogt Syndrome. Am. Neurol. Assoc. Trans. 98, 3–7.

Arnold, T.W., Faught, E., Falcon, J.C., Herrerea, G.A. & Linton, J. (1987). Adult onset neuronal ceroid-lipofuscinosis (Kuf's disease) with autosomal dominant inheritance. Neurology 37, 138–139.

Augustine, A., Fricchione, G., Woznicki, R., Broberg, D. & Holt, J. (1993). Adult neuronal ceroid-lipofuscinosis presenting with psychiatric symptoms; A case report. Int. J. Psychiat. Med. 23, 315–322.

Beal, M.F. (1992). Does impairment of energy metabolism result in excitotoxic neuronal death in neurodegenerative disease? Ann. Neurol. 31, 119–130.

Berkovic, S.F., Carpenter, S., Andermann, F. Andermann, E. & Wolfe, L.S. (1988). Kuf's disease: a critical reappraisal. Brain 111, 27–62.

Bizzozero, O.A. (1997). The mechanism and functional roles of protein palmitoylation in the nervous system. Neuropediatrics 28, 23–26.

Boehme, H.D., Cottrell, J.C., Leonberg, S.C. & Zeman, W. (1971). A dominant form of neuronal ceroid-lipofuscinosis. Brain 94, 745–760.

Boustany, R-M. (1992). Neurology of the neuronal ceroid-lipofuscinoses; late infantile and juvenile types. Am J. Med. Genet. 42, 533–535.

Braak, H. & Goebel, H.H. (1978). Loss of pigment laden stellate cells: A severe alteration of the isocortex in juvenile neuronal ceroid-lipofuscinosis. Acta Neuropathol. (Berlin) 42, 53–57.

Braak, H. & Goebel, H.H. (1979). Pigmentoarchitectonic pathology of the isocortex in juvenile neuronal ceroid-lipofuscinosis: axonal enlargements in layer IIIab and cell loss in layer V. Acta Neuropathol. (Berlin) 46, 79–83.

Bronson, R.T., Lake, B.D., Cook, S., Taylor, S. & Davisson, M.T. (1993). Motor neuron degeneration of mice is a model of neuronal ceroid- lipofuscinosis (Batten disease). Ann. Neurol. 33, 381–385.

Broom, M.F., Zhou, C., Broom, J.E., Barwell, K.J., Jolly, R.D. & Hill, D.F. (1998). Ovine ceroid lipofuscinosis: a large animal model syntenic with the human neuronal ceroid-lipofuscinosis variant CLN6. J. Med. Genet. In Press.

Buzy, A., Ryan, E.M., Jennings, K.R., Palmer, D.N., & Griffiths, D.E. (1996). Use of electrospray ionization mass spectrometry and tandem mass spectrometry to study binding of F_0 Inhibitors to ceroid-lipofuscinosis protein, a model system for subunit c of mitochondrial ATP synthase. Rapid Commun. Mass Spectrom.10, 790–796.

Campbell, M.J. & Morrison, J.H. (1989). Monoclonal antibody to neurofilament protein (SMI-32) labels a sub-population of pyramidal neurons in the human and monkey neocortex. J. Comp. Neurol. 282, 191–205.

Cavanagh, J.B. (1994). Is Purkinje cell loss in Leigh's disease an excitotoxic event secondary to damage to inferior olivary nuclei? Neuropathol. Appl. Neurobiol. 20, 599–603.

Choi, D.W. (1988). Glutamate neurotoxicity and diseases of the central nervous system. Neuron 1, 623–634.

Clark, J.B., Hayes, D.J., Morgan-Hughes, J.A. & Byrne, E. (1984). Mitochondrial myopathies: disorders of the respiratory chain and oxidative phosphorylation. J. Inherit. Metab. Dis. 7 (suppl. 1), 62–68.

Das, A.M. & Kohlschütter, A. (1996). Decreased activity of the mitochondrial ATP synthase in fibroblasts from children with late infantile and juvenile neuronal ceroid-lipofuscinosis. J. Inherit. Metab. Dis. 19, 130.–132.

Das, A., M., Jolly, R.D. & Kohlschütter, A. (1999). Dysregulation of mitochondrial ATP synthase in sheep with neuronal ceroid-lipofuscinosis (NCL). Mol. Genet. Metab. 66, 349–355.

Dawson, G., Kilkus, J., Siakotas, A.N. & Singh I. (1996). Mitochondrial abnormalities in CLN2 and CLN3 forms of Batten disease. Mol. Chem. Neuropathol. 29, 227–235.

Dunn, W.A. Jr., Raizada, M.K., Vagt, E.S. & Brown, E.A. (1994). Growth factor-induced neurite growth in primary neuronal cultures in dogs with neuronal ceroid-lipofuscinosis. Int. J. Dev. Neurosci. 12, 185–196.

Dyken, P. & Wisniewski, K. (1995). Classification of the neuronal ceroid-lipofuscinoses: Expansion of the atypical forms. Am. J. Med. Genet. 57, 150–154.

Elleder, M., Drahota, Z., Mares, V.I., Mandys, V., Müller, J., & Palmer, D.N. (1995).Tissue culture loading test with storage granules from animal models of neuronal ceroid lipofuscinosis (Batten disease): testing their lysosomal degradability by normal and Batten cells. Am.J. Med.Genet. 57, 213–221.

Ezaki, J., Wolfe, L.S., Higuti, T., Ishidoh, K., & Kominami, E. (1995a). Specific delay of degradation of mitochondrial ATP synthase subunit c in late infantile ceroid-lipofuscinosis. J. Neurochem. 64, 733–741.

Ezaki, J., Wolfe L.S., Ishidoh, K., & Kominami, E. (1995b). Abnormal degradative pathway of mitochondrial ATP synthase subunit c in late infantile ceroid-lipofuscinosis (Batten disease). Am. J. Med. Genet. 57, 254–259.

Ezaki, J., Wolfe, L.S., & Kominami, E. (1996). Specific delay in the degradation of mitochondrial ATP synthase subunit c in late infantile neuronal ceroid-lipofuscinosis is derived from cellular proteolytic dysfunction rather than structural alteration of subunit c. J. Neurochem. 67, 1677–1687.

Faust, J.R., Rodman, J.S., Daniel, P., Dice, J.F. & Bronson, R.P. (1994). Two related proteolipids and dolichol linked oligosaccharides accumulate in motor neuron degeneration mice (*mnd/mnd*), a model for neuronal ceroid-lipofuscinosis. J. Biol. Chem. 269, 10150–10155.

Fearnley, I.M., Walker, J.E., Martinus, R.D., Jolly, R.D., Kirkland, G.J., Shaw, G.J. & Palmer, D.N. (1990). The sequence of the major protein stored in ovine ceroid-lipofuscinosis is identical to that of the DCCD-reactive proteolipid of mitochondrial ATP synthase. Biochem. J. 268, 751–758.

Goebel, H.H., Braak H, H. & Seidel, D. (1982). Morphological studies on adult neuronal ceroid-lipofuscinosis (NCL). Clin. Neuropathol. 1, 151–162.

Graydon, R.J. & Jolly, R.D. (1984). Ceroid-lipofuscinosis (Batten's disease): Sequential electrophysiologic and pathologic changes in the retina of the ovine model. Invest. Ophthalomol.Vis. Sci. 25, 294–301.

Griffiths, R. (1990). Cysteine sulphinate (CSA) as an excitatory amino acid transmitter candidate in the mammalian central nervous system. Prog. Neurobiol. 35, 313–323.

Grisolía, S., Hermández-Yago, J. & Knecht, E. (1985). Regulation of mitochondrial protein concentration: a plausible model which may permit assessing protein turnover. Curr. Top. Cell. Regul. 27, 387–396.

Hall, N.A., Jolly, R.D., Palmer, D.N., Lake, B.D., & Patrick, A.D. (1989). Analysis of dolichol pyrophosphoryl oligosaccharides in purified storage cytosomes from ovine ceroid-lipofuscinosis. Biochem. Biophys Acta. 993, 245–251.

Haltia, M. (1982). Infantile neuronal ceroid-lipofuscinosis: neuropathological aspects. In: Ceroid-lipofuscinosis (Batten's Disease) (Armstrong, D., Koppang, N. & Rider, J.J. eds.), pp.105–115. Elsevier Biomedical Press, Amsterdam.

Haltia, M., Tyynelä, J., Baumann, M., Vesa, J., Peltonen, L.& Järplid, B. (1996). Congenital ovine ceroid-lipofuscinosis [abstract 35, p. 5], The Sixth International Congress on Neuronal Ceroid-lipofuscinoses (NCL-96), Gustavelund, Finland.

Haltia, M., Tyynelä, J., Hirvasniemi, A., Herva, R., Ranta, U.S., & Lehesjoki, A.-E. (1999). CLN8-Northern Epilepsy. In: The neuronal ceroid Lipofuscinoses (Batten disease) (Goebel, H., Mole, S. & Lake, B.D., Eds.), pp. 117–121. IOS Press, Amsterdam.

Hansen, E. (1979). Familial cerebro-macular degeneration (the Stengel-Batten Mayou-Spielmeyer-Vogt-Stock disease). Evaluation of the photoreceptors. Acta Ophthalmol. (Copenh.). 57, 382-396.

Haynes, M.E., Manson, J.I., Carter, R.F., Robertson, E. (1979). Electron microscopy of skin and peripheral blood lymphocytes in infantile (Santavuori) neuronal ceroid lipofuscinosis. Neuropediatrics 10, 245-263.

Hiraiwa, M., Martin, B., Kishimoto, Y., Conner, G.E., Tsuji, S. & O, Brien, J.S. (1997). Lysosomal proteolysis of prosaposin, the precursor of saposins (sphingolipid activator proteins): Its mechanism and inhibition by ganglioside. Arch. Biochem. Biophys. 341, 17–24.

Hofmann, S.L., Lee, L.A., Lu, J.Y., & Verkruyse, L.A. (1997). Plamitoyl- protein thioesterase and the molecular pathogenesis of infantile ceroid-lipofuscinosis. Neuropediatrics 28, 27–30.

Holt, I.J., Harding., A.E., Petty, R.K.H. & Morgan-Hughes, J.A. (1990). A mitochondrial disease associated with mitochondrial DNA heteroplasmy. Am. J. Hum. Genet. 46, 428–433.

Horowski, R., Wachtel, H., Turski, L. & Loschmann, P-A. (1994). Glutamate exocitotoxicity as a possible pathogenic mechanism in chronic neurodegeneration. In: Neurodegenerative Diseases (Calne D.M., ed.), pp. 163–175. W.B. Saunders, Philadelphia.

Houser, C.R., Vaughn, J.E., Hendry, S.H.C., Jones, E.G. & Peters, A. (1984). GABA neurons in the cerebral cortex. In: Cerebral Cortex: Functional Properties of Cortical Cells (Jones, E.G. & Peters, A. eds.), Vol 2, pp. 63–89. Plennum Press, New York.

Ikeda, K. & Goebel, H. (1979). Ultrastructural pathology of lymphocytes in neuronal ceroid-lipofuscinosis. Brain Dev. 1, 245–292.

International Batten Disease Consortium, (1995). Isolation of a novel gene underlying Batten disease. Cell 82, 949-957.

Järplid, B. & Haltia, M. (1993). An animal model of the infantile type neuronal ceroid-lipofuscinosis. J. Inherit. Metab. Dis. 16, 274–277.

Järvelä, I. (1991) Infantile neuronal ceroi lipofuscinosis (CLN1): Linkage disequilbrium in the Finnish population and evidence that variant late infantile form (Variant CLN2) represents a non allelic locus. Genomics 10, 333–337.

Järvelä, I., Saino, M., Rantamäki, T., Olkkonen, V.M., Carpén, O., Peltonen, L. & Jalanko, A. (1998). Biosynthesis and intracellular targeting of the CLN3 protein defective in Batten disease. Hum. Mol. Genet. 7, 85–90.

Johnson, J.L. & Wadman, S.K. (1995). Molybdenum cofactor deficiency and isolated sulfite oxidase deficiency. In: The Metabolic and Molecular Bases of Inherited Disease (Scriver, C.R., Beaudet, A.L., Sly, W.S. & Valle, D.), 7th ed., pp. 2271–2283. McGraw-Hill, New York.

Jolly, R.D. (1995). Comparative biology of the neuronal ceroid-lipofuscinoses (NCL): an overview. Am. J. Med. Genet. 57, 307–311.

Jolly, R.D. (1997). The ovine model of neuronal ceroid lipofuscinosis (NCL): its contribution to understanding the pathogenesis of Batten disease. Neuropediatrics 28, 60–62.

Jolly, R.D. & Palmer, D.N. (1995). Review—the neuronal ceroid-lipofuscinoses (Batten disease): comparative aspects. Neuropathol. Appl. Neurobiol. 21, 50–60.

Jolly, R.D., Shimada, A., Craig, A.S., Kirkland & Palmer, D.N. (1988). Ovine ceroid-lipofuscinosis II: pathologic changes interpreted in light of biochemical observations. Am.J. Med. Genet. Suppl. 5, 159–170.

Jolly, R.D., Janmaat, J., Graydon, R. J. & Clemett, R.S. (1982). Ceroid-lipofuscinosis: the ovine model. In: Ceroid-lipofuscinosis (Batten's Disease) (Armstrong, D. Koppang, N. & Rider, J.A., eds.), pp.219–228. Elsevier Biomedical Press, Amsterdam.

Jolly, R.D., Shimada, A., Dopfmer, I., Slack, P.M., Birtles, M.J. & Palmer, D.N. (1989). Ceroid-lipofuscinosis (Batten disease): pathogenesis and sequential neuropathological changes in the ovine model. Neuropathol. Appl. Neurobiol. 15, 371–383.

Jolly, R.D., Palmer, D.N., Studdert,V.P., Sutton, R.H., Kelly, W.R., Koppang, N., Dahme, E., Hartley, W.H., Patterson, JS, & Riis, R.C. (1994). Canine ceroid-lipofuscinosis: a review and classification. J. Small Anim. Pract. 35,299–306.

Katz, M.L., Gao, C-L., Prabhakaram, M., Shibya, H. Liu, P-C. & Johnson, G.S. (1997). Immunochemical localisation of the Batten disease (CLN3) protein in the retina. Invest. Ophthalmol.Vis. Sci. 38, 2375–2386.

Klockars, T., Savukoski, M., Isosomppi, J., Laan, J., Järvalä, I., Petrukhin, K., Palotie, K. & Peltonen, L. (1996). Efficient construction of a physical map by fiber-fish of the CLN5 region: refined assignment and long range contig covering the critical region. Genomics 35, 71–78.

Kohlschütter, A., Gardiner, R.M., & Goebel, H.H. (1993). Human forms of neuronal ceroid-lipofusciosis (Batten disease): Consensus on diagnostic criteria. J. Inherit. Dis. 16, 241–244.

Kohlschütter, A., Laabs, R. & Albani, M.(1988). Juvenile neuronal ceroid-lipofuscinosis (JNCL): quantitative description of its clinical variability. Acta Paediatr. Scand. 77, 867–872.

Koppang, N. (1973/74). Canine neuronal ceroid-lipofuscinosis—a model for human ceroid-lipofuscinosis. Mech. Aging Dev. 2, 421–445.

Koppang, N. (1992). English setter model and juvenile ceroid-lipofuscinosis. Am. J. Med. Genet. 42, 599–604.

Lai, J.C.K. (1992). Oxidative metabolism in neuronal and non neuronal mitochondria. Canad. J. Physiol. 70, S130–S137.

Lake, B.D. (1997). Lysosomal and peroxisomal diseases. In: Greenfield's Neuropathology, Vol.1, 6th ed. (Graham, D.I. & Lantos, P.L. eds.), pp. 658–753. Arnold, London.

Lake, B. & Cavanagh, N. (1978). Early-juvenile Batten's disease—a recognizable sub-group district from other forms of Batten's disease. Analysis of 5 patients. J. Neurol. Sci. 36, 265–271.

Lane, S.C., Jolly, R.D., Schmechel, D.E., Alroy, J. & Boustany, R-M.(1996). Apoptosis is the mechanism of neurodegeneration in Batten disease. J. Neurochem. 67, 677–683.

Lu, J.Y., Verkruyse, L.A. & Hofmann, S.L. (1996). Lipid thioesters derived from acylated proteins accumulate in infantile neuronal ceroid-lipofuscinosis: correction of the defect in lymphoblasts by recombinant palmitoyl-protein thioesterase. Proc. Natl. Acad. Sci. U. S. A. 93, 10046–10050.

Luo, X.G., Hevner, R.F. & Wong-Riley, M.T.T. (1989). Double labeling of cytochrome oxidase and gamma-amino butyric acid in central nervous system neurons of adult cats. J. Neurosci. Method. 30, 189–195.

March, P.A., Walkley, S.U. & Wurzelmann, S. (1995). Pathogenesis of brain dysfunction in Batten disease. Am. J. Med. Genet. 57, 204212.

Mark, R. J., Lovell, M. A., Markesbery, W. R., Uchida, K. & Mattson, M. P. (1997). A role for 4-hydroxynonenal in disruption of ion homeostasis and neuronal death induced by amyloid β-peptide. J. Neurochem. 68, 255–264.

Marsh, D., Watts, A., Maschke, W. & Knowles, P.F. (1978). Protein-immobilized lipid in dimyristoylphosphatidylcholine-substituted cytochrome oxidase: evidence for boundary and trapped lipid. Biochem. Biophys. Res. Commun. 81, 397–402.

Martinus, R.D., Harper, P.A.W., Jolly, R.D., Bayliss, S.L., Midwinter, G.G., Shaw, G.J. & Palmer, D.N. (1991). Bovine ceroid-lipofuscinosis (Batten disease): the major component stored is the DCCD-reactive proteolipid, subunit c of mitochondrial ATP synthase. Vet. Res. Commun. 15, 85–94.

Mattson, M. P. (1998) Modification of ion homeostasis by lipid peroxidation: roles in neuronal degeneration and adaptive plasticity. Trends Neurosci. 21, 53–57.

Mattson, M. P., Barger, S. W., Cheng, B., Lieberburg, I., Smith-Swintosky, V. L. & Rydel, R. E. (1993a). β-Amyloid precursor protein metabolites and loss of neuronal calcium homeostasis in Alzheimer's disease. Trends Neurosci. 16, 409–415.

Mattson, M.P., Zhang, Y. & Bose, S. (1993b). Growth factors prevent mitochondrial dysfunction, loss of calcium homeostasis, and cell injury, but not ATP depletion in hippocampal neurons deprived of glucose. Exp. Neurol. 121, 1–13.

Mayhew, I.G., Jolly, R.D., Picket, B.T. & Slack, P.M. (1985). Ceroid-lipofuscinosis (Batten disease): pathogenesis of blindness in the ovine model. Neuropathol. Appl. Neurobiol. 11, 273–290.

Medd, S.M., Walker, J.E. & Jolly, R.D. (1993). Characterization of the expressed genes of subunit c of mitochondrial ATP synthase in sheep with ceroid-lipofuscinosis. Biochem. J. 293, 65–73.

Mitchison, H.M., Hofman, S.L., Becerra, C.H., Munroe, P.B., Lake, B.D., Crow, Y.J., Stephenson, J.B., Williams, R.E., Hofman, I.L., Taschner, P.E.M., Martin, J.J., Philippart, M., Andermann, F., Mole, S.E., Gardiner, R.M., & O'Rawe, A.M. (1998). Mutations in the palmitoyl-protein thioesterase gene (PPT; CLN1) causing juvenile neuronal ceroid lipofuscinosis with granular osmiophilic deposits. Hum. Mol. Genet. 7, 291–297.

Mitchison, H.M., Taschner, P.E.M., Kremmidiotis, G., Callen, D.F., Doggett, N.A., Lerner, T.J., Janes, R.B., Wallace, B.A., Munroe, P.B., O'Rawe, A.M., Gardiner, R.M. & Mole, S. (1997). Structure of the CLN3 gene and predicted structure, location and function of CLN3 protein. Neuropediatrics 28, 12–14.

Novelli, A., Reilly, J.A., Lysko, P.G. & Henneberry, R.C. (1988). Glutamate becomes neurotoxic via the N-methyl-D-aspartate receptor when intracellular energy levels are reduced. Brain Res. 451, 205–212.

Olney, J.W., Misra, C.H. & de Gubareff, T. (1975). Cysteine-s-sulfate: Brain damaging metabolite in the mammalian central nervous system. J. Neuropathol. Exp. Neurol. 4, 167–176.

Palmer, D.N., Husbands, D.R., Barnes, G. & Jolly, R.D. (1986a). Ceroid-lipofuscinosis II: The major component of lipopigment in liver, kidney, pancreas and brain is low molecular weight protein. J. Biol. Chem. 261, 1773–1777.

Palmer, D.N., Husbands, D.R., Winter, P.J., Blunt, J.W. & Jolly, R.D. (1986b). Ceroid-lipofuscinossis in sheep.I. Bis(monoacylglycero)phosphate, dolichol, ubiquinone, phospholipids, fatty acids and fluorescence in liver lipopigment lipids. J. Biol. Chem. 261, 1766–1772.

Palmer, D.N., Martinus, R.D., Barns, G., Reeves, R.D. & Jolly, R.D. (1988). Ovine ceroid-lipofuscinosis 1: lipopigment composition is indicative of a lysosomal proteinosis. Am. J. Med. Genet. Suppl. 5, 141–158.

Palmer, D.N., Martinus, R.D., Cooper, S.M., Midwinter, G.G., Reid, J.C. & Jolly, R.D. (1989). Ovine ceroid-lipofuscinosis: The major lipopigment protein and the lipid binding subunit of mitochondrial ATP synthase have the same NH_2 terminal sequence. J. Biol. Chem. 264, 5736– 5740.

Palmer, D.N., Fearnley, I.M., Medd, S.M., Walker, J.E.,Martinus, R.D., Bayliss, S.L., Lake, B.D., Wolfe, L.S. & Jolly, R.D. (1990). Lysosomal storage of the DCCD reactive proteolipid subunit of mitochondrial ATPsynthase in human and ovine ceroid-lipofuscinosis. In: Lipofuscin and Ceroid Pigments (Porta E.A. ed.), pp.211–223. Plenum Press, New York.

Palmer, D.N., Fearnley, I.M., Walker, J.E., Hall, N.A., Lake, B.D., Wolfe, L.S., Haltia, M., Martinus, R.D. & Jolly, R.D. (1992). Mitochondrial ATP synthase subunit c storage in the ceroid-lipofuscinoses (Batten disease). Am. J. Med. Genet. 42, 561–567.

Palmer, N.N., Bayliss, S.L., Clifton, P.A. & Grant V.J. (1993). Storage bodies in the ceroid-lipofuscinoses (Batten disease): low-molecular-weight components, unusual amino acids and reconstruction of fluorescent bodies from non-fluorescent components. J. Inherit. Metab. Dis. 16, 292–295.

Palmer, D.N., Jolly, R.D., van Mil, H.C., Tyynelä, Y. & Westlake, V.J. (1997a). Different patterns of hydrophobic protein storage in different forms of neuronal ceroid-lipofuscinosis (NCL, Batten disease). Neuropediatrics 28, 45–48.

Palmer, D.N., Tyynelä, J., van Mil, H.C., Westlake, V.J. & Jolly, R.D. (1997b). Accumulation of sphingolipid activator proteins (SAPs) A and D in granular osmiophilic deposits in miniature Schnauzer dogs with ceroid-lipofuscinosis. J. Inherit. Metab. Dis. 20, 74–84.

Pampiglione, G. & Harden, A. (1977). So-called neuronal ceroid- lipofuscinosis. Neurophysiological studies in 60 children. J. Neurol. Neurosurg. Psychiat. 40, 323–330.

Pedersen, W.A., Fu, W., Markesbery, W.R., Appel, S.A., Smith R.G., Keller, J.N. & Mattson, M.P. (1998). Protein modification by the lipid peroxidation product 4-hydroxynonenal in spinal cord tissue of ALS patients. Ann. Neurol. 44, 819–824.

Pullarkat, R.K., Norris, G.N., Pullarkat, P.S., LaBadie, G.U. & Zawitosky, S.E. (1992). Biochemical studies on the juvenile form of ceroid-lipofuscinosis. Am. J. Med. Genet. 42, 575– 579.

Purpura, D.P. & Suzuki, K. (1976). Distortion of geometry and formation of aberrant synapses in neuronal storage disease. Brain Res. 116,1-21.

Raininko, R., Santavuori, P., Heiskala, H., Saino, K. & Palo, J. (1990). CT findings in neuronal ceroid-lipofuscinoses. Neuropediatrics, 21, 95–101.

Rider, J.A. & Rider,D.L. (1988). Batten disease: past, present and future. Am. J. Med. Genet., Suppl. 5, 21–26.

Santavuori, P. (1973). EEG in the infantile type of so called ceroid-lipofuscinosis. Neuropediatrics 4, 357–387.

Santavuori, P., Rapola, J., Saino, K. & Raitta, C. (1982). A variant of Jansky-Bielschowsky disease. Neuropediatrics 13, 135–141.

Sharp, J.D., Wheeler, R.B., Lake B.D., Savukoski, M., Järvelä, Peltonen, L., Gardiner R.M. & Williams, R.E. (1997). Loci for classical and a variant late infantile neuronal ceroid lipofuscinosis map to chromosomes 11p15 and 15q21-23. Hum. Mol. Genet. 6, 591–595.

Siakotas, A., Goebel, H., Patel, V., Watanabe, I. & Zeman, W. (1972). The morphological and biochemical characteristics of ceroid isolated from cases of neuronal ceroid-lipofuscinosis. In: Sphingolipids, Sphingolipidoses and Allied Disorders (Volk, B.W. & Aronson, S.M. eds.), pp. 53–56. Plenum Press, New York.

Sleat, D.E., Donnelly, R.J., Lackland, H., Liu, C.G., Sohar, I., Pullarkat, R.K. & Lobel, P. (1997). Association of mutations in a lysosomal protein with classical late-infantile neuronal ceroid-lipofuscinosis. Science 277, 1802–1805.

Smith, R.I.E., Sutton, R.H., Jolly, R.D. & Smith K.R. (1996). A retinal disorder associated with ceroid-lipofuscinosis in adult miniature Schnauzers. Vet. Comp. Ophthalmol. 6, 187–191.

Stengel, C. (1826). Account of a singular illness among four siblings in vicinity of Roraas. In: Ceroid-lipofuscinosis (Batten's Disease) (Armstrong, D., Koppang, N. & Rider, J.A. eds.), pp.17–19. Elsevier Biomedical Press, Amsterdam, translated from Eyr (Christiania) 1, 347–352.

Tyynelä, J. (1997). Sphingolipid activator proteins(SAPs) and glycosphingolipid metabolism in fibroblasts of patients with infantile neuronal ceroid-lipofuscinosis (INCL). In : Academic dissertation, University of Helsinki, IV, 1–16.

Tyynelä, J., Palmer, D.N., Baumann, M. & Haltia, M. (1993). Storage of saposins A and D in infantile neuronal ceroid-lipofuscinosis. FEBS Lett. 330, 8–12.

Tyynelä, J., Suopanki, J., Santavuori, P., Baumann, M. & Haltia, M. (1997). Variant late infantile neuronal ceroid-lipofuscinosis: pathophysiology and biochemistry. J. Neuropathol. Exp. Neurol. 56, 359–375.

Varilo, T., Savukoski, M., Norio, R., Santavuori, P., Peltonen, L. & Jarvela, I. (1996). The age of human mutation: genealogical and linkage disequilibrium analysis of the CLN5 mutation in the Finnish population. Am. J. Genet. 58, 506–512.

Vercruyessen, A. Martin, J.J., Ceuterick, C., Jacobs, K. & Swerts, L. (1982). Adult ceroid-lipofuscinosis: diagnostic value of biopsies and of neurological investigations. J. Neurol. Neurosurg. Psychiat. 45, 1056–1059.

Verkruyse, L.A. & Hofmann, S.L. (1996). Lysosomal targeting of palmitoyl-protein thioesterase. J.Biol. Chem. 271, 1583 –15836.

Vielhaber, G., Hurwitz, R. & Sandhoff, K. (1996). Biosynthesis, processing, and targeting of spingolipid activator protein (SAP) precursor in cultured human fibroblasts. J. Biol. Chem. 271, 32438–32446.

Walkley, S.U. (1995). Pyramidal neurons with ectopic dendrites in storage diseases contain elevated levels of GM_2 ganglioside. Neuroscience 68, 1027–1035.

Walkley, S.U. (1998). Cellular pathology of lysosomal storage disorders. Brain Pathol. 8, 175–193.

Walkley, S.U. & Wurzelmann, S. (1995). Alterations in synaptic connectivity in cerebral cortex in neuronal storage disorders. Ment. Retard. Dev. Disabil. Res. Rev. 1, 183–192.

Walkley, S.U., Baker, H.J., Rattazzi, M.C., Haskins, M.E. & Wu, J-Y. (1991). Neuroaxonal dystrophy in neuronal storage disorders: Evidence for major GABAergic neuron involvement. J. Neurol. Sci. 104, 1–8.

Walkley, S.U., March, P.A., Schroeder, C.E., Wurzelmann, S. & Jolly, R.D. (1995). Pathogenesis of brain dysfunction in Batten disease. Am J. Med. Genet. 57, 196–203.

Walkley, S.U., Siegel, D.A. & Dobrenis, K. (1996). Batten disease: a typical neuronal storage disease or a genetic neurodegenerative disorder characterized by exocitotoxicity. In: Neurodegenerative Diseases, (Fiskum, G., ed.), pp. 217–224. Plenum Press, New York.

Westlake, V.J., Jolly, R.D., Bayliss, S.L. & Palmer, D.N. (1995). Immunocytochemical studies in the ceroid-lipofuscinoses (Batten disease) using antibodies to subunit c of mitochondrial ATP synthase. Am. J. Med. Genet. 57, 177–181.

Wheeler, R.B., Sharp, J.D., Mitchell, W.A., Bate, S.L., Williams, R.E., Lake, B.D. & Gardiner, R.M. (1999). A new locus for variant late infantile neuronal ceroid lipofuscinosis-CLN7. Mol. Genet. Metab. 66, 337–338.

Williams, R.S., Lott, I.T., Ferrante, R.J. & Caviness, V.S. (1977). The cellular pathology of neuronal ceroid-lipofuscinosis. Arch. Neurol. 34, 298–305.

Wisniewski, K.E., Kida, E., Patxot, O.F. & Connell, F. (1992). Variability in the clinical and pathological findings in the neuronal ceroid-lipofuscinoses: review of data and observations. Am J. Med. Genet. 42, 525–532.

Wong-Riley, M.T.T. (1989). Cytochrome oxidase: an endogenous metabolic marker for neuronal activity. TINS. 12, 94–101.

Zeman, W. (1974). Studies in neuronal ceroid-lipofuscinosis. J. Neuropathol. Exp. Neurol. 33, 1–12.

Zeman,W. (1976). The neuronal ceroid-lipofuscinoses. In: Progress in Neuropathology, Vol. 3, (Zimmermann, H.M., ed.), pp. 203–223. Grune and Stratton, New York.

Zeman, W. & Dyken, P. (1969). Neuronal ceroid-lipofuscinosis (Batten's disease): relationship to amaurotic family idiocy. Pediatrics 44, 570–583.

INDEX

www.ingramcontent.com/pod-product-compliance
Ingram Content Group UK Ltd
Pitfield, Milton Keynes, MK11 3LW, UK
UKHW020229290926
15470UKWH00001B/429